ALUMINA AS A
CERAMIC MATERIAL

Compiled and Edited by

WALTER H. GITZEN, retired

Alcoa Research Laboratories

East St. Louis, Ill.

Special Publication No. 4

The American Ceramic Society
735 Ceramic Place
Westerville, Ohio 43081

ABSTRACT

This comprehensive survey has been compiled to provide a sourcebook for aluminum oxide ceramic technology. The features and characteristics of high-temperature, transition, and hydrated aluminas are presented in detail, and are adequate for both engineering and scientific reference. Subjects discussed include nomenclature, natural occurrence and associations, beneficiation and preparation techniques, crystallographic features, and mineralogical properties. Mechanical, thermal, sonic, electrical, magnetic, optical, radiation (nuclear), chemical, and colloidal properties are critically reviewed. The aspects of grinding, forming, and sintering pertinent to the production of refractories and other applications including uses in glass, cement, cermets, electrical insulation, thermal coatings, and airborne equipment are given. An extensive bibliography is provided.

ACKNOWLEDGMENT

This work is an extension of a report entitled "Alumina Ceramics" (AFML-TR-66-13) prepared for the Materials Information Branch, Materials Support Division, Air Force Materials Laboratory, Wright-Patterson Air Force Base, Ohio.

The contribution of Aluminum Company of America in helping to prepare the manuscript for publication by the American Ceramic Society is gratefully acknowledged.

TABLE OF CONTENTS

Page

TABLE OF CONTENTS—Continued

Page

LIST OF FIGURES

LIST OF TABLES

1 INTRODUCTION

The investigation of alumina as a ceramic material was undertaken to provide information under the following specifications: a review dealing with alumina both from a theoretical and a practical viewpoint, and including information on the nomenclature, properties, and applications of alumina. Specifically discussed, but not limited to, are the three major classifications: the hydrous aluminas, the transition forms, and the high-temperature forms. The following information has been gathered: occurrence in nature, methods of beneficiation or preparation, crystal and mineralogical characteristics, and mechanical, thermal, sonic, electrical, magnetic, optical, radiation (nuclear), chemical, and colloidal properties.

The production of ceramic grades of alumina and the practical aspects of grinding, forming, and sintering operations involved in the preparation of alumina ceramics are discussed. A comprehensive review of ceramics made essentially from alumina is provided. It includes applications in abrasives, cement, cermets, electrical insulation, glass, refractories, thermal coatings, and others. An extensive bibliography is appended. The information is presented in a format which makes the data readily available to the reader, showing relationships between applications and properties.

Data on alumina in several of these categories have been collected within recent years. Among these, *Oxide Ceramics* (1960) by Ryshkewitch* provides a good account of the

early work on sintered alumina, in large part representing his own pioneering research. Briefer coverage is given in *Industrial Ceramics* (1964) by F. and S. Singer, in the English translation of *Keramik* (1958) by H. Salmang, in *Introduction to Ceramics* (1960) by Kingery, and in *Alumina Properties* (1960) Alcoa Research Laboratories, edited by Newsome et al. Of these, only the last is devoted solely to alumina.

Ceramists are not in close agreement as to the substances included in the term "ceramics," nor do they seem to have devised a simple, consistent definition of the term that is entirely satisfactory. Kingery defined it as "the art and science of making and using solid articles which have as their essential component, and are composed in large part of inorganic nonmetallic materials." L. Mitchell defined ceramics as "all high-temperature chemistry and physics of nonmetallic materials, and the techniques of forming products at high temperatures." The first definition allows inclusion of materials having melting points below room temperature, as, for example, ice; while the second does not exclude certain organic substances that may be produced at high temperatures, such as carbon disulfide. Although materials of all kinds, including organic substances, are involved in the preparation of ceramics, it is believed that these definitions are too broad to cover the ceramic applications of alumina. Contrary to Ryshkewitch and the European practice, however, applications of alumina in glasses, glazes, and enamels are considered within the scope of this review. Also, certain ceramic operations in which Bayer-type alumina has been used to produce aluminum compounds or refractory products of ceramic interest, such as mullite, spinel, or calcium aluminate cement, are included.

---

*References are tabulated in alphabetical order for the last name of the first author, or for the title or under "anonymous" when no author is given, as indicated in the text. Multiple references to the same authors are identified by date or otherwise in the text.

2 NOMENCLATURE

The word "alumine" was suggested by de Morveau in 1786 as the proper name for the basic earth of alum. This was Anglicized to "alumina" in England, while in Germany "tonerde" is still used, meaning clay earth. The term "alumina" is presently used rather indefinitely in ceramic literature to denote (1) aluminous material of all types taken collectively, (2) the anhydrous and hydrous aluminum oxides taken indiscriminately, (3) more frequently, the calcined or substantially water-free aluminum oxides, without distinguishing the phases present, and (4) corundum or alpha alumina, specifically. It is often used interchangeably with the molecular formula, Al_2O_3. The true meaning is sometimes hard to determine from the context. In this review, as far as possible, the term is used mainly in the sense of the second definition.

More than 25 solid phases of alumina have been described in recent years, but it is doubtful whether some really exist. These phases, or forms, include amorphous hydrous and anhydrous oxides, crystalline hydroxides and oxides, and aluminas containing small amounts of oxides of the alkalies or alkaline earths, designated as beta aluminas. The beta aluminas were initially thought to be pure oxide phases by Rankin and Merwin, later disproved by Ridgway, Klein, and O'Leary. The convention still persists, although these phases should be classified more logically as aluminates.

The phases found in nature, and a few of the artificial types, have common or mineralogical names. Most are also identified by Greek-letter formulas. Unfortunately, in the United States and other countries, investigators have not been in complete agreement on nomenclature, and there is confusion particularly in the Greek-letter formulas.

Corundum, emery, sapphire, and ruby are more or less pure forms found in nature and known from antiquity as abrasives and gem stones (Greville). All consist of the phase designated alpha alumina (a-Al_2O_3). Another native mineral, described by Haüy in 1801, was named "diaspore" by him, from the Greek for "scatter," because it flew apart upon heating. Vauquelin in 1802 gave its formula as $Al_2O_3 \cdot H_2O$. Dewey named a well-crystallized mineral "gibbsite" for G. Gibbs an American mineralogist. It corresponded with the formula $Al(OH)_3$ or $Al_2O_3 \cdot 3H_2O$ (Torrey). This name is more acceptable than "hydrargillite," suggested by Rose in 1839 for an identical mineral found in the Urals. It is the principal phase of the "trihydrate" bauxites. Berthier examined a mineral from Les Baux in southern France, containing about 52% Al_2O_3 and 20% bound water, from which it was supposed that the mineral was $Al_2O_3 \cdot 2H_2O$. The mineral was named "bauxite" by St. Clair-Deville.

When X-ray diffraction became generally available as an analytical tool, Böhm and Niclassen discovered a new diffraction pattern for an aluminum hydroxide precipitated from aluminum sulfate at 100°C. Böhm subsequently identified this pattern in pressure-digested aluminum hydroxide and in pure samples of the French bauxite, both having about 15% bound water ($Al_2O_3 \cdot H_2O$). He concluded that the phase was an isomer of diaspore. The name "boehmite" was suggested by de Lapparent for both the mineral and the artificial product. It is of historical interest that Mitscherlich had anticipated both the synthetic preparation by Böhm's method and its identity as an alumina monohydrate. No phase corresponding to $Al_2O_3 \cdot 2H_2O$ has been found in natural sources by X-ray analysis. L. Milligan was unable to show an arrest at this point in the heating curve of Bayer process alumina. The preparation of this phase has been claimed by Bulashevich and Tolkachev by reacting aluminum with sodium hydroxide solution at 95 to 97°C.

In America, the word "bauxite" has come to mean any highly aluminous ore composed mainly of one or more of the phases, gibbsite, boehmite, and diaspore. Bauxite was considered by Harder to have originated mainly as a transported, weathered product. Transport of soluble or colloidal aluminous components was probably favored by environmental conditions of periodic high rainfall and low, acid pH. The related laterites are believed to have concentrated generally by chemical weathering and transport of non-aluminous components, particularly silica (Konta, S.S. Roy).

In 1925, Haber devised a system of nomenclature for the trivalent alumina phases then known. The "alpha" series included diaspore and corundum (alpha alumina); the "gamma" series included hydrargillite (gibbsite), bauxite (boehmite), and gamma alumina. The classification was obviously based on the end product of calcination, but this is somewhat arbitrary because gamma alumina also transforms to alpha alumina.

Böhm identified a new phase in aluminum hydroxide precipitates that had aged moist for several months at room temperature. The water content was about that of gibbsite, but the X-ray pattern was different, indicating an isomer of gibbsite. This phase was called "bayerite" by Fricke in 1928 on the erroneous supposition that it was the normal product from the Bayer process. L. Milligan had already shown in 1922 that the Bayer product is predominantly gibbsite. Bayerite was claimed to have been found in nature by Gedeon, and more recently (1963) by Gross and Heller. In the first instance, at least, this has been refuted by Sasvari and Zalai.

The failure of the Haber system to distinguish between bayerite and gibbsite prompted the devising of the Alcoa system of nomenclature (Frary). In this system, the choice of Greek letters was initially based on the relative abundance of the phase in nature. Gibbsite was called alpha alumina trihydrate; boehmite, alpha alumina monohydrate; bayerite, beta alumina trihydrate; and diaspore, beta alumina monohydrate. Gamma alumina and alpha alumina had the same significance as in the Haber system.

Coblentz in 1906, van Arkel and Fritzius, and Fricke and Severin found no water of hydration by infrared investigation of gibbsite, boehmite, or diaspore, but only the hydroxyl bands at 3 microns. These phases are hydroxides rather than hydrates. In present terminology, alumina trihydrate is used interchangeably with aluminum trihydroxide, and $Al_2O_3 \cdot 3H_2O$ with $Al(OH)_3$. Similarly, alumina monohydrate is used interchangeably with alumina monohydroxide or aluminum oxide hydroxide, and $Al_2O_3 \cdot H_2O$ with $AlO \cdot OH$. In commercial usage the trihydrates are also designated as hydrated aluminas.

Moiseev claimed that fibrous alumina obtained by action of mercuric chloride solution on aluminum upon heating at 80 to 100°C produces $(AlO \cdot OH)H_2O$, which he called "bauxite dihydrate."

Weiser and W. O. Milligan in 1934 adopted the Haber series, and classified bayerite as alpha alumina trihydrate. Van Nordstrand et al. reported a third alumina trihydrate in 1956 for which the names bayerite II and randomite had been suggested. Nordstrandite has also been suggested, and this has been widely accepted. The phase is thought to be intermediate in lattice structure between bayerite and gibbsite.

Cliachite, zirlite, alumogel and other names were applied to supposedly amorphous bauxitic structures in some cases long before the availability of X-ray diffraction for appraisal of the extent of crystallinity. All fine-crystal bauxites exhibit characteristics of a colloidal state at one state of their development. Aging converts the amorphous precipitates to crystalline forms, but colloidal properties may be retained.

Shishakov reported a third alumina monohydroxide formed as hexagonal $AlO \cdot OH$ on aluminum surfaces, identical to the lattice attributed by Steinheil to a cubic form. Yamaguchi et al. prepared "tohdite," which they claimed had the formula $5Al_2O_3 \cdot H_2O$, by heating an aluminum hydroxide at 450–500°C for 20–100 hours with AlF_3 or $Ti(SO_4)_2$ as a mineralizer. This is apparently identical with a transition phase described by Torkar and Krishner as Kappa-l-Al_2O_3.

Frary in 1946 and Stumpf et al. in 1950 continued the Alcoa series by naming five anhydrous or nearly anhydrous phases: delta, eta, theta, kappa, and chi aluminas. Ulrich had named gamma alumina in 1925, and the term had been used rather vaguely for aluminas calcined in the low-temperature range, and in the oxidation of aluminum. Hägg and Soderhölm, and Verwey in 1935 had applied the name to the spinel structure of alumina. Jellinek and Fankuchen in 1949 computed new dimensions for the unit cell, which were not compatible with the dimensions derived earlier by Brill for a spinel-type lattice. They concluded that crystal growth affected the diffraction pattern. Stumpf restricted the limits of gamma alumina to the product whose X-ray pattern is like that in the alpha trihydrate and alpha monohydrate dehydration sequences at 500°C. New phases include rho alumina of Tertian and Papée, xi and xi' aluminas of Cowley, and iota alumina of P. A. Foster in 1959. Some of these assigned transition phases, whose detection is based mainly on somewhat subtle X-ray diffraction differences, are of doubtful existence.

Standardization of the nomenclature for aluminas is very desirable particularly to avoid the confusion in the hydrous phases. Ginsberg reported the conclusions of a symposium held in 1957, in which an attempt was made to devise a universal standard nomenclature. Some features of the proposed system are improvements, as for example, the substitution of "hydroxide" instead of "hydrate," namely, aluminum trihydroxide for alumina trihydrate; aluminum oxide hydroxide for alumina monohydrate. Also, it was agreed to use the Alcoa nomenclature for the transition aluminas, but to designate some of them as "forms" rather than "phases," to imply the present uncertainty about them. The confusion in naming the hydroxide phases has not been resolved, however.

The designations alpha alumina (α-Al_2O_3) for corundum and beta alumina (β-Al_2O_3) assigned by Rankin and Merwin in 1916 to a supposedly monotropic form, have been retained in all systems. Following Ridgway's disclosure that beta alumina contains alkali or alkaline earth atoms, the foreign cation is now made part of the name, i.e., sodium beta alumina. Barlett called that form crystallizing from melts containing lithia, zeta alumina. Belyankin disputed its separate existence from gamma alumina radiographically as well as in refractive index and in density. Indeed, it matches eta alumina very closely, except in refractive index and, of course, composition.

Two reduced oxide forms of alumina are known, Al_2O and AlO, but they have no special designations.

The Haber system of nomenclature is in general use in Europe, the Alcoa system in America. A comparison of the major nomenclature systems most frequently confused in the literature is shown in Table 1.

In the present compilation, the Alcoa assignment of Greek letters is used.

Table 1

Nomenclature of Crystalline Aluminas

Mineralogical Name	Phase or Form Name					
	Symposium (1)	Alcoa (2)	Haber (3)	British (4)	French (5)	Other
Hydroxides						
Gibbsite (6) Hydrargillite	$Al(OH)_3$	$a\text{-}Al_2O_3\cdot 3H_2O$	$\gamma\text{-}Al(OH)_3$	$\gamma\text{-}Al_2O_3\cdot 3H_2O$		
Bayerite (7)	$Al(OH)_3$	$\beta\text{-}Al_2O_3\cdot 3H_2O$	$a\text{-}Al(OH)_3$	$a\text{-}Al_2O_3\cdot 3H_2O$		
Nordstrandite (1) Randomite (8) Bayerite II (8)	$Al(OH)_3$					
Bauxite (9)						$Al(OH)_2$
Boehmite (10)	$AlOOH$	$a\text{-}Al_2O_3\cdot H_2O$	$\gamma\text{-}AlO_2H$	$\gamma\text{-}Al_2O_3\cdot H_2O$		
Diaspore	$AlOOH$	$\beta\text{-}Al_2O_3\cdot H_2O$	$a\text{-}AlO_2H$			
Tohdite (11)						$5Al_2O_3\cdot H_2O$ $\epsilon\text{-}Al_2O_3$ (12)
Aluminas	Chi Eta Gamma Delta	Chi Eta Gamma Delta	— — Gamma —	Chi + Gamma Gamma Delta Delta + Theta	Rho Chi + Gamma Eta Gamma Delta	
	Kappa	Kappa Iota (17)	—	Kappa + Theta	Kappa + Delta	Xi^1, Xi^2 (13)
Corundum, Sapphire	Alpha AlO Al_2O $M_2O\cdot 11Al_2O_3$ (14) $M_2O\cdot 6Al_2O_3$ (15) $MO\cdot 6Al_2O_3$	Alpha	Alpha	Alpha	Alpha	
Zeta Alumina	$Li_2O\cdot 5Al_2O_3$ (16)					

Ref.:
(1) Ginsberg, Hüttig, Strunk-Lichtenberg
(2) Edwards, Frary, Stumpf, et al.
(3) Haber, Weiser, and Milligan
(4) Rooksby, Day, and Hill
(5) Thibon, Tertian, and Papée
(6) Dewey
(7) Fricke
(8) Teter, Gring, and Keith
(9) Böhm
(10) de Lapparent
(11) Yamaguchi
(12) Steinheil
(13) Cowley
(14) Rankin and Merwin
(15) Scholder
(16) Barlett
(17) P. A. Foster

3 PREPARATION OF ALUMINA PHASES

3-1 BAUXITE

The native aluminous minerals, corundum, emery, and the clays, have been associated with ceramic operations from antiquity. Exclusive of these raw materials, bauxite is the principal source of alumina for ceramic purposes. Bauxite mining began at Villeveyrac, France, in 1873. Since then, bauxite mining has spread around the world, particularly in the tropical areas, in southern United States, in the Guianas, in the Caribbean islands and Brazil, along the west coast of Africa, in the Mediterranean countries, in India, Malaya, Indonesia, and Australia, but also in Russia and China. Bauxite is generally believed to have been produced mainly by weathering, favored by abundant and intermittent rainfall. Moses and Michel conclude that a second step occurred in the case of Guiana bauxites, in which an uplift of the coastal plain, following weathering permitted bauxitization through leaching of kaolin by surface water.

Gibbsite is the most stable phase in bauxite at normal temperatures and pressures found in the tropics. As it is likely that temperatures above 100°C were involved in the genesis of bauxite, the formation of diaspore and boehmite probably resulted from high pressure.

In 1966, world production of bauxite was about 40 million long tons, of which over 14 million represented consumption of domestic and imported bauxite in the United States. Domestic ceramic uses as abrasives and refractories consumed about 509,000 tons. About 2.12 dry long tons of bauxite are required to produce one short ton of Bayer process alumina, and an average of about 3.93 tons to produce one short ton of aluminum.

Even in the primitive era of alumina production, some preparation or beneficiation was required. The South American bauxites are subjected to hydraulic cleaning operations to remove overburden and sand contamination. The mined deposits are then blended on the basis of their chemical analysis into suitable grades for various uses. The principal impurities include SiO_2, Fe_2O_3 and TiO_2.

Bauxite is used directly in adsorbents, abrasives, and refractories, aside from its use in preparing metal. For these applications, specific grades must be sized, dried, and calcined, partially at about 980°C for fused abrasive use, and more completely at about 1540°C for refractory use.

Bauxites from British (Demerara) and Dutch Guiana (Suriname) are particularly suited to ceramic uses, because of their controlled low iron and low silica contents. Properties of typical ceramic grades from each source are shown in Table 2.

3-2 PREPARATION OF BAYER ALUMINA

The major application of pure alumina is in the production of metallic aluminum. Metallurgical-grade Bayer alumina is the cheapest source of high-purity alumina available. Production in the United States in 1966 was over 5.9 million short tons of 99% purity. In view of the influence of metallurgical-grade alumina on the economics of ceramic alumina applications, the preparation of this grade will be given first consideration.

The Bayer process is the principal production method in use today. Many other production methods have been proposed, however. The subject is too voluminous to cover more extensively than by a brief outline of the processes. *Alumina and Aluminum*, by Fulda and Ginsberg (1957) is probably the most complete discussion of the commercial production for the aluminum industry. A second edition (in German) was published in 1964. *Extractive Metallurgy of Alluminum*, Volume I, Alumina, edited by Gerard and Stroup (1963) contains excellent monographs on various phases of modern alumina production. *The Aluminum Industry*, Volume I, by Edwards, Frary, and Jeffries gives very complete information on the extractive processes prior to 1930. Most of the basic procedures had been proposed prior to that date.

The incentive for many of the proposed procedures has been to use more available and cheaper raw materials than bauxite. These include common clay, leucite, alunite, and andalusite, among others.

The general processes, exclusive of metallurgical processes for winning the metal directly, include the following:

1. Wet alkaline processes: (a) autoclave Bayer, (b) tower percolation.
2. Wet acid processes: H_2SO_4, HCl, HNO_3, H_2SO_3, acid salts.
3. Alkaline furnace processes (a) Na_2CO_3, $CaCO_3$, or mixtures, (b) sulfates, chlorides or other salts and reducing agents.

Table 2

Bauxite – Ceramic Grades

	Suriname[1]			Demerara[2]		
	Dried Chemical	Abrasive	Refractory	Abrasive AAC	Refractory RAC	Refractory RASC
Al_2O_3	61-62%	87.5%	89.0%	87.75%	87.75%	88.00%
Fe_2O_3	1.3-1.7%	5.4%	1.6%	1.50%	1.50%	1.50%
TiO_2	2.7-3.2%	3.5%	3.3%	3.25%	3.25%	3.25%
SiO_2	2.0-2.5%	2.9%	6.0%	7.00%	7.00%	7.00%
Loss on Ignition	31-32%	0.7%	0.1%	0.50%	0.50%	0.25%
Free Moisture	1.5-2.5%	nil	nil	nil	nil	nil
Size Range	–	–	–	1.5 in.	lumps to fines	
Bulk Dens.	–	–	–	90	90	120

[1] Courtesy Suriname Aluminum Company.
[2] Courtesy Aluminum Limited Sales Inc.

4. Carbothermic furnace processes.
5. Electrolytic processes.

3-2-1 Wet Alkaline Processes

The wet alkaline processes, in particular the Bayer process, are still the most economical of modern methods. The Bayer process is most suitable for low-silica bauxites containing gibbsite and boehmite.

Gibbsitic bauxites are easiest to dissolve. Porter has recently described the application of the Bayer process to aluminum phosphate ores, and Shakhtakhtinskii and Nasyrov its use on alunite. The Bayer process is optionally continuous or batchwise. The dried, ground bauxite is digested in an autoclave with a solution of sodium hydroxide and sodium carbonate, containing a flocculant (starch) and lime added to causticize the soda and to act as a filter aid. This requires a concentration of about 100 to 120 grams NaOH per liter and a pressure of 40 to 50 psi (at about 145°C) to dissolve the contained gibbsite and obtain a stable supersaturated sodium aluminate liquor. Liquors having a ratio of Al_2O_3 : NaOH of about 0.8 by weight are stable against autoprecipitation at clarification temperatures around 105°C. This was the conventional American continuous process until the advent of bauxites containing appreciable amounts of boehmite.

Boehmitic European bauxites, and particularly some hard Greek bauxites, require more drastic digest conditions. Digestion at 280 to 700 psi and at above 300 grams NaOH per liter is the usual practice. A continuous process requires operation at 285°C (1400 psi) to obtain good recovery of alumina. With the advent of bauxites containing appreciable boehmite in American practice, a compromise digestion at higher pressure, 400 to 450 psi (about 230°C) has been developed.

After digestion, the slurry is cooled by flash heat interchange to near boiling at atmospheric pressure and the waste solids (red mud) are separated by sedimentation and filtration. The red mud comprises mainly iron oxides, titania, carbonated lime, and desilication product. The lime represents makeup causticization of soda lost in previous cycles and in some cases, filter aid. The desilication product represents mainly soda and alumina lost by combination with reactive silica to form zeolitic compounds resembling sodalite. The clarified (green or pregnant) liquor is cooled by heat exchange with spent liquor to about 55°C, and a portion of the dissolved alumina is precipitated by introducing fine seed alumina trihydrate. Solubilities and densities of Bayer trihydrate, bayerite, and hydrothermal boehmite (coarse) in sodium hydroxide solutions for the Bayer-process range have been reported by Russell, Edwards, and Taylor; and subsequently, Panasko and Yashunin (1964). Gibbsite is the normal precipitated phase, designated Bayer trihydrate, but Fricke showed that other phases are possible at different temperatures and with different seeding phases present, as long as the solution is supersaturated with respect to these phases. The precipitated trihydrate is separated from the spent liquor, washed, and for the most part, is calcined for ceramic applications. The spent liquor plus washings is evaporated and returned to process. Complete separation of the hydrate from the green liquor can be realized by gassing with carbon dioxide, but this then requires complete causticization as a separate external operation, as distinguished from the relatively lesser causticization required for makeup for losses in the seeding precipitation. Strokov observed that the properties of the product were markedly influenced by the temperature, duration, and speed of mixing during carbonation. At 80°C, the product contained from 0.5 to 0.6% alkali (in Al_2O_3). Tower leaching of partially calcined bauxite by sodium hydroxide under pressure is described by Fulda and Ginsberg, and by Zalar.

The conditions of seeding, temperature, agitation and time during precipitation are controlled to obtain the most economic separation of the Bayer trihydrate, and the most advantageous particle size distribution for subsequent operations. Factors which affect the purity and texture of the precipitated trihydrate also influence the quality of the calcined alumina, even though the product has passed through several transitions.

Typical properties and specifications of commercial grades of hydrated aluminas, prepared in a modern Bayer plant, are shown in Table 3.

3-2-2 Wet Acid Processes

Wet acid processes for ceramic aluminas have so far not approached the wet alkaline process in economy of operation. Difficulties have been mainly with corrosion of equipment, poor recovery of acid for reuse, and the high pickup of iron and other impurities in the product. The acid processes should be more adaptable to low-iron clays, kaolins, and high-silica bauxites, which give poor extraction or difficulty with desilication by the Bayer process. There is generally little contamination from silica in acid processes. Exhaustive references to acid processes prior to 1930 are given by Edwards, Frary and Jeffries. The following recent references (since 1958) show the continuing interest in potential methods for recovering metallurgical-grade alumina from clays and other alumina sources. Bakr preferred acid methods to alkaline or thermal reduction methods for clays. Peters et al. at the U.S. Bureau of Mines evaluated three sulfuric acid processes and one nitric acid process for calcined clay, and estimated capital and operating costs. Scott combined hydrolysis of sulfate liquor at 220°C and the use of bauxite to achieve partial neutralization of digestion liquors to recover a pure product from clays and bauxites. Kretzschmar roasted clay to make it more soluble in dilute sulfuric acid, and converted the resultant silicic acid to calcium hydrosilicate before separating the purified aluminum sulfate. Alternatively, the clay can be roasted with $(NH_4)_2SO_4$ and an addition of Na_2SO_4 at 400 to 550°C, and leached in 0.5% H_2SO_4 solution before hydrolyzing. By treating the pulp additionally with an alkaline solution, a yield of 94 to 96% Al_2O_3 was achieved (Skobeev). Loss of ammonia probably occurs above 400°C (Riedel). Mixed H_2SO_4-$(NH_4)_2SO_4$ has been used as the leaching solution and a yield of 68.4% alumina was obtained by hydrolysis with NH_3 (Bretznajder). Tucker extracted alumina from clay by dissolving it in nitric acid, and obtained an ironfree product by adding calcium chloride to

Table 3

Typical Properties and Specifications of Hydrated Aluminas-Series C-30

	C-31(1)	C-31 Coarse	C-33(1)	C-35(2)
Typical Properties:				
Al_2O_3 . . . . %	65.0	64.9	65.0	65.2
SiO_2 . . . . %	0.01	0.01	0.008	0.008
Fe_2O_3 . . . . %	0.003	0.004	0.002	0.004
Na_2O . . . . %	0.16	0.20	0.16	0.04
Moisture (110°C) . . . . . %	0.04	0.04	0.04	0.04
Bulk density, loose, lb/ft³	60-70	70-80	60-70	70-80
Bulk density, packed, lb/ft³	75-85	90-100	75-85	85-95
Specific gravity	2.42	2.42	2.42	2.42
Sieve analysis (cumulative)				
On 100 mesh . . . . %	0-1	0-10	0-1	0-2
On 200 mesh . . . . %	5-10	40-80	5-10	5-20
On 325 mesh . . . . %	30-55	85-97	30-55	30-60
Through 325 mesh . . . . %	45-70	3-15	45-70	40-70

[1] Finer material is available.
[2] Tentative properties.

Courtesy Aluminum Company of America

the filtered solution, and by separating the iron in methyl isobutyl ketone.

In 1945, a development plant was built by Defense Plants Corporation at Salem, Oregon, using the acid ammonium sulfate (Kalunite) process (Libbey).

Anaconda Aluminum Company in May 1963 was operating a sulfuric acid process on Idaho clays on a pilot scale (6200 lb. of metal per day). The process has been claimed to cost about $40 per ton.

North American Coal Corporation also operated a sulfate process on coal processing wastes, to recover both alumina and sulfate values.

Some recent references relating to improvements in Bayer plant operations are of interest. Kuznetsov et al. found that Bayer trihydrate, calcined at 250°C to boehmite structure, was the best nongibbsitic seed crystals for precipitating trihydrate, the grain size of which corresponds to industrial requirements. Lyapunov, Khodokova, and Galkina observed an increase in solubility of gibbsite in sodium aluminate solutions containing 6.9% added NaCl. Holder and Thome purified the filtered sodium aluminate liquor by treating with $KMnO_4$ and passing it through a bed of lime. Dunay et al. claimed advantages in two-fold washing, the use of hydrogen peroxide, and filtering through broken electrocorundum to reduce floating contaminants, Fe_2O_3, TiO_2, and SiO_2. Scott discussed the effects of seed and temperature on the particle size of Bayer hydrate. Reese and Cundiff described the continuous process production of alumina from Jamaican bauxite in detail. Scandrett and Porter discussed the continuous Bayer process for trihydrate bauxites containing significant amounts of boehmite.

3-2-3 Furnace Processes

Furnace processes have apparently had the most success in competing with the Bayer process in tonnage operations. Some furnacing methods require wet processing as a purification step. The ore is fused or sintered with an alkali or alkaline earth carbonate or oxide, or mixtures, to produce the corresponding aluminate. The product is then extracted and purified by wet alkaline procedures similar to the Bayer process. Sulfides, sulfates or chlorides may be substituted for the oxides. The Pedersen process is of this type. A mixture of iron ore, coke, lime, and bauxite, or other aluminous material is smelted to produce molten calcium aluminate slag containing 30 to 50% Al_2O_3 and only 5 to 10% SiO_2. Low-sulfur, high-grade iron is a by-product. Commercial operation of this process began in 1928 at Høyanger, Norway. Andalusite has been used as the aluminous material in a somewhat similar operation at Sundsvale, Sweden, producing at the rate of 4,000 tons per year during the war years.

3-2-4 Carbothermic Processes

In carbothermic furnace processes the ore is fused in contact with carbon under conditions to separate the impurities as molten metals, leaving the alumina in suffi-

ciently high purity that wet separation methods can be avoided. The Hall and Haglund processes are of this type, and are intended mainly to produce metallurgical-grade alumina. The Haglund process is economically operable when its ferrosilicon by-product is in demand. The Hall process is not presently in use.

Similar procedures have been developed for purification of fused alumina prepared from bauxite in the electric furnace for fused abrasives (Allen, Saunders, and Ridgway and Glaze). For abrasives, a mixture of bauxite, coke, and iron borings is fused in either a Higgins or a Hutchins furnace. Both comprise a water-cooled steel shell containing the charge, into which carbon electrodes are positioned. The shell of the Higgins furnace is a truncated cone, open at both ends, and with the smaller diameter up. The shell fits over a round carbon block. At the end of a furnacing cycle, the shell can be lifted off. The Hutchins furnace is crucible-shaped and closed at the bottom. The cooled product is massive and requires crushing to obtain the desired sizes.

Another type of grain is made with an addition of sulfides to the fusion. The product consists of single crystals of aluminum oxide in a water-soluble matrix.

Fused white aluminum oxide is prepared from a special grade of Bayer-process calcined alumina. Reduction to remove impurities is not required. In batch production, fusion is usually done in Hutchins-type furnaces. The impure discolored beta alumina migrates to the center of the pig, and can be hand-separated by visual inspection. White abrasive is also produced by a semicontinuous process in a pouring furnace. During pouring, micro and macro capillary pores form, thus aiding in the subsequent crushing operation (Ridgway).

During the war years the Defense Plants Corporation built two pilot plants, one for treating clay by a lime-sinter process, and the other for treating anorthosite by a soda-lime-sinter process. The processes are not presently used. R. W. Brown developed a sintering process using a mixture of soda ash and limestone, applicable to high-silica aluminous ores. The combination of the standard Bayer process and this sintering process applied to the Bayer waste product (red mud), has made practical the utilization of large reserves of low-grade bauxite in the United States.

Much of the furnacing investigation has been applied to alunite, $K_2SO_4 \cdot Al_2(SO_4)_2 \cdot 4Al(OH)_3$. Gad and Barrett; Knizek and Fetter; Nekrich; Bretsznajder and Pysiak; Asahi; and Ponomarev et al. describe roasting operations involved either in direct removal of sulfate by volatilization or in opening up the material to wet purification processes.

3-2-5 Electrolytic Processes

Roberts and Schwerin proposed electrolytic methods of separating alumina from aluminate solutions. For many years no interest was shown in these processes. Recently, Guareschi has described a process of roasting aluminosilicas, dissolving in sulfuric acid, and electrolyzing the filtered and purified liquor. Reheis electrolyzes an aqueous solution of aluminum chloride in a 3-compartment cell. The center compartment is fed a slurry of aluminum hydroxide and dilute hydrochloric acid, which combines with any chlorine migrating from the cathode. A selective membrane prevents passage of chlorine to the anode.

3-3 AMORPHOUS AND GEL ALUMINAS

Gelatinous aluminum hydroxides are of interest because of their technical importance in preparing sorbents and catalysts. Surprisingly, they have seen only limited use elsewhere, as for example, as plasticizing additives in compositions for forming sintered alumina ware. Willstätter and Kraut and their coworkers investigated various types of gels, in most cases, precipitated from alum or aluminum sulfate solutions by ammonia at temperatures below 60°C. They distinguished three types, which they designated a, β, and γ, and which have been the subject of continuing investigation to the present time. Unaged C_a is amorphous to X-rays, is readily soluble in 0.1% HCl, and dewaters to about 35% H_2O when treated with dried acetone and ether. C_a transforms within a few hours to C_β when aged in water, but loses almost one mole of bound water. C_β contains diffuse bands of boehmite, peptizes in dilute HCl, and dissolves slowly. Kohlschütter and Beutler found that aging for several hundred hours in water, or for a few hours in dilute ammonia, converts C_β into C_γ, a mixture containing boehmite and crystalline trihydroxides, almost insoluble in dilute HCl. Geiling and Glocker showed that even when the initial product is maintained at low temperature (12°C) throughout the test (up to 48 hours), the product shows the boehmite pattern.

Gels can be prepared by neutralizing other aluminum salts (for example, the chloride or nitrate) in the alkaline direction with ammonium hydroxide or alkalies. They can also be prepared by neutralizing sodium aluminate solutions in the acid direction by adding carbon dioxide, sodium bicarbonate, or an acid. Havestadt and Fricke showed that rapid precipitations of sodium aluminate by CO_2 at O° are initially amorphous, but eventually transform into bayerite. Precipitations at room temperature to about 60°C form as bayerite. Gels can be formed by reacting water with "activated" surfaces of aluminum.

A gel in fibrous form is produced by action of mercury and moisture on aluminum (Cossa). At room temperature bayerite lines develop. This gel and the previously cited preparations are generally impure because of sorbed reactants. A method of producing gels of potentially much higher purity, with respect to ionic contaminants, consists of dropping distilled aluminum triethyl, $Al(C_2H_5)_3$ into water (Thiessen and Thater). Alternatively, the aluminum triethyl can be vaporized in a stream of nitrogen. Opalescent gels prepared by this method and stored for 2 months under water lost water of hydration at 17°C to an eventual composition of about $Al_2O_3 \cdot 3H_2O$, upon drying. They developed a faint bayerite pattern. Aluminum isopropylate, $Al(C_3H_7O)_3$, prepared by reacting aluminum and alcohol in the presence of an aluminum-mercury couple, has also been used to prepare very pure hydrolyzed gels. Torkar et al. hydrolyzed the aluminum alcoholate with 3 to 30% (by weight) hydrogen peroxide at O to 22°C.

Calvet et al.; and Imelik, Mathieu, Prettre, and Teichner found that amorphous gels precipitated from aluminum salts by a base at about pH 7 are impure and contain significant amounts of anions. Attempts to purify the gel convert it to crystalline products with a marked change in specific surface. Precipitations at a more alkaline pH (about 9) are purer, but are not strictly amorphous. Papée, Tertian, and Biais claimed to make a pure amorphous gel by precipitating aluminum nitrate at pH 8.0 with ammonium hydroxide or soda, followed by rapid washing and drying at 25°C. The product had excess hydration ($Al_2O_3 \cdot 3.45H_2O$), contained less than 0.1% NO_3^-, and was considered more amorphous than glass, resins, or liquids. It had a specific surface of 170 m^2/g, stable after a year. These products aged to "pseudoboehmite," having the same pattern as well-crystallized boehmite, but more diffuse (Imelik et al.).

The preparation method of the gels became influenced by the morphology. The amorphous gel is composed of small spherical particles about 30 A in diameter as seen under the electron microscope. Turkevich and Hillier observed that the spheres were conbined into fibers in precipitates from aluminum nitrate by ammonia. On aging, these fibers arranged to form rectangular plates. The fibrous nature was retained on drying at 80°C and even on heating to 350°C. On further heating, particularly in steam, the granular fibers gradually disappeared to show lath-like surfaces. Similar results were shown for gels prepared by reaction of aluminum metal and acetic acid. Sawamura observed the fibrils in the C_a transition within two hours of aging. After 24 hours, most of the crystals were arranged in fibers of 200 to 500 A diameter and of indeterminate length. Souza Santos et al.; Watson et al.; Ohta and Kagami; Suzuki; and Moscou and van der Vlies investigated the gels in more detail. Souza Santos thought the fibers were boehmite, but Moscou claimed they were not pure and contained some bayerite. On further aging the fibers rearranged to form "somatoids," a name given by Kohlschütter to describe the characteristic particles in C_γ. The sequence is from spherical amorphous particles to fibers, to boehmite somatoids, to bayerite or other trihydroxide somatoids in hexagonal plates. Wislicenus in 1908 had described fibrous alumina formed on the surface of amalgamated aluminum. This material is composed of bundles of long, fragile fibers of aluminum hydroxide. It ages rapidly in water forming products that are morphologically analogous to those obtained from the C_a gel.

Kohlschütter had found that solid aluminum compounds react with ammoniacal solutions to yield products of easy filtration and washability. The reaction proceeds topochemically, that is, the external shape of the particles is retained. Moscou and van der Vlies state that sheet-like gelatinous boehmite forms directly without formation of amorphous hydroxide. Plate-shaped hydroxide particles are obtained from plate-shaped solid aluminum sulfate, while needle-shaped particles are obtained from solid aluminum chloride.

Cramer et al. prepared alumina gels for catalytic treatment of petroleum by reacting in aqueous solution basic aluminum chloride or nitrate, urea, and ammonium or alkali metal salts of acetic and glycolic acids to form a hydrosol at pH of about 3.5 to 7. The hydrosol was set in a water-immiscible liquid at about 52 to 100°C for about 10 hours to decompose the urea and raise the pH to about 8.

Hsu and Bates found that the molar ratio of NaOH to Al^{3+} is a significant factor in preparing gels from sulfate or chloride solutions. When the ratio was not over 2.75, the products were amorphous, and remained so for at least 6 months. When the ratio was 3.0 to 3.3, bayerite, nordstrandite, gibbsite, or mixtures were obtained within several hours.

Patrick prepared gels substantially free of water-soluble salts by mixing aluminum sulfide with water acidified with a volatile organic acid (acetic or formic) at from 0 to 80°C, and gelling the sol in dilute acid.

References to activators for accelerating the formation of aluminum hydroxide from metallic aluminum include the following: Hervert and Bloch used 100 to 10,000 parts per million by weight of the aluminum selected from the metals, tin, lead, and germanium; alkylamines, or arylalkylamines; or water-soluble alkanolamines reacting at freezing to 705°C; Evans used aniline at its boiling point; E. F. Smith claimed 0.1 to 0.2% gallium; Lefrancois any metal compound below aluminum in the electromotive series together with a strong acid having an ionization constant of at least 10^{-2}, but preferably from the group, mercury, zinc, and cadmium.

Bergman and Torkar subjected the aluminum to electrolytic attack in water containing only carbon dioxide or hydrogen peroxide. Welling applied an aqueous solution of hydrogen peroxide to amalgamated aluminum; Csordas exposed the amalgamated aluminum to a gas stream of oxygen and water vapor at 25 to 75°C.

Kandyakin states that the primary amorphous particles are oval, 0.01 to 0.05-micron in diameter, in agreement with size as determined from specific surface measurements. Papée, Tertian, and Biais found a small maximum in the X-ray pattern for pure amorphous gel between 1.5 and 3 A. Stumpf concluded that amorphous material is characterized by a band at 4.5 A, clearly distinguishable from crystalline material. Alumina prepared by hydrolysis of aluminum isopropoxide is initially completely amorphous. Wilsdorf found that a model containing two Al_2O_3 molecules agrees with the electron diffraction rings of amorphous alumina produced by 75-hour air-oxidation of aluminum. The structure consists of an octahedron of six closely packed oxygen ions with four aluminum ions in a tetrahedral arrangement and ionic distances of 2.80 and 3.95 A for 0-0; 1.72 and 3.28 A for 0-Al; and 2.80 A for Al-Al. Robson and Broussard applied radial distribution analysis to determine the most probable interatomic distances of amorphous silica-alumina compositions.

3-4 PREPARATION OF THE ALUMINA TRIDYDROXIDES

3-4-1 Gibbsite, α-$Al_2O_3 \cdot 3H_2O$, Alpha Alumina Trihydrate

Gibbsite is rare in nature in pure well-crystallized form, although it is the predominant phase of American bauxites.

As indicated in Sec. 3-3, the crystalline trihydroxides can be prepared by aging water suspensions of amorphous aluminum hydroxide. It is more expedient, however, to precipitate under conditions of temperature, concentration, and rate that favor direct formation of gibbsite. A suitable laboratory method for gibbsite consists of partly neutralizing a sodium aluminate solution having a density of about 1.32 g/ml by rapid addition of carbon dioxide at about 50 to 80°C (Fricke). The crystals vary from pseudohexagonal plates to prisms having a diameter of about 0.3 to 3 microns.

Very pure, fine gibbsite is obtained directly by slow, spontaneous hydrolysis of alkali aluminate solution at room temperature (Fricke and Jucaitis). Single crystals can be grown to large size if the precipitation continues for months.

Autoprecipitation from supersaturated sodium aluminate liquors (80 g Al_2O_3/l, 120 g NaOH/l) at about 60 to 50°C in the presence of fine seed trihydrate is the well-known Bayer process. This process yields as much as two-thirds of the charge, but the method is not very suitable for laboratory preparation.

Alkali-free gibbsite is claimed by Hauschild for hydrolysis of aluminum triethylate, $Al(OC_2H_5)_3$ in the presence of from 1 to 20% ethanolamine, $NH_2C_2H_4OH$, at 20 to 60°C, and by slow aging of the gel formed for several months. The product contains only 0.001% Na_2O, 0.001% CaO, 0.12 to 0.23% N, and only traces of organic matter.

Michel and Papée claim a high-purity gibbsite by precipitating aluminum hydroxide, washing the precipitate to obtain a cake containing about 35% by weight of Al_2O_3 and 0.1 mole of a univalent acid ion per mole Al_2O_3. The suspension is aged for at least 2 days at 60°C. The cake is immersed in an alkaline medium containing ammonium hydroxide or amines at 120°C for sufficient time to allow crystal growth to about 100 to 200 A.

3-4-2 Bayerite, β-$Al_2O_3 \cdot 3H_2O$, Beta Alumina Trihydrate

Bayerite can be produced by moderately slow carbonation of sodium aluminate solutions at 30 to 35°C. It is difficult to obtain pure bayerite in this way because of the presence of boehmite, gibbsite, and adsorbed alkali (Fricke, Fricke and Wever, Fricke and Wulhorst). The method of aging amorphous gels precipitated from aluminum salts by ammonium hydroxide in the cold (Kraut, Flake, Schmidt, and Volmer) usually does not give a uniform X-ray pattern.

The treatment of aluminum or aluminum alcoholates with water at temperatures below 40°C (Fricke and Jockers) is favored for preparing bayerite. The modification of Schmäh involving the reaction of water with finely divided or amalgamated aluminum foil is well suited to laboratory-scale preparation. Lippens obtained patterns only for bayerite for this method, upon aging the product for one week at room temperature and pH 7.9. The product conformed to $Al_2O_3 \cdot 3.09 H_2O$.

Stewart claimed bayerite, 95% pure, in which a dilute aqueous solution of aluminum chloride or nitrate is added in increments to a sodium-free, 0.1- to 4-molar aqueous basic hydroxide having a nonmetallic volatile cation. The temperature is maintained between 0 to 49°C, and the pH is not reduced below 9.0.

Doelp disclosed a process for bayerite, 96% pure, by treating boehmite, gibbsite, or amorphous gelatinous alumina in an aqueous solution of 1 to 10 moles of tetramethyl ammonium hydroxide at below 40°C, and by precipitating with carbon dioxide below 40°C.

Hauschild and Nicolaus claimed a pure bayerite by treating aluminum cuttings with water containing 7 to 15% alkylamines at below 100°C. Monomethylamine at 80°C is preferred.

Bye and Robinson prepared gels by hydrolysis of aluminum s-butoxide dissolved in benzene, using controlled mixtures of water and ethanol to provide only about twice the theoretical water for hydrolysis. The gels aged for 72 hours in the mother liquor had the composition $Al_2O_3 \cdot 1.8\text{-}1.9 H_2O$, and revealed a diffuse X-ray pattern conforming to the reactive phase called pseudoboehmite. Bayerite appeared within 6 weeks. Aging in water was faster, bayerite appearing within one week, but several additional weeks were required to attain the composition $Al_2O_3 \cdot 3.04 H_2O$. They concluded that pseudoboehmite results from condensation of OH groups, but bayerite forms by dissolution and recrystallization processes. Feitknecht stated that freshly precipitated hydroxides of the trivalent metals can be amorphous, can possess one of two modifications of a partly ordered layer structure, a partly ordered double-layer structure, or a fully crystalline double-layer structure.

Kohlschütter observed that the bayerite somatoids appear as truncated pyramids or cones under magnification of about 2000. The electron micrographs of Watson and Moscou showed these forms to consist of plates generally layered perpendicularly to the long axis. Lippens found that this arrangement conforms with the trihydroxide double-layer structure.

Tertian and Papée claimed that very pure bayerite can be produced by rehydration of rho alumina in water at about 25°C. The rho alumina is produced by dehydrating gibbsite in a vacuum at 200°C. The success of this rather complicated process depends on the original gibbsite being precipitated from sodium aluminate liquor on gelatinous aluminum hydroxide in a very fine size.

3-4-3 Nordstrandite, Bayerite II, Randomite

Kraut observed three X-ray diffraction modifications of bayerite; de Boer, Fortuin and Steggerda described two

modifications, designated bayerite I and II. Van Nordstrand et al. identified a pattern in gels precipitated from aluminum chloride or nitrate in the presence of ammonium chloride, and aged at pH 7.5 to 9.0. They suggested that it conforms to a third phase of $Al(OH)_3$. Papée claimed to have obtained pure nordstrandite, presumably by long-time aging at pH 13 of a gel formed by the cited process. Schlanger claimed that the phase occurs in limestone deposits on Guam.

Hauschild described a method for obtaining pure nordstrandite consisting of aging freshly prepared aluminum hydroxide gel in an aqueous solution of an alkylene diamine (preferably $NH_2CH:CHNH_2$ in concentrations below 35%) at 40 to 70°C. The gel is prepared by any of the usual neutralization methods, by reacting activated or finely powdered aluminum with water, or by precipitating the gel in the presence of the amine.

Keith prepares nordstrandite by precipitating a gel from aluminum chloride with ammonium hydroxide at 27 to 49°C and at pH 7 to 9. Upon washing and aging (2 to 5 days) at 43 to 49°C, nordstrandite is obtained; aging at 24 to 32°C gives gibbsite. Keith, Keith et al., and Teter et al. describe the preparation of commerical products for catalytic uses containing nordstrandite.

Ginsberg, Hüttig, and Stiehl state that conversion of bayerite to gibbsite begins with inclusion of alkali ions in the layer lattice, sodium being more effective than potassium at 20°C. Pseudoboehmite, the precursor of boehmite, is present in precipitates formed at 60°. Use of ammonium hydroxide as a precipitating agent gives distorted bayerite (nordstrandite), which, with sodium hydroxide, transforms to gibbsite.

3-5 PREPARATION OF THE ALUMINA MONOHYDROXIDES

3-5-1 Boehmite, $a\text{-}Al_2O_3 \cdot H_2O$, Alpha Alumina Monohydrate

Boehmite occurs in nature in the European bauxites. It can be prepared by aging aluminum hydroxide gels, by thermal dehydration of gibbsite, bayerite, or amorphous aluminum hydroxide, and by rehydration of the higher oxides. Although the products from these different paths of transition may have substantially the same X-ray diffraction pattern, they differ markedly in properties. This is ascribed to the ease of preparation of boehmite in a wide range of surface area or activity, from about 15 m²/g for the hydrothermal product to about 400 m²/g for some of the aged gels.

As indicated in Sec. 3-3, the normal sequence of transition of amorphous aluminum hydroxide at atmospheric pressure, and above about 20°C is to gelatinous boehmite, to bayerite, to gibbsite. Tosterud, Ginsberg, and Oomes each found that gibbsite reverts to coarse-crystalline boehmite at 60 to 100°C in sodium hydroxide solutions.

Lippens favored the method of Schmäh for preparing the surface-active form of boehmite. Fibrous aluminum hydroxide was formed by room-temperature treatment of amalgamated aluminum foil (99.99% Al) with water vapor. After aging the amorphous material in twenty times its amount of water for about 300 hours, the product was dried in vacuum over P_2O_5 to a composition corresponding to $Al_2O_3 \cdot 1.6 \ H_2O$, and having a surface area of about 400 m^2/g. The diffuse X-ray pattern of pseudoboehmite was given, but when the aging time was extended to 290 hours, bayerite lines began to appear. About the same results were obtained by topochemical conversion of alum at pH 9.2 within 22 hours.

A very suitable method for preparing well-crystallized boehmite of low surface area is by hydrothermal conversion of Bayer alumina trihydrate in water (Hüttig and von Wittgenstein) or in dilute sodium hydroxide solution at 150 to 300°C. The charge of Bayer trihydrate may be as concentrated as 400 g/l. The Bayer trihydrate usually provides a small amount of alkalinity to aid the conversion. About 6 hours is required for conversion at 180°C in water, but the product has excess water (about $Al_2O_3 \cdot 1.125 \ H_2O$). Schwiersch obtained the theoretical water content by prolonging the digest time to 106 hours at 300°C. Hüttig, Peter, and von Wittgenstein obtained boehmite by heating dried alumina trihydrate in a bomb at 200°C, without a water medium. Massive Bayer alumina trihydrate (precipitator scale), cut to geometrical forms, converts to boehmite in place without disintegration under these conditions.

The conversion to boehmite is faster with increasing concentration of alkali within limits, and with increasing temperature and increasing fineness of the charge. Ginsberg and Köster found that the conversion is about as fast for 0.6 g Na_2O/l as for 140 g/l at 180°C. They concluded, contrary to Bauermeister and Fulda, that boehmite forms by dehydration in place. Sato agreed with this. Present evidence indicates that while formation in place is the normal mode, it probably occurs by solution and reprecipitation. Precipitation can be induced on boehmite seed out of contact with the charge, and at an accelerated rate. Moreover, the rate of conversion is not the same as for dry dehydration. The equation for simple dehydration has the form: $\ln H = Kt$, in which H is the molar concentration of combined water in the solid product at time t, and K is a constant. The equation for the hydrothermal conversion satisfies the form:

$$\frac{dM}{dt} = \frac{KnM^2}{3tK+1}$$

in which M is the molar amount of boehmite formed per mole Al_2O_3 at time t, and K and n are constants. The equation indicates that the rate is proportional to the amount of boehmite present.

Residual alkali may be reduced below 0.05% Na_2O in water digests by heating at 180 to 220°C (Sablé). Boehmite forms in well-crystallized rhombs from alkaline solutions, but generally as thin curve-edged plates, often as coarse as 5 microns, when formed in water. This process has been

investigated to some extent as a potential source of alumina for ceramic purposes (Cera). It will probably receive more attention in the future.

Bloch claimed a boehmite process comprising heating finely divided metallic aluminum in water at 250 to 375°C. LePeintre obtained boehmite by electrolysis of a solution of aluminum sulfate and aluminum chloride at pH 4.0. Keith stabilized gel precipitates in the boehmite phase and prevented peptization by gassing the aqueous suspension with carbon dioxide.

Bugosh has developed a fibrous hydrothermal boehmite having discrete fibrils of average length, 0.1 to 0.7 micron, and surface area of 50 to 450 m^2/g. Bugosh et al., and Bruce investigated the ceramic possibilities of this material in binders, coatings, and thickening, emulsifying, and suspension agents.

Boehmite forms as packets of fibers, oriented in preferential crystallographic directions on albite $(Si_3AlO_3)Na$, when the albite is subjected to hydrothermal treatment in water containing CO_2 at 200°C, according to Tchoubar and Oberlin.

Boehmite prepared by thermal dehydration of Bayer trihydrate and bayerite at about 170 to 500°C is likely to contain mixtures of other phases, depending on the particle size, the rate of heating, the ambient pressure, and impurities present. This form has high surface area.

3-5-2 Diaspore, β-$Al_2O_3 \cdot H_2O$, Beta Alumina Monohydrate

Diaspore occurs rarely in well-crystallized form in some Russian sources, but more abundantly mixed with flint clay in deposits in Missouri, and in high-iron nodular deposits in central Pennsylvania. Its main ceramic use has been in refractory brick. With gradual depletion of the refractory grades of diaspore, calcined bauxite is a replacement. Diaspore is of little use in the Bayer process of purification because of its poor solubility.

Laubengayer and Weisz were the first to prepare diaspore synthetically. The preparation required hydrothermal digestion of aluminum hydroxide in water at high pressure, and in the presence of seed diaspore upon which the diaspore was grown. Diaspore is the stable phase between 280 and 450°C, and at water pressures above 140 atmospheres, a rather formidable set of conditions for prospective commercial preparation.

Krishner and Torkar prepared diaspore without seeding by reacting aluminum turnings with steam, and in a lower pressure region (about 100 atm). F. Freund noted a spontaneous formation of disoriented diaspore in the hydrothermal treatment of rho and chi aluminas formed by the vacuum dehydration of gibbsite.

3-6 TRANSITION ALUMINAS

3-6-1 Dehydration Mechanism

The transition aluminas include the group of phases which fall between boehmite and corundum. The group originally had been designated loosely as gamma alumina. None has been found in nature, but all have been prepared by thermal transformations, and in some cases, by hydrothermal transformations of the aluminum hydroxides, and by oxidation of aluminum. Although designated as oxides, some are probably hydrous. These structures have present ceramic uses, particularly as adsorbents, catalysts, coatings, and soft abrasives.

Stumpf, Russell, Newsome, and Tucker described the X-ray patterns, transition sequences, and temperature ranges of seven modifications, arbitrarily designated as chi, eta, gamma, delta, kappa, theta, and alpha aluminas. They found that the sequence of phases is influenced by the starting material. The transition temperatures are influenced by the amount of water vapor in the atmosphere and by impurities. They found that Bayer alumina trihydrate transforms successively to boehmite, to chi, to gamma, to kappa, to alpha alumina. Bayerite transforms to boehmite, to eta, to theta, to alpha alumina. Hydrothermal boehmite transforms to gamma, to delta, to theta, to alpha alumina. The designations were based primarily on differences in X-ray diffraction patterns.

Other investigators have since discovered new phases, or disagreed with the X-ray patterns or even the existence of the present phases.

Before proceeding with the preparation methods for the transition phases, it seems desirable to review early investigations of the dehydration process. Peculiarities in the dehydration phenomena of the aluminum hydroxides have been revealed by static and dynamic heating methods. These peculiarities are doubtless related to the double layer structure found by Megaw in 1934 for gibbsite, and which persists in other transition forms.

In 1932, Fricke and Severin determined the equilibrium water contents versus temperature of the trihydroxides and monohydroxides (Figure 1). Long periods (weeks) were required to establish equilibrium in some portions of the temperature range, 100 to 400°C; in others the time was relatively shorter. Gibbsite dehydrated to about the water content of boehmite at 165°C, and bayerite at 120°C, when both were held at a pressure of 100 mm Hg, but the actual transition points were vague. In the region of dehydration below residual water content corresponding to $Al_2O_3 \cdot H_2O$, both gibbsite and bayerite dehydrated to lower water levels than either hydrothermal boehmite or diaspore for the same dehydration conditions.

At atmospheric pressure, the course of dehydration of gibbsite and bayerite (Figures 2 and 3) takes place in two stages. The first stage dehydrates to a composition having a water content below $Al_2O_3 \cdot H_2O$. The lower the heating temperature used to attain equilibrium, that is, the slower the dehydration rate, the more nearly the structure approaches $Al_2O_3 \cdot H_2O$. Even at 200°C, the rate of dehydration is very slow at atmospheric pressure.

Blanchin decomposed alumina trihydrate at 100°C in vacuum. Blanchin, Imelik, and Prettre; and Tran-Huu et al. found that for an infinitely slow rise in temperature, the two stages of dehydration are separated by a temperature interval of about 100°C, during which the boehmite phase is stable, but very close to the composition $2Al_2O_3 \cdot H_2O$. Prettre et al. confirmed that for heating periods of 20

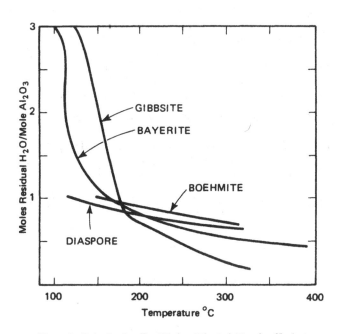

Figure 1. Dehydration Equilibria of Heated Alumina Hydrates (After Fricke and Severin)

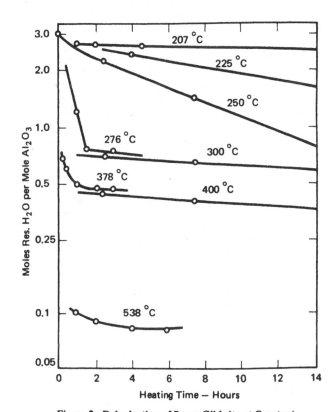

Figure 2. Dehydration of Bayer Gibbsite at Constant Temperature

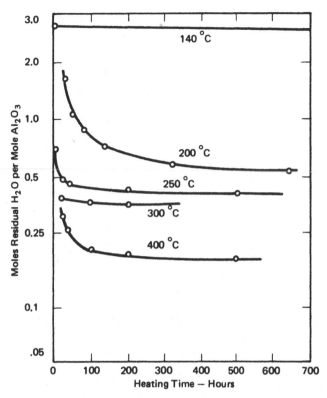

Figure 3. Dehydration of Bayerite (CO₂ Precipitated) at Constant Temperature

hours, the water content decreases uniformly in a 70°C interval from $Al_2O_3 \cdot 3H_2O$ at 100°C in vacuum (or 150°C at one atmosphere) to $2Al_2O_3 \cdot H_2O$, where an arrest for an 80° interval was observed before loss of water was again appreciable. Dehydration was essentially complete at 400°C in a vacuum, and at 500°C at atmospheric pressure. Chatelain found that gibbsite, when dehydrated in air or vacuum at about 230°C, becomes a mixture of boehmite and imperfectly crystalline alumina. Courtial and co-workers found that boehmite and a more or less amorphous solid are produced either simultaneously or successively at 230°C, and that the composition conforms to $Al_2O_3 \cdot 0.69$ H_2O. Based on thermogravimetric (TGA), calorimetric, and chemical methods, the water contents were computed, and the relative amount of each phase present was determined.

Brown, Clark, and Elliott were the first to deduce that gibbsite decomposes by two mechanisms (1) to gamma alumina directly, and (2) to boehmite, which subsequently decomposes to gamma alumina. Day and Hill thought that gibbsite and bayerite decompose directly to anhydrous alumina, and that this subsequently rehydrates partially to form boehmite.

Papée and Tertian explained the two paths of transition as the result of buildup of steam pressure in the coarser particles to create a hydrothermal condition favoring boehmite formation. At about 200 to 220°C, the internal pressure causes fracturing of the crystals to terminate further formation of boehmite. Also, very fine particles might show less boehmite formation (Tertian and Papée).

Eyraud, Goton, and co-workers found that after a short induction period, the rate of decomposition was constant to about $Al_2O_3 \cdot H_2O$ at specific temperatures, beyond which the rate decreased (Figures 2 and 3). They investigated dehydration rates in the temperature range 206 to 243°C and in the pressure range 0.001 to 15 mm Hg water-vapor pressure. They obtained activation energies ranging from 31 to 63 kcal/mole. Brindley and Nakahira obtained activation energies of 31 to 35 kcal/mole for the gibbsite to boehmite reaction, and 46.5 kcal/mole for the gibbsite to chi alumina reaction. They used a thermo-balance to measure the weight loss, and applied an Arrhenius-type relation, $\ln K = \ln A - 0.4343 E/RT$, in which T is the absolute temperature (°K), R the gas constant, and A a frequency constant.

Papée and Tertian, and de Boer derived differential heating curves from static thermal dehydration curves similar to Figure 1. The derived curves resemble actual curves obtained by differential thermal analysis (DTA).

Orcel; Jourdain; Norton; and Speil applied DTA to gibbsite and diaspore. Gibbsite heating curves reveal peaks of heat absorption (endothermic) at about 225, 300, 550°C, and a peak of heat release (exothermic) at about 980°C. Diaspore has a single endothermic peak at about 500°C. Both gibbsitic bauxite and synthetic gibbsite reveal marked differences in intensity between the first and third peaks with respect to the second peak; in some cases, the first and third are absent. Steggerda demonstrated that these differences are related to particle size. DTA curves are shown in Figure 4 for gibbsite, bayerite, boehmite, and diaspore. Also shown are two curves for two fine commercial grades of gibbsite, Hydral 705 (0.4 micron) and Hydral 710 (0.9 micron), which show the marked reduction in intensity of the peaks involved. De Boer, Fortuin, and Steggerda concluded that the first peak represents boehmite formed by hydrothermal pressure within the crystals that are coarser than about one micron. The second peak represents the formation of an anhydrous phase (chi alumina) following crystal disruption at about 200 to 220°C. The third peak at about 550°C represents the dehydration of the boehmite formed under the first peak.

Boersma and de Jong discussed the quantitative aspects of DTA. Smothers et al., and Schmitter identified and classified various patterns. Alexanian applied DTA to the gibbsite dehydration in the range to 1400°C, and established the temperature stability for the following phases: gibbsite (to 220°C), boehmite (280°), a phase at 380° (chi?), gamma (400°), delta (560°), kappa (900°), alpha alumina (from 1300°). Calvet, Thibon, and Gambino obtained characteristic thermograms in a microcalorimeter for amorphous gel, pseudoboehmite, normal light boehmite, and a fine boehmite.

LeChatelier observed the sudden heat release at about 850°C during calcination of alumina prepared from nitrates and chlorides, but not for alumina from sodium aluminate. Heat evolution in this case can be induced by small amounts of fluorides. The method is used practically to increase mineralization of the alumina, and to conserve fuel

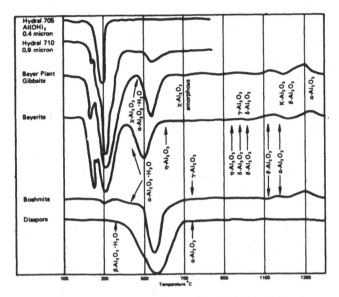

Figure 4. Differential Thermal Curves for the Hydrated Aluminas

containing up to 0.6% H_2O, including the phases, rho, chi, eta, and gamma; (b) high-temperature, nearly anhydrous, obtained at 900 to 1000°C, and including the phases, kappa, theta, and delta. Several other phases have been proposed.

3-6-2 Sequence of Transition

The following sequences of transitions (shown graphically in Figure 5), mainly due to Stumpf et al. and Tertian and Papée, are generally accepted, although there is confusion about the X-ray identification of some phases and the existence of others. The sequences are affected not only by the starting materials but also by their coarseness of crystallinity, heating rates, and impurities. The approximate temperature ranges of stability of the phases are included.

Gibbsite Transitions

In vacuum, coarse or fine material: gibbsite transforms to rho (100 to 400°C), to eta 270 to 500°C), to theta (870 to 1150°C), to alpha alumina (1150°C).

Instantaneous dehydration at 800°C: gibbsite to eta, to theta, to alpha alumina.

In air, fine gibbsite: gibbsite to chi (300 to 500°C), to kappa (800 to 1150°C), to alpha alumina.

In air, coarse gibbsite: (1) gibbsite to chi, to kappa, to alpha alumina. (2) gibbsite to boehmite (60 to 300°C), to gamma (500 to 850°C), to delta (850 to 1050°C), to theta, to alpha alumina.

Bayerite Transitions

The bayerite transitions are substantially of the same type as the gibbsite transitions except that fine bayerite in

by reducing the calcination temperature required to make the alumina nonhydroscopic (Pechiney). Maximum crystal size is attained for an aluminum fluoride concentration of about 1.4% (Lindsay).

Following the work on sequence of the transition phases of alumina by Stumpf et al., more or less complete surveys have been made by Rooksby; Day and Hill; Newsome et al.; Thibon, Charrier, and Tertian; Tertian and Papée; Ginsberg, Hüttig and Strunk-Lichtenberg; Stirland, Thomas, and Moore; de Boer and associates; and Sato. Of these, Tertian and Papée have furnished the most detailed description.

Lippens classified the transition aluminas into two groups: (a) low-temperature, dehydrated below 600°C and

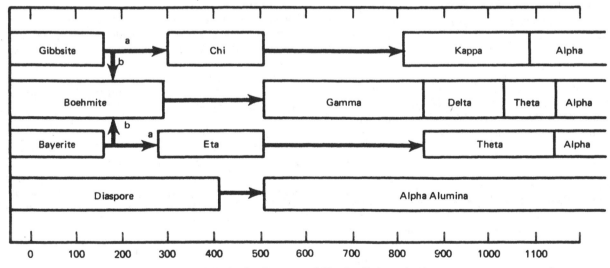

Figure 5. Dehydration Sequence of Alumina Hydrates in Air

Note: Enclosed area indicates range of stability. Open area indicates range of transition.
Path b is favored by moisture, alkalinity, and coarse particle size (100 microns);
path a by fine crystal size (below 10 microns).

air transforms as follows: bayerite to eta (250 to 500°C), to theta (850 to 1150°C), to alpha alumina.

The difference between bayerite and gibbsite transitions is the substitution of eta for chi alumina, and the absence of kappa alumina in the bayerite transitions.

Boehmite Transitions

Fine boehmite (pseudoboehmite, 350 m²/g): boehmite to bayerite, to eta (100 to 350°C), to theta, to alpha alumina.

Coarse hydrothermal boehmite (less than 15 m²/g) prepared from gelatinous aluminum hydroxide, gibbsite, bayerite, or higher transition phases by digestion in H_2O at above 150°C: boehmite to gamma (360 to 860°C), to delta, to theta, to alpha alumina.

Diaspore Transitions

Deflandre in 1932 had shown that diaspore transforms to corundum by an ordered process, without intermediate products, at about 450 to 600°C. Herold and Dodd observed that diaspore clay forms corundum and gamma alumina at 400°C, but that between 400 and 750°C there is a gradual expansion in the size of the corundum unit cell, as determined by X-ray. Upon continued heat treatment at higher temperatures, the unit cell gradually reduces to the true dimensions.

Rho Alumina

Rho alumina is primarily a vacuum-produced phase. It is substantially amorphous, as its X-ray pattern shows only a few diffuse bands, the most pronounced having a spacing of about 1.40 A. It has a surface area of about 200 m²/g, and micropores not accessible to nitrogen. Rehydration of rho alumina by water at 25°C gives very pure bayerite (Tertian and Papée). It is therefore possible to prepare all the transitions from gibbsite, either by dehydration, or rehydration.

Chi Alumina

Chi alumina is distinguished by an X-ray spacing of 2.12 A, and is probably cubic, but not of the spinel type. Tertian and Papée claim that the Alcoa structure for chi alumina is a mixture of chi and gamma aluminas.

Gamma Alumina

Gamma alumina, according to Stumpf, is the cubic structure obtained by heating Bayer alumina trihydrate or hydrothermal boehmite at 500°C. Rooksby concurred with Hägg and Verway in designating gamma alumina as a spinel. This is probably the eta alumina of Stumpf. Gamma alumina (Stumpf's designation) is distinguished from eta alumina by differences in relative intensity and sharpness of the lines for eta, by the line at about 4.6 A for eta, and by splitting, for gamma, of a line appearing at 1.97 for eta. In other respects, the two patterns are closely related.

Kimberlin and Gladrow prepared eta alumina for use as a catalyst support by hydrolysis of an aluminum alcoholate at zero to 21°C, aging the water suspension for 20 to 65 hours to form beta trihydrate, and calcining at 345 to 650°C. Kirshenbaum and Hinlicky treat similarly, but heat finally at 200 to 800°C.

Hunter and Fowle obtained eta alumina by the oxidation of aluminum near its melting point, as well as by electrolytic oxidation. Caglioti and d'Agostino produced transition alumina by blowing air through molten aluminum. Filonenko, Lavrov, Andreeva, and Pevzner obtained an alumina spinel, $AlO \cdot Al_2O_3$, by intense reduction during fusion of alumina.

Houben and de Boer suggest that the crystalline structure of gamma alumina is probably due to the presence of about 3.4% water in the lattice (one mole H_2O per five moles Al_2O_3) to fill all open spaces in the spinel lattice. Additional water may react with the surface until the surface formula might be described as $AlO \cdot OH$, while the internal structure is HAl_5O_8.

MacIver et al. differentiated gamma from eta alumina by its higher water content. This was regarded as consisting of molecular water strongly adsorbed on the surface of the gamma alumina.

Yanagida and Yamaguchi prepared gamma alumina by hydrothermal treatment at 500°C and 50 atmospheres pressure for 20 hours of boehmite previously dehydrated in air at 700°C. It contained 7.5% water upon drying at 150°C, and had a tetragonal axial ratio of 1.0357. An eta alumina specimen of lesser tetragonal deformation was obtained from dehydration in air of bayerite. It contained 3.8% water. Successive heat treatments at 400, 600, and 800°C, with cooling to room temperatures after each heat treatment, reduced the water contents of each alumina. The axial ratio and lattice dimensions of the gamma alumina were also reduced. Yanagida, Yamaguchi, and Kubota distinguished between two types of water contained in transient aluminas an "non-lattice water" in which the typical d spacings remain substantially unchanged for heating curves in air under pressure of one atmosphere water vapor, and "lattice water" for that water representing change in the lattice dimensions.

Cowley ascribed the extra spots and line segments, observed in electron diffraction patterns of gamma alumina on heated aluminum foil, to faults in the cubic close-packed stacking of the oxygen atoms, and arbitrarily designated the new phases xi, xi', nu, and mu. Alexanian claimed that these phases are merely mixtures of two real successive phases. He points out that traces of SiO_2, K_2O, Na_2O, and Li_2O combine with alumina above 600°C to give X-ray lines foreign to those of the actual phase.

Ando prepared gamma alumina by vaporization in an electric arc of an alumina-carbon electrode containing a mixture of alpha and gamma alumina bonded with low-ash carbonaceous material.

Delta Alumina

The initial patterns for delta alumina, of Stumpf et al., were obtained from low-soda hydrothermal boehmite that had been calcined in steam at 1000°C for one hour. Rooksby, and Rooksby and Rooymans suggested other preparation methods, some involving the presence of impurities to develop the structure. Improved structure was claimed for the delta phase by heating hydrothermal boehmite for 120 hours at 930 to 950°C. The phase is also obtained by rapidly calcining a mixture of 50 parts ammonium alum and 10 parts ammonium molybdate at 950°C, and holding for one hour at temperature. Delta is the form obtained by flame-spraying commercial-grade corundum on graphite substrate. The corundum should contain both lithium and sodium ions as prominent impurities. The product converts to delta alumina upon heating at 1000 to 1100°C. The necessity for so much impurity in these latter compositions is not very convincing for the method. Rooymans found Rooksby's product to be tetragonal. Tertian and Papée considered delta phase truly amorphous.

Plummer found that passage of gamma or alpha alumina through the oxyhydrogen flame forms well-developed crystals of delta and theta aluminas, providing the final particles are less than 15-micron spheres. Larger particles were alpha alumina. Juillet similarly hydrolyzed aluminum chloride vapors to obtain spherical globules only 150 A in diameter, mainly delta alumina, and with a specific area of 100 m^2/g. Delta alumina has also been obtained by combustion of aluminum in air or oxygen, or in the presence of carbonaceous material, or by combustion of aluminum carbide in oxygen (Wartenberg 1952; Schneider and Gattow 1954; and Foster, Long, and Hunter 1956). Cuer and co-workers found that the flame reactor using an oxygen-hydrogen mixture is particularly suitable for preparing nonporous alumina chiefly in the delta phase and in spherical form.

Kappa and Theta Alumina

Kappa and theta aluminas, and some alpha alumina are the principal components of metallurgical-grade alumina. Kappa alumina has received some attention as a dental polish (Broge). Both structures are well-crystallized and of markedly lower surface area than the lower transition phases. Kappa is orthorhombic and theta is monoclinic. Kappa is the normal phase obtained by calcining fine-crystal gibbsite. It is prepared in fair purity from Bayer trihydrate by atmospheric dehydration in the narrow range, 1000 to 1050°C. Tertian and Papée claimed that the Alcoa structure for kappa is a mixture of true kappa' and delta aluminas.

Theta alumina is the normal product obtained by calcination of fine-crystal bayerite. It is prepared by heating bayerite for one hour at 1100°C in moving dry air.

Funaki and Shimizu (1959) determined that the thermal decomposition of $Al(NO_3)_3 \cdot 9H_2O$, $AlCl_3 \cdot 6H_2O$, and $Al(OAc)_3$ at 500 to 700°C is to amorphous alumina. On thermal decomposition, $Al_2(SO_4)_3 \cdot 18H_2O$ and $NH_4Al(SO_4)_2 \cdot 12H_2O$ yield gamma alumina at 850 to 900°C, but the product contains some amorphous alumina. All of the mentioned salts yield upon further calcination gamma, delta, theta and alpha aluminas, and in that order.

3-6-3 Phases Formed on Aluminum

An instantaneous amorphous film about 20 A thick forms on aluminum in air (Hass). Growth to about 90 A continues and stops in about one month. Heated films become crystalline at about 500°C. Aylmore, Gregg, and Jepson investigated the effects of exposures of 170 to 400 hours at atmospheric pressure and at 50°C intervals from 400 to 650°C. Oxidation at 400°C follows a parabolic-rate law. At higher temperatures the gain in weight rates consists of three distinct types in which the rate decreases, becomes constant, and again decreases.

Greenblatt found that the structure of oxide films formed on aluminum after exposure to water at 150 to 350°C consists of zones which increase in number as the exposure becomes more extreme and with ease of corrosion of the aluminum type. Columnar structures were observed joining the outer crystalline layer to the metal surface at the site of second-phase particles. Taylor, Tucker, and Edwards found that the oxide from electrolytic oxidation at less than 100 volts is amorphous, but is eta alumina at higher voltages. Lichtenberger noted that the oxidation of films at 20 to 80 volts originated as nuclei at the grain boundaries, then spread radially and coalesced until a barrier layer formed. Cass found that the film is anisotropic and biaxial, the bisector of the optical axes being always perpendicular to the surface. Koenig measured the lattice constant and found that it varies from 7.73 to 8.06 A, but is about 7.89 A to 800°C for all formation methods. Burnham and Robinson stated that the electrolytic oxidation product of aluminum at high voltages in boric, citric, and tartaric acids is nonporous, platelike, gammatype alumina of hexagonal symmetry. Bergmann prepared bayerite by electrolyzing aluminum in dilute solutions of hydrogen peroxide, and obtained mixed boehmite-bayerite in solutions containing carbon dioxide.

Steinheil prepared face-centered cubic "epsilon alumina" having a space constant of 5.33 A by oxidation of aluminum. Upon heating to melt off the metal, gamma alumina was obtained. The contact of aluminum with the oxidation product appears to lower the transition temperature to alpha alumina to about 660°C; if the metal is dissolved away from the alumina the transition occurs at 900 to 1000°C in air, and at 1100 to 1200°C in vacuum (Beghi et al.). Continental Oil Company claimed an epsilon alumina having a surface area of 300 m^2/g, prepared by hydrolysis of aluminum alcoholate at 80 to 100°C, and by calcining almost exclusively to the phase.

Hannon patented the method of preparing aluminum oxide by grinding aluminum pellets, while maintaining the pellets free from contact with substances other than aluminum.

3-6-4 Rehydration

Rehydration of the intermediate, rho alumina, was mentioned previously as a method for preparing pure bayerite. Rehydration is a factor of significance in connection with the formation of active aluminous sorbents, their continued cyclic use, and gradual changes in their sorptive capacity. Nieuwenberg and Pieters, and Hüttig and Kölbl investigated rehydration of activated alumina. The latter prepared curves of isobaric dehydration and hydration. Bentley and Feichem demonstrated the rehydration of anhydrous intermediate alumina to boehmite in steam. Day and Hill claimed that boehmite is formed in dehydrating both gibbsite and bayerite by secondary reactions between the original dehydration products and the water released.

Calvet and Thibon investigated the rehydration of activated aluminas prepared from gibbsite and bayerite sources in a microcalorimeter (Calvet). They found that alumina activated in a vacuum is more easily rehydrated than that activated at atmospheric pressure. Increase in the temperature of activation has an unfavorable effect on the tendency of the alumina to rehydrate. Alumina which has been activated at 180 to 200°C under vacuum and then heated in air at about 350 to 450°C does not rehydrate as readily as unheated alumina.

Tertian and Papée found that vacuum-dehydrated Bayer trihydrate (rho alumina) rehydrated to bayerite at 25°C, but any contained hydrothermal boehmite remained unchanged. Rho alumina prepared from bayerite was influenced by the conditions of water vapor release during the initial dehydration. Slow release and long contact of the water with the active surface greatly reduces the rate of rehydration. Rehydration to boehmite occurs at 100°C but requires days. Amorphous alumina produced by calcination of aluminum salts at 500 to 700°C rehydrates at room temperature to bayerite, and then gibbsite, independently of bayerite formation (Funaki and Shimizu 1959). These investigators in 1952 found no transformation of bayerite into gibbsite at room temperature during a 10-year aging period.

Ervin and Osborn found no rehydration for alpha alumina of low surface area. Krishner and Torkar prepared an active form of alpha alumina (pseudocorundum, by analogy with boehmite nomenclature), having a surface area of 40 to 70 m^2/g, probably representing primary particles without micropores. The active alpha alumina hydrated to boehmite and diaspore, the latter in the form of small rods.

3-7 ALPHA ALUMINA

3-7-1 Preparation

The principal sources of alpha aluminum oxide are native corundum and manufactured production derived from bauxite. Mining of a corundum deposit in South Carolina was in operation during the war. Corundum from South Africa (Northern Transvaal) is still used, mainly for abrasives. Manufactured abrasives are, in general, superior. The South African product, according to Palmour, Waller et al., analyzes about 85% Al_2O_3 4% TiO_2 8% SiO_2 and 1% Fe_2O_3.

The principal methods for preparing alpha aluminum oxide are: calcination of the aluminum hydroxides, transition aluminas, and aluminum salts, and solidification from melts. Less common methods are: hydrothermal synthesis at high pressure, vapor-phase transition, and burning aluminum in oxygen (Wartenberg 1952). Alpha alumina can be prepared in a wide range of properties insofar as they are affected by crystal size, crystal habit and purity. Crystal sizes within a range from about 0.03 micron to 30,000 microns or more have been available commercially. Crystal shapes throughout the gamut of habit from thin plates through equants to needles and whiskers can now be produced for various commercial applications.

The most important method for preparing alpha alumina for ceramic purposes, as well as for metal, is calcination of Bayer alumina trihydrate. Table 4 shows typical data for six commercial grades of ceramic alumina, designated "calcined aluminas." They are prepared at temperatures below the sintering level. They include variations in crystal size and in soda content. Also included in the table are four grades of tabular aluminas. The tabular aluminas are massive low-shrinkage forms that have been sintered without added permanent binders. The tabular aluminas are available in ball forms and in crushed and graded granular sizes from a top size of about one-half inch for dense forms (T-60 and T-61) to minus 325-mesh. A porous grade (T-71) having about 35 to 50% porosity is also available.

These aluminas are typical of the range of properties obtainable in Bayer process plants designed for the production of ceramic aluminas. With the exception of the grade designated A-1, all are substantially alpha aluminum oxide. The principal variations in these grades are crystal size, surface area and residual soda content. The grades, designated Calcined and Low-Soda Aluminas are soft-bonded and easily disrupted to ultimate crystallites.

The stated temperature of formation of alpha alumina from the transition phases varies from the range of 1100 to 1150°C (Stumpf et al.; Rooksby; and Tertian and Papée) to 1200 to 1300°C (Gorbuna and Vaganova, Arakelyan and Chistyakova). The transition is influenced by impurities and particle size. Dawihl and Kuhn prepared fine pure corundum containing less than 4 ppm Na_2O by calcining aluminum methylate, $AlO(C_2H_5)_3$, at 1200°C. Brundin and Palmqvist reacted a water-soluble aluminum salt and a water-soluble carbonate to obtain a microporous calcined oxide.

Linde Company has furnished two commercial types of submicron aluminas prepared from ammonium alum. Type A is calcined in the alpha alumina range, is 99.9% pure, and is useful as an abrasive, a catalyst carrier, and a raw alumina for fine sintering. Type B has cubic structure and is suitable for the Verneuil process for making jewel alumina. Other references to processes involving decomposition of sulfates include; Bretsznajder; Hegedüs and Fukker; Antonsen; Dingman et al.; and Stokes and Secord. Chirnside and

Table 4

Typical Properties of Calcined and Low-Soda Aluminas

Typical Properties	A-1	A-2	A-3	A-5	A-10	A-14
Al_2O_3 .... %	98.9	99.2	99.0	99.2	99.5	99.6
SiO_2 .... %	0.02	0.02	0.02	0.02	0.08	0.12
Fe_2O_3 .... %	0.03	0.03	0.03	0.03	0.03	0.03
Na_2O .... %	0.45	0.45	0.45	0.50	0.10	0.04
Loss on ignition (1100°C) .... %	0.6	0.4	0.4	0.2	0.2	0.2
Total H_2O (by Sorption-Ignition Test) .. %	1.	0.3	0.7	0.3	0.3	0.3
Bulk density, loose, lb/ft³	55.	52.	55.	48.	60.	63.
Bulk density, packed, lb/ft³	68.	68.	68.	63.	80.	83.
Specific gravity	3.6-3.8	3.7-3.9	3.6-3.8	3.7-3.9	3.8-3.9	3.8-3.9
Sieve analysis (cumulative)						
On 100 mesh................... %		4-15		2-10	4-15	
On 200 mesh.................... %		50-75		40-65	50-75	
On 325 mesh.................... %		88-98		75-95	88-98	
Through 325 mesh .... %		2-12		5-25	2-12	

Typical Properties of Tabular Alumina

Typical Properties	T-60	T-61	T-71	Balls 3/8 Inch
Al_2O_3 .... %	99.5+	99.5+	99.5+	99.5
SiO_2 .... %	0.06	0.06	0.04	0.04
Fe_2O_3 .... %	0.06	0.06	0.06	0.06
Na_2O .... %	0.20	0.02	0.01	0.05
Bulk density, packed lb/ft³				
Converter discharge	125.	125.	85.	
Granular — Minus 14 mesh..........	135.	135.	—	
Powder — Minus 325 mesh	140.	140.	—	
Specific gravity	3.65-3.8	3.65-3.8	3.65-3.8	
Apparent porosity .... %	5.	5.	30.50	
Water absorption .... %	1.5	1.5	15.25	

74° to 212°F (Btu/lb/°F)	0.200
74° to 932°F (Btu/lb/°F)	0.233
74° to 1832°F (Btu/lb/°F)....	0.244
Apparent porosity .... %	3.0
Water absorption .... %	0.9
Specific gravity	3.7-3.9
Balls per pound	202
Bulk density, lb/ft³........	135

Courtesy Aluminum Company of America

Dauncey; Barnes; Burdick and Jones; Ikornekova and Popova; Nikitichev et al.; and Zemlicka discuss phases of the production of synthetic jewel corundum as a boule grown in the oxyhydrogen flame.

Ali specified two ranges of hydrothermal transition as (1) a subcritical (two-phase) range in which the pressure required is determined by a given temperature, and (2) a hypercritical (monophase) range in which both pressure and temperature can be varied independently.

The hydrothermal system Al_2O_3-H_2O has been investigated by Laubengayer and Weisz; Cooke and Haresnape; Ervin and Osborn; Roy and Osborn; Newsome; Druzhinina; Kennedy; and Krishner and Torkar. Ervin and Osborn used aluminum hydroxide gels in water to confirm and extend the phase diagrams of the earlier investigators. They delineated regions of stability for gibbsite, boehmite, diaspore, and corundum. Gibbsite transforms to boehmite at 140°C in room air or in water, at least up to 680 atm (10,000 psi).

Torkar and Krischner substituted formation reactions for dehydration reactions and extended the diagram to new fields (Figure 6). At pressures above atmospheric the reaction of metallic aluminum of high surface with water was found sufficiently rapid and without the use of activators such as iodine or mercury. Besides forming diaspore at pressures of about 100 atm, other phases prepared included: metastable reactive alpha alumina (a^*-Al_2O_3), formed between 320 and 400°C, and at steam pressures from 15 to 100 atm; "autoclave" gamma alumina, similar to eta or chi aluminas obtained by calcination; kappa-l-alumina, a new form that transforms into kappa alumina and finally stable alpha alumina. The temperature range found for hydrothermal alpha alumina agrees well with the range 400 to 550°C specified by Newsome. The reactive alpha alumina, of particle size only 300 to 800 A, had a lower density (3.7 g/cc) than for stable corundum (3.98). The thermodynamic relationship between the reactive form and corundum is analogous to that between the pseudoboehmite and hydrothermal boehmite. The investigators considered the role of entropy and enthalpy in establishing reaction equilibrium and stability, a subject also discussed by Searcy. Krishner and Torkar determined the free enthalpy or "activity," assuming that the displacement of the equilibrium curve is due solely to the activity. Maximum attainable activity was taken as the disorder of the molten state. The excess of total enthalpy of the active alpha alumina was computed at 4 to 5 kcal/mole, and the entropy at 3.5 to 3.9 clausius higher than in stable alpha alumina. They concluded that reactions with active forms begin sooner and proceed faster than with stable substances. Sintering occurs sooner, and reactions can take place at lower temperatures, a desirable feature for ceramic applications.

Newsome obtained relatively large crystals of alpha alumina at 450°C and 300 atm steam pressure. Yanada et al. claim that the dimensions of the crystals depend on the grain size of the starting aluminum hydroxide. Kryukova

and Nozdrina found that about 65% of the pressure-crystallized material is boehmite, 10% is diaspore, and only 20% is corundum. Yamaguchi, Yanagida, and Soejima determined that the equilibrium solubility of corundum is independent of pressure. The gram equivalent ratio of dissolved corundum with respect to aqueous sodium hydroxide is 2.46 at 350°C, and 2.95 at 500°C. The activation energy is 3.8 kcal. Nucleation without seed starts at a supersaturation ratio of 0.12. Kuznetsov calculated the activation energy of hydrothermal crystallization of corundum to be 17,500 kcal/mole for prism ($11\bar{2}0$) planes and 32,000 for rhombohedral ($10\bar{1}1$) planes. Nozdrina and Tsinobar found that corundum crystallized spontaneously in aqueous Na_2CO_3 solutions either as plates or equants; in $NaHCO_3$ solution prisms elongated along the C-axis are formed.

Ballman described a method for growing synthetic rubies on seed crystals in an aqueous medium containing corundum and a soluble chromium compound in contact with solid corundum held at 415 to 520°C, and not more than 20° higher than the seed temperature.

The preparation of fused alumina from electric furnace melts was mentioned in Section 3-2-4. It is also possible to grow alumina crystals from fusions other than alumina, but saturated with alumina. Remeika used mixtures of the oxides of lead and boron. Barks and Roy found four mechanisms of growth from various modified borate fluxes: lateral growth, screw dislocation, vapor deposition, and twin-plane reentrant edge. E.A.D. White crystallized corundum from lead fluoride (PbF_2) at 1100 to 1300°C.

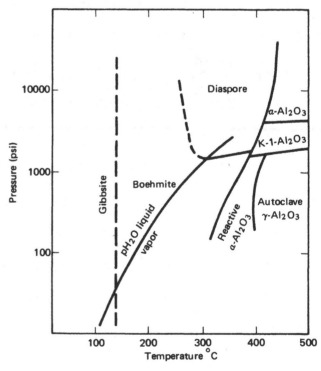

Figure 6. System Al_2O_3-H_2O
(After Ervin and Osborn, and Krishner and Torkar)

Timafeeva and Voskanyan; and Giess found a marked increase in solubility of alumina in lead fluoride between 900 and 1250°C. Giess found the weight ratio of Al_2O_3 to PbF_2 was 0.093 at the lower temperature, and 0.151 at the higher. The principal faces of corundum crystals grown from these melts were (0001) and (10$\bar{1}$1), with minor (01$\bar{1}$2) and very small faces of (22$\bar{4}$3). Timofeeva discussed the growth of corundum from such melts. Sysoev and Obukhovskii used supersaturated melts of sodium, potassium, or lithium cryolite. A. B. Chase used La_2O_3 as a growth modifier for single-crystal corundum grown from PbF_2-Bi_2O_3 melts.

Gaudin attempted to prepare synthetic corundum for jewel applications in 1837, by heating a mixture of powdered alumina and potassium sulfide in a crucible covered with charcoal. Only very small crystals (probably beta alumina) were obtained. Many investigators have since made the attempt. Fremy and Feil were the most successful until the advent of the Verneuil process in 1902. They fused alumina with lead oxide in a fireclay crucible; the lead aluminate was attacked by the fireclay, and the liberated alumina crystallized as small plates. The addition of 2 to 3% chromium salts imparted a ruby color. Watch jewels were made from the crystals.

The Verneuil flame-fusion process for preparing artificial sapphires consists of dropping finely divided pure alumina onto a preoriented crystal of corundum, 2 to 3 mm in diameter, surrounded by an oxy-hydrogen flame furnace. The container for the alumina supply is equipped with a tapping device for feeding the powder uniformly to the area of impingement with the growing crystals (boule). The feed alumina is calcined ammonium alum. The boule is rotatable, and can be withdrawn from the heating zone in a slow continuous manner as the growth proceeds (Popov). Rods as slender as 1.5 mm may be grown to avoid excessive loss in cutting operations. Single crystals of 100 to 150 carats (20 to 30 grams) in 2 to 3 hours have been obtained as expanded boules.

The purest alumina is required for colorless sapphire. Metal oxides are added to the alum for colored crystals. Red (ruby) crystals contain about 1 to 7% chromic oxide; blue crystals a mixture of about 1% titania and 2% iron oxide; green crystals a mixture of vanadium and cobalt oxides; and alexandrites vanadium oxide only. Green crystals are also produced if the chromium content exceeds 8 mole %. Alumina and chromia are completely miscible in both the liquid and solid states, and do not form compounds (Bunting, 1931). A supposed discontinuity in lattice change with increasing chromia content, noted by Thilo, has been discussed by Zen; Spriggs and Bender; and Sarver. Fang cited evidence to disprove the discontinuity, and to confirm Vegard's law relating the lattice parameters of solid-solution binary systems linearly with the atomic percentage of one component.

Single-crystal growth of sapphire using a plasma torch in plasmas of air, argon, nitrogen, or oxygen at radio-frequencies was described by Alford and Bauer. The densities obtained were slightly higher than those for flame-fused crystals. Almost perfect crystals 100 mm long by 15 mm diameter have been grown by vertical pulling from the melt at pull rates of 6 to 50 mm/hr (Cockayne et al.).

Spoerry prepared corundum from aluminum dross collected when aluminum scrap is melted. Wendell prepared the oxide by blowing an oxygen-rich flame against a pool of molten aluminum.

Coble prepared semi-transparent polycrystalline alumina having an in-line transmission of not less than 0.5%/mm thickness of the body in the wavelength range 0.3 to 6.6 microns, and not less than 10% at one wavelength. Increased transmission is attained by adding not more than 0.5% magnesia as spinel to reduce internal voids. St. Pierre and Gatti improved the transparency by firing the polycrystalline compact in a hydrogen atmosphere at 1650 to 1750°C for 50 to 300 minutes to remove gas-containing pores, and then firing at 1800 to 2000°C for not less than 15 minutes additional.

Worel and Torkar hydrolyze aluminum fluoride at 700 to 1000°C in air or oxygen containing from 10^{-16} to 3 mg H_2O/liter. Locsei hydrolyzes the fluoride in a fluidized reactor at 800 to 1000°C to obtain laminarly crystallizing alpha alumina. Schaffer devised a chemical vapor deposition technique for the epitaxial growth of alpha alumina crystals by reacting aluminum chloride with a mixture of hydrogen and carbon dioxide at 1550 to 1800°C.

May observed the growth of microcrystalline corundum platelets from vapor on objects heated in a resistance furnace with an aluminum oxide furnace tube and a hydrogen atmosphere. There was evidence for growth by the screw dislocation mechanism. Sears deposited nearly perfect single crystals from the vapor phase upon a suitable substrate. The degree of supersaturation of the vapor was adjusted by varying the temperature according to the equation

$$\log_n \, a \; = \; (\pi a \theta^2 M)/\left(\rho k T^2 R \; \log \frac{B}{N}\right)$$

in which a is the ratio of the partial pressure of the vapor to the equilibrium partial pressure of the vapor at the deposition temperature, a is the interlayer spacing of the material, ρ is the density of the material, θ is the surface free energy of the material, M is the molecular weight, k is Boltzmann's constant, R is the molar gas constant, B is equal to 10^{20} sec^{-1}, and T is the deposition temperature (°K). Bauer, and Sears and DeVries investigated the growth process by the vapor-phase mechanism.

3-7-2 Factors Affecting Alumina Transitions

A few instances of factors affecting either the temperature of transition or the sequence of phases have already been cited. Some of these factors are: the starting material, particle size, the extent of disorder or activity, gases in the calcining atmosphere, impurities, and additives that either promote or suppress crystal growth or affect other properties. In general, the factors discussed here are those which affect loose powder transitions. Factors which affect sintering will be treated under that subject.

Alkaline impurities in Bayer trihydrate agglomerates generally increase in amount with increasing size of the agglomerates and with prolonged aging in the precipitating liquor. These impurities and pseudomorphic factors (Section 4-2) influence the crystal size of the calcined product. The effective crystal size distribution of some calcined grades may extend through a range of 10 to 15 times (8 to 9 hypothetical $\sqrt{2}$ sieve intervals).

Ginsberg, Hüttig, and Strunk-Lichtenberg discussed the influence of the starting material on the transitions. They observed that Bayer alumina trihydrate, dried in air at $105°C$, always contains a small amount of boehmite. Potassium and sodium ions in the lattice increase the temperature of transition of both gamma and delta aluminas to alpha alumina. The transition temperature is normally about $1150°C$ for Bayer alumina, but may be as low as $1050°C$ for reactive alumina containing less than 0.08% Na_2O (MacZura). The stability of boehmite is independent of the alkali content, however.

Layng; Huffman; and Bailey and Bittner prepared alumina catalyst supports containing as much as 12% silica, coprecipitated with the alumina to repress the transition to alpha alumina. Eliasson produced fine crystalline alumina by addition of 10% $(NH_4)_2PO_4$ to Bayer alumina trihydrate and by calcination for from one to five hours at 1000 to $1300°C$. Pouillard found that trivalent iron reduces the transition temperature from gamma to alpha alumina.

Noda and Isihara found that fluorides, chlorides, and borax accelerate crystal growth of alpha alumina. Hüttig, and Hüttig and Markus showed that HCl atmospheres induce formation of alpha alumina at $850°C$, while SO_3, HBr, Cl_2, NO_2, and SO_2 produce only gamma alumina. Murray and Rhodes converted amorphous alumina to alpha alumina by heating in an atmosphere containing oxides of nitrogen and a small amount of sodium nitrate at only $400°C$. The addition of solid boric acid inhibited the formation of corundum. Edling et al. found that the low conversion temperature does not apply to crystalline alumina. Hedvall and associates investigated the effect of different atmospheres on the chemical reactivity of gamma and alpha aluminas with lime. Oxygen, nitrogen, and sulfur dioxide were activators, but not sulfur trioxide.

Voltz and Weller patented the use of 0.5 to 1.5 cation percent of thoria, or 0.2 to 2.6 cation percent of hafnia to inhibit transformation to alpha alumina. Iler (1964) claimed that the addition of from 2.7 to 20% silica stabilized calcinations of fibrillar boehmite in the theta phase at $1150°C$ for at least 10 hours at the lower concentration, and in gamma phase for 24 hours at the higher concentration.

Wakao and Hibino found that the transition temperature from gamma to alpha alumina was reduced by additions of from one to 10% MgO, NiO, CuO, MnO_2, Fe_2O_3, TiO_2, or

SiO_2. The maximum reduction was $250°C$ with 10% additions and $150°$ with 1%. The larger the metallic ion of the oxide and the higher the vapor pressure of the oxide at the transformation temperature of alumina, the larger the effect. Roberts and Jukkola used the Pechiney method of addition of a volatile fluoride to reduce the conversion temperature to alpha alumina during fluid calcination of finely divided alumina.

3-7-3 Special Ceramic Aluminas

When the German ceramic ware "Sinterkorund" was developed around 1930 (Kohl), the only types of pure alumina available in significant commercial amounts for ceramic purposes were a single metallurgical grade of Bayer alumina, fused white alumina fines, and small amounts of calcined, recrystallized ammonium alum used in the Verneuil flame-fusion process for sapphire (1902). The Bayer alumina was produced as a mixture of phases, about 40% higher transition phases and 60% alpha alumina, the most significant impurity of which was about 0.5% Na_2O. The fused alumina fines was about of the same purity as that of the Bayer alumina from which it had been prepared, except for a lower free-soda content.

At that time slip casting appeared to be the most feasible forming method, particularly in the highly acid suspensions developed by Reichmann to improve the forming operation. The method had objectionable features in that the high acidity was corrosive on mold materials, and the slip-casting process was not well adapted to large-volume, commercial production of small items such as spark-plug porcelains. The slip-cast product was exceptionally good, however, when prepared from Bayer-process alumina, partly because the free alkali was converted to a more volatile salt, and much of it had been withdrawn during the casting operation. The residual alkali in sintered ware was considered detrimental mainly to electrical properties, without, however, definite information on tolerable amounts.

Improvement in the quality of ceramic grades of alumina has developed along two different paths: (1) the reduction of impurities in normal Bayer alumina, and (2) the development of particle distributions having better technical properties.

Attempts to reduce the soda content of Bayer trihydrate by leaching in dilute mineral acids were not very rewarding because of the dispersion of the alkali within the crystal structure and the slow rate of diffusion (Stowe 1946). Upon calcination, the Bayer structure opens sufficiently to enable the practical release of about one-half the contained soda by dilute acid leaching (Gitzen, 1934, Yamauchi and Kondo). Ozment claimed a reduction to less than 0.03% Na_2O by water-leaching a rehydratable transition alumina at 80 to $98°C$ in a column. Leum et al. reduced the sodium oxide content to about 0.12% by mixing calcined alumina in water suspensions with Amberlite IR-120 ion-exchange resin.

Fessler (1937) reduced the soda to less than 0.05% by calcining in the presence of small amounts of boric acid.

Thompson restricted crystal growth during this process by preventing the recycling of sodium tetraborate in the rotary kiln. Riesmeyer, and Riesmeyer and Gitzen leached with fluoboric acid or mixed acids to remove the soda effectually from alumina that had been precalcined without excessive crystal growth. Alternatively, fine crystal size and low soda content could be achieved by calcining Bayer trihydrate with boric oxide or with boric oxide and a fluoride, at a temperature below the effective volatilization temperature, and then leaching out the solubilized sodium salt (Gitzen, 1963). These methods leave a small amount of boron in the alumina, which acts as a fluxing agent in ceramic compositions. Lindsay and Gailey (1965) claimed the method of heating the alumina to above 1000°C with gaseous Cl and AlF_3.

Holder, Helmboldt and Vogt (to Giulini, 1963) claimed a process in which Bayer trihydrate is calcined in the presence of any of various silica-containing minerals of coarser particle size than the alumina. The soda in the alumina combines to form insoluble sodium aluminum silicates on the siliceous material, which can be separated substantially from the calcined alumina by sieving. Watson, Lippman, and Royce claimed by this process to reduce the sodium oxide content below 0.02%. The residual silica was as low as 0.07%, and the crystal size below 2 microns for alumina calcined at 1260 to 1540°C. Holder, Vogt, and Helmboldt also claimed that a soda content below 0.1% Na_2O can be attained merely by calcining Bayer hydrate at 1300°C for a sufficiently long period (determined by a dye test with alizarine red), and by removing by air separation the fraction substantially coarser than about 60 microns in diameter.

Fenerty calcined the alumina in the presence of sufficient aluminum fluoride (about 3%) at a temperature high enough to volatilize the sodium fluoride formed. Alternatively, unvolatilized soda could be leached out with dilute acid or alkaline solutions. Products containing less than 0.04% Na_2O were claimed, but the crystallinity of the calcined alumina was increased into coarser size distributions. Nixon and Davis removed alkali to about 0.2% Na_2O by treating with sulfamic reagents. Giulini (1965) patented a process of calcining hydrated alumina in the presence of halides under reducing conditions to vaporize ZnO and Na_2O. The preferred halides were NH_4Cl or $AlCl_3$; fluorides were not considered desirable. Pechiney, Netherlands Application 6,611,170, claimed the removal of silica, iron oxides, and titania in a 2-step treatment: (a) raising the temperature to 1200°C in the presence of C and S to reduce the oxides to metals and sulfides, and (b) reacting at 500 to 700°C with Cl_2.

The low-soda transition of Bayer hydrate to boehmite by hydrothermal pressure digestion is also a feasible method for producing ceramic alumina containing less than 0.05% Na_2O (Barrett and Welling). High-purity alpha monohydrate, obtained by hydrolysis of aluminum alkyl (Continental Oil Company), and calcined in the range 800 to 1200°C, is available in tonnage amounts and purity of 99.97% Al_2O_3. The sintering reactivity of the calcined

monohydrates has so far not been quite as satisfactory as that obtainable by calcining gibbsitic types, perhaps because of the extreme platiness of the crystallites.

Some methods for producing ceramic alumina of a higher purity than is normally attainable in the Bayer process have been alluded to previously. These are relatively high-cost procedures because they generally involve retreatment of Bayer alumina by solution in acid or alkaline processes, or solution of high-purity aluminum (99.99%) in acids, alkalies, or anhydrous organic liquids. The calcination of recrystallized salts (sulfates, chlorides, nitrates, acetates, etc.) results in the formation of finely divided, high-surface transition aluminas, but small amounts of the anions may be retained under calcination conditions that produce reactive size distributions. Suehiro, et al. roasted basic alum to prepare alkali-free, high-purity alumina for refractories. A laser grade having a purity of 99.997% is claimed for triple recrystallized alum, calcined to submicron particle size (Linde). The calcination product of pure aluminum isopropylate has been shown to contain less than 4 ppm sodium oxide (Dawihl and Kuhn).

Residual alkali below 10 ppm is claimed for a Solvay process involving the vaporization and oxidation of aluminum chloride, but 0.02% residual SiO_2 is about in the same range as for Bayer alumina. Thiele, Schwartz, and Dettmann burned triethyl aluminum in the O-H flame, and heated the product to 1200°C to obtain spherical particles 0.02 to 0.2 micron in size. Iwatani alumina, another finely divided product, is prepared by passing a high-frequency spark discharge through a suspension of coarse metallic aluminum in water or other nonconducting liquid, and by calcining the hydrolized fine product (Ishibashi). The purity of the product is dependent on the purity of the metal, which may run as high as 99.995 to 99.999% in metal purified by a special electrolytic process or perhaps by zone refining.

Iron-group impurities have been removed as carbonyls by treatment at 300 to 600°C with carbon monoxide and moisture (G. Free), or by heating at 600 to 800°C in carbon tetrachloride (Lehman), or by passing the acid aluminum salt solutions through a strongly basic anion-exchange resin column (Nagumo and Murakoshi).

Hayes removed calcium and magnesium salts from aluminum salt solutions prior to filtration by complexing with ethylene diamine tetraacetic acid (EDTA). Chromium was removed by oxidizing in sulfuric acid solutions and by separating in Amberlite LA-l resin (Minamiie and Toshiro).

The analyses and, in some cases, attainable forming properties, are shown for some commercial high-purity reactive aluminas in Table 5.

The term "reactivity" is currently used to describe the new approach to improved pure-oxide ceramics. The minimum time-temperature conditions required to sinter or otherwise mature the ceramic body is stressed. An added criterion is optimum compaction or minimum fired shrinkage to mature. The phenomena involved in sintering are discussed to some extent in Section 16. The mechanism of

Table 5

Typical Properties of Low-Soda Ceramic Aluminas

	Republic Foil Type 9902209	Cabot Alon	A-10	A-14	A-15	RC-152	A-16	Linde A	Gulton Alucer MC
Chemical Analysis (%)									
Na_2O	.02		.10	.04	.06	.05	.07	.02	.01
SiO_2	.05		.08	.12	.02	.07	.02	.01	< .01
Fe_2O_3	.003		.03	.03	.006	.03	.006	.002	< .003
TiO_2	< .001		.002	.002	< .001	.002	< .001	—	< .0008
CaO	.08		.04	.04	.04	.03	.04	—	.001
Ga_2O_3	.00	See note 3	.02	.02	.004	.02	.003	—	—
B_2O_3	< .001		.1	.06	< .001	< .001	< .001	—	—
MnO	—		.0003	< .0003	.001	< .0001	.001	—	—
Cr_2O_3	.0005		.0002	.0002	.0002	< .0005	.0002	—	.007
MgO	.020		.001	.001	.001	.002	.05	—	—
ZnO	—		.003	.003	.001	.005	.002	—	< .001
CuO	—		.000	.000	.0002	< .0005	.0002	—	.0003
V_2O_5	—		.001	< .0001	.0003	< .0005	.0005	—	—
F	—		< .03	< .03	.01	< .03	< .01	—	—
Crystal Size (μ)									
Median	< 0.1	.01-.02	7.0	4.0	2.5	2.0	0.7	0.1	1.5-2.0
Ceramic Properties[1]									
Density (g/cc)									
Pressed at 5000 psi	—	—	2.36	2.21	2.51	2.33	2.11	—	—
Fired	—	—	3.37	3.62	3.89	3.88	3.93[2]	—	—
Linear Shrinkage (%)	—	—	11.5	15.2	13.5	15.6	18.7	—	—

[1]Alumina ground to ultimate crystals, compacted at 5000 psi and fired 1 hr at 1700°C.
[2]A-16 fired 1 hr at 1550°C.
[3]Alon C & R-13 are α and only 95% pure. Alon C contains Fe (.2%), Cl (<1%), H2O (2-3%). Alon R-13 has residual sulfate of 2%.

sintering has been investigated mainly on the basis of uniformly sized spheres (Coble 1958), or available size distributions which do not necessarily satisfy the criterion for good compaction.

Ideality of size distribution has been considered from the standpoint of selecting size distributions so that the finer particles fill the voids created in packing the coarser particles (Furnas, 1931; Lewis and Goldman, 1966). The mathematical determination of the optimum size distribution in this manner has provided improved compaction and lower porosity in coarse size distributions in which it is possible to sieve mechanically the size increments. The treatment still requires some experimental determination of the best combinations because the particles themselves do not satisfy the assumed ideal shapes, or fail to settle into the necessary compaction positions.

In the range of ceramic reactivity of present interest, about an average particle size from 0.05 to 5 microns, no mechanical separation methods are presently feasible for obtaining ideal size distributions commercially. Moreover, present techniques have not developed economical methods for altering the usual platy Bayer crystals to equants or spherical shapes, which might provide more suitable compaction properties. The reactivity has been improved by the development of control of calcination so as to obtain two or more crystal size distributions within the desired range, from which combinations may be selected for control of ceramic reactivity (Gitzen and MacZura, 1966). In this way the reactivity is improved so that fired densities of cold-pressed bodies (99.5% Al_2O_3) above 3.90 g/ml are attainable within one hour at 1500 to 1550°C. Former alumina types required about 4 hours at 1700°C to attain the equivalent fired properties. Densities as high as 3.93 g/ml have been achieved at shrinkages below 11% when fired for one hour at 1650°C. Shrinkages as high as 20 to 23% were obtained with former types having the same equivalent reactivity.

Increasing the fineness of ceramic-grade alumina further in the submicron range has so far not been very advantageous. This is shown in low compaction densities by the usual forming methods, in excessive crystal growth, and in defective sintered ware.

3-8 BETA AND ZETA ALUMINAS

Beta alumina is the phase observed as an impurity in white fused alumina, representing the lower-melting fraction of the ingot. The soda in the Bayer alumina raw charge migrates to this fraction. The beta alumina fraction is usually separated from fused abrasive alumina because of its poorer hardness relative to alpha alumina. Its main ceramic usefulness at present is as fused cast refractories. Smith and Beeck prepared active beta-alumina catalysts by heating mixtures of gamma alumina with about 5% Na_2O as sodium nitrate at 1050°C for six hours. Fusion in the electric furnace is a suitable method for preparing well-crystallized massive beta alumina.

Yamauchi and Kato found beta alumina in Bayer trihydrate (0.87% Na_2O) heated at 1300°C. By using the correct ratio of sodium carbonate to alumina, conversion can be initiated at 900°C and is completed at 1200°C. Excess Na_2O reacts to form normal $Na_2O \cdot Al_2O_3$ which melts at 1650°C. If no excess Na_2O is present, beta alumina transforms to alpha alumina at 1730°C. Sodium, magnesium, and calcium beta aluminas are stable at 1700°C if they have more than 5% of the corresponding oxides present (Funaki). According to Gallup, sodium beta alumina converts to alpha Al_2O_3 at 1300°C in vacuum or in hydrogen, but 1650°C is required in air or argon. Conversely, Saalfeld in 1956 found that alpha alumina converts to beta alumina at about 1000°C on treatment with hydrofluoric acid. Chiolite or cryolite are also effective. P. A. Foster in 1962 precipitated beta alumina from melts of sodium fluoride in the presence of solid sodium aluminate.

Ridgway, Klein, and O'Leary tentatively concluded that the formula for the alkali beta aluminas conforms to $Na_2O \cdot 12Al_2O_3$. Yamauchi and associates in 1943 prepared products having molar ratios conforming closely with this formula. Beevers and associates, by correlating the hexagonal structure with a revised value for density, 3.25, obtained the formula $Na_2O \cdot 11Al_2O_3$, rather than $NaAl_{23}O_{35}$, as suggested by Bragg, Gottfried, and West. DePablo-Galan and Foster confirmed the Beevers formula. In the belief that beta alumina is a stable phase, they revised the phase diagram, Na_2O-Al_2O_3-SiO_2, to indicate its field. Analogous compounds that have been described, include $K_2O \cdot 11Al_2O_3$, $Rb_2O \cdot 11Al_2O_3$, $Cs_2O \cdot 11Al_2O_3$, and $MgO \cdot 11Al_2O_3$. The related compounds in the alkaline earth series, however, have the formulas $CaO \cdot 6Al_2O_3$, $BaO \cdot 6Al_2O_3$, and $SrO \cdot 6Al_2O_3$. These also belong to the hexagonal crystal system. The lithium aluminate, zeta alumina, is cubic, and has the formula $Li_2O \cdot 5Al_2O_3$.

Brownmiller prepared potassium beta alumina at 1550°C. Kato and Yamauchi prepared potassium beta alumina at 1640 to 1700°C, and magnesium beta alumina at 1600 to 1800°C. Foster and Stumpf demonstrated the analogies between the alumina and gallia beta structures. Roth and Hasko claim a structure in the same crystal group

for $La_2O_3 \cdot 11Al_2O_3$. Cirilli and Brisi prepared solid solutions of beta alumina and beta ferric oxide.

Scholder and Mansman reinvestigated the beta alumina structure in 1963 and concluded that the structure is $Na_2O \cdot 6Al_2O_3$. They claimed that all reaction products having a molar ratio of $Na_2O \cdot 11Al_2O_3$ are mixtures of $Na_2O \cdot 6Al_2O_3$ and Al_2O_3. They prepared isomorphous compositions of $Na_2O \cdot 6Al_2O_3$, $K_2O \cdot 6Al_2O_3$, $Rb_2O \cdot 6Al_2O_3$, and $Cs_2O \cdot 6Al_2O_3$. Rolin and Pham (1965) presented data from cooling-curve arrests that confirm the formula, $Na_2O \cdot 6Al_2O_3$. Thery and Briancon (1964), on the basis of X-ray interpretation of phases in specimens prepared in a mirror furnace, claimed that the phase, $Na_2O \cdot 11Al_2O_3$ exists, but is stable only above 1650°C, and that a new phase, $Na_2O \cdot 5Al_2O_3$, can be prepared by prolonged heating in suitable stoichiometric proportions of the components at 1100 to 1200°C. It is still indefinite whether the beta aluminas have fixed ratios of alkali to alumina.

Bor found that beta alumina recrystallizes in corroded glass tank firebrick in thin sharp-edged perfect hexagons embedded in glass. This is also the habit of crystallization from Bayer alumina.

Toropov and Stukalova found that sodium in beta alumina can be substituted by calcium, strontium, and barium by a double fusion with a 6-fold stoichiometric amount of the corresponding alkaline earth chloride. Filonenko and Lavrov prepared calcium beta alumina by prolonged heating of the oxide mixtures. Wisnyi synthesized $CaO \cdot 6Al_2O_3$ by flame fusion. Massazza prepared $SrO \cdot 6Al_2O_3$.

Barlett precipitated zeta alumina, $Li_2O \cdot 5Al_2O_3$, as octahedra from melts containing about 0.35% Li_2O. Akiyama added 2% lithium oxide to alumina and heated at 1700°C. Long and Foster in 1961 found that fused alumina containing 3.9% nitrogen, added as aluminum nitride, has an X-ray powder pattern identical with that of lithium zeta alumina, a spinel structure closely similar to eta alumina.

Kordes proposed the spinel structure for zeta alumina. Datta and Roy find that zeta alumina exists in many forms; it transforms at about 1295°C to a low-temperature cubic form. Hafner and Laves claimed intermediate forms between these two.

3-9 SUBOXIDES AND GASEOUS PHASES

Alpha aluminum oxide is one of the most stable compounds. Lanyi concluded that there was no reduction to metal in free hydrogen up to 900°C, and only slight formation with atomic hydrogen. In contact with carbon in the electric furnace at its melting point, alumina forms two oxycarbides, Al_4O_4C and Al_2OC with some metal (Foster, Long, and Hunter). Beletskii and Rapoport claimed to have prepared the suboxide Al_2O (crystallizing as needles from the vapor phase) by reduction with silica in a carbon monoxide atmosphere at 1800°C and 1 mm Hg pressure. The crystals contained no carbide or nitride. The spinel form, $AlO \cdot Al_2O_3$, claimed by Filonenko, Lavrov, Andreeva, and Pevzner is a partially reduced form. Carnahan, Johnston, and Li found that the contact angle of liquid aluminum on alumina at about 1240°C and 10^{-6} mm Hg pressure decreased within 20 minutes to a steady value of 60 degrees. On sapphire at 1200°C, the drops spread and then suddenly contracted to a contact angle of 80 degrees. The spreading and contraction were repeated cyclicly. The behavior was ascribed to the formation of aluminum suboxide gas.

Arndt and Hornke observed that pure alumina that had been sintered at 1700°C volatilized appreciably at 1500°C. Zintl et al. claimed that alumina and boron react when heated to form the volatile monoxides, AlO and BO.

Bevan, Shelton, and Anderson; and Sandford and Ericsson showed that alumina loses oxygen when heated at 1400 to 1500°C in water vapor, but retains its weight unchanged in air or nitrogen. The loss amounted to as much as 0.2% in excess of possible Na_2O volatilization in 8 hours. Ordinary kiln gases are dissolved in mullitic material at 800 to 1200°C. Their solubility increases with increasing temperature. Rhodes found that alumina vapor diffuses in hydrogen or at low oxygen pressures to induce contamination of objects being furnaced at temperatures around 1200°C. Roy and Coble (1967) derived an equation for solubility of hydrogen in porous polycrystalline aluminum oxide: $\ln X = (-2.28 \pm 0.5) - [(8024/T) \pm 1500]$, in which X is the atomic ratio of H to Al, and T is the absolute temperature. Readey and Kuczinski attributed the control of sublimation of sapphire in dry hydrogen to gaseous diffusion of either AlO or H_2O vapor through a stagnant boundary layer. Poluboyarinov, Andrianov, et al. stated that the comparative rates of vaporization of porous oxide ceramics were: MgO, ZrO_2, Al_2O_3, and BeO.

Brewer and Searcy concluded from vapor pressure measurements in Knudsen effusion cells that alumina volatilizes as AlO when heated alone, but as Al_2O when heated with aluminum. They found vapor pressures of 2×10^{-39} mm Hg at 1200°C and 5×10^{-19} mm Hg at 1830°C. The tests were made in tungsten cells with tantalum top shields. Von Wartenberg (1952) obtained higher values, around 8×10^{-4} mm Hg at 1950°C. The stable phases when aluminum and alumina are heated together are Al and Al_2O_3 below 1100°C, Al_2O between 1100 and 1600°C, and AlO above 1500°C (Hoch and Johnston). Yanagida and Kroger found no evidence of solid-state existence of either Al_2O or AlO.

Wartenberg and Moehl; Ackermann and Thorn; Navias; and Sears and Navias found that there is a reaction between alumina and the metals, tungsten and tantalum at high vacuum, even though not in contact. In tantalum equipment, aluminum is formed at 1600°C, and in tungsten equipment at 1900°C. The evaporation rate of alumina at 1900°C out of contact with possible reactants was less than 5.5×10^{-9} g/cm²/sec, about three orders of magnitude less than the evaporation rate obtained by Brewer and Searcy in contact with a tungsten cell.

Walker, Efimenko, and Lofgren found that the calculated vapor pressures (assuming AlO to be the vapor species) in the temperature range 1493 to 1614°C were about five times as high as the least-squares extrapolation of the effusion-cell determinations of Brewer and Searcy. They found no reaction between sapphire crystals and water vapor up to 1600°C and a partial pressure of H_2O of 10 mm. There was severe reaction between water vapor and the platinum suspension. Vaporization was extensive, and not inconsistent with the possible formation of AlOH molecules in the vapor.

The seeming inconsistency in these rates of vaporization suggests the possibility of minor impurities, either present in the sapphire raw material or picked up during the formation of the boule, having an effect on the evaporated phase.

Burns, Jason, and Inghram found that the rate of evaporation of polycrystalline alumina, heated in vacuum in an arc-image furnace, increased by a factor of 3.3 as the phase changed from solid to liquid. A similar discontinuity was observed for mullite.

Juillet et al., Arghiropoulos et al., and Teichner et al., described the preparation of oxygen-deficient aluminum oxide (approximately $Al_2O_{2.96}$) by heating alumina phases of high surface area (e.g. hydrolyzed $AlCl_3$) in a vacuum at 500 to 1000°C. The alumina assumes a black color and becomes a semiconductor. The normal white color is restored when the nonstoichiometric form is reheated in air or oxygen. Krieger computed the thermodynamics of the Alumina/Aluminum-Oxygen system to 6000°K and pressures to 10^3 atmospheres, on the basis of both a pure gas-phase system and a mixed gas- and condensed-phase system containing the 12 gaseous species: Al, Al^+, e^-, O, O^-, O^+, O_2, O_2^+, AlO, Al_2, Al_2O, and Al_2O_2, and the condensed species, Al_2O_3 (crystal or liquid). The two sets of data were plotted as a Mollier diagram of specific enthalpy vs. specific entropy in two mutually exclusive contiguous regions.

The form of the alumina structure produced from vapor-phase transport is variable, depending on the conditions, and is significant for certain ceramic applications. Discrete platelets, fibers (whiskers), and spheres are common forms obtained. The vapor-phase method entails the presence of controlled amounts of moisture to facilitate formation of volatile aluminum suboxide. Hegedus and Kurthy obtained amorphous, spherical alumina, deficient in oxygen by calcining aluminum ethylate in hydrogen at 1600°C. Hargreaves, from thermodynamic considerations, concluded that the vapor-phase growth of whisker and platelet crystals of alpha alumina is by transport of Al_2O but not AlO. Timofeeva and Yamzin, and Shternberg and Kuznetsov grew corundum crystals on a seed from the gas phase by heating aluminum fluoride at 900 to 1250°C.

The preparation and applications of fiber alumina are discussed more fully in Section 26.

4 STRUCTURE AND MINERALOGICAL PROPERTIES

The remarkable range of properties of the hydrous and nonhydrous crystalline phases of alumina has induced much scientific curiosity about their structure. Examples of these structural peculiarities are the factors determining the chemical reactivity of the hydrous forms, the high surface phenomena of the transition phases, and the exceptional strength and hardness of corundum. Besides the ideal crystal structures, which are a rarity in actual ceramic systems, defect crystal structure and microstructure are significant. Gross structure beyond the crystal lattice is also of ceramic interest. The symposium, *Microstructure of Ceramic Materials*, NBS Misc. Pub. No. 257, and *Defect Solid State*, Gray et al. discuss some of these factors.

4-1 STRUCTURE OF THE ALUMINA PHASES

The properties of the aluminas are determined mainly by the crystal structure. In general, the phases of most significance in alumina are those produced by pseudomorphic dehydration.

Powder X-ray methods for determining crystal structure are not as suitable as single-crystal methods. In many cases, however, the lack of single crystals of the alumina phases as coarse as one or two tenths of a millimeter has restricted the development of exact information. Lack of this information on the atom parameters has thrown some doubt on the named structure or crystal system for some phases.

The crystal structures of the alumina phases are shown in Table 6. Mineralogical properties of the various phases are shown in Table 7.

The crystalline modifications of alumina are, without exception, classed as ionic (Verhoogen). For the arrangement of ions in the lattice, the coordination number of the cations is of paramount importance. The relatively small Al^{3+} cation is usually 6-coordinated with respect to O^{2-} or $(OH)^-$ anions, and is located in the interstices of octahedral anion groups. It may appear at the center of tetrahedral groups of 4-coordination, as in glass and in mullite. The Pauling principle of coordination is generally satisfied within each octahedron, even with different kinds of octahedral coupling, but in any case, neutralization is obtained at least for each separate unit cell.

In the crystal lattices of the alumina phases, both cubic and hexagonal close packing may take place. In cubic close-packing of the densest arrangement, the ions of one plane occupy the hollows formed by three ions of the next plane. Every fourth plane of superposition in the crystal lattice repeats (ABC ABC) to yield the face-centered cubic cell.

In hexagonal close packing, the sequence of close-packed anionic planes recurs every third plane (ABAB). The linking of octahedra and tetrahedra differs from that of cubic close-packing. In the direction parallel to the hexagonal c-axis, adjacent forms are linked through faces; in other directions through edges. In both crystal structures, the ions situated at octahedral corners belong to six adjacent octahedra. In general, the octahedra and tetrahedra will be distorted by the cations. Empty octahedra are larger, while those filled with aluminum ions become smaller.

Gibbsite has been the most precisely determined of the trihydroxide phases because of the availability of well crystallized coarse specimens. Gibbsite is designated as monoclinic by Dana. It usually occurs both native and in commercial production as twinned, pseudohexagonal tablets. Saalfeld (1960) observed triclinic crystals dispersed in the monoclinic structure of the native form.

Megaw determined the structural arrangement of gibbsite to consist of double layers of hydroxyl ions of closest packing enclosing the aluminum atoms within their interstices. Each aluminum atom is enclosed by three anions above and three below to form an anionic octahedron. The double layers are stacked in the direction of the c-axis. The aluminum atoms occupy only two-thirds of the available interstices in their plane and are arrayed in a pattern of hexagonal rings surrounding the vacant sites. There are no aluminum ions in the plane between adjacent double layers, and since the anions are superposed in these contacts, rather than lying in depressions, this plane represents a cleavage plane of weakness.

The stacking is deformed slightly in the direction of the a-axis to give the monoclinic lattice. The triclinic form observed by Saalfeld is obtained by a possible displacement in a direction perpendicular to the (110)-plane (60° with the a-axis). The open structure, in particular the hole through the centers of the hexagonal rings, allows paths for

Table 6

Crystal Structure of the Aluminas

Phase	Formula	Crystal System	Space Group	Mole-cules	Unit Cell Parameters Angstroms a	b	c	Angle	Ref.
Hydrated Aluminas									
Gibbsite	α-$Al_2O_3 \cdot 3H_2O$	Monoclinic	C_{2h}^5	4	8.641	5.070	9.720	85°26′	1
Bayerite	β-$Al_2O_3 \cdot 3H_2O$	Monoclinic	C_{2h}^5	2	4.716	8.679	5.060	90°07′	3
Nordstrandite	$Al_2O_3 \cdot 3H_2O$	Monoclinic		8	8.63	5.01	19.12	92°00′	4
Boehmite	α-$Al_2O_3 \cdot H_2O$	Orthorhombic	$D_2^{17}h$	2	2.868	12.227	3.700		5,12
Diaspore	β-$Al_2O_3 \cdot H_2O$	Orthorhombic	$D_2^{16}h$	2	4.396	9.426	2.844		5
Transition Aluminas									
Chi		Cubic		10	7.95				9
Eta		Cubic (spinel)	O_h^7	10	7.90				9,8
Gamma		Tetragonal			7.95	7.95	7.79		6,7
Delta		Tetragonal		32	7.967	7.967	23.47		10,11
Iota		Orthorhombic		4	7.73	7.78	2.92		9
Theta		Monoclinic	C_2^3h	4	5.63	2.95	11.86	103°42′	13
Kappa		Orthorhombic		32	8.49	12.73	13.39		9
Corundum	α-Al_2O_3	Rhombohedral	D_3^6d	2	4.758		12.991		2
	Al_2O	Cubic			4.98 (1100°C)				20
					5.67 (1700°C)				
	$AlO \cdot Al_2O_3$	Cubic (spinel)	O_h^7		7.915				22
Beta Aluminas (21)	$Na_2O \cdot 11Al_2O_3$	Hexagonal	D_6^4h	1	5.58		22.45		14
	$K_2O \cdot 11Al_2O_3$	Hexagonal	D_6^4h	1	5.58		22.67		14
	$MgO \cdot 11Al_2O_3$	Hexagonal	D_6^4h	1	5.56		22.55		16
	$CaO \cdot 6Al_2O_3$	Hexagonal	D_6^4h	2	5.54		21.83		15
	$SrO \cdot 6Al_2O_3$	Hexagonal	D_6^4h	2	5.56		21.95		15
	$BaO \cdot 6Al\ O$	Hexagonal	D_6^4h	2	5.58		22.67		17
Zeta Alumina	$Li_2O \cdot 5Al_2O_3$	Cubic	O_h^7	2	7.90				18,19

Ref.: (1) Megaw
(2) Swanson, Cook, Isaacs,
& Evans
(3) Unmack
(4) Lippens
(5) Swanson & Fuyat
(6) Saalfeld
(7) Brindley & Nakahira
(8) Verwey

(9) Stumpf
(10) Tertian & Papée
(11) Rooymans
(12) Reichertz & Yost
(13) Kohn
(14) Beevers & Brohult
(15) Lägerqvist
(16) Bragg
(17) Adelsköld

(18) Kordes
(19) Braun
(20) Hoch & Johnston
(21) See Scholder & Mansmann for claimed structure $Na_2O \cdot 6Al_2O_3$
(22) Filonenko, Lavrov, Andreeva & Pevzner

Table 7

Mineralogical Properties of the Aluminas

Phase	Index of Refraction n_d			Cleav-age	Mohs Hard	Micro Hard kg/mm^2	Dens. Measured g/ml	Ref.
	α	β	γ					
Hydrated Aluminas								
Gibbsite	1.568	1.568	1.587	(001)	2.5-3.5		2.42	1,2
Bayerite	1.583*	1.568					2.53	3,2
Boehmite	1.649	1.659	1.665	(010)	3.5-4		3.01	4,5,14
Diaspore	1.702	1.722	1.750	(010)	6.5-7		3.44	1,14
	ϵ	ω	*					
Transition Aluminas								
Chi							3.0†	19
Eta		1.59-1.65					2.5-3.6	6
Gamma							3.2†	20
Delta							3.2†	
Iota			1.604				3.71†	7
Theta			1.66-1.67				3.56	6
Kappa			1.67-1.69				3.3†	6
Corundum								
Al_2O	1.7604	1.7686		none	9.0	2150	3.96-3.98	21,15
$AlO \cdot Al_2O_3$			1.77-1.80			2070	3.84	13
Beta Aluminas								
Sodium Beta	1.635-1.650	1.676					3.25-3.33	8,10,18
Potassium Beta	1.642	1.675						17,18
	1.640	1.668						
Magnesium Beta	1.629	1.665-1.680						16,8
Calcium Beta	1.752	1.759						
	1.754	1.763					3.731	11,12
Barium Beta	1.694	1.702					3.69	10
Lithium Zeta			1.735				3.61	9

*Average.
†Estimate.

Ref.: (1) Dana
(2) Roth
(3) Montoro
(4) Ervin & Osborn
(5) Bonshtedt-Kupletskkaya
(6) Thibon
(7) Foster, P. A.
(8) Rankin & Merwin

(9) Kordes
(10) Toropov
(11) Filonenko & Lavrov
(12) Wisnyi
(13) Filonenko, Lavrov, Andreeva
     & Pevzner
(14) Fricke & Severin

(15) Coble
(16) Bragg, Gottfried, & West
(17) Kato & Yamauchi
(18) Beevers & Brohult
(19) Stumpf
(20) Ginsberg, et al.
(21) Biltz & Lemke

ionic diffusion (Barrer 1941). Gibbsite crystals grown as pseudohexagonal prisms in the presence of potassium ions, but often as platelet layers with spiral growth or with distortions in the base layer in the presence of sodium ions (Wefers).

Bayerite was classified in the hexagonal system by Montoro, and Yamaguchi and Sakamoto. The latter rejected X-ray reflections that could not be indexed in a hexagonal lattice. Also, a number of reflections that should coincide in a hexagonal lattice were found to be clearly separated (600, 330 and 601,331). Unmack classified bayerite in the monoclinic system. Wefers states that bayerite forms wedge-shaped somatoids composed of loosely packed $Al(OH)_3$ layers a few angstroms thick, and with their main axis perpendicular to the layers.

The interplanar spacings of nordstrandite have been furnished by Papée, Tertian, and Biais; and by Lippens. The latter found a monoclinic lattice most acceptable to explain the structure. Saalfeld and Mehrotra, however claim a triclinic unit cell.

The three trihydroxides have substantially the same layer structure and the same dimensions except in the direction of the c-axis. The differences between the lattices in the length of the c-axis is about proportional to the number of $Al(OH)_3$ molecules per unit cell.

The two monohydrates, boehmite and diaspore, are closely related. Boehmite is characterized by cubic close-packing of the anions, which carries over into the transition aluminas. Diaspore is characterized by hexagonal close-packing, which accounts for its direct transition to corundum at relatively low temperature (Ewing, Hoppe, Ervin 1952).

Achenbach established the orthorhombic symmetry of boehmite; confirmed by Dana, who found rhombic lamellae in French bauxites. Reichertz and Yost, and Milligan and McAtee determined the structure, based on a comparison with the isomorphous iron hydroxide, lepidocrocite. Similarly, diaspore was correlated with its iron oxide counterpart, goethite. Van Oosterhout provided a schematic model that explains the structure of the aluminum hydroxides, consisting of an anti-parallel coupling having the form:

$$HO-Al-O$$
$$| \quad |$$
$$O -Al-OH$$

The couples can form chains with others in which it may be assumed that the hydrogen ions act as bonds uniting the oxygens in each of the octahedral layers with those in the adjacent octahedral layers. Molecular water is not present (Imelik, Glemser). In boehmite the couple is oriented with the longer axis parallel to the b-axis; in diaspore, the couple is oriented more compactly on the bias so that the aluminum ions are joined by lines parallel with the b-axis. The OH-directions form zig-zag chains between the planes of the oxygen ions in boehmite (Kroon and Stolpe). The short unshared 0-0 distance is 2.70 A.

Busing and Levy could determine by neutron diffraction that the positions of the hydrogen ions in diaspore are at an angle of 12.1° with the line of centers of the two different types of oxygen atoms, and are closest to the oxygens farthest from the aluminum ions.

Rho alumina, obtained by vacuum heating of gibbsite, is almost completely amorphous. Its X-ray powder pattern shows only a few diffuse bands, the most pronounced of them having a spacing of about 1.4 A.

Chi alumina, the lowest transition phase from gibbsite in air (or nitrogen) also has a diffuse pattern, but is distinguishable from the other low-temperature aluminas by a line with spacing of 2.11 A. Stumpf concluded that it is cubic, but not a spinel, with a unit cell dimension of 7.95 A.

Both Saalfeld (1960) and Brindley and Choe ascribe to chi alumina a close-packed hexagonal structure with a = 5.57 A, c = 13.44 or 8.64 A, respectively.

Eta alumina is described as a cubic spinel. The unit cell of spinel, $MgAl_2O_4$, is formed by a cubic close stacking of 32 oxygen atoms, 16 aluminum atoms occupying one-half of the available interstices, and 8 magnesium atoms in tetrahedral holes. The formula is $Mg_8[Al_{16}]O_{32}$: the brackets are conventional for indicating octahedral sites. According to Verwey, and Hägg and Söderholm, gamma alumina (eta alumina) is a spinel with 32 oxygen atoms and, consequently, only 21-1/3 aluminum atoms. The formula is $Al_8[Al_{40/3}\square_{8/3}]O_{32}$; the square indicating that vacancies are distributed in octahedral sites. Jagodinski claimed that the vacancies are exclusively in the octahedral sites, on the basis of Fourier analysis.

Gamma alumina, obtained by anodic oxidation of aluminum, supposedly has its vacant places distributed at random over all possible cation positions.

Stumpf's gamma alumina, conforming to a one-hour heat treatment of Bayer trihydrate at 500°C, was thought to be cubic, but of too many interplanar spacings to conform with the spinel structure. Saalfeld (1958) found a tetragonal cell for the gamma alumina derived from gibbsite, confirmed by Brindley and Nakahira. A slow dehydration leads to the ordered tetragonal form, a fast, more drastic, heat treatment and less favorable starting material may give rise to a "statistically" cubic form (Tertian and Papée 1958). At about 430°C boehmite is no longer detectable in heated Bayer trihydrate. Glemser and Rieck observed that all gamma alumina has water strongly bound as hydroxyl ions. It was concluded by de Boer and Houben that gamma alumina (eta alumina) is a hydrogen spinel having the formula $5Al_2O_3 \cdot H_2O$ or HAl_5O_8, analogous to the zeta-alumina spinel $LiAl_5O_8$. There is little evidence for such a hydrate. Fortuin determined that all OH-groups are located at the surface.

It was concluded by Saalfeld (1960) that the dehydration scheme of the boehmite series is based on the cubic oxygen layer sequence ABC ABC, which is found in gamma alumina. In gamma alumina the aluminum atoms are all probably located in octahedral sites of the spinel.

Delta alumina, prepared by heating pure boehmite for only one to two hours at 1000°C, fits an orthorhombic structure, according to Stumpf. Rooksby obtained more sharply crystalline material by prolonged heating or by addition of mineralizers. Rooymans found a relationship between the interplanar spacings and those found by von Oosterhout and Rooymans for gamma ferric oxide, regarded as a defect spinel. The data fit a tetragonal cell having an axial ration of c/a = 2.946:1. The cell contains three spinel cells, that is, a whole number of Al_2O_3 molecules (32). A fourfold screw axis parallel to the c-axis was also found (Lippens). Saalfeld, on the other hand, denied the existence of delta alumina.

Foster and Stumpf found that theta alumina and beta gallia are isomorphous. Indeed, all the gallia phases have exact counterparts in the alumina system. Roy, Hill, and Osborn (1953) established a series of solid solutions between theta alumina and a gallia isomorph. Kohn, Katz, and Broder obtained beta gallia in large single crystals by vapor phase deposition, and were able to establish the structure of theta alumina as a distinct form of alumina from the analogy in structure. Theta alumina is monoclinic. Saalfeld (1960) obtained substantially the same unit cell dimensions as Kohn et al. In contrast with the spinel lattice, the aluminum atoms preferentially occupy tetrahedral positions. Theta alumina occupies an intermediate structural transition between the spinel lattice with the cubic oxygen arrangement and the corundum lattice with a dense hexagonal oxygen arrangement and the aluminum atoms exclusively in the octahedral positions.

Stumpf established the orthorhombic crystal structure for kappa alumina. According to Saalfeld (1960), kappa alumina has the layer sequence ABAB-BABA . . . , with the aluminum atoms in octahedral and tetrahedral sites. In common with gamma, theta, and chi aluminas, the preferred orientations of its crystallites confirm the fact that the gibbsite lattice is not completely destroyed upon calcination, but that special directions are preserved. The strongest lattice direction is the a-axis of gibbsite.

Sasvari and Zalai explained the mechanism of lattice change during dehydration of gibbsite by assuming that the changes occur by collapse and disappearance of specific anionic planes of the layers. The cations lying between the single planes so produced and the double layers realign into octahedral environment by cubic close packing. In going to boehmite, the original crystal layering arrangement AB BA AB BA . . . loses every third single layer to give the sequence A BA B B AB A A BA . . . before collapse, and B BA BA A . . . finally. If all the water is expelled at once, the disappearance of both planes from every second layer appeared to be the more likely method. Several different paths were conceivable for the dehydration of boehmite and diaspore.

The structure of corundum was approximately determined by X-ray methods by Bragg and Bragg. Pauling and Hendricks established the structure of rhombohedral alpha-aluminum oxide in 1925. The best present values for cell dimensions are probably the values of Swanson and Fuyat. Newnham and de Haan have recently investigated the crystal structure of Verneuil process specimens by zero-layer intensity data collected about the (100) face with a Weissenberg camera. The oxygen positions approximate hexagonal close packing, with the trivalent cations occupying two-thirds of the octahedral interstices. Hoch and Johnston claimed that the crystal structure of corundum at 2000°C is the same as that at room temperature. The theoretical density calculated from the unit cell constants is: 3.987 g/ml (Swanson and Fuyat), and 3.996 g/ml (Sedlacek, 1964). The measured theoretical density determined by Coble (1962) is 3.98±0.01 on pure low-color sapphire. Biltz and Lemke (1930) found 3.96.

Alpha aluminum oxide crystals, as exemplified by Bayer process ceramic types have the shape of thin hexagonal, or more rarely, three-cornered plates. Small differences in habit are observed, mainly relating to the thickness of the plates, the edge faceting, and the size and degree of perfection of the crystals. Crystal habit is affected by several factors, among which are the kind and concentration of mineralizer present (fluorides, borates, phosphates, etc.) and the temperature of calcination. Crystal thickness may, for example, be altered by the choice of fluoride; cryolite produces thin, sharpcornered plates in comparison with thick, rounded crystals produced by calcium fluoride.

Dana mentions and illustrates many examples of the crystal habit of native corundum, ranging from tabular to elongated hexagonal prisms and pyramids, and as rounded or tabular aggregates, as found in emery. The characteristic lamellar and striated structures showing rhombohedral and basal parting planes in hexagonal prisms are common (Schroder).

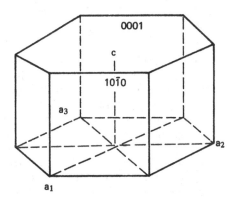

Alumina ceramic products are manufactured under conditions quite different than those under which the native form is produced, and their crystal habits are different, but there are analogies.

Baumann has described the crystal habit and microstructure of alpha alumina as it occurs in massive sintered and electrofused forms. In sintered alumina well developed crystal faces (euhedral to subhedral) having the tabular habit are usual. Spherical pores both within the crystals, often separated from the boundaries by a margin, and pores at the boundaries are numerous.

In fused melts from bauxite and Bayer alumina, the crystallization increases in size from the wall to the center of the ingot, depending on the rate of cooling. Crystal size of several millimeters in diameter is attained from melts containing a small amount of titania. The crystallization is usually anhedral. In ingots prepared from Bayer alumina, occasionally alpha alumina crystals are free to grow in open spaces, forming well developed rhombohedra. On surfaces, particularly of shrinkage pipes or canals, skeletal needles and dendritic forms project, often growing along the direction of the c-axis.

Filonenko and Borovkova found that titanous compounds, in amount not over 4 to 6%, cause anomalous thermal expansion of fused alumina, with formation of a network of cracks on the surface. Other impurities have a marked effect on the structure and properties of fused alumina. Silica up to 3%, inhibits the development of beta alumina in white fusion alumina, but the rhombohedral crystallization persists (Schroeder and Logan).

X-ray examination shows that no natural sapphires approach the perfection of structure of synthetic sapphires. In the early days of production by the Verneuil process, defects in structure were observed, however, among which are strain and longitudinal splitting. Longitudinal splitting does not necessarily indicate twinning, but the release of strain, and is not generally in a principal crystallographic plane (Brown, Chirnside et al.), contrary to the opinion of Saucier. The optic axis and the plane of splitting lie close together, but both may be at any angle to the axis of the boule. Stofel and Conrad found that mechanical twins, both (0001) and $(0\bar{1}11)$ types, were produced by compression in the range 1100 to $1300°C$. Cracks propagated more readily along the twin boundaries than in the nontwinned crystal, suggesting a lower fracture strength than that of nontwinned crystals.

Conrad et al. observed twinning in 60 degree-oriented sapphire at temperatures as high as $1500°C$. Gumilevskii observed surface grain markings indicating the crystal axes, and skeletal evidence in the form of particles at the grain boundaries and dendrites. Belova and Ikornikova reported that synthetic sapphires have a mosaic structure intermediate between the ideal monocrystal and polycrystalline growth. Amelinckx (1952) observed triangular growth spirals on the (0001) plane of synthetic rubies. This is the plane in which the aluminum atoms are located in the octahedral interstices of the oxygen atoms. Lima-de-Faria noted that intermediate stages in the dehydration process of diaspore are revealed by satellite X-ray reflections, which occur close to developed corundum spots at certain temperatures. The satellites may correspond to a sinusoidal structure-amplitude modulation along the c-axis of the corundum crystallite before dehydration is complete. The wavelength of the modulation was 39 A. Nonuniform line broadening was ascribed to stacking faults.

Webb and associates, and Dragsdorf and Webb grew sapphire whiskers and platelets by heating aluminum in moist hydrogen at 1300 to $1450°C$ for 2 to 24 hours. The whiskers were 3 to 50 microns in diameter and 1 to 30 mm long; they had hexagonal cross sections and [0001] axes. Evidence of screw dislocations were observed in the whiskers; none was found in the platelets. Many whiskers were hollow, the axial cavities having diameters of 0.5 to 3 microns. Axial twists, determined by X-ray and confirmed by optical observation, were ascribed to a screw-dislocation Burgers vector equal to integral multiples of the c-axis, 12.97 A. Theories of Eshelby and some others require the presence of screw dislocations. Sears, DeVries, and Huffine have observed the Eshelby twists on the tapered tips of alumina whiskers. The twist and associated screw dislocation can be removed by nonuniform bending, occurring during the breakage of the whisker from its base. May (1959,a) found that the platelets grown in moist hydrogen

atmosphere also show evidence of screw dislocation (Frank mechanism).

McCandless and Yenni; Smith and Yenni; and Hart, by applying heat and force, altered the shape of corundum unicrystals. Rod-shaped crystals were bent in a plane which makes an angle between zero and 45 degrees with the plane defined by the c-axis and the longitudinal axis of the rod. The stretched portion became opaque, while the body remained a unicrystal. Kronberg observed the arrangement of the tiny crystals that formed in the plastically bent regions, a phenomenon designated "polygonization." The dislocations tend to line up in particular arrangements. Just above an edge dislocation where an extra plane of atoms is inserted, there is a compressive stress, and below the dislocation a tensile stress. Dislocations of the same sign (positive for those with the extra plane inserted above the slip plane) tend to repel one another if they are in the same slip plane. Dislocations of the same sign in different slip planes tend to line up above each other to form low-angle grain boundaries. A mosaic structure forms upon annealing. May investigated the kinetics of the process over the range 1750 to $1850°C$, and found an activation energy of 140 kcal/mole.

Scheuplein and Gibbs found triangular etch pits corresponding to the intersection of individual edge dislocations with the surface, after immersing the corundum crystals in boiling phosphoric acid for about 5 minutes. Dislocations of both the (0001), $\langle 11\bar{2}0 \rangle$ and $\{1\bar{2}10\}$, $\langle \bar{1}010 \rangle$ slip system (where reference is made to the slip plane and slip direction, respectively) were detected. Crystals, bent while cooling, had about twice as many dislocations as are predicted by the Nye formula, and these were distributed randomly. A special method of deformation, which favored the participation of both the basal (0001), $\langle 11\bar{2}0 \rangle$ and the prismatic $\{1\bar{2}10\}$, $\langle \bar{1}010 \rangle$ slip systems, resulted in highly stressed crystals which fractured spontaneously, once the outside surface was scratched.

Palmour, Du Plessis, and Kriegel observed that basal-, prism-, and rhombohedral-plane specimens cut and polished from flame-grown sapphire boules developed fine surface textures during heat-treatment in the range of $1700°$ to $1900°C$. Microscopic examinations of such surfaces revealed low-angle grain boundaries, dislocations, and other crystal imperfections. Thermally etched textures apparently were generated by three overlapping processes; annealing, etching, and decoration. Annealing commenced at 1500 to $1700°C$ and was essentially complete at $1850°C$. The cold-worked surface was progressively readjusted by surface and volume diffusion. At higher temperatures, the surface began to etch by evaporation. Surface decoration was indicated by a deposition of aluminous material on favorable sites in the form of low rounded bumps, occurring, presumably, during cooling. Thermal etching gave evidence of high-temperature relaxation of residual strain energy introduced at low temperature (analogous to cold work in metals) including that introduced by fracture, microindentation, and cutting and polishing during surface preparation. Voruz et al. observed dislocations in thin

34

specimens of sapphire platelets directly by electron transmission techniques. The observations indicate the possibility that corundum has a high stacking fault energy and that slip may occur by the movement of perfect dislocations or that dissociation may occur over a distance of several Burgers vectors.

Fuchs claimed that even brittle corundum powders exhibit lattice distortion during milling on the basis of measurement of X-ray line broadening. Stephens and Alford found that the average density of edge dislocations lying in prism planes was 3.0×10^5 per cm^2, which could be changed only slightly by additions of chromium and by annealing at $2000°C$. An average basal dislocation density of 2×10^6 per cm^2 decreased by 35 to 80% upon annealing. Crystal orientation (the angle between the c-axis and the axis of growth) showed no effect on dislocation density but a pronounced effect on the subboundary arrangement and density. The substructure of 60-degree crystals is intermediate between that of 0-degree and 90-degree crystals. Prismatic and basal slip were observed on all as-grown crystals; profuse basal slip polygonized readily upon annealing; the densities of dislocations in flux-grown crystals were lower than those of Verneuil crystals; and basal twinning on the plane $\{\bar{1}010\}$ was observed on all flux crystals. Lommel and Kronberg, using a high-resolution Berg-Barrett X-ray technique, compared dislocation data with etch patterns. Thermal etching data were in good agreement with the chemical etching data of Scheuplein and Gibbs. Correlation between X-ray data and etch figures was good for sapphire but not for a light ruby containing 0.75% Cr_2O_3.

Paladino and Reuter grew sapphire single crystals of moderate size by the Czochralski method of continuous pulling from the melt. A pickup of less than 0.01% W from the containing tungsten crucible and 0.0001% Ni was observed from melts of Linde alumina. Impurities that became less or disappeared by spectrographic analysis include: B, Na, Mn, Fe, Ga, In, and Pb. Impurities that remained the same include: Mg, Si, Cr, Cu, and Ag. The defect structure by this method was decidely less than that of the Verneuil process.

Laudise and Ballman hydrothermally synthesized even more perfectly crystalline corundum by seeding in dilute sodium carbonate solution at 395 to $520°C$ and 12,000 to 55,000 psi. The crystal growth was about 0.006 in/day. Kryukov found that boehmite and gibbsite are by-products from the hydrothermal synthesis of corundum.

4-2 PSEUDOMORPHOSIS

Achenbach (1931), Damerell et al. (1932), and Tertian (1950) noted that the dehydration of gibbsite crystals is pseudomorphic, that is, the external shape of the crystals is retained and there is an orientation relationship of the crystal axes of the new phase to those of the original. The crystals lose transparency and smoothness upon heating, and fine-grained fibers develop parallel to the hexagonal surface. The gross form of the Bayer agglomerate is retained

even to $1200°C$ substantially unchanged. Void space resulting from the loss of water from the gibbsite and increasing density of the transition phases is distributed in microporosity of very high surface area in the porous skeleton (Weitbrecht and Fricke). Pseudomorphosis is of considerable importance because of its effect on surface area of the intermediate phase structures, and on crystal size and size distribution of the fully calcined aluminas for ceramic forming processes.

Hüttig and Ginsberg concluded from micrographs of the heated Bayer trihydrate, that dehydration takes place in a continuous process preceding lattice change. The transition phases are fragments in the crystal structure. During boehmite formation, the retention of form was best when the transformation was in alkali; next best in water; and poorest when in air. Transparency was lost from the interior of the crystal, at least in dry dehydration. Nahin and Huffman prepared electronmicrographs of heated aluminum hydroxides. Voids, apparently generated by dehydration are shown in theta alumina and other high-temperature phases. Bayer trihydrate calcined at $1040°C$ was sharply microcrystalline in the cubic system. A diffuse X-ray pattern was shown for crystals ranging in size from less than 100 to 250 A. The pore size was about 40 to 60 A. Stirland, Thomas, and Moore concluded from electronmicrographs of thermal decomposition of hydrated aluminum chloride, that at about $300°C$, striated flakes 0.5 to 5 microns form, containing a few holes, 400 to 500 A in diameter. The holes generally are partially occupied by dense inclusions. The holes and inclusions increase in number with increasing temperature. At $1000°C$ the inclusions appear as hexagonal or rectangular forms. At $1100°C$, the inclusions loose sharpness, and coalesce, apparently by engulfing the striated structure, to form relatively large open pores.

Thibon, Charrier, and Tertian; de Boer, Fortuin, and Steggerda; Steggerda; Saalfeld; Ervin; and Lippens and de Boer were also interested in pseudomorphic changes. Ervin observed that the X-ray powder diffraction patterns of all of the transition aluminas have in common the strong line at 1.39 A. This reflection belongs to the (440) plane for the spinel structure. It passes through all the oxygen ion centers and the possible aluminum ion centers, whether tetrahedral or octahedral. The less intense (400) plane of spinel at about 1.99 A passes through every anionic and octahedral cationic position. On this basis, all the transition aluminas have oxygen ions in approximately cubic close packing. The differences in their patterns represent changes in intensities of reflections resulting from differences in distribution of the aluminum ions. The initial cationic disorder of the low-temperature phases depends upon the source of the alumina. The transitions become more ordered with increasing heat treatment.

4-3 SURFACE AREA OF ALUMINA

Gelatinous alumina pastes, prepared by the methods described in Section 3-3, especially the methods of Kraut and Schmäh, exhibit high surface area and high sorptive capacity for water vapor and other gaseous phases merely upon drying at about 120°C. Harris and Sing obtained initial surface areas as high as 1100 m^2/g for gels formed by hydrolysis of aluminum isopropoxide. Storage of these unstable gels in the presence of water vapor caused a loss in surface area to about 500 m^2/g. The adsorption isotherms of nitrogen, determined at -196°C on the outgassed products were of the reversible S-type, characteristic of physical adsorption on nonporous solids. Gels that had been dehydrated at room temperature approximately to the formula $Al_2O_3 \cdot 3H_2O$ showed no X-ray evidence of crystalline hydrate structure.

The initial gel loses water to about $Al_2O_3 \cdot 1.1\text{-}1.9H_2O$ upon drying at room temperature. Bye and Robinson (1961) obtained peak surface areas of about 480 m^2/g for samples aged about three hours in water at room temperature into the pseudo-boehmite structure. Upon continued aging, the development of crystalline bayerite caused a reduction in surface area to about 250 m^2/g within 18 hours, and to as low as 50 m^2/g within 96 hours in the presence of a crystal-growth accelerator, NH_4OH (Lippens).

Tertian and Papée (1958) showed that the reactive pseudo-boehmite can be recrystallized in various stages of microcrystal growth to normal boehmite having a surface area of only 40 m^2/g, by hydrothermal digestion at 300°C for about 18 hours. Calvet and Thibon observed that the surface of well crystallized hydrothermal boehmite may be less than 10 m^2/g. Pseudoboehmite is not merely finely divided or poorly crystallized boehmite similar to hydrothermal boehmite (Papée, Tertian, and Biais). Differences in structure were shown by variations in the interplanar spacings, 6.11 A for well-crystallized boehmite, but 6.6 to 6.7 A for the pseudoboehmite. The water in excess of the theoretical amount in well-crystallized hydrothermal boehmite may be more than 0.25 mole, and in pseudoboehmite more than 0.6 mole. Goton considered this to be a trihydrate phase located around the surface of the crystallites because of the heating conditions required to decompose it. Tertian and Papée rejected this view, and pointed out that the amount, if present as trihydrate, could be detected by X-ray. Also, the loss of up to 12% water at about 300°C is followed by neither a significant increase in surface area (from 3 to only 3.5 m^2/g) nor a structural change. The excess represented by $Al_2O_3 \cdot 1.6H_2O$, if present as a monolayer, would cover about 265 m^2/g, essentially the total active surface of the pseudoboehmite.

Russell and Cochran measured the change in surface area of the hydrated aluminas with the dehydration temperature for heating periods of one hour at temperature. The surface area was measured by adsorption of n-butane. Gelatinous boehmite (dried at 140°) having an initial surface area of 340 m^2/g, commenced to lose surface area at about 400°C, but retained about 70 m^2/g at 1200°C. Coarsely crystalline hydrothermal boehmite, having an initial surface area of about 2.5 m^2/g, developed only 120 m^2/g, peaking at about 500°C, pointing up the relatively poorer surface properties of this type. The surface area of the heated gibbsite did not reach as much as 2 m^2/g until the weight

loss was 8%, that is, a product equivalent to $Al_2O_3 \cdot 2.5H_2O$. The relatively low value for surface area at this point implies that the new surface was largely concealed in the interior of the particles, or, more probably, that it represented the formation of low-surface hydrothermal boehmite. The maximum surface area attained was about 340 m^2/g, peaking at 400°C in dry air, and thereafter falling to less than 25 m^2/g at 1200°C. Fine-crystal gibbsite and bayerite developed 425 and 380 m^2/g, respectively, the higher values reflecting less susceptibility to coarse boehmite formation. The bayerite peaks of area occurred at about 300°C.

Blanchin found a substantial decrease in the surface area of gibbsite held for twenty hours at 400°C. Fricke and Jockers, and Fricke and Eberspächer confirmed the losses in surface area for bayerite and boehmite above 400°C. Emolenko and Efros determined the loss for activated aluminas produced by calcination of aluminum oxychloride. For 2-hr calcinations at 500, 600, 700, and 800°C, the specific surfaces were about 255, 174, 160, and 127 m^2/g, respectively.

Gregory and Moorbath obtained curves for variation in surface area with temperature resembling those of Russell and Cochran, but by another technique, the emanation of thoron from the specimens. The method, devised by Hahn, has been improved in sensitivity by the techniques of electronic counting and recording. Upon heating gibbsite, impregnated with thorium salt during precipitation, a large increase in emanation occurred at 300 to 450°C. Minor changes were found at 850, 890, 1020, 1070, and 1090°C. A large decrease occurred at 1130°C. The changes did not correlate very well with structural changes observed photographically (Moorbath).

Carruthers and Gill investigated the surface area and structural changes occurring in an industrial trihydrate and a monohydrate in the heating range from 1200 to 1720°C. The surface areas for heating periods of two hours at 1200, 1400, 1600, and 1700°C were 5, 3, 1, and 0.3 m^2/g, respectively. The surface area of alpha aluminum oxide obtained by calcination of the lower phases is usually less than 1 m^2/g. Russell and Cochran obtained a surface of 85 m^2/g for Missouri diaspore clay upon heating at 600°C. Only alpha alumina was shown by X-ray analysis. A strained, high-surface structure might account for the peculiarities observed by Herold and Dodd in this calcined product. Krishner and Torkar obtained 40 to 70 m^2/g for hydrothermally produced alpha alumina having a particle size range of 300 to 800 A, and a pore distribution of 100 to 50 A.

4-4 POROSITY

Porosity, and its special case, permeability significantly affect the properties of alumina ceramics, and in a wide range of magnitude. Porosity is generated in sintered alumina structures for various reasons, some of which are: to improve permeability to gases and liquids for porous diaphragms and diffuser plates, to increase the thermal insulation of refractories, and to improve the fuel combustion in radiant heaters. Volatile or combustible burn-outs (sawdust, naphthalene) have been used to generate pores. Gas generators include: hydrogen peroxide (Ryshkewitch 1953), and aluminum powder with acids or alkalies.

Gross porosity beyond 50% by volume can be developed by calcining mixtures of ground and unground Bayer alumina at high temperatures. Uniformly distributed porosity is attained by "bisque" firing fine-ground alumina in the undersintered range 1000 to 1400°C.

Torkar (1954) pointed out that the property relationship of a porous or mixed body to those of its components applies to magnetic permeability, the dielectric constant, gaseous and liquid diffusion, the velocity of sound, the refractive index, elasticity, and thermal and electrical conductivity. The application of the mixed body theory to the "conductivity properties" of a porous material, that is, a solid-air system, leads to the expression: $y/y_1 = Kx_1/[K+(1-x_1)]$, in which y and y_1 are the values for the mixed body and for the solid component; x_1 is the volume proportion of the solid component; $1-x_1$ is the porosity, p; and K is the "structure factor." The value of K varies from infinity for parallel orientation to zero for series orientation. It has the same value for all properties. The ratio y/y_1 is known as the β value of the property. When β is less than 0.7, the curves deviate markedly from the constant value.

Among the applications for sintered alumina in which porosity is undesirable are vacuum tube envelopes and ceramic lathe tools. Before firing such a compacted form, the porosity is almost entirely open or permeable. Upon sintering, the open pores are gone when the total porosity has decreased to 5%. At 95% of theoretical density, the structure is gas-tight (O'Neil, Hey, and Livey).

Hayes, Budworth, and Roberts; Fryer et al.; and Budworth investigated the permeability of extruded sintered alumina tubes (total porosity 4 to 9%, purity 99.3 to 99.8% Al_2O_3). These tubes were impermeable to oxygen, nitrogen, and argon at temperatures below 1500°C. At 1500 to 1750°C, the specimens showed appreciable permeation to oxygen, presumably by a surface diffusion process. The diffusion coefficient was about $100\mu^2$/sec. Very slight or no permeation was found for nitrogen, and none for argon. After continued exposure for 100 hours at 1700°C, permeation by normal channel-flow developed suddenly and swamped the earlier phenomena. The permeation of nitrogen through hot-pressed sintered alumina (4 to 14% total porosity) was predominantly by Knudsen flow. An activation energy of 70 kcal/mole was found for the transport process involving oxygen in polycrystalline alumina at 1500 to 1750°C. M. O. Davies explained an observed change in the activation energy in about this temperature range from about 60 kcal/mole to 135 kcal as a change from extrinsic to intrinsic transport. Volk and Meszaros, in an investigation of alumina as a coating for the high-temperature protection of graphite and refractory metals, concluded that sapphire is the best oxygen barrier of all the refractory oxides, and polycrystalline alumina compares favorably with zirconia, hafnia, and thoria.

The effect of porosity on the mechanical strength and other properties of sintered alumina is discussed in those sections.

Surface area alone is not sufficient to qualify a porous desiccant or catalyst. De Boer, Heuvel, and Linsen; de Boer and Lippens; Barrer, McKenzie and Reay; Cranston and Inkley; Innes; and others have attempted to determine the shape and distribution of the pores from the shapes of the hysteresis curves obtained in sorption and desorption of the activated aluminas, and by application of the Kelvin equation for vapor pressure. The vapor pressure over a liquid with a curved surface is given by the equation:

$$\ln x = \ln p/p_0 = \frac{-2\,\sigma\,V\cos\psi}{RT\,r_k},$$

in which x is the relative vapor pressure over the liquid, σ is the surface tension of the liquid, ψ is the contact angle between the liquid surface and the solid, V is the molecular volume of liquid, and r_k is the Kelvin radius. In empty pores the sorbed liquid may condense at a higher relative pressure than it evaporates from the fully filled pores. This is the phenomenon called hysteresis.

Barrett, Joyner, and Hallenda gave a method for cylindrically shaped pores, Steggerda and Innes for slit-shaped pores.

Five general types of hysteresis loops have been distinguished, from which fifteen capillary shapes could be deduced. The adsorption isotherms of the activated forms of alumina fit the three main types, A, B, and E, all of which have steep desorption curves. Type A has a steep sorption branch, type B a gradual sorption branch, with a broad hysteresis range, and type E a gradual sorption branch with a narrow hysteresis range. The pore shapes are mainly open and closed tubular capillaries, ink bottle shapes, and slit shapes.

Sheet-like gelatinous boehmite shows a hysteresis loop intermediate between types A and E. The crystalline hydroxides give type B loop upon heating. Many of the narrow pores formed in activated gibbsite do not allow capillary condensation to occur. Two types of pores are believed to form in succession: platelike pores having a width of about 30 A, lying in the cleavage plane of the gibbsite crystals; and pores having a width of about 10 A, which divide the alumina structure into parallel rod-shaped crystals. Hydrothermal boehmite, upon heating develops slit-shaped pores, generally with type-B characteristics, and with openings of about 25 A.

Fricke and Jockers (1951), and Fricke and Eberspächer found that the pore diameter of dried gelatinous boehmite was about 60 A at room temperature for a surface area of about 200 m²/g. The pore diameter began to increase above 400°C, reaching a value of 140 A at 1100°C, for a surface area of only 37 m²/g Fine bayerite, having a pore diameter of 14 A at room temperature to 400°C, increased in pore size to 144 A at 1100°C with an accompanying drop in surface area from 431 m²/g to only 33 m²/g. Boreskov et al. found that the porosity as determined by sorption and mercury porosimeter methods was unaffected by heating up to 600°C. On heating to 1200°C, the surface area decreased by a factor of 60, but the pore volume decreased by less than a factor of 2. Several investigators have provided electron micrographs showing that the pores become fewer and larger with increased heating.

A. G. Foster determined a Gaussian distribution for heated alpha trihydrate with a maximum at 26 A, by applying the Kelvin equation to water sorption results. Drake and Ritter measured the size distribution of the coarser porosity by compressing mercury into the pores. The total pore volume was 0.388 ml/g, of which over 0.082 ml/g represented porosity coarser than 1000 A in diameter. Some gross permeability is essential, however, to enable easy access of the sorbate to the micropore structure.

Chrétien and Papée determined the relative amounts of micro and macroporosity from the difference between total volume by immersion and the sorption of water and some organic liquids. Wosniczek showed that the conditions under which the alumina gels were produced determine to a large extend the internal structure and particularly the volume and size distribution of the pores. Sanlaville found two distinct maxima in pore size distribution of transition alumina: (1) 7 to 18 A, having a surface area of 235 m²/g; and (2) 18 to 300 A, having a surface area of 36 m²/g.

Bielanski and Burk found that water vapor is adsorbed in three stages at 25 and 50°C on alumina that has been activated at 450°C. Diffusion in oxide micropores is one mechanism. At pressures above 4.6 mm Hg, surface diffusion probably predominates. Water-vapor adsorption was measured at 40°C on samples of alumina that had been activated at temperatures from 250 to 1000°C. Isotherms were obtained from which the specific surface of the samples (180 m²/g at 450°C, falling to 46 m²/g at 1000°C) and pore structure could be determined. The smaller the diameter of the pores, the lower the activation temperature for reaching a maximum surface area.

As examples of the importance of control of pore size in sorptive and catalytic operations, the preparation of molecular sieves is cited. Basmadjian et al. controlled the pore volume and size distribution in alumina by the addition of water-soluble organic polymers (polyethylene glycols and oxides, methyl cellulose, and polyvinyl alcohols) to suspensions of the alumina prior to or during precipitation. Konoval'chikov et al. adjusted the pore structure of a precipitated Al_2O_3-Cr_2O_3 catalyst by displacement of intramicellar water from the initial gel with isoamyl alcohol. Kimberlin and Gladrow subjected dried hydrous aluminum oxide to an elevated pressure in the range of 15 to 1000 psi and a temperature in the range of 212° to 550°F, and dried the product by releasing steam from the reaction zone while maintaining the elevated pressure. A catalyst of increased pore volume and diameter was obtained.

4-5 SORPTIVE CAPACITY

Adsorption is the phenomenon of attraction and fixing of gaseous and liquid molecules to the surface of a solid, a property possessed by all solids to some degree. Adsorption results from unbalanced surface forces or unsatisfied valences of the surface molecules. The amount of adsorption depends upon the attractive force of the adsorbent and the amount of surface area available. Although all molecules are adsorbed to a certain extent, molecules having high polarity, such as water, methylchloride, and some olefins, are likely to be more strongly held. Adsorption is accompanied by the liberation of heat and, conversely, desorption requires the application of heat. In physical adsorption, the heat released equals the heat of condensation of the adsorbed material plus the heat of wetting.

The strong desiccating action of activated alumina has been known at least since 1879 (Cross). The properties that make the activated aluminas particularly suitable for desiccant use are: the ability to develop high surface area during formation or dehydration; a high degree of chemical inertness; resistance to softening, swelling, and disintegration when immersed in water or other liquids; high resistance to shock and abrasion; and the ability to return to the original highly adsorptive form by a suitable thermal regenerative treatment.

The activated bauxites provide similar properties, but, in general, to a lesser degree and in frailer structures, subject to deterioration in moist condition (Heinemann). Ramaswamy et al. used activated bauxite in drying wet chlorine gas, and reactivated without impairing its efficiency. Bell nodulized flame-activated, ground bauxite. The nodules were cured to firm, spherical shapes upon aging until rehydration took place.

Commercial types of activated alumina are prepared: (1) by partially dehydrating massive alumina trihydrate, a by-product of the Bayer process, designated F-1*; (2) by forming amorphous gels according to the methods described in Section 3-3, type H-151. A special type, F-6, that indicates visually when the adsorbent is spent (at about 20% relative humidity) contains cobaltous chloride. This grade is used in breather-equipment on tank cars, in transformers and storage tanks, on pneumatic and electrical control equipment, in laboratory desiccators, between multiple-glazed windows, and like applications where a visual indication of the degree of adsorption is desirable. A second type, F-5, is impregnated with calcium chloride, which about doubles the sorptive capacity or "one-shot" rapid desiccating applications at somewhat higher dewpoints. Reactivation above about 180°C causes progressive decomposition of the calcium chloride. Another type for controlling atmospheres in heat-treating furnaces contains about 2.5% nickel formate, F-7. Many other types, in which silica or metal oxides or salts are co-precipitated with or impregnated on the activated alumina, have catalytic applications. In these applications the alumina may be in the form of a powder for fluid bed use, or in a variety of macrosize forms for tower packing use. The forms may be crushed irregular shapes, or extruded, pelletized, or otherwise shaped. Spheres are commonly produced by spray

---

*Aluminum Company of America designations.

drying, or by the ingenious method of suspending droplets of alumina hydrogels or hydrosols in water-immiscible liquids (hydrocarbon oils) under conditions to induce set (Archibald; Messenger; Hoekstra). A list enumerating over 180 representative applications of catalytic processes using alumina directly or as a support for other catalysts is compiled in Alcoa Product Data Brochure "Activated and Catalytic Aluminas" June 1, 1967.

Alumina is also used in the special case of adsorption called chromatography, in which the identification and separation of adsorbed ions are usually based on a visual, spatial order of adsorption. While the method has been applied to a large number of organic substances, the main applications of ceramic interest are probably restricted to the separation and purification of inorganic salts (Strain; Zechmeister and Cholnocky; E. and M. Lederer; Bobbitt).

Some types of activated alumina are shown in Table 8. Sorptive capacities of two grades are shown in Figure 7.

Ehman prevented dusting of granular activated alumina in refrigeration systems by coating the particles with 0.05 to 5% by weight of phosphoric anhydride. Smirnov et al. prepared mechanically strong activated alumina by decomposing the aluminum carbonates $(Na,K)_2O \cdot Al_2O_3 \cdot 2CO_2 \cdot nH_2O$ at about $100^\circ C$.

A partial list of gases and liquids which can be dried by activated alumina (Alcoa brochure, June 1, 1967) includes the following:

Gases

Acetylene, air, ammonia, argon, carbon dioxide, chlorine, cracked gas, ethane, ethylene. Freon, furnace gas, helium, hydrogen, hydrogen chloride, hydrogen sulfide, methane, natural gas, nitrogen, oxygen, propane, propylene, and sulfur dioxide.

Liquids

Benzene, butadiene, butane, butene, butyl acetate, carbon tetrachloride, chlorobenzene, cyclohexane, ethyl acetate, Freon, gasolines, heptane, hexane, jet fuel, kerosene, lubricating oils, naphtha, nitrobenzene, pentane, pipe-line products, propane, propylene, styrene, toluene, transformer oils, vegetable oils, and xylene.

Activated alumina, produced from massive Bayer trihydrate (F-1 type), adsorbs moisture to 14 to 16 percent of

Table 8

Typical Properties of Desiccant, Chromatographic, and Catalytic Aluminas

	F-1	H-151	F-20	T-71	F-110	F-7
Typical Properties						
Al_2O_3 .... %	92.0	90.0	92.0	99.5+	92-94	84.0
Na_2O .... %	0.90	1.6	0.90	0.01	0.08	0.90
Fe_2O_3 .... %	0.08	0.13	0.08	0.06	0.03	0.08
SiO_2 .... %	0.09	2.2	0.09	0.04	0.01	0.09
Loss on ignition ($1100^\circ C$) .. %	6.5	6.0	6.2	0.0	6.0-8.0	12.1
SO_3 .... %				0.09		
CaO .... %				0.06		
Nickel formate .... %						2.5
Form	Granular	Ball	Granular	Granular	Balls	Granular
Surface area, m^2/g	210.0	390.0	210.0	0.5	180-280	
Bulk density, loose, lb/ft^3	52.0	51.0	58.0	76.0	50.0	52.0
Bulk density, packed, lb/ft^3 ...	55.0	53.0	68.0	85.0	55.0	55.0
Specific gravity	3.3	3.1-3.3	3.3			
Static sorption at 60% RH	14-16	22-25				
Crushing strength	55.0	75.0				
Pore volume, ml/gm				0.15-0.20	0.38	
pH			9.0			
Sieve analysis						
on 80 mesh .... %			2 max.			
through 270 mesh ..... %			5 max.			

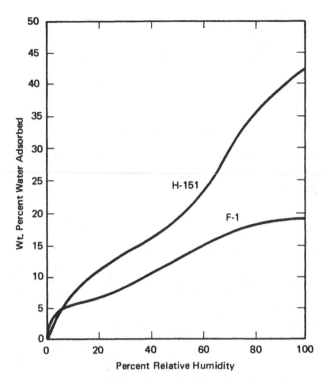

Figure 7. Equilibrium Curves for Water After Normal Activation at 400 °F

its dry weight at 60 percent relative humidity. Dew points of minus 75°C are attainable. Activated alumina (H-151 type) is especially suited for the drying of foods and drugs. Its static sorptive capacity at 60 percent relative humidity is 22 to 25 percent of the dry weight.

Kipling and Peakall observed that the adsorption of vapors by solid oxides was only partially reversible. Water vapor and the vapors of the lower aliphatic alcohols are held by chemisorption if oxide ions are present in the surface. Physical adsorption can take place on the chemisorbed material. De Boer, Fortuin, Lippens and Meijs stated that the chemisorbed water is proportional to the specific surface area, amounting to about 24.8 mg H_2O/100 m^2. This suggests a binding of one water molecule to two oxygen atoms in the surface. The surface covered with chemisorbed water adsorbs more water by physical sorption. The monolayer capacity is also approximately proportional to the specific surface area, and is about 33 mg H_2O/100 m^2. Lauric acid can be adsorbed on the water layer adsorbed on the alumina surface, and can displace water beyond a critical water content of about 50 mg H_2O/100 m^2. Actually, alumina that has been calcined at 800 to 1200°C picks up and retains a significant amount of water which resists expulsion at temperatures as high as 500°C.

The decrease in effective surface area by chemisorption is a factor in the cyclic use of activated alumina in service as a sorbent. Applications of adsorption generally take place under dynamic conditions, that is, under conditions of gas flow in which the desiccant is not allowed to come to

equilibrium as it becomes spent and its efficiency decreases. There may be an initial loss of as much as 30 percent of the original dynamic sorptive capacity depending upon the conditions selected, after which the loss becomes exceedingly slow. The service can be carried through hundreds of cycles; an original charge (F-1) in a unit for drying annealing-furnace atmosphere performed satisfactorily for more than 25 years.

This stability against loss of activity is possible only if there is no chemical reactivity between the active surface and the sorbed substances. Activated alumina reacts chemically with some fluorides, hence cannot dry them. Advantage has been taken of this, however, to diminish the level of fluorine in municipal water treatment to useful concentrations for dental protection. Activated alumina also provides practically complete removal of fluorine from the product of the hydrogen fluoride alkylation process. Savinelli investigated the fluoride-ion exchange capacity of activated alumina regenerated by dilute alum solutions. Satisfactory exchange of fluorides was obtained from waters of high carbonate hardness, but not from waters of high sodium or potassium bicarbonate alkalinity.

Some liquids react with activated alumina; for example, acetone is catalytically oxidized to mesityl compounds upon heating. Some components may be deposited as solids or viscous liquids on the active surface, as for example, sulfur. Miller and Weir claimed improved drying of paraffinic hydrocarbons containing sulfur compounds by treating the activated alumina with dilute mineral acid. Bienstock et al. removed sulfur oxides from flue gas by absorption at 330°C in alkalized alumina. The absorbent was then regenerated by heating with hydrogen or steam-reformed natural gas.

Unsaturated organic compounds sometimes polymerize to high-boiling compounds which may foul the desiccant. Carbonaceous deposits form from natural gas and unsaturated hydrocarbons, which act to reduce the active surface. High-temperature regeneration (500 to 600°C) of H-151 type of activated alumina under oxidizing conditions restores almost the original sorptive capacity. Krieger and Heinemann purged with steam at 370°C and finally reactivated at 400 to 700°C.

Common methods for determining the sorptive capacity of desiccants involve the isothermal pickup of water at controlled humidities and temperatures, either under static or dynamic conditions of flow. In many commercial operations, however, the adsorption is to a large extent adiabatic. Bower compared the efficiency of several desiccants (barium oxide, lime, anhydrous calcium chloride, and calcium sulfate, and silica gel) with that of activated alumina under dynamic flow of moist air at 30°C. The activated alumina reduced the moisture content of the air to only 0.0018 mg/l, and was exceeded in efficiency of removal only by barium oxide on the weight basis.

Stowe determined that about 18.0 cal/g of activated alumina represents the heat release when activated alumina (F-1) is immersed in liquid water. About 24.4 cal/g is released for activated gelatinous, amorphous alumina (H-51). Robert found that the heat of wetting of 15 organic liquids on activated alumina is proportional to their degree of adsorbability. The heat of adsorption varied from 10 kcal/mole water adsorbed when the level of water adsorbed was 3% by weight to over 105 kcal/mole when the level was about 0.1% (Cornelius et al.). Sanlaville found that the differential heat of adsorption is 6 to 7 kcal/mole for 0 to 30 mg H_2O/g Al_2O_3 (the first layer), 0.9 kcal/mole for 130 mg H_2O/g Al_2O_3 (the second layer), and only 0.25 kcal/mole for more than 150 mg H_2O/g Al_2O_3.

Freymann concluded that the water adsorbed on alumina is in the liquid state, inasmuch as it has a microwave absorption at 3.15 cm. Liquid water has an absorption in this region, but not hydrated alumina.

Getty, Lamb, and Montgomery; Miller and Roberts; and Getty and Armstrong investigated the dynamic sorption of water vapor from gases, to improve the understanding of industrial dynamic drying conditions. In drying gases under pressure, the heat of adsorption is dissipated easily because of the high heat capacity of the gas. Under certain conditions, a temperature wave precedes the sorption wave. The adiabatic operating capacity of the desiccant bed is much less than that calculated from isothermal equilibrium data. Equations for adiabatic capacity were determined for use in the design of large industrial gas dryers which operate without internal cooling. Fleming, Getty, and Townsend applied these principles to the design of a 4-tower natural gas dehydration section in a helium extraction facility at Liberal, Kansas. It uses about 350,000 lb of granular activated alumina to dry over 840,000,000 scf of natural gas daily.

Winter followed the exchange between an oxygen atmosphere containing 1.2% ^{18}O and transition alumina by means of a mass spectrograph. At 500°C, an initially fast first-order exchange had a heat of activation of 37 kcal/mole. A slow reaction followed. A similar exchange was observed for heavy-oxygen water vapor (Whalley and Winter). The exchange between water vapor and alumina attained 20, 30, and 50% equilibrium at 200, 400, and 600°C, respectively (Karpacheva and Rozen). The initial first-order exchange was independent of the oxygen pressure, and was considered to result from surface oxygen ions. The rate constant increased from 0.0067 min^{-1} at 485°C to 0.0146 min^{-1} at 592°C. The subsequent slow exchange was ascribed to diffusion of oxygen ions from an interior layer to the surface. After exchange of ^{18}O from water vapor, further exchange could be effected with ethyl alcohol vapor at 200 and 400°C, probably by the same mechanism. Dontsova observed a similar exchange in CO_2 at 1100°C. Rozen et al. concluded that the mechanisms involved are the heterogeneity of the surface, polydispersion, and a combination of adsorption and solution.

Greaves and Linnett determined the coefficients of recombination for the removal of oxygen atoms at the surface of alumina. The decay of atom concentration was exponential (first order), but the activity could not be related to any single parameter associated with the surface material.

Holm and Blue found that pure alumina of high surface area has high activity for hydrogen-deuterium exchange. This is greatly increased by raising the calcination temperature from 400°C to 800°C. Silica in the catalyst decreased the exchange but increased the rate of hydrogen transfer reactions from naphthenes to olefines. Bernstein observed that the rate of exchange of ^{18}O was appreciably less than that of deuterons from $D_2^{18}O$ to boehmite powder (12.2 m^2/g surface area) from 100 to 230°C. This suggested that the diffusion of oxygen-bearing carriers (such as OH$^-$, H_2O or O^{2-}) is independent of the protons (deuterons) through the lattice. Apparently only one-half the oxygen atoms in the boehmite are exchangeable. The activation energy for the proton (deuteron) diffusion coefficient is approximately 12 kcal/mole. The deuteron exchange between D_2O and bayerite at 25°C was more rapid than with the smaller particles of boehmite at the same temperature. The proton mobility in bayerite appears to be an order of magnitude greater.

The active surface of the alumina spinel may be assumed to contain covalent bonds locally between incompletely coordinated aluminum atoms and the surrounding oxygen atoms. These aluminum atoms are associated with acid sites on the surface. They behave as electron pair acceptors (Lewis acids) and will strongly bind water molecules with the formation of protonic acids (Brönsted acids), as is shown by the two structures:

$$\overset{\cdot\cdot}{:}\overset{\cdot\cdot}{O}:Al:\overset{\cdot\cdot}{O}:$$

$$O$$

Lewis acid

$$H \quad H^+$$
$$:\overset{\cdot\cdot}{O}:$$
$$O:Al^-:O$$
$$:\overset{\cdot\cdot}{O}:$$

Alumina-silica cracking catalysts owe their activity to such acid sites (Trambouze and Perrin). Mixtures of neutral silica sols with neutral alumina sols develop strong acid. One equivalent of acid is produced for each atom of aluminum consumed in this reaction. The reaction becomes less effective with aging of the gels (Tamele).

De Rossett, Finstrom, and Adams claimed that the adsorption of hydrogen sulfide from mixtures of hydrogen sulfide and hydrogen at 260 to 560°C was like that of water and ammonia in reacting as a base at Lewis acid sites formed by stripping oxygen anions from the spinel surface. Over a range of H_2S partial pressure from one micron to 4.4 mm, 39 to 107 micromoles H_2S/g Al_2O_3 were observed. Heats of adsorption ranged from -25 to -38 kcal/mole, depending on the degree of predrying of the alumina.

Statistical calculations made by Clark and Holm for the adsorption of NH_3 in the region of fixed adsorption indicate a broad distribution of adsorption energies. The adsorption data fit the integral form of the Langmuir equation. Entropies were several times smaller than those for uniform sites. Below a coverage of 0.36×10^{14} moles/cm^2, the adsorbed NH$_3$ molecules have lost all translation and rotation. Above this value one degree of rotational freedom may have been regained.

Smith and Metzner developed a molecular model of the surface transport rate process, containing only one arbitrary parameter. Experimental measurements of the migration rates of propane, butane, propylene, perfluoropropane, and Br$_2$CF$_2$ over a wide range of pressures and a moderate range of temperatures on an alumina catalyst surface support the theoretical model.

Mizutani, Sakaguchi, and Iizuka used the adsorption of alcohol and aromatic compounds as a measure of adsorbent activity for use in petroleum refining. Topchieva et al. found that during the dehydration of ethyl alcohol with activated alumina, the surface of the alumina became hydrated, and this increased its catalytic activity. It was possible to adsorb methyl alcohol both on hydrated and dehydrated surfaces.

Roach and Himmelblau investigated the adsorption of calcium, strontium, and thallium ions from molten mixtures of zinc and potassium chlorides by both activated alumina and silica. No adsorption for any of the cations could be obtained on the silica. Thallium was not adsorbed on alumina, but strontium and calcium were adsorbed, the rate being twice as fast for strontium. The calcium isotherm at 250°C fitted the Freundlich and Langmuir equations. No effective method was found to elute the cations from the alumina column.

Aleixandre-Ferrandis determined the quantitative sorption on activated alumina of many organic acids dissolved in common organic solvents. He preferred activated alumina prepared by the method given in German Patent 561,713.

Fridman claimed that activated alumina having a surface area of about 180 to 370 m^2/g and a pore radius of 25 to 55 A has a higher sorptive capacity for the acidic compounds in oxidized transformer and turbine oils than silica gel. This is particularly true for acids of low molecular weight. For a period of 370 days, 0.15% of activated alumina in the oil sorbed the same amount of acids as is sorbed by 2% of silica gel. Fairly stable oils with an acid number as high as 0.10 mg KOH were regenerated in two weeks to the specifications of fresh oil. Aluminum Company of America offered instructions in a manual "Activated Alumina Maintenance Program: Power System Oils" (1955) for preventing deterioration of oils used in transformers, oil circuit breakers, and hydroelectric powerhouse equipment by continuous circulation of the oil through bypass filters in which the alumina adsorbs moisture and acids and prevents sludging and corrosion of metal parts. Sargent and Kipp, and Sawyer, Keefer, Kipp, and Shaw demonstrated the excellent service of activated alumina in maintaining lubricants in gas burning, in-line, internal combustion engines and in large radial gas engines.

Engel and Krijger impregnated a porous alumina with an aqueous solution of an aluminum salt in an amount sufficient to fill the pores without causing agglomeration, and heated at above 200°C in order to obtain an active contact decolorizer of hydrocarbon oils.

Czaplinski and Zielinski claimed that it is very effective to separate helium, neon, and hydrogen by adsorption in alumina under pressure of a few atmospheres at the temperature of liquefied nitrogen. Hydrogen is the most strongly adsorbed, followed by neon; helium is the least adsorbed. At pressures of about 30 atmospheres, helium and neon are about equally adsorbed. Menon found that adsorption of carbon monoxide in alumina at temperatures of 0 to 50°C and pressures up to 2940 atm is fully reversible, and the alumina suffers no permanent change.

5 MECHANICAL PROPERTIES OF ALUMINA

5-1 GENERAL CONSIDERATIONS

Alpha aluminum oxide, both as single crystals and in polycrystalline sintered form, has remarkable mechanical properties in comparison with conventional porcelains and other single oxide ceramics. None of the likely refractory single-oxide contenders (BeO, CaO, CeO, MgO, Nb_2O_5, SnO_2, ThO_2, TiO_2, ZrO_2, UO_2, and the rare-earth oxides) approaches pure sintered alumina in bending and tensile strengths at room temperature, and only ZrO_2 and ThO_2, are comparable in compressive strength. At about 1000°C, alumina is not approached by them in tensile strength, is matched only by BeO and MgO in bending strength, and is exceeded only by stabilized ZrO_2 in compressive strength (Hague et al., 1963). Many of the advantageous strength characteristics are retained to lesser extent by the high and low-alumina porcelains (Austin, Schofield, and Haldy; Haldy et al.).

The interest in mechanical properties stems from several modern applications such as the possible substitution of alumina ceramics for refractory metal parts in air-borne equipment, or fabrication forms in which high mechanical strength, hardness, or thermal shock resistance is important. The various structures are classified as brittle ceramics.

The mechanical tests of particular significance include: flexural, compressive, tensile torsional, and impact strengths; moduli of elasticity and rigidity; Poisson's ratio and bulk modulus; fatigue, creep, internal friction, thermal shock resistance, and flaw detection; and hardness. Binns (1965) discussed test methods for general mechanical, electrical, and thermal properties of alumina ceramics used in engineering applications. Shook gave a critical survey of mechanical property test methods, representing the present state of the art (1963). He discussed the probabilistic nature of brittle failure in terms of Weibull's statistical theory, and Griffith's theory of the existence of microcracks to explain the discrepancy between the theoretical and observed strengths. Orowan developed this theory further. Weibull proposed a semiempirical function of the form:

$$S = 1 - \exp \left[-V \left(\frac{\sigma - \sigma_u}{\sigma_0} \right)^m \right],$$

in which σ_0, σ_u are the classical flawless strength and the lower limit stress below which a fracture cannot occur, respectively, m is the flaw density, V is the volume of component subjected to stress, and S is the rupturing stress.

The theoretical strength can be deduced from consideration of the amount of work required to form the new surfaces resulting from fracture of the atomic planes. For alpha alumina this is about 5.5×10^6 psi, but in practice one-tenth of this is usually beyond realization. Under careful conditions of preparation, particularly with respect to elimination of surface flaws, strengths of single-crystal ceramics may approach the theoretical E/lO, where E is Young's modulus. In the case of polycrystalline ceramics, strengths may be less than 0.01 (E/lO). The highest strengths are obtained with materials of high density, fine crystal size, and usually high purity of a single phase. High purity in alumina, however, appears to favor crystal growth during sintering. Since the strengths of polycrystalline ceramics, in general, do not approach the values computed from theoretical considerations, much attention has been directed to the appraisal, analysis, and production factors governing microstructure (*Ceramic Microstructures*, ed. Fulrath and Pask). Although specific issues in these investigations have been to obtain a better understanding of the factors affecting forming and sintering and the perfection of equations typifying sintering models, in the case of alumina ceramics at least, the main objectives are to improve strength and other properties and to make these properties more reproducible as a practical consideration.

Sintered alumina shows the closest relationship between the calculated and the theoretical strength values of the general pure oxide ceramics (Matveev and Kharitonov, 1966). A large amount of test data in the literature has limited value because of inadequate description (size, surface preparation, ambient test conditions, chemical and physical structure, statistical treatments, grain boundaries, impurities, etc.).

Since a porosity of only 10% may reduce strength by as much as 50%, this has necessitated the computation of strength at zero porosity for comparative purposes. Knudsen derived an empirical equation showing that the fracture strength of completely brittle polycrystalline materials increases with decreasing grain size and porosity, having the following form: $\sigma_F = K_x d^{-a} e^{-bP}$, in which d is grain diameter, P is porosity, K_x a factor, and a and b the coefficients of grain size and porosity, respectively. Carniglia, however, claimed that when adjustment had been made for porosity, fracture strength more closely approaches the equation: $\sigma_F = K_x d^{-1/2}$ for large grain size, and $\sigma_F = \sigma_y + K_y d^{-1/2}$ for small grain size, relating the latter equation to a similar one developed earlier by Petch to explain slip-band failure of semibrittle materials. Factors affecting crystal growth are discussed in the chapter on sintering.

Congleton and Petch evaluated the surface energy of running cracks, derived from crack-branching measurements in alumina ceramics. The surface energy at low stress values approached the intrinsic surface energy, but rose well above this as the stress on the crack increased, typical values being 2.5 and 15 X 10 ergs/cm^2, respectively. As mentioned, significant improvement in mechanical strength of polycrystalline alumina is achieved by surface polishing, as with diamond powder (Rigby and Hesketh).

The strength of alumina ceramics falls off substantially as the temperature increases beyond 1/2 the melting point. This is discussed under specific mechanical strength tests.

The testing procedures are also subject to criticism because they may not represent pure application of the supposed test. Tensile tests are the least used to evaluate strength of brittle materials. Duckworth observed that the tensile strengths of brittle materials are much lower than the compressive strengths. Bending tests fail in tension, and can be used as a precise method for determining tensile strength. Compressive strength tests usually involve failure in tension and are also surface-sensitive.

Other relevant literature includes articles by Gurney; Salmassy, Schwope, and Duckworth; Glenny and Taylor; D. Weyl; Duckworth and Rudnick; and Bravinskii and Reshetnikov; WADC and other reports by Duckworth, Schwope, Salmassy, Carlson, and Schofield; Salmassy, Duckworth, and Schwope; Salmassy, Bodine, Duckworth, and Manning; Pulliam; and Bortz et al. Books on the subject include, among others, *Property Measurements at High Temperature*, W. D. Kingery; *The Mechanical Properties of Engineering Ceramics*, edited by Kriegel and Palmour; *Mechanical Behavior of Materials at Elevated Temperatures*, edited by Dorn; and *Fracture of Solids*, edited by Drucker and Gilman. The handbook, Refractory Ceramics of Interest in Aerospace Structural Applications ASD-TDR-63-4102 (October 1963) by Hague, Lynch, Rudnick, Holden, and Duckworth shows selected mechanical and thermal properties of aluminum oxide in comparison with other ceramic materials.

Duckworth proposed that the "strength" of brittle materials used in design structures should be based on a stress corresponding to a given probability of fracture, as derived from a statistical analysis of the test data, rather than on the mean strength obtained from the test data and to which a factor of safety has been applied. The stress corresponding to zero probability of fracture, the "zero strength" is particularly significant.

King found that the microstructure was quite influential in changing the mechanical properties of hot-pressed polycrystalline alumina, having a density of at least 3.964 g/ml. The wear resistance increased but impact strength decreased with decreasing temperature of compaction. Samples which had been hot-pressed at 1450 to 1500°C and annealed at 1500°C had marked grain growth after annealing. This growth, which normally does not occur at 1500°C, suggests that residual strain energy from plastic deformations exists after hot pressing.

Weil, Bortz, and Firestone undertook a program to determine the applicability of statistical fracture theories to inorganic ceramics and to define the major parameters affecting fracture strength. The found that the prior thermal history, the specimen finish, the test temperatures, and specimen size had a primary influence on fracture strength. Environmental effects (water content) were significant only for specimens tested at 20°C; at 1000°C, all environmental influences became negligible. Fracture at room temperature is governed by surface-induced failure mechanisms. Both Weibull constants, the flaw-density parameter (m) and zero strength (σ_u), are sensitive to surface treatment and thermal history. Grinding increases the value of m but leaves σ_u unchanged; annealing increases the value of m, but reduces σ_u to zero. Both weaken the material. Weakening is particularly pronounced in annealed specimens. This is thought to result from destruction of beneficial residual stress distribution.

Ryshkewitch (Oxydkeramik, page 192) claimed that rapid cooling by application of an air blast increased the strength of sintered alumina (95% Al_2O_3) by 35% and impact resistance by 30%. Smoke, Illyn, and Koenig; Preist and Talcott; Phillips and DiVita; and Insley and Barczak confirm this effect. Insley and Barczac perform the rapid cooling following a reheating treatment. The proposed mechanism for the strength increase is the prevention of exsolution of components soluble in either a major or minor crystalline phase. Contrariwise, in the case of single-crystal sapphire ceramics, annealing in the temperature region, 1000 to 1800°C improved tensile strength (Kvapil; Davies), and was considered more effective than mechanical polishing (Heuer and Roberts, 1966). Chemical strengthening of polycrystalline alumina was attained by application of surface layers of Al_2O_3-Cr_2O_3 composition (Kirchner and Gruver, 1966).

Structural applications of aluminum oxide in the high-temperature field require a knowledge of the effect of temperature on the mechanical properties. Kingery's book, *Property Measurements at High Temperature*, provides much information on the mechanical properties of aluminum oxide and other high-temperature ceramic materials.

Data on the mechanical properties of alumina are collected in Table 9 and in Figures 8 through 12. The data in the table include information taken from the ceramic literature, as well as average values for commercial production, taken from the standards of the Alumina Ceramic Manufacturers Association and the literature of several of

Table 9

Mechanical Properties of Alpha Aluminum Oxide

Bending Strength (Modulus of Rupture)

		Temp. °C	psi	
Sapphire	Flame-fused, oriented 0° between optic axis and bar axis[a]	25	102,000	(1)
		600	27,000	
		1000	45,000	
Sapphire	Flame-fused, oriented 45° between optic axis and bar axis[a,b]	25	72,000	
		600	47,000	
		1000	85,000	
Ruby	Flame-fused, oriented 45° between optic axis and bar axis[a,b]	25	50,000	
		600	33,000	
		1000	85,000	

Polycrystalline Alumina

$$S_{25°C} = 142,500\ e^{-11.83P_G-0.60+3.33P} \quad (18)$$
$$S_{1200°C} = 73,000\ e^{-11.33P_G-0.60+3.33P} \quad (18)$$

Polycrystalline Alumina (99.9% Al_2O_3, 98% theoret. density, hot pressed) (2)

Crystal Size (microns)	1-2	10-15	40-50
25°C	67,000 psi	48,000 psi	35,000 psi
400	52,000	37,000	34,000
1000	49,000	37,000	31,000
1350	37,000	16,000	14,000

Commercial Grades of Polycrystalline Alumina

Nominal % Al_2O_3	99.9 (5)	99 (3)	94 (3)	85 (3)
25°C	62,000	52,000	46,000	46,000
980	–	23,000	17,000	12,000

Compressive Strength (psi)

	Sapphire	Polycrystalline			
		100% Al_2O_3	99% Al_2O_3	94% Al_2O_3	85% Al_2O_3
	(6)	(7)	(3)	(3)	(3)
25°C	443,000 to	560,000[c]	300,000	300,000	240,000
25°C	495,000	426,000[d]			
400		213,000			
800		185,000			
1000		128,000			
1200		71,000			
1400		35,600			
1600		7,100			

Tensile Strength (psi)

	Single Crystal Orientation 45° to optic axis	Filaments Uncoated	Coated	Polycrystalline (e)	94% Al_2O_3	85% Al_2O_3
	(9)	(10)	(10)	(7)	(3)	(3)
30°C	71,000 psi	70,000	210,000	37,600	26,000	17,500
300	52,500			36,400		
800	52,500			34,000		
1050				33,800		
1100	88,000			31,400	9,500	8,500
1200				18,500		
1400				4,250		

[a]Loading rate 14,200 psi/minute; [b]minimum creep at 45°; [c]zero porosity; [d]less than 5% porosity; [e]rupture time less than one minute.

Ref.: (1) Wachtman & Maxwell (1959)
    (2) Spriggs, Mitchell & Vasilos (1964)
    (3) Coors Porcelain Data Sheet 0001, August 1964
(5) Frenchtown Porcelain Co., Bull. 5462
(6) Pavlushkin (1957)
(7) Ryshkewitch (1941)
(9) Wachtman & Maxwell (1954)
(10) Berezhkova & Rozhanskii
(18) Passmore, Spriggs & Vasilos (1965)

Table 9 – Continued

Modulus of Elasticity (E), $X10^6$ psi

$$E \text{ (polycrystalline)} = 59.49 \ X10^6 \ e^{-3.95P}, \text{ where P = fractional pore volume} \qquad (8)$$

	Single Crystal (12)		Polycrystalline		
	Sapphire	Ruby (0.75% Cr_2O_3)	(11)	94% Al_2O_3 (3)	85% Al_2O_3 (3)
25°C	52.6	54.1	59.30	40.2	31.9
500	48.1	49.5	57.27		
1000	43.5	45.1	54.89		
1200	41.9	43.5	53.65		

Elastic Constants ($X10^6$ psi)		Elastic Compliances ($X10^{-13}$ cm^2/dyne)		(13)
C_{11}	72.05	S_{11}	2.353	
C_{33}	72.24	S_{33}	2.170	
C_{44}	21.40	S_{44}	6.940	
C_{12}	23.73	S_{12}	−0.716	
C_{13}	16.08	S_{13}	−0.364	
C_{14}	−3.41	S_{14}	0.489	

Modulus of Rigidity (G), $X10^6$ psi

		Polycrystalline Alumina			
	Sapphire (13)	Hot-pressed Zero porosity (14)	Cold-pressed Zero porosity (14)	94% Al_2O_3 Density 3.62 (3)	85% Al_2O_3 Density 3.42 (3)
25°C	23.29 (Reuss) 24.07 (Voigt)	23.26	23.89	17	13

Poisson's Ratio (μ)

$$\mu = 0.257 - 0.35P, \text{ where P = fractional pore volume} \qquad (16)$$

	Polycrystalline Alumina			
	Cold-pressed Zero porosity (16)	Cold-formed 98% Theoret. density (15)	94% Al_2O_3 Density 3.62 (3)	85% Al_2O_3 Density 3.42 (3)
25°C	0.257	0.32	0.21	0.22
1000		0.32		
1400		0.45		

Impact Resistance (In-lb)

	Polycrystalline Alumina			
	Slip-cast 92% Theoret. density (17)	96% Al_2O_3	90-95% Al_2O_3 Charpy Test Method (4)	85-90% Al_2O_3
25°C	1.2	7.0 to 7.6	6.5 to 6.8	6.3 to 6.5
800	1.0			
1000	0.55			
1600	0.32			

Ref.: (3) Coors Porcelain Data Sheet 0001, August 1964
(4) Diamonite Products Manufacturing Company (1963)
(8) Knudsen

(11) Crandall, Chung, & Gray (1961)
(12) Wachtman & Lam (1959)
(13) Wachtman, Tefft, Lam, & Stinchfield (1960)

(14) Lang (1960)
(15) Ryshkewitch (1951)
(16) Spriggs & Brissette
(17) Kingery & Pappis

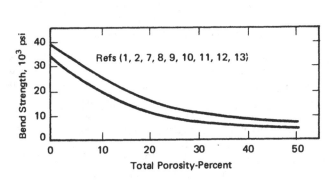

Figure 8a. Effect of Porosity on Room-Temperature Bending Strength of Polycrystalline Aluminum Oxide.

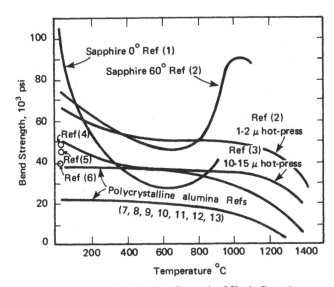

Figure 8b. Bending Strength of Single-Crystal and Polycrystalline Aluminum Oxide

Single-Crystal Alumina

Ref 1—Wachtman and Maxwell (WADC-TR, 1957). Sapphire rod, 0-deg. see oriented (crystallographic axis to rod axis); centerless-ground and flame-polished.

Ref 2—Wachtman and Maxwell (WADC-TR, 1954). Wachtman and Maxwell (1957). Sapphire rod, 60-deg. see orientation; centerless-ground and flame-polished; 0-10-inch diameter, 2.3-inch gauge length, quarter-point loaded at 17000 psi/min.

Polycrystalline Alumina

Ref 3—Spriggs, Mitchell, and Vasilos (1964). Linde A-5175 alumina 99.9%; hot-pressed at 1700°C; density $>$98%; specimen size 0.25 by 0.15 by 1.75 inch (1.5-inch span); 4-point loaded at third points, load rate 0.0001 inch per second.

Ref 4, 5, 6—Alumina Ceramic Manufacturers Association, 96% plus Al_2O_3, 70 to 96% Al_2O_3, and 80 to 90% Al_2O_3, respectively.

Ref 7—Coble and Kingery (1956).

Ref 8, 9—Jackman and Roberts (1953), Pearson (1956). Commercial recrystallized alumina; 94 to 95% density; average grain size about 20 microns; bars 0.25 inch in diameter by 3 inches, 4-point bending, loaded at 30,000 psi/minute.

Ref 10—Roberts and Watt (1949). Alumina 99.1%, slip-cast and sintered; 87 to 96% density; 0.16 by 3.2-inch bars, 4-point bending (load rate not specified).

Ref 11—Ryshkewitch (1956). Alumina 99.9%; pressed and sintered; 93 to 96% density; 0.14 by 0.14 by 2.75-inch bars, 3-point bending (no load rate specified); time to fracture, one to several minutes.

Ref 12—Tacvorian (1955). High-purity alumina; sintered; about 96% density; bending of 0.16 by 0.16 by 2.66-inch bars (load rate not specified).

Ref 13—Tunison and Burdick (1961). Norton 38X alumina; slip-cast and sintered (sieved naphthalene added to control porosity); average grain size about 23 microns; bars 1/4 inch in diameter by 3.0 inches gauge-length, 4-point bending, loaded at 900 psi/minute.

Figure 8.

its members. The values are typical rather than representative of the highest that have been reported. In much of the ceramic literature, insufficient data have been provided for comparative appraisal of test results by different investigators. In some cases the specimens have been prepared by unsatisfactory forming methods or by excessive heat treatments for best strength. Such factors as crystal size, porosity, and vitreous impurities, have contributed to the wide range of strength properties reported. The influence of crystal size appears to be fairly consistent, however, in compositions prepared by different forming methods.

5-2 BENDING, COMPRESSIVE, TENSILE, AND TORSIONAL STRENGTHS

Bending strength involves the placing of some portion of a specimen beam in pure bending according to the expression: $\sigma = Mc/I$, in which σ is the modulus of rupture, M is the bending moment at the point of rupture, c is the distance from the neutral axis of the beam to the extreme fiber, and I is the moment of inertia of the cross-section about the neutral axis.

The bending strength of sintered alumina specimens is usually determined on several different bar sizes, ranging from the "needle" size of about 80 mils to about 3/8 inch in cross sectional dimensions. The microsize was suggested (H.B. Barlett) to enable determinations on specimens cut from sections of forms that may develop strengths of different values peculiar to the conditions of size and orientation in the form. L. H. Milligan recommended a ratio of span to depth of specimen greater than 10:1, and a loading at the 1/3 or 1/4 point rather than midpoint. Other conditions, particularly symmetrical loading, are commonly used. The microsize specimens usually receive surface polishing to eliminate flaws. Yurchak claimed that the best form is a cylindrical rod 6.5 to 8.6 mm in diameter.

Probably the first determinations of fracture strength on sintered alumina were made in 1933 by H. Gerdian. A flexural strength of about 17,200 psi was obtained on "Sinterkorund," that had been fired at about 1770°C, and having a porosity of about 5% (3.98 g/ml theoretical density of a-Al_2O_3). Ryshkewitch reported in *Oxydkeramik* values of about 50,000 psi on cold-pressed bars containing less than 2% porosity, measured at room temperature. The flexural and other strength measurements fell off remarkably when determined at temperatures above 1000°C. In another elaborate 4-point loading, specimens, fired at 1640 to 1780°C (3.48 g/cc bulk density, 12% porosity), developed only 35,500 psi at room temperature, which fell to 22,200 psi at 1000°C.

Ryshkewitch in 1953 found that the strength decreases in an exponential manner with increasing porosity; a 10% increase by volume causes a 50% decrease in strength from initial values. Coble and Kingery confirmed this, and also found that other mechanical properties were similarly affected. Extrapolation of Ryshkewitch's data to 100% theoretical density increases the flexural strength to about 52,500 psi, but this is far from the ultimate strength. Pores perpendicular to the direction of application of pressure decreased strength more than those parallel to the pressure direction.

Ryshkewitch found that the flexural strength of sintered alumina is affected by the crystal size, the finer crystalline material having the higher strength. Cutler used extruded, low-soda alumina specimens containing additives to develop grain growth to different sizes by heating at temperatures below and above the crystallization region. He concluded that grain growth caused a minor decrease in strength, but that porosity was a major factor in decreasing strength in his tests.

Crandall, Chung and Gray; and Spriggs and Vasilos came closer to overcoming the objection of simultaneous changes in porosity and grain size by applying hot-pressing to the forming. The latter's specimens ranged in grain size from about one to 260 microns, and in porosity from about 0.7 to 9.5% by volume. Their relation: Bending strength = $86,000G^{-1/3}$ (psi), fits the room temperature data well, after correcting to zero porosity. G is the grain size in microns. The data indicate that strengths around 100,000 psi are attainable by hot-pressing to obtain a median crystal size of about one micron. At a crystal size of 1000 microns, the strength drops to about 10,000 psi.

Wachtman and Maxwell in 1954 obtained modulus of rupture on synthetic sapphire of 43,000 to 131,000 psi. The scatter is partly dependent on the specimen orientation with respect to the crystallographic axes. Klassen-Neklyudova obtained values of 44,300 to 54,500 psi for bars cut with the length parallel to the base plane (0001), and 99,700 to 115,500 psi for bars cut perpendicular to the base plane. Annealing at 1500 to 1900°C, followed by slow cooling to room temperature, increased the strength.

Ryshkewitch (1941) observed that fracture occurs through the crystal as well as at the boundaries, and concluded that the strength of polycrystalline alumina should be substantially the same as that of single-crystal sapphire. Roberts, and Roberts and Watt confirmed Ryshkewitch's conclusion, and observed that transcrystalline failure is more common than intercrystalline. The nature of the fracture changes from transcrystalline to intercrystalline (grain boundary) at the higher temperatures at which strength decreases rapidly. The impurities even in specimens of better than 99% Al_2O_3 probably concentrate at grain boundaries, in effect to lower the temperature of deformation and fracture. Tunison and Burdick analyzed the fracture of hot-pressed alumina by means of the electron microscope and determined that the type of fracture obtained in bending rupture depends upon the relation of crystalline to intercrystalline bond strengths. These strengths were equal at 1200°C, the equicohesive temperature at which the fracture was entirely intergranular. It was speculated that for high-pressure grinding wheels, agglomerated, fine-grained alumina could give a lower breakdown rate than the same sized monocrystalline grit used in a bonded abrasive wheel.

Roberts investigated the influence of temperature on the flexural strength of both single-crystal and polycrystalline corundum. A loss in strength of the single-crystal specimens between 300 and 600°C, which was regained at 1000°C, was attributed to relief of stresses by microscopic plastic deformations. Wachtman and Maxwell (1959) determined the strengths of single-crystal sapphire and ruby under conditions of orientation of the crystal axes with respect to the direction of application of the pressure which either favored or hindered plastic flow. The data are consistent with the hypothesis of plastic deformation. The initial bending strengths obtained were about 100,000 psi for zero-degree sapphire, 70,000 psi for 45-degree sapphire, and 50,000 psi for 45-degree ruby. These values are less than one-tenth the theoretical strength.

Rasmussen, Stringfellow, and Cutler (1965) found no intrinsic effect on the strength of polycrystalline alumina by the formation of solid solutions of Cr_2O_3, Fe_2O_3, or TiO_2. The additions were made in amounts to 2% TiO_2, 20% Fe_2O_3, and 50% Cr_2O_3. The data were corrected to 3.0% porosity and 17.3 microns average grain size.

Passmore, Springgs, and Vasilos (1965), from experimental measurements of the coincident effects of grain size G and porosity P on the transverse bending strength S of alumina at 25°C and 1200°C, have derived the following relations:

$$S_{25°C} = 142,500e^{-11.83P}G^{-0.60+3.33P} \text{ and}$$

$$S_{1200°C} = 73,000e^{-11.33P}G^{-0.60+3.33P}, \text{ respectively.}$$

Brittle fracture in bending at both temperatures appears to be controlled by a Griffith-Orowan mechanism in which

preexisting surface defects, produced during grinding, are propagated to fracture at a critical level of stress,

$$\sigma \cong \sqrt{\frac{E\gamma}{C_0}},$$

in which C_0 is the critical crack length, E is the elastic modulus, and γ is the effective surface energy for crack propagation. The specimens were prepared by hot-pressing aluminum oxide (99.9 + % pure) of initial average particle size of 0.3 micron at about 1400°C. By heat-treating at

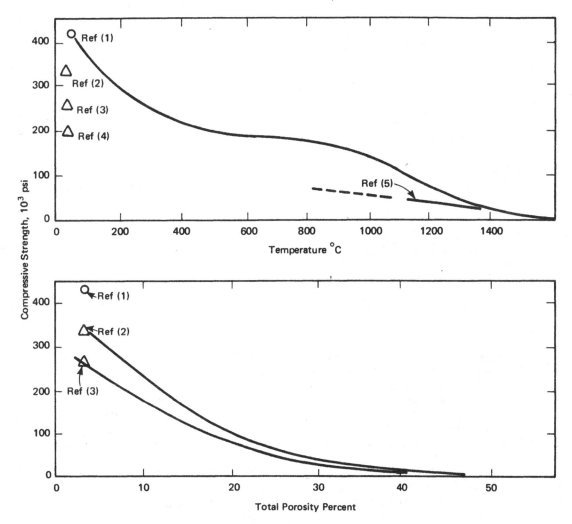

Ref 1–Ryshkewitch (1948). High-purity; 96 to 98% density; 0.25-inch cubes, ground, polished, and graphite coated; rapid loading in protective atmosphere; load duration less than 1 minute.
Ref 2, 3, 4–Ryshkewitch (1953). Slip-cast and sintered; H_2O_2 added to control porosity, 0.5 wt. % MgF_2 added to control grain size; 97% density:
    Ref 2–Pores parallel to loading direction
    Ref 3–Pores perpendicular to loading direction
    Ref 4–Same as preceding, but 90% density; average of parallel and perpendicular pore alignments.
Ref 5–Schofield, Lynch, and Duckworth (1949). Pressed and sintered; 78 to 93% density; average grain size 11 to 74 microns.

Figure 9. Effect of Temperature and Porosity on Compressive Strength of Polycrystalline Alpha Aluminum Oxide

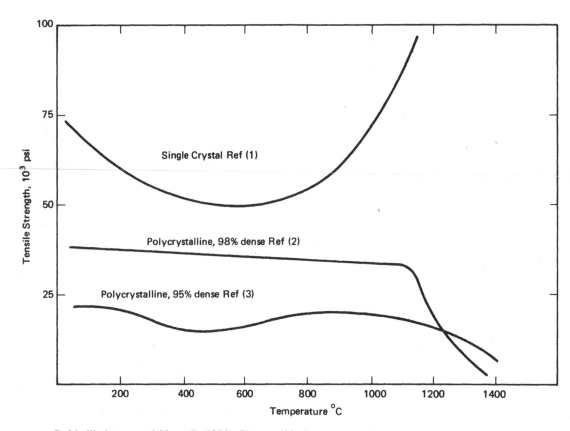

Ref 1–Wachtman and Maxwell (1954). Flame-polished, single-crystal sapphire rods, 45-degree angle between crystal c-axis and rod axis.
Ref 2–Ryshkewitch (1941). High-purity, pressed and sintered, about 98% dense.
Ref 3–Schwartz (1952). Slip-cast, 99% pure, 95% dense; 4-point bending of 5-1/2 by 3/8 by 3/16-inch
bars loaded at 13.2 lb/min in flexure.

Figure 10. Tensile Strength of Aluminum Oxide

1600 to 1700°C a range of grain sizes up to 100 microns
was achieved. The porosity ranged from 2.7 to 6.6% depending on the heat treatment. The surfaces were diamond-
machined and annealed for 24 hours at 900°C to final surface finishes of 0.2 to 2 microns (centerline averages). The
defects were thought to have been created by the pulling-
out of individual grains during the surface preparation by
grinding. Values for the critical Griffith crack length, C_o,
were calculated from values of the elastic modulus, E, for
each temperature and porosity (Mitchell, Spriggs, and
Vasilos, 1963), modified by data shown by Knudsen, and
the analytical equations of Spriggs (1961) and of Hasselman.
The ratios of crack length to grain size, C_o/G, were plotted
as functions of grain size for each temperature and nominal
porosity. Interactions between spherical pores and the
propagating crack front during brittle fracture inhibited
crack propagation apparently by reducing local stresses at
the crack front. Increasing pore size, at least up to 5
microns, favored the reduction of stress concentration.
Characteristic patterns on the fracture surface were produced by this porosity-crack interaction, which was considered to be responsible for the decreasing dependence of

strength with increasing grain size, and the decreasing dependence of grain size with increasing porosity.

Guard and Romo showed by use of a double-crystal
X-ray spectrometer that the fractured surfaces of hot-
pressed, 20-micron, polycrystalline alumina exhibit two
zones of distortion. Zone 1, of high distortion, extends to a
depth of about 10 microns; zone 2, having much lower
distortion, extends over 50 microns (in excess of one grain
diameter). Both zones are observed in specimens fractured
at 20° and at 1700°C. Distortion in zone 1 was believed to
result from basal and nonbasal slip in these grains through
which a fracture crack passes. Zone 1 has much higher
misorientation in the 1700°C fractures than in the 20°C
fractures; zone 2 is quite similar at the two temperatures.
Zone 2 is observed only in (0330) and (2240) reflections,
and was believed to result from basal slip only. The plastic
work of fracture was estimated at about 1500 ergs/cm²,
assuming that 1° of misorientation is equivalent to 1% of
strain. A fracture stress was assumed to be a tensile yield
stress under impact of 35,000 psi at 20°C and 20,000 psi at

1700°C. The value of the critical crack length for the Griffith-Orowan criterion is about 15 microns (about the grain size).

Compressive strength is determined from the expression

$$\text{Stress at fracture } \sigma = \frac{\text{Applied load P}}{\text{Cross-sectional area A}}$$

in which the area A is transverse to the load.

Ryshkewitch reported the measurement of compressive strength of sintered alumina in 1941, in 1953, and also in 1960 in *Oxide Ceramics-Physical Chemistry and Technology*. As in the case of flexural strength, compressive strength is temperature-dependent and decreases from a room-temperature value of about 425,000 psi to 128,000 psi at 1000°C, and only 7100 psi at 1600°C. The strength at 1000°C is quite impressive, however, in comparison with the strengths of the common structural metals, with which alumina might compete in high-temperature applications. For synthetic sapphire, the room temperature value was 300,000 psi. The influence of controlled porosity on compressive strength was measured on 1.2 cm cubes ranging in porosity from 3 to 60% by volume (Ryshkewitch 1953). The pores were generated by hydrogen peroxide and had an average diameter of 0.4 mm. The spatial arrangement of anisomorphic pores, formed during pressing, influenced the strength, inasmuch as pores that were parallel to the pressure direction imparted a higher strength than when perpendicular to this direction. The compressive strength decreased with increasing porosity in the same exponential manner as was found for flexural strength.

Pavlushkin in 1957 found the room-temperature strength of polycrystalline alumina to range from 443,000 to 495,000 psi. In one isolated test, a strength of 1,140,000 psi was claimed.

Ryshkewitch (1941) determined the tensile strength of sintered alumina on slender rods, 2 to 4 mm in diameter, expanded on the ends for easy gripping. The rods were made long to avoid error resulting from axial strain. Such thin specimens might suffer from inhomogeneities resulting from unequal heat treatment or marked influence of surface crystallization. His tensile strengths agree well with more recent values obtained on larger specimens, however. He obtained a tensile strength of about 37,500 psi at room temperature, which fell off gradually to 33,800 psi at 1050°C. The strength thereafter decreased rapidly to about 1560 psi at 1460°C. By comparison, however, nickel, as an example of a high-temperature metal, having almost twice the room-temperature tensile strength, retains 1400 psi in tensile strength only to about 700°C.

Roberts and Watt in 1951 used strain gauges on the surface of thicker specimens to enable adjusting the loading axis to avoid strain. The attained room-temperature tensile strength was about 17,400 psi. Pears et al. determined tensile strength on rods 3/8 in. O.D. by 5 in. length, prepared from Wesgo Al-995 (99.5% Al_2O_3) sintered alumina by centerless grinding of formed specimens. (This is a commercial product of Western Gold and Platinum Company, having a bulk specific gravity of 3.89 g/cc, less than 3% porosity, and an average crystal size of 22 microns). Gas bearings were used to provide uniaxial loading. The tensile strength was 35,500 psi at 21°C, and 1100 psi at 870°C.

Sedlacek applied isostatic pressure to the inside of a sintered alumina ring (Wesgo Al-995) in order to rupture it in tension by an ingenious method that avoids the use of nonyielding constraints in supporting the specimen. He found that the measured values for tensile strength increase with increased rate of applying the load within the range of rates from 70 psi/sec to 4500 psi/sec. At a loading rate of 3000 psi/sec, the maximum room temperature tensile strength was 30,920±1175 psi. Halden and Sedlacek claimed that it should be possible to attain a batch-to-batch reproducibility in tensile tests on commercial high-alumina shapes within 10% in standard deviation, and a piece-to-piece reproducibility of about 5%.

Wachtman and Maxwell (1954) obtained tensile strengths on flame-polished sapphire rods having a 45-degree angle between the axis of the rod and the c-axis of the crystal. The values obtained were considerably higher than those for polycrystalline alumina, amounting to 71,000 psi at 30°C. The specimens exhibited loss in strength above room temperature, gradually recovered by heating at about 800°C, and an unusual increase in strength at 1100°C of 88,000 psi.

Berezhkova and Rozhanskii found that the surfaces of filamentary corundum crystals grown from the gaseous phase were covered with twinned corundum coatings which hindered the propagation of cracks through brittle fracture. The room temperature strengths of filaments with a cross-section of 0.1 micron was 21,000 psi in the presence of the coating, and about 70,000 to 140,000 in the absence. The twinning occurs along the $(11\bar{2}1)$ and $(\bar{1}126)$ faces.

In determining torsional strength, a twisting moment is applied to a cylindrical rod, fixed at one end, or equal and opposite twisting moments are applied. The maximum stress is called torsional strength or modulus of rupture in torsion.

$$t_{max} = \frac{16T}{\pi d^3}$$

$$\sigma_{max} = \frac{16}{\pi d^3}(M + \sqrt{M^2 + T^2}) \ ,$$

in which σ_{max} is the tensile stress at the surface, d is the rod diameter, M is the bending moment, and T is the twisting moment.

Ryshkewitch; Decker and Royal; and Stavrolakis and Norton investigated the measurement of torsional properties of sintered alumina throughout a range of temperatures. Decker pointed out the advantages of torsion as a quick method for obtaining other strength tests because of the freedom from thermal expansion errors, the absence of the need for close dimensional tolerances, the simplicity of specimen installation in grips, and the ability to make deformation measurements outside the furnace on the cool end of the specimen. Stavrolakis found that the ultimate shear stress conforms with the ultimate tensile stress. Strength in torsion decreases in a slow linear manner from a value of about 29,400 psi at 25°C to 29,300 psi at 500°C. In the temperature region beyond 800°C, torsional strength drops at a rapid rate to about 3350 psi at 1500°C.

5-3 IMPACT STRENGTH

Poor impact resistance of brittle ceramics has limited their practical use. Factors affecting the strength of polycrystalline oxides are the internal microstresses, porosity, stress corrosion in damp air, and surface defects. Experience with glass has shown the importance of surface scratches in reducing fracture strength.

Moore measured the impact strength of slip-cast and extruded forms of sintered alumina by dropping a weight from predetermined heights onto 0.5-inch cubes until fracture. The strength was computed from the formula $f = \sqrt{2Ewh/v}$, in which E is Young's modulus, w is the weight of the falling load, h is the height of fall, and v is the volume of the specimen cube. The strength for both types of forms was about 43,000 psi.

Kingery and Pappis determined the influence of temperature on the impact resistance of slip-cast cylinders of sintered alumina having about 8% total porosity. The cylinders, 0.5 inch in diameter, were supported across a 4.5-inch span, and were impacted by an alumina-faced pendulum having a velocity of about 41 inches per second at impact. The impact energy of about 1.2 inch-lb at room temperature, fell only slightly until in the region 700 to 1000°C, whereupon it dropped to about 0.5 inch-lb, and continued to decrease slowly until at 1600°C it has reached 0.3 inch-lb.

5-4 MODULI OF ELASTICITY (E), AND RIGIDITY (G)

The modulus of elasticity (Young's modulus) may be determined by measuring a beam deflection for various applied loads within the elastic limits, according to the expression: $E = PL^3/48\delta I$, in which E is the modulus of elasticity, P is the applied load, L is the span between supports, δ is the deflection, and I is the moment of inertia of the cross-section about the neutral axis.

More recently, sonic and ultrasonic resonant frequency methods, and ultrasonic pulse techniques have been applied. Substantially the same results are obtained by each method at low and intermediate temperatures; the static moduli decrease faster at high temperatures, probably indicating a transition from elastic to plastic strain.

Some factors affecting the modulus of elasticity and the related modulus of rigidity of alpha aluminum oxide are shown in Figure 11. The exposition of these factors appears in the text.

Ryshkewitch (1942); Schwartz; Majumber; Schofield, Lynch, and Duckworth; Duckworth, Johnston, Jackson and Schofield; and Coble and Kingery measured the temperature dependence of Young's modulus of polycrystalline alumina by static bending methods. Ault and Ueltz; Crandall and Bryant; Wachtman and Lam; Lang; Spriggs; Spriggs and Vasilos; Chung; and Kovalev applied either damped or sustained flexural vibration methods. Schreiber and Anderson determined the influence of high pressure on the elastic properties of polycrystalline alumina by ultrasonic interferometry, and Soga, Schreiber, and Anderson estimated the influence of high temperature. Soga and Anderson computed the properties of powdered alumina from the compressibility and Debye temperature data. Values agreed within a few percent of those determined by standard resonance methods.

Ryshkewitch found that the moduli of elasticity of single-crystal sapphire and polycrystalline alumina are about the same at room temperature, from which he concluded that the intercrystalline bond is about as strong as the intracrystalline bond. Wachtman, Teft, Lam, and Stinchfield determined the zero-porosity modulus of elasticity of polycrystalline alumina to be 59.22×10^6 psi at 4084 kilobars (Voight method), and 57.59×10^6 psi at 3972 kilobars (Reuss method), based on their determinations of the elastic constants of single-crystal sapphire.

Scheetz stated that extrapolation of elastic moduli data for porous ceramic bodies (porosities of 5% or less) lead to values for the fully dense body which lie between those computed by space-averaging of single-crystal elastic stiffness and compliance constants. The elastic properties of composite bodies, including multiple-phase ceramics, may be predicted (with small error for bodies which remain continuous solids) by a method developed by Kerner. The values of elastic moduli for fully dense polycrystalline ceramic bodies should be used rather than the lower values associated with porous ceramics.

The earlier investigators used cold-forming, slip-casting or extrusion for forming the test specimens. Spriggs and Vasilos, and later investigators applied hot-pressing to obtain fine-grained polycrystalline bodies of density approaching the theoretical. Spriggs, Mitchell, and Vasilos (1964) found that the elastic modulus is essentially independent of grain size within the range 1 to 100 microns, at least, and within a temperature range to at least 1500°C.

The curve relating elasticity to temperature for polycrystalline alumina is similar to those found for flexural and compressive strengths (Figure 11). It decreases in a relatively slow linear manner to above 800°C, followed by a more rapid nonlinear decrease at higher temperatures in the range of usefulness. Coble and Kingery suggested that the

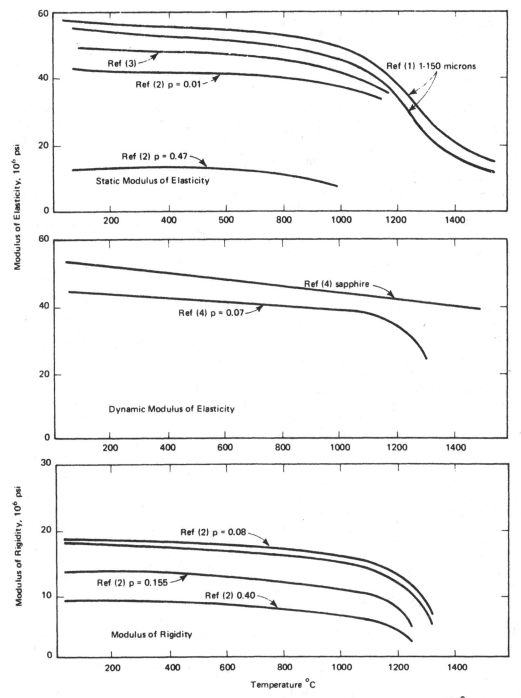

Ref 1—Spriggs, Mitchell, and Vasilos (1964). Linde A-5175 alumina 99.9%; hot-pressed at 1700°C; density >98%; specimen size 0.25 by 0.15 by 1.5-inch span; 4-point loaded at third points, load rate 0.0001 inch/sec.

Ref 2—Coble and Kingery (1956). Norton 38X alumina, slip-cast and sintered (sieved naphthalene added to control porosity), average grain size 23 microns; porosity varied from 0.01 to 0.47 for elastic modulus tests, and from 0.08 to 0.51 for rigidity tests; bars 1/4 inch in diameter by 3.0 inches gauge length, 4-point static bending.

Ref 3—Schwartz, B. (1952). Alumina plus 99% pure, slip-cast and sintered; 95% density; modulus of elasticity determined in 4-point bending (5-1/2 by 3/8 by 3/16-inch bars); modulus of rigidity probably determined similarly.

Ref 4—Wachtman and Lam (1959). Flame-polished sapphire rods, axis parallel to optic c-axis; elastic modulus determined by dynamic sonic method on 0.10-inch diameter by 6-inch specimens. Similar tests on polycrystalline, commercial bars (Norton's La 7365, 3.7 g/ml density = 9.07% porosity, 0.2 by 0.1 by 6 inches.

Figure 11. Moduli of Elasticity and Rigidity of Alpha Alumina

rapid decrease in mechanical properties around 1000°C is caused by grain-boundary slip. This appears plausible, since Wachtman and Lam found that single-crystal sapphire exhibits only linear decrease in Young's modulus from room temperature to the highest temperature measurement.

Spriggs proposed a general equation relating porosity of polycrystalline alumina to Young's modulus, having the form: $E = E_o e^{-bP}$, in which E is Young's modulus for the porous polycrystalline specimen, E_o is Young's modulus for the nonporous body, e is the Napierian number 2.71828 . . . , b is an empirical constant, and P is the fractional pore volume. Knudsen found that the equation: E(in kilobars) = 4102 $e^{-3.95 P}$ approximates the data well except for some, but not all, of the tests in which porosity was obtained by the use of a fugitive particulate filler. The zero porosity value at room temperature is 59.48 × 10^6 psi. Spriggs corrected the equation to compensate for differences in the effects of open and closed porosity. Hasselman, and Piatasik and Hasselman have since proposed a mechanical model for showing the effect of porosity on strength, based on the postulation that a fraction of the material is stress-free owing to the shielding action of the pores.

Lang found that the greatest variation in modulus of elasticity occurs in hot-pressed alumina compositions (mainly commercial specimens), although the averages were higher than for other forming methods. The dynamic elastic constants determined by the sonic vibration method are more sensitive than bulk density measurements as indicators of homogeneity.

Rao; Bhimasenacher; Mayer and Hiedemann; Wachtman, Tefft, Lam, and Stinchfield; Tefft; and Ramamurthy determined the six elastic compliances and elastic constants of single-crystal corundum. Wachtman et al. and Tefft applied a resonance technique in the range 1 to 45 kcps in the evaluation. Ramamurthy used a piezoelectric oscillator. Values are shown in Table 9. The variations of Young's modulus and of the shear modulus with crystal orientation were determined from the elastic compliances.

The modulus of rigidity (shear modulus) may be visualized as the distortion resulting from a compression of one diagonal of a specimen and the stretching of the other. The determination is most readily achieved by placing a cylinder in torsion, and is measured by the expression: $G = TL/\theta J$, in which T is the twisting moment, L is the specimen length in strain, θ is the angle of twist in radians, and J is the polar moment of inertia of cross section.

The modulus of rigidity is related to the modulus of elasticity by the expression: $G = E/2(1+\mu)$, in which G is the modulus of rigidity, E is the modulus of elasticity, and μ is Poisson's ratio (dimensionless). This is applicable only to isotropic bodies.

The resonant frequency of a beam in its fundamental mode of torsional vibration is related to the modulus of rigidity by: $G = \rho(WL/\pi)^2$, in which G is the modulus of rigidity, ρ is the mass density, W is the resonant frequency of the fundamental vibration, and L is the beam length.

Ryshkewitch in 1942, and in 1951; Stavrolakis and Norton; Coble and Kingery; and Lang obtained data on polycrystalline alumina. Wachtman, Tefft, Lam, and Stinchfield obtained data on single-crystal sapphire at room temperature. The modulus of rigidity of polycrystalline alumina changes with temperature in the same manner as the modulus of elasticity, that is, a slow linear decrease to about 800°C, followed by a very rapid decrease at high temperatures. The change with increasing porosity is also similar. This follows from the Poisson's ratio relationship between the two moduli.

Spriggs and Brisette computed the zero-porosity modulus of rigidity for polycrystalline alumina from the values of Lang, and Coble and Kingery on specimens prepared by cold-pressing, slip-casting, and hot-pressing, by use of an exponential expression similar to that used in determining the influence of porosity on modulus of elasticity. The expression has an empirical constant that varies with the fabrication technique. An average value of 23.68 × 10^6 psi was obtained, which agrees well with the average of the values calculated by Wachtman et al. from the single crystal constants, 23.29 × 10^6 by the Reuss theory, and 24.07 × 10^6 by the Voigt theory.

Spinner and Valore found that the empirical determination of the relationship between shear modulus and the fundamental torsional resonance frequency, mass, and dimensions of bars of rectangular cross-section is lower than the theoretical approximation given by Pickett by an amount increasing to about 1.75% as the cross-sectional width-to-depth ratio of the bars approaches 10.

5-5 POISSON'S RATIO (μ)

When a material is strained in one direction, it undergoes strain of opposite sign in the transverse directions. The ratio of transverse strain to principal strain is Poisson's ratio (μ). The changes can be shown as volume changes.

Ryshkewitch (1960) computed Poisson's ratio for sintered alumina from the relationship: $\mu = E/2G - 1$, and obtained a value of 0.32 for temperatures up to about 1000°C, increasing above that temperature to about 0.5 at 1900°C. Alumina exhibits pronounced creep under these higher temperatures. Lang, Coble and Kingery, and Knudsen found that Poisson's ratio is porosity-dependent, decreasing with increasing porosity. Spriggs and Brissette derived a linear relation between Poisson's ration and porosity, calculated from the exponentially dependent relations of elastic modulus (E) and of shear modulus (G). The equation has the form: $\mu = \mu_o - mP$, in which μ_o is Poisson's ratio of the nonporous specimen, P is the volume fraction of porosity, and m is an empirical constant which must be derived for each type of specimen fabrication. For cold-pressed, sintered alumina the equation $\mu = 0.257 - 0.350 P$, provides a close fit.

5-6 CREEP CHARACTERISTICS

Creep is the slow and progressive deformation of a material with time under constant stress. Related phenomena include stress relaxation, internal friction, dynamic elastic modulus relaxation, and grain-boundary relaxation of polycrystalline materials.

Bridgman showed that aluminum oxide as well as other hard, brittle materials creep at room temperature under pressures of 30,000 atm. Ryshkewitch (1957) observed evidence of plastic flow on a microscale for the coiled chips of sapphire, ruby, and quartz glass that had been cut with a diamond tool at room temperature. Gorum predicted that the most likely brittle materials to possess ductility are the ionic solids having cubic crystal structures. Weinig, and Parker and associates investigated atmospheric and impurity factors affecting the limited ductility of brittle oxides. Sintered alumina does not presently appear to have useful room-temperature ductility.

McDowall and Vose discussed methods for determining pyroplastic deformation during heating of ceramic bodies. They found a relationship for sag of a rod specimen in which Sd^2/L^4 is a characteristic constant. S is the sag at the center of the rod of circular section, d is the diameter, and L is the length between suspensions.

Wygant (1951) stated that polycrystalline alumina has superior creep resistance to hydrostatically-pressed magnesia, thoria, zircon, and mullite. Alumina probably has the best creep resistance (around 0.13×10^{-5} at 1300°C, 1800 psi) of all the oxides, but is exceeded at high temperatures by covalent materials such as silicon carbide and graphite (Coble 1960). Poluboyarinov and Kalliga measured the deformation under load for alumina-silica compositions containing from 39 to 99% Al_2O_3, and from 0.5 to 4.7% fluxes. The temperature of deformation under controlled loading increased with increasing content of Al_2O_3. Partridge found that the creep rate of polycrystalline refractories containing a glassy phase is governed by the viscosity of the glassy matrix.

Stavrolakis and Norton found that the recovery of short-time creep of torsional stress in polycrystalline alumina was complete at 1200°C within 12 hours after removal of the load (unspecified); at 1300°C the deformation was permanent.

Wachtman and Maxwell investigated the creep rates of both polycrystalline and single-crystal alumina. Dense sintered alumina and sintered alumina containing 1.4% Cr_2O_3 developed an instantaneous strain, ϵ_o, under load (8530 psi), which continued to increase with time, as shown in Figure 12. When the strain was released after 10 hours, an instantaneous recovered strain ϵ'_o, was obtained, which was about equal to the initial, instantaneous strain. This was followed by a partial time-dependent recovery within 10 hours. In the case of the single crystals, plastic deformation was attributed to slip of the base plane (0001) in the $[11\bar{2}0]$ direction. The creep curves under constant load consisted of an initial period of increasing creep rate commencing above 900°C, followed by a period of decreasing creep rate, a period of constant creep rate, and finally, a period of increasing rate. The stress required to initiate creep fell uniformly from about 11,000 psi at 900°C to about 1850 psi at 1400°C.

Creep has been conceived to occur by several methods. The Peierls stress is the force necessary to move a dislocation along its slip plane, thus creating an edge dislocation. In grain boundary sliding, the creep rate, ϵ, is proportional to the applied stress, σ, and inversely proportional to the grain diameter, d, in polycrystalline materials. Dislocation climb is the motion into an adjacent plane of atoms. Weertman's theory of steady-state creep assumes that the rate-controlling process is the diffusion of vacancies in dislocation climb between dislocations that are creating vacancies and those that are destroying them, and that the elastic energy is measured by the change in free energy caused by the change in the number of vacancies. The steady-state creep rate, $\dot{\epsilon}$, is proportional to

$$\frac{\sigma^4 \, e^{-U/kT}}{T}.$$

The Herring, or Nabarro-Herring diffusional flow theory states that deformation results from diffusion away from a boundary under compressive stress toward boundaries having tensile stress. This makes the steady-state creep rate proportional to

$$\frac{\sigma \, D_0 e^{-U/kT}}{dT}.$$

In these equations, U is the activation energy of the process, D_0 the diffusion constant, and k and T have the usual significance.

Chang (1960) found that the dislocation climb mechanism fits the steady-state creep of sapphire crystals. The ratio of Peierls force to shear modulus was only about 0.00002, apparently too weak to be rate-controlling. Kronberg, Westbrook, and May; and Kronberg stressed the function of temperature and strain rate in controlling the flow properties. The tensile deformation of sapphire between 1200 and 1700°C and at a strain rate of 0.001 to 0.01 in/in-min was relatively sharp. The specific temperature of transition from brittle fracture to massive plastic flow in this rate change increased from about 1270°C to 1520°C. The stress-strain relation for plastic flow was characterized by a pronounced drop in the yield-point. The stress required for starting macroscopic flow was about double that required for subsequent flow. Both the upper and lower values for yield stress were temperature-sensitive; both decreased approximately exponentially with increasing temperature for a given strain rate. The fracture stress before yielding, however, was essentially independent of both temperature and strain rate, falling within a range of about 16,000 to 20,000 psi.

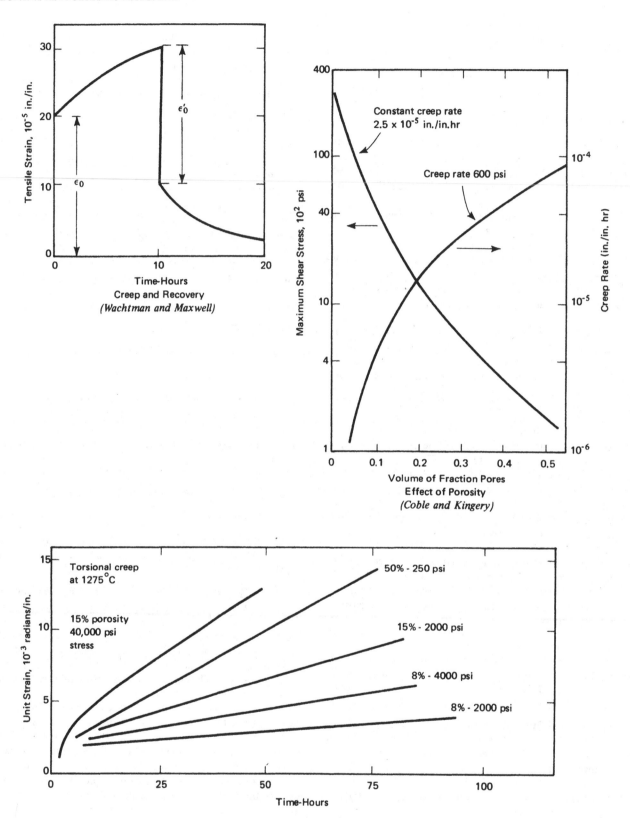

Figure 12. Creep, Creep Recovery, Torsional Creep and Influence of Porosity on Creep
(Coble and Kingery)

Conrad, Stone, and Janowski found that the deformation rate (γ) of sapphire from 900 to 1700°C fits either of two equations:

$$(1) \quad \gamma = (A/T) \, (\tau)^n \exp - (H_o/RT)$$

$$(2) \quad \gamma = \nu \exp \left[- H(\tau)/RT \right]$$

A, n, and ν are constants. In equation (1), n is approximately 3 to 7 and H_o is 90 to 125 kcal/mole. In equation (2), H is 120 kcal/mole and $\nu = - dH/d\tau$, and is $110b^3$ at $\tau = 1$ kg/mm^2, decreasing with stress. Equation (1) agrees with the dislocation climb mechanism; equation (2) supports the Peierls-Nabarro mechanism.

Seigle; Folweiler; and Warshaw and Norton also found the Nabarro-Herring mechanism to be the effective rate-controlling action in creep of polycrystalline alumina. Creep was found to vary linearly with stress between 1500 and 1800°C at an activation energy of 130 kcal/mole. Warshaw and Norton found that grain size is an important factor. Fine-grained aluminum oxide (3 to 13 microns) is viscous and creep rate is by the Herring equation; coarse-grained material (50 to 100 microns) deforms plastically, the plastic flow probably being by a dislocation or glide mechanism. Orlova claimed that *unfired* alumina deforms at a rate which is linearly dependent on the applied load: at 1300°C the rate was linear for loads of 200 to 2300 g/mm^2; at 1400°C for loads of 135 to 200 g/mm^2; at 1500°C for loads of 35 to 100 g/mm^2; and at 1600°C for loads up to 25 g/mm^2.

Passmore and Vasilos (1966) found no temperature region in which creep of dense, pure, fine-grained alumina conformed to the steady-state conditions, in contrast to these cited earlier references. They found that the creep rate continued to decrease with increasing creep strains up to at least 1.5% deformation and at a lower temperature range (1630 to 1740°K) than that of Warshaw and Norton (up to 1800°C). The transition from brittle to ductile behavior corresponds to the temperature at which the upper yield stress equals the fracture stress. For fine-grained sintered alumina it occurs at 1300 to 1400°C. An associated increase in modulus of rupture (about 19% at 1425°C) was ascribed to the strengthening effect of plastic deformation.

Coble and Guerard obtained good agreement with the diffusion coefficients obtained from data of Warshaw and Norton. They also found good agreement between measured and calculated aluminum-ion diffusion coefficients. The observed grain-size dependence for creep is satisfied by a lattice-diffusion model. Addition of Cr_2O_3 reduces the temperature at which the transition to a multiple slip process occurs. Orlova and Kainarskii found that the rate of deformation of polycrystalline alumina containing 0.25% talc or 0.07% MgO was similar to that of the pure corundum, and the rate approached linearity with time, suggesting the diffusion mechanism.

Particularly at high temperatures but also for rapidly applied stresses, the strain does not instantaneously come to equilibrium. This is called anelasticity, and the time required to attain equilibrium is the relaxation time. Chang, and Chang and Graham measured the activation energies for both steady-state creep and grain-boundary relaxation in polycrystalline aluminum oxide; the values were approximately 200,000 cal/mole and 120,000 cal/mole, respectively.

In both single-crystal and polycrystalline specimens, a controlled amount of solid solution was claimed to cause increased hardening and greater creep resistance. The addition of about 1 to 2% (by weight) of Cr_2O_3 or La_2O_3 to the alumina introduced additional grain-boundary relaxation peaks at lower temperatures (800 to 900°C), reduced the grain-boundary viscosity, and improved the ductility at temperatures around 1100°C. Evidence that grain boundary effects are the predominant factors challenges the diffusional-creep theory.

Beauchamp; and Beauchamp, Baker, and Gibbs measured the deformation of polycrystalline sintered compacts of doped Gulton Alucer MC alumina in 3-point beam loading in the range 1000 to 1350°C. The creep consisted of a transient deformation superposed on a steady-state deformation. The steady-state creep rate of specimens doped with 50 to 5000 ppm (wt) of MgO or MnCO$_3$ was fitted to the expression: A exp $(- E/KT)$, in which the activation energy (E) is 130 kcal/mole, independent of the added impurity. The constant (A) is independent of the MnCO$_3$ addition, but is decreased by a factor of 3 to 5 in samples doped with MgO. Gibbs, Baker et al. concluded that the steady-state creep is controlled by diffusion of vacancies. Hewson and Kingery (1967) found that the addition of 1000 to 2000 ppm MgO to polycrystalline alumina decreases the apparent diffusion coefficient by a factor of about 20, explained by the entry of magnesia into alumina by solid solution. The addition of MgTiO$_3$ however, did not change the diffusion constant.

Coble and Kingery observed the influence of porosity on the torsional creep rate of polycrystalline alumina having porosity in the range from 5 to 50%, induced by addition of different amounts of crushed naphthalene during the forming operation. The induced pore size was about 100 microns, in comparison with an average grain size of 23 microns. The creep rate (at constant stress and constant temperature), following a suitable relaxation interval at temperature, increased with increasing porosity.

McClelland derived a linear expression relating porosity to creep of sintered alumina, based on the Mackenzie-Shuttleworth plastic flow model, having the form:

$$p_E = p \, \frac{1}{1 - P^{2/3}} \, ,$$

in which p_E = effective pressure in closing pores, p = the externally applied pressure, and P = the volume fraction porosity. Spriggs and Vasilos (1964) have derived an explicit equation to show the relation in terms of the Nabarro-Herring diffusional creep equation, having the form:

$$\epsilon = \text{strain rate} = \frac{10\sigma D\Omega_o \dfrac{1}{1-P^{2/3}}}{r^2 k\, T},$$

in which σ = the stress, D = the lattice diffusion coefficient, Ω_o = the vacancy volume, P = the volume fraction porosity, r = the grain radius, k = Boltzmann's constant, and T = the absolute temperature.

5-7 THERMAL SHOCK

Poor resistance to thermal shock in comparison with the refractory metals limits the usefulness of the brittle ceramic oxides. Thermal stresses are particularly harmful during sharp cooling, in which the cooler exterior of the ceramic develops tensile stresses resulting from differential volume changes of the different parts of the structure. Since the compressive strength of ceramics may be about eight times the tensile strength, failure from compressive stresses is not important.

General reviews of the factors affecting thermal shock resistance of ceramics are given by Kingery (1955), and by Manson. Manson has shown good correlation between the theoretical and experimental values of thermal shock resistance of some brittle materials. The theoretical solution is based on the concept of an infinite plate at uniform temperature that is immersed in a cooling medium. The cooling is expressed in terms of a dimensionless heat-transfer coefficient β termed Biot's modulus, defined as: $\beta = ah/k$, in which a = 1/2 the plate thickness, h = the heat transfer coefficient, and k = the thermal conductivity of the plate. In general, the thermal stress resistance determining the temperature difference between initial and final conditions that just induces failure at different rates of thermal shock is represented by the equation:

$$\Delta T_f = \frac{k\,(ts)\,(1-\mu)S}{E\,a},$$

in which ΔT_f is the temperature difference just causing failure, k is the thermal conductivity, ts is the tensile strength, μ is Poisson's ratio, E the elastic modulus, a the coefficient of expansion, and S a shape factor. The thermal conductivity, k, enters the equation only at relatively low values of heat transfer ratio β.

Present heat-shocking methods include: the plasma arc; the arc imaging furnace; a gas-air blast directed onto the specimen, rocket motor exhaust; fused quartz heating lamps; and exposure to heating or cooling fluids. The increasingly severe demands for heat-resistant materials in high-duty vehicles or for reentry into the earth's atmosphere have taxed the present technical capability.

Manson and Smith found that the failure in thermal shock materials is not entirely the result of attainment of a definite critical stress but is dependent upon the stress distribution within the body, and can be accounted for on the basis of Weibull's statistical theory of strength. J. White concluded that the largely theoretical consideration of the thermal shock behavior of brittle materials, the factors involved, and the basis of a quantitative treatment are complicated by the fact that the properties of the material vary with the temperature and because there is no unequivocal definition of strength of a brittle material. The maximum stress criterion of failure was compared with the statistical concept of strength, based on the assumed presence of flaws in the material. White concluded that stresses set up under slow conditions of heating or cooling are also influential. Wenger and Knapp derived a stress formula based on work by Lidman and Bobrowsky, and concluded that (1) prestressing increases the relative resistance to thermal shock, and (2) the internal flow theory offers an explanation of the failure by tension upon cooling. Hummel found that the low-expansion materials suffer from one or more of the disadvantages of high cost, insufficient refractoriness, poor thermal conductivity, poor strength, and anisotropic expansion behavior.

Schwartz selected a hollow cylindrical shape heated uniformly from the inner surface for testing sintered alumina specimens in thermal shock. The resistance to heat shock was found to be a function of (1) the physical properties of the alumina, (2) the temperature range of testing, and (3) the temperature distribution within the specimen. Crandall and Ging described a method and the theoretical considerations involved in using a solid spherical shape (Coors grinding balls, type AB-2). The test consisted of plunging the heated sphere into a medium (salt bath) at a different temperature, and determining the time to fracture and the temperature difference causing 50% of the spheres to fail. These agreed well with the theoretical calculations based on the determined physical properties: Poisson's ratio, Biot's modulus, Young's modulus, the coefficient of thermal expansion, the breaking tensile stress, the surface heat-transfer coefficient, and the sphere radius. Hasselman determined the thermal shock on balls of this same composition by radiation heating. The emissivity of the material was shown to be as important as the thermal and mechanical properties usually considered in thermal-shock theory. The calculated and experimental values of the maximum radiation temperature to which the body could be subjected without fracture (T_{max}) at 95% stress level were in good agreement, as for example, 1255 and 1250°K for the 3-inch ball diameter, respectively, and 1564 and 1644°K for the 1.25-inch balls, respectively. Buessem and Bush used the ring shape, 2 in. in outside diameter by one in. in inside diameter, by 1/2 in. in length, and stacked rings for test, heating from the inside and cooling from the outside by a calorimetric chamber. Upon increasing the gradient slowly, failure eventually occurred. The thermal stress resistivity constants were determined from the maximum gradient at thermal fracture and from the thermal conductivity.

Baroody et al. characterized the thermal fracture of hollow cylinders in terms of two factors: M, the material, and S, the shape. Changing the dimensions (but not the ratios of dimensions) by as much as 1.6 times did not affect the shape factor. Using theorems derived by Biot, it was shown that the product relationship of the two factors S and M hold for a considerably wider variety of shapes than the simple circular tube.

Tacvarian claimed that a correlation exists between the ratio of the transverse strength before and after thermal shock and the difference in temperature of the heat shock, and proposed this as a simple method for evaluating thermal shock resistance.

Glenny and Royston found reasonable agreement between theory and experiment in calculating the transient thermal stresses in sintered alumina cylinders produced by rapid heating and cooling in fluidized beds.

Rubin found that for moderate cooling rates, the high thermal conductivity of beryllia gives a higher thermal shock resistance than that of alumina, but at high cooling rates, the higher strength of alumina provides greater resistance. Walton and Bowen claimed that the thermal shock resistance of slip-cast fused silica was superior to that of Pyroceram, which in turn was found to be superior to sintered alumina in the thermal shock environment provided by the exhaust of a small O-H rocket motor.

Hasselman and Shaffer investigated the material properties which affect the thermal shock resistance of polyphase ceramic systems composed of a continuous phase of high Young's modulus containing a dispersed phase of low Young's modulus. The model system investigated was zirconium carbide-graphite. The principal effect of the dispersed phase (graphite) on the thermal shock resistance of the zirconium carbide was to reduce the extent of damage resulting from fracture. The graphite, while decreasing the strength and Young's modulus, decreased the elastic energy stored at fracture, and at higher volume fractions of graphite, increased the extensibility, that is, the strain at fracture. Emissivity (absorptivity) was introduced as a variable affecting the thermal shock resistance of ceramic materials. Some metal-reinforced ceramics, ceramic fiber-reinforced plastics and metals are combinations that offer possible improvements over single component materials.

Reinhart claimed to improve the thermal shock resistance of sintered alumina by incorporating a small amount of the boride of zirconium or titanium.

There has been a general opinion that increased porosity improves thermal stress resistance of ceramic materials because greater flexibility or extension before fracture is obtained. Much of the early experience related to the effect on strength or spalling resistance of refractory brick. Variations due to forming methods and other causes probably were greater than those due to porosity. Coble and Kingery (1955) investigated the effect of porosity on the thermal shock resistance of sintered alumina under conditions believed to represent only this variable. The porosity was obtained by adding ground naphthalene to a slip-casting composition. All samples were fired together to the same temperature so that crystal growth and grain boundaries would be equally developed. Pore size was uniform. Within the range from 4 to 50% porosity, the thermal stress resistance was found to decrease with increasing porosity; resistance to thermal stress at 50% porosity was only about one-third of that estimated for zero porosity.

5-8 INTERNAL FRICTION

In a perfectly elastic system (Hookean) with stress proportional to strain, cyclic energy loss occurs only under dynamic adiabatic conditions. Kelvin established the difference between adiabatic and isothermal conditions of elasticity, $E_a - E_i$, by the expression, $a^2 T E_i E_a / \rho S$, in which a is the coefficient of thermal expansion, T is the absolute temperature, ρ is the density, and S is the specific heat. Real solids depart from the Hookean concept. Differences larger than predicted from the formula (about 0.01%) are ascribed to internal friction.

The relaxed (static) strain is greater than the instantaneous strain, and the relaxed modulus of elasticity is correspondingly higher, that is, the material is more rigid under rapid loading. Under sinusoidal loading, the energy loss per cycle is proportional to a function of the phase angle θ by which the strain lags the stress; it reaches a maximum when the relaxation times matches the period of vibration, so that the energy absorption is sensitive to frequency. The width of a natural resonant peak at 0.707 maximum amplitude, f, is measured, since $\Delta f/f$ approximately equals tangent θ, defined as the measure of internal friction. Internal friction may also be determined by measuring the decay in amplitude of free vibration (logarithmic decrement).

Dew, Wachtman and Lam; and Huber investigated the internal friction of polycrystalline alumina. Dew found a rapid broadening of the resonant peak widths above 800°C, which he regarded as an indication of increasing internal friction. Dew ascribed this to plastic deformation, on the basis of torsion tests made at about 20 cycles per second. Wachtman and Maxwell (1957) and Chang concur in the more likely belief that the cause is grain-boundary slip. Turnbaugh and Norton (1968) found two relaxation mechanisms operative in polycrystalline, but not monocrystalline, aluminum oxide at low frequencies (15 to 25 cps), (a) a high-temperature effect (above 1150°C), believed to be a diffusion-controlled, grain-boundary motion, independent of impurity content, and (b) a lower temperature effect (at 900°C) which was apparently related to impurity content but was not frequency-dependent.

Asbury and Davis showed the influence of porosity on the internal friction in terms of the phase difference between stress and strain in cyclic deformation. The phase displacement increases with increasing porosity but decreases with increasing frequency of vibration. For perfect elasticity, the phase angle is zero. For ordinary electrical porcelain, the value ranges from 0.3 to 1.0 radian; for polycrystalline alumina, a value of about 0.01 to 0.03 radian is found; and for single-crystal alumina and quartz, below 0.01 radian.

5-9 FATIGUE

The hysteretic energy absorbed during cyclic stress of a body is largely absorbed in regions of local stress concentration. Failure may occur at some critical stress level. The endurance limit is the greatest stress that can be applied repeatedly without causing failure; its measurement, statistically, has a band width of about one logarithmic order of magnitude. Fatigue fracture of the cyclic type is rare in ceramics.

Static fatigue, or delayed fracture, is more common in ceramics. There is a marked dependence upon the ambient atmosphere. Brittle fracture is of most significance. Williams; Pearson; Roberts and Watt; Wachtman and Maxwell (1954); and Charles and Shaw investigated the embrittlement by moisture, or other factors affecting the delayed failure of both polycrystalline and single-crystal aluminum oxide. Mountvala and Murray found that exposure of sapphire to moisture at elevated temperatures (above 180°C) caused the formation of hydrates, which impaired the strength measured at room temperature. The strength was recovered when the moisture-exposed specimens were heated to 400°C. Polycrystalline alumina (Lucalox) having an average crystal size of 30 microns did not decrease significantly in bend strength (from 33,300 to 28,000 psi) following steam spraying at 180°C for 52 hours. The strength of single-crystal alumina was unaffected by exposure to nitrogen at 1000°C, or hydrogen at 800°C.

Delayed fracture (steady-load) data have been correlated with fatigue (cyclic-load) results and with the short-time determination of modulus of rupture. Both types of corundum exhibit strong delayed failure characteristics which depend upon the atmosphere, temperature, and strain rate. In normal atmosphere the effect was found to depend on the time for which the stress was applied. Under constant load, the average time to fracture increased from one second to about 11 days when the stress was reduced by 22%. This type of delayed fracture could be largely eliminated by heat-treating and testing in a vacuum. The delayed-fracture effect reappeared, however, when the vacuum heat-treated material was again exposed to atmosphere. The effect was absent at liquid nitrogen temperatures, and decreased at temperatures above room temperature until it disappeared around 900°C.

A fatigue theory involving stress corrosion of an elastic continuum was applied to the low-temperature data on sapphire to obtain reasonable concordance.

5-10 HARDNESS AND ABRASIVENESS OF ALUMINA

Hardness has been defined as the resistance of the surface layers of a body to penetration under the effect of external mechanical forces concentrated in a restricted area of the body (Koifman). A clear understanding of the hardness range of minerals existed at least 200 years before Mohs devised his arbitrary scale about the year 1820 (Gradzinski). On the Mohs scale, corundum is fixed at 9.0 and diamond at 10. The scale has since been revised by Ridgway, Ballard, and Bailey, and by Markes, to show quartz 8, topaz 9, garnet 10, fused zirconia 11, corundum 12, silicon carbide, 13, boron carbide 14, and diamond 15 or 15.1.

A more quantitative evaluation of hardness than the Mohs scale has been determined in terms of valence number and atomic volume or the interatomic distances (Friederich; Plendl and Gielisse; Kick; and E. Meyer. The different surface-indentation tests (Vickers, Knoop, Tukon, Rockwell) are based on the relation between hardness and the load required to produce a definite indentation. The methods differ in the shape of the diamond indenter and in the loading (Ludwik; Smith and Sandland; Brinell; and Knoop, Peters, and Emerson).

A distinction is made between macrohardness and microhardness determinations made with the diamond indenter. The former requires from 10 to 20 kilograms of loading, and the indentation diagonal usually covers many crystals. The determination is more or less independent of the load value. Microhardness determines the indentation on single crystals usually, and at loadings of about 5 to 100 grams, the hardness is dependent upon the loading (Hanemann and Bernhardt).

Hardness (H) is determined as the loading (P) per unit of superficial area of penetration derived from the length of the diagonal of the indentation (d): $H = kP/d^n$, in which n is 2 for macrohardness determinations, and varies from about 1.7 to 2.5 for microhardness determinations. The factor k relates the length of the diagonal of penetration to the area of identation, and has a value of about 1.8544 for the Vickers indenter.

The Bierbaum equipment is sometimes used for measuring the hardness of ceramic materials. It measures the width of scratch produced by a diamond cube corner under standard loading, usually three grams.

Tabor found a correlation between the Mohs hardness and the indentation hardness, based on the closely similar ratios (about 1.6) between steps on the Mohs scale.

Knoop et al. obtained a hardness of alpha aluminum oxide of about 3000 kg/mm^2. Ryshkewitch claimed that fused alumina has a distorted structure which weakens its mechanical strength. Thibault and Nyquist obtained an average value for abrasive alumina of 2050 by the Knoop indenter at 100 gram loading, in comparison with 820 for quartz, 1340 for topaz, 2480 for silicon carbide, 2760 for boron carbide, and 8000 to 8500 for diamond.

Attinger, and Stern determined the directional hardness of synthetic corundum. The faces parallel to the optic axis were harder than those perpendicular to it. Sapphires with pores normal to the optic axis were the softest. Fissures were usually at $45°$ to the optic axis. Maximum hardness is claimed to be at about 60 degrees to the c-axis (Klassen-Neklyudova) and both minima occur at $0°$ and $90°$ to this axis. The abrasion resistance is also greater in the direction parallel to the optic axis. The average hardness was about 2000 kg/mm^2 (Knoop$_{200}$).

The hardness of corundum containing about 3% Cr_2O_3 in solid solution is higher by about 10% than that of pure sapphire, but decreases at higher chromia contents. The hardness is strongly anisotropic. Bradt claimed that the concentration of Cr_2O_3 increase in microhardness, according to the empirical relation: Hardness at 500g load (DPH$_{500}$) = 1945 + 15.23 (mole % Cr_2O_3). Schrewelius found that for an expansion in the unit cell of fused alumina from 5.126A to 5.130A, caused by a change in the titania content of 0.6 mole%, the microhardness (Knoop$_{100}$) dropped from 2300 kg/mm^2 to only 1800 kg/mm^2. Titania is added to fused bauxite compositions to about 2 to 3% by weight to increase the toughness of the grain (Parché). Baumann showed that such grain has large crystalline areas of alpha alumina with slag-like inclusions of siliceous ground-mass and numerous included microcrystals of titanium and aluminum compounds. He ascribed the toughness of the abrasive grain to this residual matrix. Hsu, Kobes and Fine observed that Verneuil-process sapphire, doped with 0.5 cation % Ti as $Ti_2 (C_2O_4)_3$, upon aging at $1500°C$ to precipitate acicular crystallites, increased in hardness from 2400 kg/mm^2 for undoped specimens, to as much as 3300 kg/mm^2. Fracture strength was also increased, probably by interference of the acicular crystals with local plastic deformation. The toughness is also improved by additions of 5 to 15% by weight ZrO_2 to fused alumina, by solidifying the melt rapidly, and by breaking the massive product by impact rather than by crushing (Carborundum Co. British Patent 993,891).

Schrewelius obtained only 2000 to 2200 kg/mm^2 for beta alumina in comparison with 3000 for alpha alumina by the Zeiss microhardness test. Jorgensen and Westbrook obtained a microhardness of 2500 kg/mm^2 for polycrystalline alumina sintered from Linde A alumina powder (0.3 micron) fired for 16 hours at $1830°C$. When sintered with 0.05 to 0.25 MgO by weight, the bulk hardness (crystal face) was about 2550 kg/mm^2 and the crystal-boundary hardness about 2800 to 2900 kg/mm^2. The adsorption of water on the surface of the polished specimens affected the hardness values. Specimens dried in oxygen for 16 hours at $200°C$ and measured under toluene gave consistent values. This effect together with the differences in crystal orientation and in solid solution of impurities perhaps accounts for the differences in hardness reported by different investigators. The microhardness of sintered alumina doped with 0.5 mole% Cr_2O_3 (25g load) increases with increasing distance from the interface (Aust et al.). This appears to be rational. The cited finding, that MgO-doped alumina decreases in microhardness with increasing distance from the interface (Jorgensen and Westbrook), is difficult to reconcile with the work of Pearson, Marhanka, et al. They found

that the wear of both undoped or MgO-doped alumina grinding balls is significantly dependent on crystal size, which implies that service wear is mainly by loss of unit crystals by impact rather than by surface attrition. They found that the effect of crystal size (G) and of porosity (P) on the wet-abrasion loss (A) is described approximately by the relation: $A = 0.00968G^{1.480}e^{0.191P}$.

Whitney attempted to measure hot hardness using a Tukon tester fitted in a temperature-controlled furnace in a hydrogen atmosphere.

Pemberton determined the hardness of alumina abrasives simply by measuring the load in grams required to crush a sufficient number of single grains, 290 to 340 microns in size for statistical significance. Abrecht et al. measured the toughness of the grain by throwing sized grain continuously (by recirculation) from a centrifugal device against an impact surface. The number of cycles required to halve the number of grains of the original size was arbitrarily taken as the measure of impact resistance. The loss in weight of the impact surface measured the abrasiveness. Mann determined the crushing resistance of synthetic abrasives by milling uniform grit sizes (2000 to 125 microns) with steel balls, and by calculating the fraction of attrition. The order of decreasing resistance for various 2000-micron abrasives was: dark silicon carbide, brown corundum, white corundum, boron carbide, and light silicon carbide, designated a,b,c,d, and e, respectively. The order for 125-mesh grits was: d,e,a,c,b, possibly indicating that strain or partial cracking upon crushing may differ for different sizings of the materials.

Cadwell and Duwell, and Stotko measured the friability by crushing with steel balls in grinding operations.

Several synthetic abrasive materials are harder than corundum. Among these are silicon carbide, boron carbide, boron, boron nitride (cubic), and, of course, diamond. While silicon carbide is harder than corundum, it is more brittle. Alumina is therefore recommended for grinding operations on materials of high tensile strength, while silicon carbide is used on materials of low tensile strength.

Giessen (1959) reviewed the available methods for determining the tenacity of abrasives and recommended the use of impact methods, and sample sizes of about 160 to 200 grams.

H. W. Wagner in 1950 stated that an abrasive yields higher grinding value when its penetration hardness is higher, when its body strength permits fracture (self-sharpening) of the grain at the proper degree of dullness, and when its attrition resistance is higher. Dulling of abrasive points is influenced by solubility of the abrasive in the work material being ground. The effect of oxidation in grinding is small compared with that of solubility. In molten steel, boron carbide and silicon carbide dissolve rapidly, but aluminum oxide is practically insoluble. In molten glass, aluminum oxide dissolves more rapidly than silicon carbide. Brown, Eiss, and McAdams confirmed that chemical mechanisms are involved in the wear of single-crystal sapphire on steel. Duwell and McDonald measured the wear and cutting performance of sixteen abrasive minerals, and found that wear may be by attrition and fragmentation. Lubricants can increase the rate of cut and decrease the rate of wear. Carbides and borides have less resistance to attritional wear on steel than oxides because of their chemical instability.

Steijn; Coffin; Sibley et al.; Riesz; and Riesz and Weber investigated the mechanism of wear of sapphire as a bearing material. A statistical correlation exists between the measured wear rates and the coefficient of friction, the thermal-stress resistance, and the thermal diffusivity of the mated materials for unlubricated wear resistance at high sliding speeds (100 to 200 fps). During high-speed sliding, wear appears to be induced by the inability of ceramic and cermet materials to resist thermal stresses produced by temperature gradients within each rubbing surface between small asperities nor hot spots in frictional contact. Corundum and Cr-Mo cermets appear to be promising materials for high-temperature, high-speed sliding bearings and seals. Friction and wear of single-crystal sapphire, sliding in vacuum from 30 to 1550°C, was influenced by surface cleanliness. High coefficients of friction and "stick-slip" sliding occurred up to 300°C. At 300 to 1000°C, lower coefficients and smooth sliding occurred. In the region from 1000 to 1350°C, stick-slip reappeared at lower coefficients. Chevron-shaped subsurface fractures of the sapphire plate were observed below 300°C, their formation being orientation-dependent. In the absence of fracture, adhesions were noted. Under high vacuum, a weld-adhesion mechanism of sliding friction was observed, particularly above 1000°C. High surface temperatures occur coincidently with the friction process, but direct evidence of an adhesion mechanism is inconclusive.

The frictional characteristics of single-crystal sapphire on its own type are highly anisotropic. Polycrystalline alumina had frictional characteristics intermediate between those for prismatic- and base-oriented single crystals (Buckley).

Dawihl and Doerre (1966) related the frictional action to the shear strength for different sintering binders.

Parker, Grisaffe, and Zaretsky used a five-ball fatigue tester to study the behavior of hot-pressed and cold-press-and-sintered alumina balls under repeated stresses applied in rolling contact. The failures that developed were shallow eroded areas unlike fatigue spalls found in bearing steels. The room-temperature load capacity of hot-pressed aluminum oxide was about one-fifteenth that of a typical bearing steel and seven times that of cold-pressed-and-sintered alumina. Preliminary tests at 1000°C with molybdenum disulfide-argon mist lubrication indicate that polycrystalline aluminum oxide is capable of operating satisfactorily under high-temperature rolling contact conditions.

Groszek found that preferential heats of adsorption of various wear-reducing agents on different types of porous solid surfaces correlated with their wear-reducing action on sliding steel surfaces.

The abrasiveness and wear resistance of alumina have been the subject of investigation because of the importance of applications as abrasives, lathe tools, bearings, gauge blocks, dies, etc. These applications involve alumina both in particulate form and in massive shapes.

The Bayer process hydrates, calcined aluminas, and tabular aluminas together with fused alumina cover a remarkable range of abrasive properties. These effects are related to a wide variation in hardness between the different alumina phases, and to an extended range of particle and crystal size. Changes in texture and in the bond holding crystallites together in the Bayer grain, particularly in the transition alumina structures, afford various degrees of abrasiveness in the polishing range. Russell and Lewis showed the range of abrasiveness of Bayer-process grades in terms of a standardized air-blast test. Tabular alumina (sintered alumina) was found to be about as abrasive as fused alumina by this "one-shot" test.

6 THERMAL PROPERTIES

Several excellent compilations of thermal properties of alumina and other materials have appeared within recent years. Among these are Goldsmith, Waterman, and Hirschhorn, *Handbook of Thermophysical Properties of Solid Materials*, The Macmillan Company, New York, 1961, 5 volumes; Bradshaw and Mathews, *Properties of Refractory Materials* (1958); Shaffer, Peter T. B., *Handbook of High-Temperature Materials*, No. 1, Materials Index, Plenum Press, New York, xx+ 740 pp (1964); Cohen, Crowe, and Dumond, *The Fundamental Constants of Physics*, Interscience Publishers, New York (1957); Krishman, R. S., Progress in Crystal Physics, Vol. I, *Thermal, Elastic, and Optical Properties*, Interscience Publishers, New York, 1960; JANAF Interim Thermochemical Tables, Vol. I; Fieldhouse, Hedge and Lang, *Measurements of Thermal Properties*, PB Rept. *151583*, furnish some thermal data on sapphire. Rossini, Wagman et al. (compilers), *Selected Values of Chemical Thermodynamic Properties*, N.B.S. (US) Circular No. 500 (Feb. 1952) provides thermodynamic properties of some alumina phases.

A selected set of thermal properties of the aluminas has been collected in Table 10.

6-1 THERMORPHYSICAL AND THERMOCHEMICAL CONSTANTS

Many investigators have determined the melting point of alpha aluminum oxide, among whom are included Kanolt; Geller and Bunting; and Trombe. In general, they have striven for high sample purity, but in some cases their estimations have been vitiated by questionable heating conditions or by the methods used to detect incipient melting. Some investigators, including Kanolt, used graphite resistor tubes as furnaces, which might have caused reduction of the alumina in the presence of reducing atmosphere and water vapor. In other cases, the direct observation of melting indicated a lack of black body conditions and a temperature gradient in the furnace (Schneider, 1963).

Geller and Yavorsky (1945) obtained apparently low values ranging from 2000 to 2030°C on fused and recrystallized alumina 99.99% pure, melted in an oxidizing atmosphere, but based on the visual observation of edge deformation, believed unsatisfactory. Ryshkewitch (1960) had pointed out that an alumina rod in the horizontal position, fastened at one end, deforms at 1400°C, and that pyrometric cone 42 is pure alumina deforming at 2000°C.

Kanolt's painstaking determination, 2050°C, made in 1913, had been erroneously corrected for changes in the International Temperature Scale (1948) to 2040°C, which was considered valid until recently. The present temperature scale, however, could not be established within three degrees at 2000°C between leading national laboratories as recently as 1957 (Kostkowski). Kanolt's determination, adjusted more reasonably on the basis of the melting point of platinum, has the value 2072°C (Schneider), which appears somewhat higher than some recent determinations. Diamond and Schneider found 2025°C for melting in a solar furnace, but do not recommend the method. Schneider and McDaniel obtained 2051°C for melting in a tungsten crucible in a vacuum (6.5 × 10⁻⁵ torr.). Gitlesen and Motzfeld found 2041°C by a drop method in argon; Urbain and Rouannet 2047°C for melting in tungsten in argon; and Sata 2037°C for small amounts of alumina supported in a tungsten ring in argon.

Lambertson and Gunzel obtained 2034 ± 16°C for the melting point of tabular alumina T-61, indicating the high purity of alumina from Bayer sources. Markovskii et al. suggested a micromethod for determining the melting point using hot-pressing equipment.

Brewer and Searcy (1951) found 3530 ± 200°C as the boiling point of alpha alumina; Wartenberg (1952) obtained 3300°C, based on extrapolation of the vapor pressure. Ruff and Konschak (1926) had obtained 2980°C. All the data are questionable because of the claimed decomposition of the alumina by tantalum containers used by Ruff (Brewer and Searcy), and by tungsten containers used by Brewer and Searcy (Wartenberg; Ackermann and Thorn). Yudin and Karklit (1966) found 3717°C for the normal components present at one atmosphere. The overall thermal effect of evaporation at this temperature was 472.3 kcal/mole.

Treadwell and Terebesi (1933) stated that the heat of formation of alpha aluminum oxide is fairly independent of temperature, citing 392 kcal/mole at 25°C, and 387 kcal/mole at 2300°C. Roth, Wolf, and Fritz (1940) obtained 402.9 kcal/mole, Snyder and Seltz 399.0, and Holley and Huber 400.29. Schneider and Gattow (1954) claimed that the value of Snyder and Seltz was too low

63

Table 10

Thermal Properties

Property	Phase	Value		Reference
Melting Point	α-Al_2O_3	$2051.0 \pm 9.7^\circ C$		(1)
Boiling Point	α-Al_2O_3	$3530^\circ C$ ($3800 \pm 200^\circ K$)		(2)
Vapor Pressure		T $^\circ$K	Atm	(2)
		2309	8.7×10^{-6}	
		2325	1.03×10^{-5}	
		2370	1.66×10^{-5}	
		2393	1.68×10^{-5}	
		2399	2.15×10^{-5}	
		2459	3.78×10^{-5}	
		2478	5.81×10^{-4}	
		2487	9.10×10^{-5}	
		2545	2.00×10^{-4}	
		2565	1.29×10^{-4}	
		2605	1.91×10^{-4}	

Heat of Formation at $298.16^\circ K$ (kcal/mole)

			Entropy at $298.16^\circ K$ (kcal/mole)	
α-$Al_2O_3 \cdot 3H_2O$		-612.8	33.51	(5,3)
β-$Al_2O_3 \cdot 3H_2O$		-609.4		(4)
Amorphous		-304.2		(3)
$\alpha Al_2O_3 \cdot H_2O$		-471.8	23.15	(4)
β-$Al_2O_3 \cdot H_2O$		–	8.43	(3)
α-Al_2O_3		-400.4	12.16	(7,6)
AlO (g)		-138	48.967	(2,30)
Al_2O (g)		-248	59.75	(2,30)

Enthalpy (E + PV) in kcal/mole   (9)

γ-$Al_2O_3 \rightarrow \alpha$-$Al_2O_3 - 5.3$ kcal/mole at $705^\circ C$

κ-$Al_2O_3 \rightarrow \alpha$-$Al_2O_3 - 3.6$ kcal/mole at $705^\circ C$

δ-$Al_2O_3 \rightarrow \alpha$-$Al_2O_3 - 2.7$ kcal/mole at $705^\circ C$

α-Al_2O_3 ($\Delta H_{298.16^\circ K}^T$)

$0.03550922T - 4.0884(10^{-7})T^2 - 11.23206 \log_{10}T + 19.63341$ (a)   (11)

$0.03549846T - 3.9085(10^{-7})T^2 - 11.2306 \log_{10}T + 17.23778$ (b)   (32)

$0.03031602T + 8.3979(10^{-7})T^2 + (2.81406 \times 10^3/T) - 12.87764$ (c)   (31)

(a) Range 400 to 1200°K.
(b) Range 678 to 1330°K.
(c) Range 1290 to 1673°K.

Table 10 — Continued

Property	Phase			Value			Reference

Heat of Immersion at 25°C (cal/m² surface area)

(Ref. 28) Type Alumina	A Tabular T-60	B Alucer MC	E Alucer HSB	F Alon C	G Alucer MA	H Act. Alumina F-20
Phase	α-Al$_2$O$_3$	α-Al$_2$O$_3$	α-Al$_2$O$_3$	γ-Al$_2$O$_3$	γ-Al$_2$O$_3$	–
Area m²/g	0.22	2.72	4.56	65.2	109	
Degassing Temp °C						
100	0.157	0.139	0.127	0.098	0.092	0.077
300	0.202	0.222	0.170	–	0.150	0.123
400	0.206	0.241	0.170	–	0.172	0.153
450	0.208	0.257	–	0.200	0.177	0.163
Immersion Liquid	α-Al$_2$O$_3$	γ-Al$_2$O$_3$	Al$_2$O$_3$/Al			(33)
H$_2$O	0.1850	0.1550	0.2408			
C$_6$H$_6$	0.0396	0.0551	0.0480			
n-BuOH	0.0570	0.1008	0.1341			
n-BuCl	0.0700	0.0712	0.0946			

Specific Heat (cal/g°K)

α-Al$_2$O$_3 \cdot$3H$_2$O		$0.2694 + 6.43 \times 10^{-4} t$		(a)	(10)
		0.2855 at 25°C			
α-Al$_2$O$_3$		$0.348264 - 8.019 \times 10^{-6} T - 47.8423/T$		(a,b)	(11)

°K	cal/g°K	°K	cal/g°K	
400	0.22545	900	0.28789	(11)
500	0.24857	1000	0.29240	
600	0.26372	1100	0.29595	
700	0.27431	1200	0.29877	
800	0.28205			
1318	0.2783			(12)
1510	0.3364			
1660	0.3814			
1787	0.4196			
2575	0.468			(13)

(a) T = °Kelvin = 273.16 + °C; t = °C.
(b) Range of equation 400 to 1200°K.

Table 10 – Continued

Property	Phase		Value		Reference
Diffusivity (cm^2/sec)					
α-Al_2O_3	°C	cm^2/sec	°C	cm^2/sec	(26,27)
	25	0.091	1000	0.015	
	100	0.055	1500	0.0063 to 0.0108	
	200	0.043	1600	0.0106 to 0.0184	
	400	0.024	1700	0.0130 to 0.0224	
	800	0.018	1800	0.0170 to 0.0294	

Property	Phase	Value	Reference
Heat of Fusion (kcal/mole)			
	α-Al_2O_3	26 at 298.16°K	(3)
Heat of Sublimation (kcal/mole)			
	α-Al_2O_3	456.0 at 298.16°K	(2)
Heat of Vaporization (kcal/mole)			
	α-Al_2O_3	150.1 at 1950°C	(8)
		443.0 at 25°C	(2)
		729.0 at 2220 to 2333°C in vacuum	(29)

Thermal Expansion, Linear ($\times 10^{-6}$/°C)

α-$Al_2O_3 \cdot 3H_2O$ (Perpendicular to specified crystal plane)

25 to 100°C (010), 10.9; (001), 15.4; (100), 13.1; (101), 39; (101), −5.6   (16)

α-Al_2O_3

Temp. Range °C	Single-Crystal (14)		Polycrystalline	
	Orientation 0°	C-axis 90°	(14)	(15) AD-94
−273 to 0	1.95	1.65	1.89	
− 73 to 0	4.39	3.75	4.10	
0 to 127	6.26	5.51	6.03	
327	7.31	6.52	6.93	
527	7.96	7.15	7.50	
927 (c)	8.65	7.80	8.08	
1127	8.84	7.96	8.25	7.95
1327	8.98	8.12	8.39	
1527	9.08	8.20	8.49	
1727	9.18	8.30	8.58	

β-Al_2O_3

25 to 200°C − 5.1 to 5.7 $\times 10^{-6}$/°C
600 to 700°C − 6.0 to 7.6

    (17)

ζ-Al_2O_3

25 to 200°C − 6.0 $\times 10^{-6}$/°C
600 to 700°C − 7.7

(c) Data obtained by extrapolation

Table 10 – Continued

Property	Phase	Value			Reference

Thermal Conductivity (cal/sec cm °C)

α-Al$_2$O$_3$ (dense)

Temp °C	°K	cal/sec cm °C	Coors AD-995	AD-94	(15,18,19, 20,21,22, 23,24,25)
-263		3			
-253		9			
-233		14			
-223		12			
25	298	0.086	0.080	0.046	
100	373	0.069			
300	573	0.038			
500	773	0.025			
700	973	0.018	0.018	0.011	
900	1173	0.015			
1100	1373	0.014			
1300	1573	0.014			
1500	1773	0.013			
1700	1973	0.014			
1900	2173	0.015			

α-Al$_2$O$_3$ (23.4% porosity)

25		0.055			
100		0.049			
300		0.029			
500		0.018			
700		0.013			
900		0.012			
1100		1.008			

α-Al$_2$O$_3$ (48.7% porosity)

25		0.0397			
100		0.0312			
300		0.0191			
500		0.0104			
700		0.0070			
900		0.0062			
1100		0.0070			

References

(1) S. J. Schneider, Nat. Bur. Stds. Private communication, Nov. 6, 1968
(2) Brewer & Searcy
(3) Rossini, Wagman, et al. N.B.S. Circ. 500, 1952
(4) Russell et al. (1955)
(5) Barany & Kelley
(6) Kerr, Johnston, & Hallett
(7) Mah
(8) Wartenberg
(9) Yokokawa & Kleppe
(10) Roth, Wirths, & Berendt (1942)

(11) Furukawa et al.
(12) Shomate & Naylor
(13) Sheindlin (1964)
(14) Wachtman, Scuderi, & Cleek
(15) Coors Porcelain Co. (AD 995)
(16) Megaw (1933)
(17) Nat. Bur. Stds. News Bull. April 1934, p. 39
(18) Frankl & Kingery (1954)
(19) Weeks & Seifert
(20) Francis, McNamara, & Tinklepaugh
(21) Adams, M. (1954)
(22) Charvat & Kingery

(23) McQuarrie (1954)
(24) Norton (1951)
(25) Berman (1951, 1952)
(26) Plummer, Campbell, & Comstock
(27) Paladino, Swarts, & Crandall
(28) Wade & Hackerman (1964)
(29) Diamond & Dragoo (1966)
(30) Panyushkin & Mal'tsev
(31) Banashek, Sokolov, Rubinchik & Fomin
(32) Sokolov, Banashek, & Rubinchik
(33) Cochrane & Rudham

because of incomplete conversion to alpha alumina in the oxygen pressure bomb. They corrected on the basis of about 25% transition alumina having a heat of transition to alpha alumina of 10 kcal/mole, and estimated 402 kcal/mole for the heat of formation of alpha alumina. The value, 400.4, obtained by Mah in 1957, appears to be more trustworthy, because the product of combustion was shown to be alpha aluminum oxide by X-ray.

Wade and Hackerman (1960) determined the heats of immersion in water of aluminas varying in surface area from 0.22 m²/g for tabular alumina to 221 m²/g for activated alumina (F-20). The data are shown in Table 10. Morimoto, Shioma, and Tanaka (1964) found that the curves for heat of immersion in water have maxima at about 600 to 650°C. The heats of immersion calculated from the heats of immersion were 15.8 kcal/mole for a-Al_2O_3 and 9.60 kcal/mole for γ-Al_2O_3. Rehydration of the dehydration sites is an important factor in immersion. Cochrane and Rudham found somewhat lower heats of immersion in common organic liquids than in water. B. Frisch noted that sintered alumina acquired a chemisorbed, monomolecular layer of OH^-, which was irreversible; the calculated area occupied by unit of OH^- was 16.7 A^2.

Kingery (1959) found the density of liquid alumina, by the pendent drop method, just above the melting point, to have a mean value of 2.97 g/ml, somewhat higher than the 2.5 g/ml reported by Wartenberg, Wehner, and Saran in 1936. Kirshenbaum and Cahill measured the density at 2100 to 2350°C (2375 to 2625°K) by use of a molybdenum crucible and a tungsten sinker. Their data fit a linear curve having the form: $D = 5.632 - 1.127 \times 10^{-3}T$, in which $T = °K$. The calculated liquid density at the melting point is 3.053 g/ml. The increase in volume found by Kingery was 20.4% and by Kirshenbaum 22%, assuming the correctness of Ebert's solid data. Wartenberg's value, 33%, seems unlikely in comparison with the expansion of other ionic solids. Hasapis, Panish, and Rosen found the viscosity at 2200°C to be only 13 poises, in comparison with 104.4 poises for fused silica at 2560°C. Kingery (1959) measured the surface tension by the pendent drop method (from a molybdenum rod) and obtained a mean value of 690 dynes/cm.

The surface energy of solid aluminum oxide was 905 ergs/cm at 1850°C, or about 885 ergs/cm² at the melting point (Norton, Kingery, Economos, and Humenik). Livey and Murray (1956) considered these values low, and probably due to the formation of a low-energy surface consisting of oxygen ions highly polarized by the submerged and powerful cations. The surface energy is low in comparison with several thousand ergs/cm² for diamond and silicon carbide. Livey and Murray (1959) reported a value of -0.2 erg/cm²/°K for the surface energy for crack propagation, γ, which would make the room temperature value of surface energy about 1270 ergs/cm².

6-2 SPECIFIC HEAT

The variation in specific heat of alpha aluminum oxide with temperature has been determined by many investi-gators, among whom are Shomate and Naylor (1945); Kerr, Johnston, and Hallet; Ginnings and Furukawa (1953); Furukawa et al.; J. I. Lang (1959); Hoch and Johnston (1961); Ewing and Baker; and Kirillin, Sheindlin, and Chekhovskoi (1964). In general, the method of dropping the specimen from an equilibrium temperature into a water or ice calorimeter has been used. The specimen usually has been sapphire. Agreement between the different investi-gators has been close in the temperature region from near absolute zero to about 1200°K. The data of Furukawa et al. (Table No. 10) is generally accepted in the temperature region 400 to 1200°K. Dawson (1963) found values only 0.3% higher than these.

In the region above 1200°K, there is increasing variance between different investigators. At 1700°K, the scatter ranges from extrapolated values of about 0.295 cal/g °K for the data of Lucks and Deem, to 0.390 cal/g °K for that of Rodigina and Gomelskii. Goldsmith, Waterman, and Hirsch-horn plotted a most probable average curve from the data prior to 1959, which conforms to the data of Ginnings and Furukawa from the absolute zero to 1200°K. The values of Shomate and Naylor continue the curve slope smoothly to about 1800°K. Sheindlin et al. found a value of 0.4677 cal/g °K for the melt at 2350 to 2800°K, indicating an upsweep to the curve, which makes Rodigina's data appear high. In their method the corundum melt, heated diather-mally in a molybdenum container in argon, was mixed in a massive metallic calorimeter. Alternatively, tests in vacuum gave closely similar values for specific heat.

Fieldhouse, Hedge, and Lang found only slightly higher results for polycrystalline alumina than for sapphire by the drop method in the temperature region 1300 to 1800°K.

6-3 THERMAL EXPANSION

The linear thermal expansion of various specimens of alpha aluminum oxide has been determined by many investigators, and using different methods. Austin deter-mined the expansion of single-crystal natural sapphire by the interferometric method. He reported that alumina expands anisotropically, with the expansion 7 to 15% higher parallel to the c-axis. Austin claimed that the expansion of an aggregate of nonisotropic crystals often shows hysteresis which is not observed in single crystals.

Schwartz (1952) obtained values ranging from 5.5 $\times$ 10^{-6} at 25°C to 10.0 $\times$ 10^{-6} at 1300°C by telemicroscopic sighting on the ends of slip-cast pure alumina rods having about 5% porosity after sintering. Smoke (1949) deter-mined expansions on six alumina spark plug porcelains by dilatometric measurements against silica on 8-inch bars. Shevlin and Hauck similarly measured the expansion of bars prepared from Alcoa T-61 to about 700°C. Whittemore and Ault obtained data to about 1500°C on 7-inch bars prepared from fused alumina 99% Al_2O_3), sintered alumina (99% Al_2O_3), and clay-bonded fused alumina (88% Al_2O_3). Their values for the expansion coefficients were: (6.7, 5.9, and 4.4) $\times$ 10^{-6}/°C, respectively, in the range 25 to 900°C; and (10.2, 9.6, and 8.7) $\times$ 10^{-6}/°C, respectively, in the range 25 to 1500°C.

Coble and Kingery (1956) found no difference in the thermal expansion of polycrystalline alumina ranging in porosity from 4 to 49%.

Shalnikova and Yakovlev (1956) reported that alumina expands isotropically, from X-ray measurements of the lattice expansions. Zimmerman and Allen reviewed the use of a simple back-reflection X-ray camera to determine the expansion, and concluded that the accuracy of measurement was adequate. Beals and Cook applied the method to reagent-grade alumina, and reported the following average values: (9.39, 9.48, and 9.50) $\times 10^{-6}/°C$ in the temperature ranges, 20 to 200°C, 20 to 600°C, and 20 to 1200°C, respectively. These values are on the high side of the average curve. They claimed that the change in the ratio of crystallographic axes, c/a, is insignificant. Mauer and Bolz measured the expansion of two orientations of sapphire and a commercial-grade alumina (0.50% Na_2O) in neon atmosphere by the X-ray method, and obtained values on the low side of the curve. They ascribed this to inadequate thermal calibration for samples of lower thermal conductivity. Klein claimed that the anisotropy of alpha aluminum oxide is low, and that within a small limit of error, the ratio c/a is constant. This was believed to confirm that the expansion is substantially isotropic.

Engberg and Zehms extended the telemicroscopic evaluation to 1600°C by use of a graphite tube in which the carbon gases were swept away from contact with the alumina specimens by a small flow of argon, introduced through a tantalum tube. Consistent values were obtained on cold-pressed, extruded, and single-crystal melts of alumina on the low side of the average of previous investigations. The average coefficients of linear thermal expansion found for alumina in comparison with several other high-temperature materials is as follows:

Material	$\bar{a} \times 10^{-6}$	Temp. range °C
Al_2O_3	7.5 ± 0.4	1000 to 1600
BeO	10.1 ± 0.6	1000 to 2000
MgO	12.6 ± 0.5	1000 to 2000
B_4C	6.02 ± 0.51	1000 to 2400
SiC	5.68 ± 0.11	1000 to 2400
TiC	8.31 ± 0.68	1000 to 2600

All BeO specimens that had been heated to above 2050°C had very large expansions. The specimens were bent and cracked. The alumina specimens exhibited irreversible behavior above 1600°C, probably representing additional sintering contraction, particularly of the specimens having a fired density of 3.83 g/ml obtained by cold pressing, and 3.81 g/ml obtained by extrusion. Peculiarly, the single-crystal alumina (3.98 g/ml) also exhibited irreversible shrinkage, and on a different slope than the ascending curve.

Campbell and Grain investigated the c/a ratio rigorously by X-ray diffractometer methods and found that alpha alumina expands anisotropically, with the expansion coefficient parallel to the c-axis about 10% higher than that parallel to the a-axis, in substantial confirmation of Austin's data. The c/a ratio determined at 25, 200, 400, 600, 800, and 1000°C was 2.7294, 2.7298, 2.7303, 2.7307, 2.7312, and 2.7316, respectively.

Burk measured the thermal expansion of several specimens of sintered alumina, Coors AD-85, AD-94-E, AD-99, and AP-100, in the low temperature range from about 73°K to 273°K. In these compositions the number signifies the approximate alumina content. In this region, the coefficient of expansion was in the range 2 to 4 $\times 10^{-6}$, in comparison with 6 to 8 $\times 10^{-6}$ from room temperature to 100°C. The specimen with almost 100% Al_2O_3 and porous, AP-100 (18 to 21% porosity), had the highest thermal expansion coefficient. The lowest coefficient was found for the specimen having the lowest Al_2O_3 content, owing to the presence of spinel or silicate phases.

Wachtman, Scuderi, and Cleek (1962) measured the linear thermal expansion of polycrystalline alumina (Lucalox containing 0.1 to 1% MgO) and of sapphire (0.01 to 0.1% Si) in two orientations by the interferometric method in air. The orientations of sapphire were for an angle (ω) between the crystallographic c-axis and the direction of measurement equal to 10 degrees, and for the angle equal to 90 degrees. Values were computed for the orientation 57.6 degrees, which is frequently used for sapphire rod dilatometers. Measurements obtained at 100 to 1100°K were fitted to Grüneisen's statistical mechanical equation using a Nernst-Lindemann energy function. The closeness of fit made it reasonable to extrapolate the values to 2000°K. The values for the polycrystalline alumina fell intermediate between those for the two orientations of sapphire, the higher value being for the 10 degree orientation. Their data agree with those of Mauer and Bolz, but are on the low side of the curve of probable values shown by Goldsmith et al.

Neilson and Leipold (1963) investigated the thermal expansion of hot-pressed, slip-cast, and isostatically pressed, polycrystalline, single-phase alumina heated in air in an automatic, recording dilatometer. Expansion measurements above 1100°C in air were made in an oxide induction furnace. The hot-pressed alumina (98.5% theoretical density) developed crystal growth during testing, increasing in size from 3 microns original size to 17 microns at 1700°C, 20 to 50 microns at 1900°C, and 80 to 200 microns at 2040°C. The isostatically pressed alumina (Norton AW1F) having a grain size initially of about 45 microns, developed growth to about 65 microns around 1900°C. The coefficient of expansion was not affected by the crystal size or by the fabrication techniques. The samples showed stability of weight and lattice parameters, but developed permanent expansion when heated above 1500°C, which is well below the melting point. It appears that an oxidizing ambient during initial heating causes a permanent expansion and correspondingly higher values for the coeffficient, while a reducing atmosphere causes permanent shrinkage and lower values for the coefficient (Engberg and Zehms). The effect repeats to some degree with continued cycling.

The marked differences between the results of various investigators probably result in part from the differences in grain growth or sintering activity of the polycrystalline aluminas. These changes are of particular significance in high-temperature applications. Further elucidation is required of the factors affecting expansion. More consideration should probably be given to data obtained under oxidizing conditions than reducing, because of the lesser likelihood of reduction of the alumina.

Stutzman, Salvaggi, and Kirschner (1960) have collected the literature on the reversible thermal expansion of crystalline ceramics.

Filonenko and Kuznetsova observed anomalous thermal expansion of electrocorundum, which they ascribed to oxidation of titaniferous ferroalloys at 400 to 600°C. This is accompanied by a considerable decrease in density. Wenzel investigated the effect of the anomalies in thermal expansion on the quality of abrasives. Titanium carbide is the most damaging impurity; titanium nitride and iron alloys, which are also detrimental, may be removed by proper calcination at 1200°C. Mischke and Smith determined the thermal conductivity of alumina catalyst pellets of different densities as a function of the macropore volume fraction. Values were measured under vacuum and also under atmospheric conditions to 150°C. Low conductivities indicated that severe temperature gradients are likely in porous catalysts. The area of contact between the powder particles and in the pellets is more significant than the conductivity of the solid phase itself.

Kingery (1957) noted that the thermal expansion of two-phase compositions is not simply the averages of the end member values, but agrees quantitatively with calculations based on the assumption of substantial residual microstresses resulting from the constraint of each phase on cooling.

Schneider and Mong investigated the thermal expansion of certain refractory castables, using a sapphire rod dilatometer. Ruh and Wallace (1963) have extended the determinations to many types of refractory brick, including alumina brick in the classes from 60% to 99% Al_2O_3, and ranging in porosity from 19.5 to 23.4%. The data for the high-purity alumina brick agree well with that of Wachtman, Scuderi, and Cleek for Lucalox. As the alumina content decreases, the thermal expansion also decreases, which indicates the importance of the chemical composition and the mineral phases present, as well as their distribution. The 99% Al_2O_3 brick had a linear expansion of about 1.2% for the range from room temperature to 1300°C (2400°F). This is less than the expansion of the magnesia types (fused, forsterite-bonded, spinel-bonded, spinel, magnesia-chrome, and chrome-magnesia), and less than the expansion of super-duty silica-alumina and superduty silica brick. Stabilized zirconia is about in the same range, while zircon and clay-bonded silicon carbide have lower expansions.

6-4 THERMAL CONDUCTIVITY

The thermal conductivity of alpha aluminum oxide is relatively high for ceramic materials, which helps to explain its good thermal shock resistance. It is exceeded, however, by several materials among which are beryllia, magnesia, and silicon carbide. The variation of thermal conductivity of alumina with temperature is typical of dielectrics. The heat conduction is apparently mainly by lattice vibration of the crystals in quantized units called "phonons." The temperature region below 100°K is of unusual interest because of the influence of the marked variation in mean free path, and the effect of various factors on scattering (Umklapp processes). Berman found a peak of very high conductivity, about 15 cal/sec cm °K, for a sapphire crystal at about 40°K. The conductivity of polycrystalline alumina varies near this maximum, dependent on the lattice defects, crystallite size, impurities, etc., in the specimen, and the relaxation rates associated with the phonon scattering. The thermal conductivity then falls to a minimum of about 0.012 cal/sec cm °K with increasing temperature to about 1700°K (the reciprocal temperature rule). Above 1700°K, radiation contributes to increased thermal conductivity.

Jaeger (1950); Francl and Kingery; McQuarrie; and Sutton have described methods and equipment for testing thermal conductivity. The usual methods involve the determination of steady-state heat flow through a cube or rod form by comparison or substitution of standards or the direct determination dynamically of heat flow or temperature rise. The heat source in some cases has been enclosed within an ellipsoid-shaped envelope (Norton et al., 1950). It may consist of a thermocouple-resistance unit embedded in the specimen symmetrically (Haupin), or an inductively heated resistor surrounding the specimen.

Determinations of the variation of thermal conductivity of alumina with temperature have been made by many investigators, among whom are Knapp (1943), Norton, Fellows, et al.; Norton and Kingery; and Kingery, Klein and McQuarrie. Weeks and Seifert compared the thermal conductivity of synthetic sapphire against Armco iron standards in a vacuum, and found that in the direction 60 degrees from the c-axis, the thermal conductivity is 0.065 cal/sec. cm °C at 90 to 100°C, in good agreement with the average curve (Goldsmith, et al.). Knapp earlier had obtained low values for native corundum and synthetic sapphire, both normal and parallel to the c-axis, by a similar method.

Kingery (1954) found that the thermal conductivity of polycrystalline alumina by the spherical envelope and cylinder methods gave values in close agreement with other methods. Adams obtained values in the region 780 to 1570°K by the prolate spheroid method on slip-cast sintered alumina (porosity 6.3 to 7.1%) that conformed closely with the average curve. McQuarrie (1954) fitted the previously published data for alumina to an empirical equation. The experimental data depart from the reciprocal temperature law at higher temperatures, believed to result from increased passage of radiant energy. Jamieson and Lawson (1958) fitted the data by an equation which was exponential in reciprocal temperature, and ascribed the deviations to the flow of excitons. Excitons are electron-hole pairs of insufficient energy to allow the electrons to leave, but which can transfer energy from point to point. Whitmore (1960) concluded that phonon electronic heat transfer, as well as transport by electron-hole pairs, excitons, and dissociated gas molecules contribute to the heat flow, and estimated the possible magnitude of these effects.

It was long known that gas-filled or vacant pores have a lower thermal conductivity than any solid phase at low temperatures. Francl and Kingery measured the influence of porosity on conductivity. At temperatures below 500°C, the conductivity of a sample with isolated pores in any given direction k_p was found to be approximately equal to the solid-structure conductivity $k_s (1-P_c)$, in which P_c is the cross-sectional pore fraction. At higher temperatures, the pore size and emissivity become significant. The effect of isometric approximately spherical pores was quite different from that of anisometric cylindrical pores. Pore orientation was found to affect the thermal conductivity remarkably. The "structure factor" for porous bodies varies from infinity for parallel orientation to zero for series orientation in the direction of heat flow (Torkar). Convection becomes important when the pores are larger than several millimeters (Kingery). Kingery, Francl, Coble, and Vasilos applied a correction factor on this basis to obtain thermal conductivity at zero porosity. Schwiete, Granitzki, and Karsh found that the rate of decrease in thermal conductivity was highest below 800°C. Charvat and Kingery found that the conductivity of polycrystalline alumina rods (9 and 12 micron average crystal size, data corrected to zero porosity) is the same as that of monocrystal sapphire, at temperatures below 300°C. At about 300°C radiant heat transfer begins. At 1000°C, the value for sapphire was about 0.018 cal sec^{-1} cm^{-1} °F^{-1}, and for polycrystalline alumina about 0.008 cal sec^{-1} cm^{-1} °F^{-1}. Tinklepaugh, Truesdale, Swica, and Hoskyns (1961) found that the thermal conductivity of sintered alumina decreased slightly but significantly as the average grain size was decreased from 10 to 4 to 2 microns, in the temperature range from 100 to 1000°C. Francis, McNamara, and Tinklepaugh determined the thermal conductivity between 200 and 700°C of a commercial high-purity sintered alumina (Wesgo Al-300, 97.6% Al_2O_3, density 3.68 g/ml). Members of the Alumina Ceramic Manufacturers Association furnish values for thermal conductivity of their high-alumina ceramics.

Hoch and Silberstein measured the thermal conductivity of commercial, sintered alumina disks only ¼ to ½ inch in diameter, heated inductively by circumferentially placed platinum resistors in the temperature region, 1010 to 1170°C. Thermal conductivity (k) was found to satisfy the equation:

$$k = [(88.3 \pm 0.4)/T] + (4.4 \pm 0.6)(10^{-12})(T^3).$$

Godbee and Ziegler determined the thermal conductivity of alumina powders compressed to various densities (volume fractions from 0.49 to 0.70) as a function of temperature from 10 to 850°C, by radial heat flow in a hollow cylinder of a perfect conductor surrounded by an infinite amount of the test powder. A theoretical expression was derived relating the effective thermal conductivity of the two-phase system to the conductivities of the pure phases, the volume concentrations of the phases, and a shape factor. Deissler and Boegli (1958) showed that the effective conductivity varies with the atmosphere; the conductivity of alumina powder in air or nitrogen was less than 50% of that in hydrogen. Zhorov et al. (1966) determined the thermal conductivity of alumina coatings, plasma-sprayed onto molybdenum plates to a thickness of 130 or 300 microns, heated in argon.

Cowling, Elliott, and Hale (1954) concluded from a large number of production control tests that there was a relationship between bulk density and thermal conductivity of specific types of refractory insulating bricks that was sufficiently accurate for appraising thermal conductivity for most practical purposes. Lasch observed an increase in thermal conductivity of firebrick containing up to 50% added corundum. Brick with high corundum content (80%) showed a comparatively rapid decrease in conductivity with increasing temperature.

Ruh and McDowell determined the thermal conductivity of many types of refractory brick, including high-alumina types, using apparatus modified from the ASTM C201-49 specifications. The measurements covered mean specimen temperatures from about 90 to 1090°C. The apparent porosities of the specimens varied from about 16 to 23.5%. The thermal conductivities increased with increasing alumina content from about 0.0034 cal/sec cm °C at 150°C mean temperature for 60% Al_2O_3 to about 0.0137 cal/sec cm °C at 150°C mean temperature for 99% Al_2O_3. At a mean temperature of 1100°C, the values were 0.0034 cal/sec cm °C and 0.0060 cal/sec cm °C, respectively.

Hansen and Livovich investigated the thermal conductivity of both insulating and dense refractory cement (Lumnite) concretes. The data for the concretes generally were in good agreement with a curve establishing a relationship between thermal conductivity and bulk density. Some deviation in excess of 5% in the direction of higher conductivity apparently indicated larger than average pore size. Well-dispersed pores provide better insulation than the same quantity in a poorly dispersed condition. Ruh and Renkey also determined the thermal conductivity of refractory castables of both dense fireclay and high-alumina types, as well as of several insulating castables. These tests, covering mean specimen temperatures of about 90 to 1090°C, indicated that the thermal conductivity of a castable material varies with its thermal history. Castables merely dried at 110°C and retaining their constitutional water had high initial thermal conductivities, but when the combined water was driven off between 200 and 800°C, usually associated with loss of strength, the thermal conductivity was also lowered. As the temperature was further raised beyond 1100°C, the formation of a ceramic bond caused an increase in the thermal conductivity. In actual castables in service, these effects take place in layers that merge into each other.

6-5 THERMAL DIFFUSIVITY

Thermal diffusivity is defined as the ratio of thermal conductivity to the product of specific heat and density. The measurement in the CGS system is in cm^2/sec. Thermal conductivity is important in those cases which involve steady-state conditions. Diffusivity is of more significance in those cases where transient heat flow is important, as in spalling of refractories. Thermal diffusivity is also of more fundamental significance with reference to phonon scattering because the mean free path of the scattered phonons is more directly related to diffusivity. Diffusivity methods which have been used are the periodic heat flow methods, Biot's modulus method, the radial flow method, and the infinite solid method (Plummer, Campbell, and Comstock).

Fitzsimmons (1950) determined the diffusivity of alumina at 400 to 700°C. Paladino, Swarts, and Crandall covered the range 1500 to 1800°C, and Plummer, Campbell, and Comstock the range from room temperature to 1000°C. Rudkin, Parker, and Jenkins (1963) have reported high-temperature measurements. The values vary from about 0.091 cm^2/sec at 25°C to 0.0063 at 1500°C, and 0.0294 at 1800°C.

7 SONIC EFFECTS IN ALUMINA

7-1 VELOCITY OF SOUND IN ALUMINA

The longitudinal velocity of sound in solids is related to Young's modulus of elasticity (E) and the density (d) by Newton's equation, $V = \sqrt{E/d}$. This elastic modulus is specifically defined for slender rods whose lateral dimensions change according to Poisson's ratio (μ) when a longitudinal stress is applied. According to Hueter and Bolt, *Sonics*, page 25, the velocity of sound in solids varies, dependent on the degree of constraint of the lateral dimensions. For an infinite solid body in which the lateral dimensions are constrained by stiffness, the stiffness (or bulk) modulus (E_b) is higher than E, from theoretical considerations, by an amount determined by $E/E_b = 1 - b$, in which $b = 2\mu^2/(1 - \mu)$. For sintered alumina of theoretical density, μ may be assumed to be 0.257, d 3.987, and E 58.5×10^6 psi (4033.3 kbars) at one atmosphere and 25°C, based on determinations of E by beam deflection. On these assumptions, the calculated velocity of sound (V) in unconstrained sintered alumina bar forms is about 33,000 ft/sec (10.058 km/sec).

In a reversal of the preceding methods, the velocity of sound is measured in order to determine the modulus of elasticity. Young's modulus is also determined by measurement of the fundamental resonant frequency of the sound waves generated in a bar of suitable geometry. This also involves the speed of sound in conformity with the expression, $N = (m^2k/2\pi L^2) \sqrt{E/d}$, in which E = Young's modulus, d = density, k = radius of gyration, L = bar length, and m is a constant representing the mode of vibration (Powers, 1938). As the frequency increases, and the wavelength of the sound diminishes to about twice the diameter of the rod, the dominant wave form changes from longitudinal to transverse (shear). The shear velocity (V_s) = $V \sqrt{1/2(1 + \mu)} = \sqrt{E_s/d}$, in which E_s is the shear modulus of elasticity. For alumina, V_s is about 20,810 ft/sec (6.343 km/sec).

Elegant reasonant methods at measured frequencies of about 100 kc/sec, and ultrasonic interference methods at about 30 Mc/sec have been used to determine the velocity as well as the elastic constants. Schreiber and Anderson, using the latter method, found the longitudinal velocity of sound in polycrystalline alumina to be 10.845 km/sec at one atmosphere and 25°C, and the shear velocity to be

6.3730 km/sec. They found a pressure derivative dV_p/dP of 5.175×10^{-3} km/sec/kbar (longitudinal wave) and a pressure derivative dV_s/dP of 2.207×10^{-3} km/sec/kbar (shear wave) for experiments up to 4 kbars at 25°C. The bulk modulus (E_b), derived from the expression, $E_b = d(V^2 - 4/3\ V_s^2)$, amounted to 2504.5 kbars (isothermal). The density of the sintered alumina was 3.972. Poisson's ratio was found to vary with pressure according to $d\mu/dP = 1.02 \times 10^{-4}$/kbar, and bulk modulus according to $dB/dP = 3.99$, in which B is the isothermal bulk modulus.

Soga, Schreiber, and Anderson determined the sound velocities at very high temperatures by applying the Mie-Grueneisen equation to establish the linear dependence of the bulk modulus on temperature. The Grueneisen constant relates the expansion coefficient (a) with compressibility (K) and specific heat (C_p) by the expression: $\gamma = a/dC_pK$. Caddes and Wilkinson observed a surprisingly large photoelastic anisotropy for longitudinal acoustic waves propagating along the C-axis of synthetic sapphire.

7-2 ULTRASONIC ABSORPTION

Ultrasound can be applied to the study of dislocations, internal friction, stress effects, relaxation, fatigue, and superconductivity of solids (Redwood, 1959).

The attenuation of sound (acoustical damping) in solids depends on imperfections which cause scattering. In crystalline materials, foreign impurities occur in the lattice as a substitute for the solvent, or in interstitial positions. A small number of foreign atoms markedly affects the physical properties. Two kinds of single point defect structure represent atoms out of normal position, that have moved to the surface (Schottky) or into an interstitial position (Frenkel). Line defects (dislocations), hysteresis, and relaxation also contribute to ultrasonic attenuation (Hutchison, 1960). The relaxation process involves a characteristic rate.

Sligh and Bixby observed that the ultrasonic energy absorbed is converted to heat. Solid particles acquire high velocity and acceleration, cavitation occurs in liquids. These effects have been applied to ceramic processes to obtain improved dispersion of slips, to provide heat during the application of glazes, to cut and drill hard ceramics, and to detect flaws in heat-fused ware, among others (Dickinson, Gibbs).

Fitzgerald, Chick, and Truell found that the ultrasonic attenuation for ruby was higher than for sapphire in the low-temperature region, conforming with the prediction from the corresponding effect of impurities on the thermal phonon relaxation time. Ciccarello and Dransfield found, contrary to some theoretical expectations, that the absorption of longitudinal waves in the low-temperature region was stronger than that of transverse waves. This was ascribed to 3-phonon processes. The intrinsic attenuation and velocity of the compressional and shear waves measured in monocrystal sapphire at one Gc/sec for a-axis propagation shows a T^4 dependence on temperature, but the fast shear mode shows a T^7 dependence below $50°K$ and a T^4 dependence at higher temperatures (Klerk).

8 ELECTRICAL PROPERTIES OF ALUMINA

8-1 INTRODUCTION

The remarkable electrical insulating properties of alumina were applied early in the technical development of sintered alumina as insulators for spark plugs. H. Gerdian and R. Reichmann (1927) prepared an insulation composition which was substantially pure alumina. The spark plug manufacturers became interested in the electrical resistivity particularly at elevated temperatures. A common criterion in appraising the resistivity was the determination of T_e, a designation for the temperature at which the resistance of a one-centimeter cube of the insulator falls to one megohm. Other electrical measurements of significance mainly in high-frequency applications include the determination of the dielectric constant, the power factor, the loss factor, and the dielectric strength or breakdown voltage.

From practical considerations, the excellent insulating properties of alumina are partly dependent on high refractoriness, high strength, and chemical inertness of the compositions.

Some data on the electrical properties of pure alumina and various commercial compositions are shown in Table 11.

8-2 ELECTRICAL CONDUCTIVITY OF ALUMINA

Many investigators have determined the variation in electrical properties of alumina with change in temperature, both on single-crystal and polycrystalline specimens. Cited among the contributors to the literature on the electrical conductivity (or its inverse, resistivity) are: Diepschlag and Wulfestieg; Wartenberg and Prophet; Podszus; Backhaus; Rochow; Shul'man; Rögener; Arizumi and Tani; Oreshkin; and Pentecost, Davies, and Ritt. Doelter (1910) classed corundum as an insulator at ordinary temperatures, and an electrolytic conductor at higher temperatures, showing noticeable polarization. Many subsequent investigations, even to the present time, were undertaken with marked variations in purity, porosity, thermal history, ambient conditions during test (air, water vapor, vacuum), contacts to the test specimens, and test methods (both ac and dc), among others. It is therefore quite understandable why the data of different investigators show a divergency as great as six orders of magnitude for both single-crystal and polycrystalline alpha aluminum oxide (Figure 13).

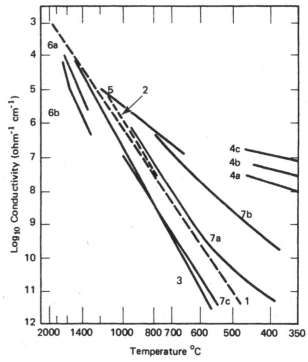

Variation in electrical conductivity with temperature. Solid lines represent polycrystalline alumina, dotted lines single-crystal sapphire; (1) Linde Air Products; (2) Wachtman and Maxwell (1957), annealed at 1800°C for 4 hours, tested in air, (3) Heldt and Haase (1954), vacuum sintered 99.97% Al_2O_3, tested in vacuum; (4) Hartmann (1936) a, b, c, repeated tests in vacuum at increasing conductivity, 4c represents equilibrium; (5) Hensler and Henry (1953), porosity 27%, acid treated specimens, ac test in air; (6) Pappis and Kingery (1961), isostatic-molded Linde powder, sintered at 1900°C for 7 hours to 3% porosity, a in oxygen at one atm., b in oxygen at 10^{-5} atm.; (7a) Coors Porcelain Company, AD-995, (7b) Coors Porcelain Company, AD-85, (7c) Coors AD-99C.

Figure 13. Electrical Conductivity of Alpha Alumina

Kose and Hamano found that adsorbed water in open pores may be a main contribution to dielectric losses. Peters, Feinstein, and Peltzer suggested that the conductivity of the ambient air is significant in high temperature measurements. Kainarskii, Karyakin, et al. pointed out that variable changes in cross section of polycrystalline alumina in different temperature ranges, particularly between 1550 to 1650°C, influenced the dielectric properties.

The conductivity in mhos is the reciprocal of the bulk resistivity. When the logarithm of electrical conductivity is plotted vs. the reciprocal of the absolute temperature, a straight line is generally obtained, the slope of which is proportional to the activation energy. A substantial amount of data has been found to obey an expression having the form: $\theta = Ae^{-E/kT}$, in which θ is the conductivity, A is a constant, k is the Boltzmann constant, T is the absolute temperature, and E is the activation energy. Ford and White summarized the fundamental knowledge relating to ionic conductors in 1952. Cohen reviewed the literature on conductivity in 1959. Florio (1960), and Pappis and Kingery (1961) also reviewed the temperature dependence of the results of conductivity of earlier investigators.

At room temperature the conductivity of alumina is extremely low, around 10^{-17} mhos. In common with other insulators, the conductivity increases with increasing temperature. At high temperatures the insulator acquires the characteristics of a semiconductor. Shul'man (1940) noted a break in the conductivity curve between 1200 and 1500°C, and stated that it was not caused by structural changes, but was a real change in conductivity. The reported activation energies vary from about 0.250 ev to above 5 ev, but are generally close to 2.8 ev. Cohen concluded, from the point of view of electrical conduction, that the distinction between an insulator and a semiconductor may be considered one of degree rather than of kind. There is no abrupt change in resistivity to separate the two classes.

Hartmann in 1936 had observed that the electrical resistance of sintered alumina at high temperatures increases with increasing oxygen in the atmosphere, and had concluded that conduction is an oxidation process. Heldt and Haase (1954) measured the electrical resistance in vacuum of very pure alumina (less than 0.03% total impurities) that had been sintered in vacuum, and compared the results with similar measurements made on less pure specimens (about 0.06% impurities) that had been sintered in a gas furnace. The "pure" material had a linear curve with an activation energy of 2.50 ev. The less pure specimens yielded curves which exhibited a break at about 1100°C. Below 1100°C the activation energy was 2.38 ev; above 1100°C it was 2.5 ev. The conductivity above 1100°C was designated intrinsic to imply real semiconduction substantially by electron transfer and mechanisms associated with the pure oxide. Extrinsic conduction, largely impurity-controlled, was believed to occur below 1100°C. Matsumura described the conduction as ionic below 830°C, and electronic at higher temperatures.

Heldt and Haase also investigated the effect of air and vacuum on resistance. When the alumina was removed from the vacuum and aged in air the resistance decreased, being the lowest for high humidity. At room temperature, the resistance in air was 10^9 ohms, at 10^{-1} mm Hg 10^{12} ohms, and in vacuum, following heating at 1500°C and cooling to room temperature, 10^{14} ohms.

Pappis and Kingery measured the electrical conductivity of both polycrystalline and single-crystal aluminum oxide at 1300 to 1750°C, and at oxygen partial pressures of 10^0 to 10^{-10} atmospheres. At 10^0 atmosphere the activation energy was 2.97 ev and the conductivity was p-type; at 10^{-5} atmosphere the activation energy was 2.62 ev below 1627°C and 5.5 ev above, with intermediate-type conductivity; and at 10^{-10} atmosphere the activation energy was 2.84 ev below 1627°C and 5.8 ev above, and the conductivity was n-type. They concluded that the conductivity of alpha aluminum oxide does not result from any single simple process over wide ranges of temperature and oxygen partial pressure. The increase in conductivity at high and low oxygen pressures indicated that alumina is a nonstoichiometric semiconductor. Kroger and Vink suggested possible structures by which changes in the conductivity can be interpreted. These are limited, since interstitial anions are unlikely in alumina. Harrop used ultraviolet reflection and absorption data to show the temperature dependence of exciton absorption in alumina, and by extrapolation of the first ionization level of Al_2O_3 (6.0 ev at 1950°K) a comparison could be made with the activation energy of intrinsic electrical conductivity found by Pappis and Kingery at this temperature. It was inferred that a carrier jump energy of 2.5 ev is involved.

Vest studied the transport mechanism by electrons and ions in various oxide compositions including Cr_2O_3-Al_2O_3. He concluded that because the bonding in metal oxides is partially ionic, there is always an ionic component of the total measured conductivity. Hensler and Henry measured the influence of controlled additions of small amounts of Cr_2O_3 on the electrical resistance of polycrystalline alumina, measured in air. The specimens were dry-pressed, 0.4 cm-thick disks fired at 1500°C for 10 hours. X-ray patterns showed the Cr_2O_3 to be in complete solid solution. With additions to about 6% Cr_2O_3, the log resistivity vs. reciprocal temperature curves behaved similarly to that of the pure alumina, being approximately linear and having about the same slope. The calculated activation energy was about 1.0 ev. Specimens containing up to 6% Cr_2O_3 had higher resistivities than the pure alumina, with a maximum at about 1%. Another peak resistivity occurred at 92% Cr_2O_3. Chiochetti and Henry found that the electrical resistance of commercial refractory brick (alumina-silica mixtures, fireclay brick, high-alumina brick, mullite-type brick, and fused-cast brick containing some chromite) are similarly straight line curves. Fused-cast alumina containing mixtures of alpha and beta aluminas exhibits a different type of curve.

Barta, Bartuska et al. investigated the influence of purity and crystal size on the specific resistance in the range 400 to 700°C. Alkalies decreased the specific resistance considerably, but small quantities of SiO_2, CaO, TiO_2, Fe_2O_3, V_2O_5, or Cr_2O_3 had no significant effect. MnO and calcium compounds increased the specific resistance. Fine-grained sinters had a higher resistivity than coarse-grained sinters.

Yamauchi and Kondo (1950) investigated the effects of several mineralizers on sintered alumina disks that had been fired at 1550 to 1700°C. Pure sintered alumina had greater resistance to dc at high temperatures than other ceramics. The electrical conductivity was claimed to result from the alkali present in the vitreous component. Iron oxide also decreased the electrical conductivity, while lime and silica increased it and decreased its temperature coefficient.

Starokadomskaya et al. appraised the electroconductivity of alundum, 90% Al_2O_3-10% BeO, and Al_2O_3 plus 1% Cr_2O_3 as materials for thermocathode preheaters at working temperatures of 1600 to 1800°C. All specimens behaved as semiconductors, with changes in the type of conductivity at 1200 to 1400°C. The sharp increase in conductivity begins at much lower temperatures for the mixtures than for the pure substances. Calcination in vacuum decreased both the conductivity and the scatter in the measurements.

Budnikov and Tresvyatskii found that corundum refractories prepared with 2.5% clay bond reduced the electrical resistivity to one-tenth that of shapes without clay. Further addition to 20% dropped the resistance only to 0.4. The resistivity depends very little on composition for a clay content of 20 to 50%. Eremenko and Beinish correlated the electrical conductivity of binary refractory oxides with their chemical analysis. Garn and Flaschen applied electrical conductivity to the detection of phase transformations.

Schwab (1962) investigated the semiconductor properties of aluminum oxide from the standpoint of its use as a catalyst in gas reactions. The conductivities of both alpha and gamma aluminas increased with doping with any foreign cations. These properties were not affected markedly by oxygen. Both p- and n-doped oxide rectification was obtained, suggesting that both alumina phases are intrinsic semiconductors.

Reported values for the electroconductivity of single-crystal alumina are about in the same range as that for polycrystalline alumina. Rochow (1938) measured the dc conductivity of sapphire from 600° to 1200°C, and obtained an activation energy of 1.8 ev. Wachtman and Maxwell (1954, 1957) determined the effect of plastic deformation on the electroconductivity of Linde sapphires and found that deformation increased the resistivity. The activation energy for the undeformed portion was 1.5 ev and for the deformed portion 2.7 ev. The resistivity curves (extrapolated) crossed at about 1800°C. Annealing the deformed specimens at 1600°C did not affect the conductivity significantly, but annealing at 1800° reduced the resistivity even lower than the original curve. This is interpreted to indicate that resistivity in the original crystal, caused by strain and subsequent resistivity caused by work-hardening of plastic deformation, is eliminated at high temperature. Linde Air Products Company obtained the same activation energy (about 1.5 ev) for annealed and unannealed sapphire.

Tucker and Gibbs observed a narrow peak in the conduction current of sapphire at about 250°C, and exponentially decaying pulses of about 10^9 electrons at 400 to 550°C, when the sapphire was exposed to water-saturated nitrogen. Both the peak and pulses were absent

on cooling from 1000°C, or on reheating in dry nitrogen. Similar impurity effects at different temperatures were observed for HCl, NaCl, and $Mg(NO_3)_2$. The Verneuil-type crystals normally contain about 10^6 dislocations/cm^2, as indicated by etch-pit studies. Bending through a radius of about one centimeter introduces about 2×10^7 dislocations in the (0001) slip plane, the length of which lies parallel to the axis of bending. Gibbs (1959) concluded that the impurities penetrate into the body of sapphire along dislocation lines. This provides high-conductivity "tubes" which surround the dislocation lines. The high-conductivity regions also cause dielectric-loss peaks, but these can be eliminated by heating to volatilize the impurity atoms.

Harrop and Creamer (1963) found an activation energy of 4.64 ev in the range 1280 to 1480°C for the low-voltage, dc conductivity of sapphire. When chromium was present as an impurity, the conductivity was lower and the activation energy was 4.26 ev. They concluded that the charge carriers were p-type, in agreement with earlier investigators, but ascribed the conductivity mechanism to impurity effects, specifically traces of iron, rather than to intrinsic behavior, as suggested by Pappis and Kingery.

Champion (1964) continued the investigation of the conductivity of alumina containing chromia in air at 20 to 850°C by a 2-terminal dc method. When first heated, all specimens showed peaks and pulses in conduction current similar to those described by Tucker and Gibbs for clear sapphire. After several heating cycles, the variation of conductivity with temperature became reproducible. The activation energy averaged 0.5 ev at 200 to 450°C, and 1.7 ev at 450 to 850°C. Conductivities of 1×10^{-9} ohm^{-1} – cm^{-1} for clear sapphire and of 2×10^{-10} ohm^{-1} – cm^{-1} for pink ruby were compatible with the values obtained by Harrop and Creamer, but their suggestion that trace impurities of iron control the conductivity was not confirmed.

Chang (1963) found that electrical resistivity decreases initially during transient creep. During steady-state creep at 1525°C and a pressure of 2175 psi, the resistivity increased almost linearly with strain. This was ascribed to the creation of carrier-trapping point defects in the wake of moving dislocations. The increase in resistance amounted to about 1.3×10^{-3} ohm^{-1} – cm^{-1} for undoped crystals, and 1.3×10^{-4} for Cr-doped (0.3% wt Cr) crystals (Chang and Graves).

The current-voltage characteristics of very thin barrier layers (20 to 40 A thick) of aluminum oxide between aluminum electrodes was investigated by Nakai and Miyazaki. Conduction results from a quantum-mechanics tunneling effect. The tunneling current at constant applied voltage increases with increasing temperature.

Peters determined the electrical conductivity through junctions in single-crystal sapphire formed by sintering together two single crystals with a 50% $CaCO_3$-50% SiO_2 bond at 1750°C in air. A large barrier to current flow was found to exist across the barrier. Peters suggested several mechanisms for the junction barrier including: (1) differences in work functions between the sintered layer and the sapphire, (2) diffusional effects, and (3) a difference in the type of charge carrier between the junction region and the sapphire region. Peters (1966) measured the thermoelectric power of single-crystal sapphire at 400 to 1000°C in air, and found that it is positive, and decreases with increasing temperature. This is typical of extrinsic conduction in which mobile carriers are released from impurity atoms.

Matiasovsky et al. (1964) found that the specific conductivities of cryolite and of melts in related systems decrease with increasing concentration of dissolved alumina in the melt. The addition of sodium chloride increases the specific conductivity of cryolite and of the system $Na_2AlF_6 \cdot Al_2O_3$ also. The influence of the sodium chloride is most pronounced at low concentrations of NaCl and at higher concentrations of Al_2O_3.

8-3 DIELECTRIC CONSTANT AND LOSS FACTOR OF ALUMINA

The dielectric constant (k) or relative permittivity of crystalline insulator materials is characterized by a combination of electronic, ionic, and dipole orientations that cause polarization of the insulator in electrical fields. For an alternating field, polarization shows as a phase retardation of the charging current. The complex dielectric constant, (k) equals $k'-ik''$, in which k' is the relative dielectric constant, i is the imaginary, $\sqrt{-1}$, and k'' is the loss factor. If θ is the dielectric phase angle, $90-\theta$ is the dielectric loss angle (δ); $\tan \delta$ is the dissipation factor (D); and k' $\tan \delta$ is called the dielectric loss factor (k''). The rate of dissipation of electrical energy lost as heat is proportional to the ac conductivity (σ); and σ equals $fk''/1.8 \times 10^{12}$, in which f is the frequency.

In the optical frequency range, electronic orientations are present. The relative dielectric constant (k'_e equals n^2, in which n is the index of refraction (Maxwell). The value is about one-third the measured value. A common method for measuring the dielectric constant is by substitution of a specific geometrical form in a capacitance bridge circuit.

The dielectric constant and the dielectric loss factor are of significance in alumina ceramics applied principally in frequency applications, as for example, capacitor and electronic insulating elements, and transparent radar windows.

Glemser (1939) determined the dielectric constants of several hydrated aluminas by an immersion method. The values ranged from 20.6 for amorphous alumina hydrate (49.6% ignition loss) as well as for alpha monohydrate (17.4% ignition loss), to 9.4 for beta trihydrate, and 8.7 for alpha trihydrate. On heating, the values either decreased or increased in the direction of 12.3, the value found by Glemser for either native or artificial corundum (Table 11).

Ebert (1962) examined the dielectric loss of gamma alumina containing sufficient water of hydration corresponding to a bimolecular layer at temperatures from −60 to +60°C and at 100 kcps to 10 Mcps. Several sharp maxima, traceable to resonance, were observed in the curves of $\tan \delta$ vs. frequency. Curves of dielectric constant

vs. temperature agreed with calculations made by modifications of the Debye relaxation equations. Dekker and van Geel found no difference in dielectric constant of the amorphous and crystalline layers formed by anodic oxidation of aluminum.

Bogoroditskii and Polyakova based the low dielectric losses for corundum on the polarization effects being ionic and electronic. Variations in dielectric properties were attributed mainly to the formation of beta alumina, in which structural polarizations are marked because of the open crystal lattice. Calcination under reducing conditions was thought to convert the beta alumina to alpha alumina, possibly by favoring the volatilization of contained soda. Bogoroditskii and Fridberg ascribed large dielectric losses in alumina insulators at temperatures below 200°C to moisture retention in submicroscopic pores.

Bowie (1957) described apparatus for measuring the dielectric constant and dielectric losses at 3000 Mcps. and at temperatures to about 1600°C. The average dielectric constants for several commercial, nominally 95% Al_2O_3 bodies in the temperature range, 250 to 850°C, were found to be between 8.4 and 8.9.

Cohen (1959) had indicated the presence of a high-resistance transition layer at the dielectric-electrode interface on sapphire as a complicating factor in bridge methods for determining the value.

According to von Hippel (1958) the dielectric constant of pure sapphire at 25°C is 9.34 perpendicular to the optic axis, and 11.55 parallel to this axis. It increases slowly with temperature to about 10.40 perpendicular to the optic axis and 13.50 parallel to it at 800°C. It is substantially independent of frequency to about 500°C, but slight increases in value are observed at frequencies below 10 kcps. The dielectric loss tangent (dissipation factor) showed no particular trend with orientation. The values were very low, being of the order of 0.00001 at 25°C. The loss was both frequency- and temperature-dependent, generally decreasing with increasing frequency, and increasing temperature to a value as high as 0.23 for 100 cps at 600°C. The dielectric constant of dense, polycrystalline alumina (99.9% Al_2O_3) fell intermediate between the values for the two orientations of sapphire. Relatively high values for both the dielectric constant and the loss tangent were obtained at temperatures above 300°C, and at 1 kcps. Differences in the value of the dielectric constant were obtained for differences in orientation of the test specimen with respect to the forming operation, indicating an influence of preferential crystal orientation in forming.

Above 900°C, the dielectric constant is far from constant. Florio used a 3-electrode guard-ring to determine the value at 900 to 1300°C. The dielectric constant of sapphire became increasingly frequency-dependent, as was also the case for the loss factor, which decreased at a rate about proportional to 1/frequency. The dielectric constant of polycrystalline alumina comtaining significant amounts of impurities (0.1% Fe, 0.1% Si, 0.08% Na, 0.1% Mg, and 0.1% Ca), confirmed von Hippel's finding that the measured

Table 11

Electrical Properties of Alumina

Resistivity (ohm-cm)

% Al_2O_3	Single-Crystal	Polycrystalline				
Temp °C	+99.5%	99.97%	99.0%	99.5%	94%	85%
	(1)	(2)	(3a)	(3b)	(3c)	(3d)
200				2.0×10^{12}	1.0×10^{15}	4.0×10^{11}
400			1.0×10^{13}	1.2×10^{10}	3.0×10^{10}	3.5×10^9
500	1×10^{11}	1×10^{13}	6×10^{11}	1.5×10^9	3.0×10^9	4.0×10^8
800	–	5×10^8	3×10^8	8×10^6	1.0×10^7	1.2×10^6
1000	1×10^6	7×10^6	8×10^6	8×10^5	5.0×10^5	
1500	1×10^4	1×10^4 (extrapolated)				
2000	1×10^3	–				

Dielectric Strength (volts/mil)

Conditions	Test Thickness	Value (v/mil)	Reference
Anodized films, tested at	<100 A	27,500	(4)
15°C	>100 A	15,000	(4)
−75°C, 33,000 cps	100-2000 A	34,060	(5)
27°C, 33,000 cps	100-2000 A	40,500	(5)
100°C, 133,000 cps	100-2000 A	42,100	(5)

		99.5% Al_2O_3	94% Al_2O_3	85% Al_2O_3
		(3)	(3)	(3)
ASTM Method D116-63	0.250 in.	230	230	230
	0.125	330	330	330
	0.050	400	500	–
	0.025	425	550	–
	0.010	450	600	–

Dielectric Constant

Phase	Preparation	Per Cent Loss on Ignition	Dielectric Constant	Reference
Amorphous	Stock	49.6	20.6	Glemser
Alpha Trihydrate	Fricke & Wullhorst	34.6	8.7	Glemser
Beta Trihydrate	Fricke & Wullhorst	34.6	9.4	Glemser
Alpha Monohydrate	Fricke & Severin	17.4	20.5	Glemser
Amorphous	Heated 13 hr 600°C	21.4	9.3	Glemser
Amorphous	Heated 13 hr 800°C	–	9.6	Glemser
Amorphous	Heated 13 hr 1200°C	–	11.3	Glemser
Beta Trihydrate	Heated 13 hr 600°C	0.88	13.1	Glemser
Beta Trihydrate	Heated 13 hr 1000°C	–	11.8	Glemser
Alpha Monohydrate	Heated 13 hr 600°C	0.13	10.6	Glemser
Alpha Monohydrate	Heated 13 hr 800°C	–	10.0	Glemser
Alpha Monohydrate	Heated 13 hr 1000°C	–	10.0	Glemser
Alpha	Synthetic or Corundum	–	12.3	Glemser

Table 11 — Continued

Temp °C	Frequency cps	Sapphire Orientation to Optic axis ⊥	∥	99.9%	99% 3.83 g/ml	94% 3.62 g/ml	85% 3.42 g/ml
		(7)		(7)	(3)	(3)	(3)
25	10^3	9.3	11.5	10.5	9.5	8.9	8.2
	10^5	9.3	11.5	10.5	9.5	8.9	8.2
	10^9				9.5	8.9	8.2
	10^{10}	9.3	11.5	9.6	9.4	8.8	8.2
300	10^3	9.6	12.1	21.6			
	10^5	9.6	12.1	11.1			
	10^9						
	10^{10}	9.6	12.1	9.9			
500	10^3	9.9	12.5	69	11.3	11.8	13.9
	10^5	9.9	12.5	13	10.0	10.1	8.9
	10^9				10.0	9.2	
	10^{10}	9.9	12.5	10.1	10.0	9.1	8.3
800	10^3						
	10^5						
	10^9				10.5		
	10^{10}	10.4	13.5	—	10.4	9.4	
			60° orient.	(8)			
1000	10^3	12.2			—		
	10^5	10.7			30		
1100	10^3	15.5			—		
	10^5	11.3			31		

Dielectric Loss Tangent (Dissipation Factor)

Temp °C	% Al$_2$O$_3$ Frequency cps	Single-Crystal Orientation ⊥	∥	Polycrystalline 99.5	99.0	94	85
		(7)		(3)	(3)	(3)	(3)
25	10^2		0.000012	—	0.0054	0.00018	0.0015
	10^6			0.0001	0.0008	0.00012	0.0009
	10^9			0.0001	0.00014	0.0008	0.0014
	10^{10}	0.00003	0.000086	0.0001	0.00019	0.0009	0.0019
500	10^3	0.005	0.0035	—	0.15	0.215	0.580
	10^6			0.0023	0.0047	0.008	0.024
	10^9			0.0002	0.0003	—	—
	10^{10}	0.00009	0.00015	0.0003	0.0002	0.002	0.003
800	10^6			—	—	—	—
	10^9			0.0003	0.0005	—	
	10^{10}	0.00043	0.00021	0.0006	0.0006	0.004	
	24×10^{10}			0.0007	0.0003	0.006	
	50×10^{10}			—	—	—	

Table 11 Continued

Miscellaneous Electrical Properties			
Secondary Electron Emission		2.5 to 3.5	(9)
Conditions: 30 to 500 microsecond pulses at 50 pulses/sec., at 500 volts; specimen 10 microns thick, spectroscopically pure			
Activation Energy for Electrical Breakdown		5 ev	(10)
Single-crystal sapphire			
200 to 450°C	0.5 ev		(11)
450 to 850°C	1.7 ev		(11)
1280 to 1480°C	4.64 ev		(12)
Single-crystal ruby			
1280 to 1480°C	4.26 ev		(12)

Ref.: (1) Linde Air Products
      (2) Heldt & Hasse (1954)
      (3) Coors Porcelain Company
          (a,b,c,d)
      (4) Lomer
      (5) Kawamura & Azuma
      (7) von Hippel (1958)

      (8) Florio (1960)
      (9) Shul'man & Rosentsveig (1959)
          Shul'man, Makeslonskii, &
          Yaroshetskii (1953)
    (10) Miyazawa & Okada
    (11) Champion (1964)
    (12) Harrop & Creamer (1963)

values become very high for low frequencies at elevated temperatures. The observed dielectric losses were ascribed to free-electron conduction mechanisms. Large thermal activation energy values suggested that the conductivity is intrinsic; the energy band gap was calculated to be about 7.3 ev. Anomalous dispersion, in which the dielectric constant decreases with increasing frequency, was encountered in the alumina insulating applications. Types of polarization which yield anomalous dispersion curves include: orientation polarization due to polar molecules, distortional polarization due to a displacement of bound ions, and interfacial or space-charge polarization produced by traveling charge carriers (Florio 1960).

Tallan, and Detwiler and Tallan observed dielectric loss maxima of clear sapphire at frequencies between 10^2 and 10^4 cps, and at temperatures between $-160°$ and $400°C$. It was observed that the loss process was more pronounced with the optic axis in the direction of the applied field than with it perpendicular to the field. With careful balancing of the bridge-guard circuit system used, no significant conduction loss was observed. They suggested that the conduction loss commonly observed in the upper part of this temperature range for unguarded sapphire samples is a surface rather than bulk effect.

Sang (1958) discussed the application of high-strength dielectric materials in different types of air-borne equipment for Mach 3 to 6 operations. Fallon (1960) described the preparation of sapphire dielectrics.

Atlas, Nagao, and Nakamura (1962) determined the dissipation factors and dielectric constants of polycrystalline alumina ceramics containing less than 100 ppm of impurities and of specimens doped with Si, Ti, Ca, Mg, Fe, and Cr ions at 25 to 875°C, and at 10^2 to 8.5×10^9 cps.

Disks, 0.2 in. thick for frequency measurements to 10^7 cps and disks 0.6 in. thick for microwave determinations at 4000 Mcps were cold-pressed from calcined aluminum chloride prepared from 99.999% pure aluminum ingot. The aluminum chloride was precalcined at 1000°C (75.6 m²/g) to 1300°C (5.7 m²/g). The impurity additions in the form of oxides or carbonates were wet-milled with the alumina portions for one or more days in plastic jars. The specimens fired at 1800 to 1950°C reached a maximum density of only 3.7 g/ml. Unresolved discrepancies in the analyses of the wet-ground alumina compositions were ascribed to impurities introduced during firing and to errors in the spectrographic analysis, but not to the prolonged wet grinding. Multiple regression analysis of the data obtained at 500°C and at 10^6 cps showed a linear relationship between the impurity concentration and tan δ with a correlation coefficient of 0.93. The greatest increase in tan δ was caused by Si ions, followed by Mg and Ti, and Ca. Cr and Fe had no significant effect. These effects decreased with rising frequency and became negligible in the microwave region. Activation energies of conduction for the pure and doped alumina compositions were calculated from measurements of the loss tangent at 10^5 cps and 500°C. Values between 1.2 and 1.6 ev were calculated for all compositions except the one containing Mg^{2+} ions, for which 2.0 ev was obtained. At low frequencies, an exponential rise of the dielectric constant (k') with temperature was interpreted as reflecting a like rise in the number of free charge carriers contributing to interfacial polarization. At higher frequencies, the temperature coefficient was not affected by low concentrations of impurities, but could be compensated effectively without excessive loss by additions of 10 to 20% strontium titanate. None of the data showed clear evidence of dipole rotation loss mechanism.

Ioffe et al. (1964) investigated the electrical conductivity of cerium aluminate ($CeAlO_3$) and some solid solutions based on it in weak and strong electric fields. Some specimens having anomalous dielectric properties showed hysteresis loops in which the electrical conductivity varied exponentially with the field intensity. Reversible dielectric constant was found to be independent of the intensity of the displacing field.

Smoke et al. (1962) determined the dielectric constant and loss tangent of alumina in comparison with other ceramic materials over a range of frequencies from 100 to 12,000 Mcps and at 25 to 816°C. Basic formulas were derived for a range of low to high loss.

Variations in loss tangent with temperature of commercial alumina compositions are shown in Figure 14b.

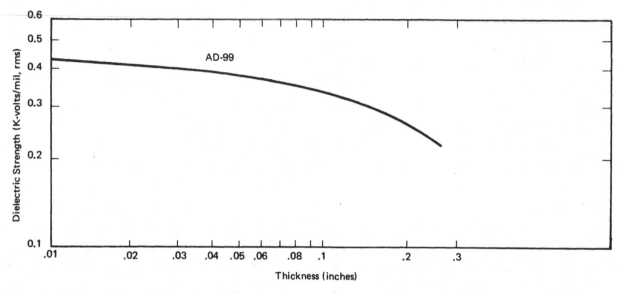

Figure 14-a

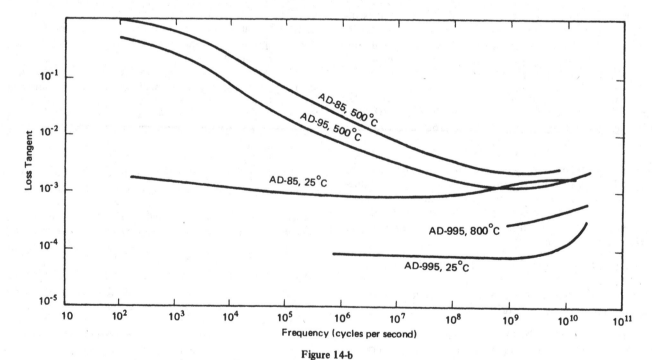

Figure 14-b

The composition number represents the nominal Al_2O_3 content. Courtesy Coors Porcelain Company.

Figure 14. Dielectric Properties of Alumina

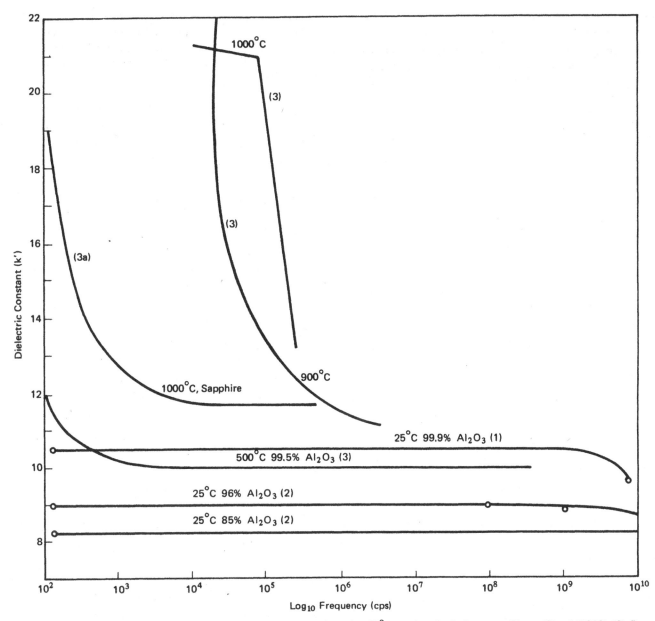

All specimens dense polycrystalline alumina except 3a, sapphire, oriented at 60° to optic axis. References: (1) von Hippel (1958); (2) Coors Porcelain; (3) Florio (1960).

Figure 15. Dielectric Constant of Alumina

Variations in dielectric constant with frequency and temperature are shown in Figure 15.

8-4 DIELECTRIC STRENGTH

Ueda and Okada obtained 2×10^6 v/cm (5000V/mil) for the dielectric strength of alumina at 50 cps. Lomer measured the dielectric strength of anodized films on aluminum at 15°C in vacuum. The value decreased from 11 megavolts/cm for films less than 1000 A thick to 6 megavolts/cm for films greater than 100 A. Kawamura and Azuma measured the breakdown of thin films 100 to 2000 A thick with a rectangular pulse of 30 microseconds.

Measurements at 198, 300, and 373°K gave breakdown voltages of 1.36×10^7, 1.62×10^7, and 1.85×10^7 v/cm, respectively, for the three temperatures.

Kainarskii et al. (1964) observed that both the dielectric strength and the mechanical strength are affected by the microstructure of the alumina. The microstructure is determined by the sintering, which in turn depends on the mode of preparation of the alumina. An increase in the area of individual crystals to above 0.02 mm² (140 microns diameter) was found to increase the breakdown voltage from 70 kv/cm (175 v/mil) to 170 kv/cm (425 v/mil). These values appear somewhat low, but unless careful precautions are observed, values in this range are likely to be obtained at room temperature.

Commercial manufacturers of high-alumina electrical insulators obtain dielectric strength values ranging from about 200 to 600 v/mil, depending on the purity of the alumina composition and the extent of sintering. Typical data are shown in Table 11 and Figure 14a. These values are much lower than are obtained from measurements on thin films. The determination is quite sensitive to the thickness of specimen tested.

9 MAGNETIC PROPERTIES OF ALUMINA

9-1 MAGNETIC SUSCEPTIBILITY

Magnetic susceptibility (X) is the ratio of the intensity of magnetization produced in a substance (M) to the magnetizing field intensity producing it (H). Magnetic permeability (μ) is the ratio of magnetic induction (B) to the field intensity (H), and $\mu = 1 + 4\pi X$. Substances having magnetic permeability less than that of air (about one) are said to be diamagnetic, in contrast to those having higher values, called paramagnetic. The pure alumina phases are diamagnetic. As shown in Table 12, the susceptibility is very small, less than 10^{-6}. Pascal (1924) investigated the magnetic susceptibility of the precipitated hydrous phases and their dehydrated products as a means of determining changes in the content of constitutional water. A definite correlation was found, confirmed by Zimens. The value changed from about -0.35×10^{-6} for the calcined product to -0.62×10^{-6} for aluminum hydroxide containing about 72% H_2O. The negative sign denotes the diamagnetic property. Selwood obtained substantially the same values. The method has been recommended for determining rates of hydration of various materials as for example, of cements.

Rao and Leela (1952, 1953) observed a slightly lower diamagnetic susceptibility for pure synthetic sapphire in the direction of orientation parallel to the optic axis than at a right angle to this axis. Native ruby, containing a maximum of about 0.27% Cr_2O_3, was also diamagnetic, and showed about the same disparity in effect of orientation. Synthetic ruby containing 0.82% Cr_2O_3, however, was strongly paramagnetic.

Cirilli (1950) precipitated a continuous series of mixtures of hydroxides of Al_2O_3 and Fe_2O_3 subsequently heat-treating at 600°C. The magnetic susceptibility increased to a maximum at a content of 90.4% Fe_2O_3, and then decreased abruptly for higher iron oxide contents. The γ-Fe_2O_3 alone changes to α-Fe_2O_3 at 300°C, but if it contains Al_2O_3 in intimate mixture, it changes at 550°C.

Gorter (1954) investigated the magnetization of various mixed crystal oxides, including $NiFe_2O_4$-$NiFeAlO_4$. Selwood (1956) presented information on the magnetic

Table 12

Magnetic Properties of Alumina

	Pascal	Zimens	Selwood	Rao & Leela
Magnetic Susceptibility ($\times 10^{-6}$)				
α-Al_2O_3 (polycrystalline)	0.35	−0.30	−0.23	
Synthetic sapphire $\perp$				−0.25
Synthetic sapphire $\parallel$				−0.21
Synthetic ruby $\perp$				+0.38
Synthetic ruby $\parallel$				+0.41
Natural ruby $\perp$				−0.20
Natural ruby $\parallel$				−0.15
γ-Al_2O_3		−0.28	−0.34	
α-$Al_2O_3 \cdot H_2O$			−0.37	
α-$Al_2O_3 \cdot 3H_2O$	−0.50		−0.43	
$Al(OH)_3$ 20% H_2O	−0.42			
$Al(OH)_3$ 71.6% H_2O	−0.62			

properties of solid solutions of Cr_2O_3 and Fe_2O_3 in alumina. Shirikov and Kirillov investigated the magnetic properties of mixtures of Al_2O_3 and NiO prepared either by repeated impregnation of gamma alumina with $Ni(NO_3)_2$ solution to different concentrations, or by grinding together NiO and Al_2O_3 pastes in HNO_3. Both series were calcined at 300, 500, and 700°C. In both cases, the susceptibility was maximum at 6 to 15 mole % NiO, and it fell off continuously up to 65 mole %.

9-2 MAGNETIC RESONANCE OF ALUMINA

Nuclear magnetic resonance (NMR) and electron paramagnetic resonance (EPR) have been investigated by Varian Associates Instrument Division (1961) and others. In nuclear magnetic resonance spectroscopy, isotopes of the elements can be identified by their nuclear gyromagnetic constants. Isotopes whose spins do not equal zero have a magnetic moment. The ratio of the value of magnetic moment to spin constitutes the gyromagnetic ratio. In application, the specimen, immersed within a magnetic field of variable field strength, is excited by a coil in the circuit of a small radio-frequency transmitter. A second coil in a sensitive receiving circuit, having substantially zero coupling with the transmitter, receives any induced signal voltage in the specimen, and detects it on a voltmeter, an oscillograph, or a graphic recorder. A suitable system amounts to a high-resolution, crystal-controlled, fixed-frequency spectrometer.

Saito (1942, 1960) determined the NMR of bayerite and gibbsite and their dehydration processes through change in the shape of the magnetic resonance line. Both were the same. The second moment for bayerite is 13 to 14 gauss², and is independent of the dehydration temperature to 220°C. It decreases gradually to 4 to 5 gauss² at 400°C, remains constant to 550°C, and then gradually decreases. This agrees with the sequential phase transformations of bayerite through boehmite and eta alumina.

Osiowski thought that NMR should be particularly useful in identifying the different forms in which water can be found in cements, and for following the kinetics of cement hydration.

Pecherskaya, Kazanskii, and Voevodskii applied electron spin resonance (ESR) to alumina catalysts containing 0.3 to 11.4% (wt) Cr_2O_3. Samples of low chromium content show a second narrow line of 30 to 40 gauss width, which was attributed to a dilute 2-dimensional solution of Cr^{3+} on the surface of the alumina. Co-precipitated catalysts do not exhibit the line. The effects were similar for $SiO_2-Cr_2O_3$ catalysts. Hall, Leftin et al. investigated the deuterium exchange of several catalyst types on a rising temperature by NMR. The deuterium was intended to distinguish the different kinds of hydrogen held by the solids. Their data showed that the temperature region (the activation energy) for exchange increases in the order: Al_2O_3, $Al_2O_3-SiO_2$, SiO_2. The magnetic resonance absorption spectra from $SiO_2-Al_2O_3$ were indistinguishable from SiO_2, qualitatively. There was no evidence to show that any portion of the hydrogen of dehydrated $SiO_2-Al_2O_3$ is acidic, but

possibly 20% of the total hydrogen was acidic AlOH. O'Reilly and Poole interpreted changes in intensity of the signals with metal concentration of adsorbed oxides of chromium, nickel, and cobalt in high-area alumina in terms of changes in the ^{27}Al spin-lattice relaxation times induced by the adsorbed oxides. Clusters of these metal ions on the surface of the alumina were found to be effective in removing the ^{27}Al nuclei near the surface.

Poole and Itzel (1964) investigated the ESR at 125 to 550°K of chromia-alumina mixtures that had been calcined between 500 and 1400°C. The chromic ions present in compositions of low chromia content were paramagnetic; at high-chromia contents the spectra exhibited antiferromagnetic behavior. This is a system analogous to the ferromagnetic state except that neighboring spins are antiparallel instead of parallel; that is, the substance exhibits paramagnetism (a low positive susceptibility) that varies with temperature in a manner similar to ferromagnetism, and passes through a Curie point. Sharp Néel points (paramagnetic to antiferromagnetic state) were obtained for Cr/Al ratios of 2/3, calcined at 1400°C. Samples with Cr/Al ratios of 1/2 gave broad Néel points, which became less well resolved with decreasing chromia content. The spectra of high Cr/Al ratio samples calcined at 900°C or lower consisted of superpositions of a paramagnetic resonance (β_w) and a fairly sharp Néel point (β_n). M. G. Townsend examined the ESR of pink single crystals of alpha alumina doped with 0.02% cobalt, grown by cooling from a $PbO-PbF_2$ melt. Measurements at 4.2°K with a spectrometer operating at 1 cm wavelength, and the use of phase-sensitive detection with 100 kcps modulation, indicated the presence of Co^{2+} in only one substitutional site. This is contrary to the measurements of Zverev and Prokhorov, which showed Co^{2+} in two different sites, those normally occupied by Al^{3+} and vacant octahedral sites present in the lattice. The latter deduced that for every two Co^{2+} substituting for Al^{3+}, a third Co^{2+} was in an interstitial site. Pink crystals with localized green areas were grown by using an alumina melt doped with 0.02% cobalt and 0.02% magnesium as oxides. Their optical spectrum was characteristic of both Co^{3+} and Co^{2+}, indicating that magnesia charge-compensates for Co^{3+}.

Krebs applied ESR to crystalline alpha alumina containing Fe^{3+} and Mn^{2+}. The broadening of the ESR lines was interpreted as a splitting of an unresolved line which is linear in the applied field. A change in the spin Hamiltonian D parameter with the field applied parallel to the c-axis $(\partial D/\partial E)$ was 1.00 ± 0.09 for Fe^{3+} and 1.03 ± 0.09 for Mn^{2+} in units of 10^{-5} gauss-cm/v. This effect implies a substitutional replacement of Al^{3+} ions in the lattice. Dixon and Bloembergen observed electrical perturbations of the NMR of ^{27}Al in single-crystal sapphire as splittings of the quadrupole satellite resonance lines in fields as high as 300 kv/cm. The splittings were ascribed to symmetrically opposed changes in the electric-field gradient tensor at crystallographic Al sites which were related to each other by inversion symmetry. By applying the magnetic field in various crystallographic orientations, the magnitudes and relative signs of five independent R-tensor elements, which completely describe the shifts, were determined.

Pace, Sampson, and Thorp measured the spin-lattice relaxation time in sapphire in the range 1.4 to 56°K at frequencies as high as 34.6 Gcps. The relaxation times were as much as five milliseconds in this range, and were found to be temperature-dependent, as was also the case for chromia-doped specimens. The relaxation times for ruby were independent of frequency.

10 OPTICAL PROPERTIES OF ALUMINA

10-1 REFRACTIVE INDEX OF ALUMINA

The pure alumina phases, in general, appear as colorless or white powders. Except for artificial corundum, none is commercially available in sufficient single-crystal size to be used as components of optical instruments. Refractive index is used as a means of identification of the alumina phases, since it can be applied to crystals of microscopic size by observation of the behavior of the Becke line when the powder is immersed in a liquid of suitable refractive index and dispersion.

Some of the hydrated and calcined alumina phases have good opacifying or hiding properties as pigments, because of their higher than average refractive indices. Other factors, which are characteristic of all powders, are also significant in affecting the pigment properties, however. Maximum reflectance is generally attained at about 0.4 to 0.7 micron particle size, maximum color brilliance with particles of about 5 microns. The alumina trihydrates, at refractive indices of 1.57 to 1.58, have no pronounced opacifying action in the usual suspension media for pigments.

The refractive indices of the different alumina phases at room temperature and for the nominal wavelength 5893 A are shown in Table 7. Some optical properties of general interest are shown in Table 13.

Synthetic corundum is now available in sizes of two inches or greater for use in windows, lenses, and other optical equipment in which transparence and moderately high refractive index are significant, and for uses at elevated temperatures, or where abrasive or corrosive action may be involved.

Alpha aluminum oxide is anisotropic with variable properties in different crystal directions. It belongs to the rhombohedral (trigonal) division of the hexagonal system. It is convenient to use the hexagonal set of axes: c, a_1, a_2, and a_3, in describing crystal directions. Corundum is uniaxial, indicating that the c-axis is the only direction that does not exhibit double refraction of light. The crystal is negative, the convention indicating that the ordinary index (n_o) is larger than the extraordinary index (n_e).

Kebler developed a Hartmann-type formula for calculating the refractive index of pure sapphire at 25°C for wavelengths within the range from the C-line (6563 A) to the Hg-line (3660 A). The equation: $n_o = 1.74453 + 101.0/(\lambda - 1598)$, in which λ is measured in angstroms, fits the data within the experimental error (about 0.00003). Malitson (1962) measured the refractive index of sapphire at selected wavelengths, and found that it varies from 1.8336 at 0.2652 micron to 1.5864 at 5.577 microns for the ordinary ray, corrected to 24°C. In the visual region, the thermal coefficient of the index is about $13 \times 10^{-6}/°C$, and it decreases with increasing wavelength. A three-term Sellmeier dispersion equation was fitted to the experimental data, as shown in Table 13. The dispersive coefficient, $dn/d\lambda$, was computed at regular wavelength intervals, from which the relative "dispersive power," $dn/d\lambda/(1-n)$, was determined. The relative dispersion decreased from a value of about 0.8 at 0.25 micron to about 0.02 near 1.3 microns, and then increased again at higher wavelengths. A plot of the quantity, $(\lambda dn/d\lambda)^{-1}$, against wavelength indicated that optimum resolution is obtained in the wavelength regions below about 0.4 micron and longer than about 2.5 microns.

Loewenstein obtained $n_o = 3.4 \pm 4\%$ and $n_e = 3.61 \pm 4\%$, in the infrared region, 170 to 500 microns. The extrapolation of Malitson's equation as the wavelength approaches infinity, n_∞, equals 8.4, the approximate dielectric constant.

Jaffe (1956) stated that the empirical rule of Gladstone and Dale: $(n-1)/d = k$, the specific refraction, holds very well. The formula of Lorenz and Lorentz is claimed to be superior. The refractive index and dispersion of light by a dielectric material is explained in terms of electric dipole movements. The dielectric material contains charge carriers that can be displaced to retard light waves. Dielectric polarization is caused by a shift in the electron cloud surrounding the atomic nucleus. In an electric field the charge displacement neutralizes part of the field. The remainder, the bound charge, is neutralized by polarization of the dielectric. The interactions of the dielectric and electromagnetic radiations cause the index of refraction to increase with decreasing wavelength in the visual range (normal dispersion), but to decrease in the region of natural frequency where resonance occurs (anomalous dispersion). The application of a stress (tensile) increases the index of alumina perpendicular to the direction of the stress, and

Table 13

Optical Properties of Alumina[1]

Refractive Index

Sapphire 24°C

$$n_o^2 - 1 = \frac{1.023798\,\lambda}{\lambda^2 - 0.00377588} + \frac{1.058264\,\lambda^2}{\lambda^2 - 0.0122544} + \frac{5.280792\,\lambda^2}{\lambda^2 - 321.3616}$$

Malitson, 1962

λ	n_o	λ	n_o	λ	n_o	λ	n_o
0.270	1.83047	0.570	1.76925	0.870	1.75845	2.600	1.72368
0.290	1.81915	0.590	1.76810	0.890	1.75800	2.800	1.71818
0.310	1.81026	0.610	1.76705	0.910	1.75756	3.000	1.71224
0.330	1.80312	0.630	1.76609	0.930	1.75713	3.200	1.70584
0.350	1.79729	0.650	1.76521	0.950	1.75672	3.400	1.69896
0.370	1.79245	0.670	1.76439	0.970	1.75632	3.600	1.69158
0.390	1.78840	0.690	1.76362	0.990	1.75592	3.800	1.68368
0.410	1.78495	0.710	1.76291	1.000	1.75554	4.000	1.67524
0.430	1.78199	0.730	1.76224	1.200	1.75216	4.200	1.66623
0.450	1.77942	0.750	1.76162	1.400	1.74880	4.400	1.65662
0.470	1.77719	0.770	1.76102	1.600	1.74535	4.600	1.64640
0.490	1.77522	0.790	1.76046	1.800	1.74169	4.800	1.63553
0.510	1.77347	0.810	1.75992	2.000	1.73773	5.000	1.62397
0.530	1.77191	0.830	1.75941	2.200	1.73344	5.200	1.61168
0.550	1.77051	0.850	1.75892	2.400	1.72876	5.400	1.59864
						5.600	1.58479

Temperature coefficient of refractive index

0.4 micron	$20 \times 10^{-6}/°C$
0.4 to 0.7 micron	$13 \times 10^{-6}/°C$
4 microns	$10 \times 10^{-6}/°C$

Infrared Spectra of Aluminas[2]

Phase	OH-Stretch	OH-Bend	Unassigned	Al-O Stretch	Reference
Alpha Trihydrate	2.765(m), 2.842(s), 2.917(vs), 2.960(m), 2.975(s)	9.85(vs), 10.37(m)	11.0(w), 12.1*, 12.58(s)	13.48(s)	Frederickson Hannah
Beta Trihydrate	2.830(m), 2.842(m), 2.895(m), 2.940(m)	9.85(s), 10.25(s)	11.6(m), 12.3(m)*, 13.9(m)*	12.88(s)	Frederickson Hannah
Alpha Monohydrate	3.065(s), 3.247(s)	8.75(m), 9.35(s)		13.5(s)	Frederickson Hannah
Beta Monohydrate	3.42(vs), 4.27(m), 4.73(m), 5.04(m)	9.35(s), 10.42(s)		13.89(vs)	Cabannes-Ott
Alpha Alumina			15.58(m), 16.61(s), 17.89(m), 20.41(w), 22.21(s)	13.16(vs), 12.84-13.89	Hannah Kolesova
Alumina "Gel"		8.5, 8.3, 9.8	6.9, 7.7, 10.7, 12.3	13.5	Imelik (1954)
Anodic Oxide	4.15	8.4-8.7	10.2-10.6		Fichter

m = Medium intensity   vs = Very strong   w = Weak   s = Strong   * = Shoulder

[1]Wavelength, microns.
[2]Taken from Alcoa Technical Paper No. 10.

decreases it parallel to the stress direction. The birefringence produced can be applied to measure the stress. Davis and Vedam measured the change in refractive index of sapphire due to strain (n_s) to 7 kbars at 22°C, and found the value to decrease quite linearly with increasing pressure with slopes of $(1.0 \pm 0.2) \times 10^{-4}$/kbar and $(1.1 \pm 0.2) \times 10^{-4}$/kbar for the ordinary and extraordinary rays, respectively. A shift of about 20 fringes was observed with each crystal. Corundum is relatively incompressible, about 9% at 300 kbars, but exhibits an increase in the rhombohedral angle of about 0.5° for this pressure change (Hart and Drickhamer).

The reciprocal dispersion of sapphire, $(\nu) = (n_d - 1/(n_f - n_c))$, has a value of 72.2 (Kebler). This is high compared with available optical glasses, and would be important in the fabrication of photographic lens elements. Strain effects in flame-fusion sapphire and hardness cause difficulties in this application.

10-2 TRANSMISSION, EMISSIVITY, AND ABSORPTION OF ALUMINA

The light transmission of sapphire is only slightly poorer than that of fluorite, lithium fluoride, and rock salt, but none of these is satisfactory at temperatures above 600 to 800°C because of surface-clouding effects. Klevens and Platt; Bauple, Gilles, Romand, and Vodar; Gaunt; Gilles; and Kebler have contributed to the literature. A sharp cutoff in transmission occurs in the Schumann region of the ultraviolet at about 0.170 to 0.145 micron, with an absorption peak at 0.184 micron. The transmission increases to about 84% at 0.30 micron, and smoothly rises to 93% at 6 microns, thereafter falling rapidly to less than 20% at 7 microns. The loss in transmission in the visual range is mainly Fresnel-type reflection. The loss to this source, determined from the formula: $R = [(n-1)/(n+1)]^2$ for normal incidence of the radiation, is about 8%. Gilles showed that the absorption at the shorter wavelengths increases with increasing temperature.

Lemonnier et al. presented reflectance data in the form of graphs for both sapphire and ruby single crystals in the region from 0.02 to 0.3 micron. The reflectance for sapphire was about 20% with a cut-off at about 0.03 micron. Ruby had a maximum of about 15% at 0.137 micron, which decreased to about 5% at 0.05 micron. Piriou derived the optical constants of corundum from reflectance values in the region 0.67 to 3.3 microns. Schatz ascribed the large changes observed for reflectance of sintered alumina and other oxides sintered at temperatures below 1550°C to loss of trace-water, decrease in surface roughness (increased reflectance), and specimen shrinkage (decreased reflectance). Surface roughness and porosity have been thought to be the cause of large discrepancies in reported emittance values of alumina (Pattison), but surface roughness variations from 20μ in. to 200μ in., and porosity variations from 9.7% to 30.0% were found to affect the emittance only slightly. Porosity in the range 0 to 10% was probably significant, however (Gannon and Linder). J. M.

Adams derived the absorption constant of molten alumina from emittance measurements on small flames at visible wavelengths.

The radiant energy emitted per unit area of a source per unit time is called the radiant emittance (W). The total emissivity of an incomplete radiator is defined as the ratio of its radiant emittance to that of a complete radiator ("black body"). Total emissivity is expressed by the equation: $e_t = Q/\sigma A(T_1^4 - T_2^4)$, in which Q is the heat-flow rate, A the area of radiating surface, T_1 and T_2 the absolute temperatures of radiating surface and external environment, respectively, and σ the Stefan-Boltzmann constant, having a value of about 5.6697×10^{-5} erg cm^{-2} deg^{-4} sec^{-1}. Spectral emissivity, e_λ, for a narrow band of wavelength of radiation is defined by the equation: $\ln e_\lambda = C_2/\lambda(1/T - 1/T')$, in which C_2 is Planck's second radiation constant (1.43879 cm deg), and T and T' are the true and brightness temperatures, respectively. (Allen, 1961).

Early investigators determined the emissivities of polycrystalline powders and massive structures used as thermal insulators. Heilman; Taylor and Edwards; Kilham; and Michaud, in general, have shown that the total emissivity of alumina decreases from about 98% at room temperature to 80% at 400°C, 50% at 750°C, 30% at 1000°C, and 18% at 1600°C. Skaupy and Hoppe observed no absorption in pure sapphire below 1230°C. Polycrystalline alumina had substantial emissivity at elevated temperatures, especially in the blue wavelengths, which was ascribed to emission centers located at crystal boundaries. Skaupy and Liebmann established a relation between the emissivity and decreasing crystal size, and found that the emissivity increased to a critical size, 1 to 2 microns. Two absorption bands were found in the infrared at 1.0 and 5 microns wavelength. At elevated temperatures, the total emissivity was about 10% from which it was concluded that alumina is substantially a "white body." Ruby was found to be very transparent at room temperature, but was almost a black body at high temperatures, having an emissivity of 80% at 1100°C. Michaud noted that the monochromatic emissivity at 0.655 micron was 15% from 1000 to 1600°C.

Kingery and Norton (1955) found that the transmissivity (corrected for Fresnel reflection) of sapphire was independent of temperature to beyond 3 microns wavelength within the temperature range from 30 to 1200°C. The transmissivity was about 99% at 2 microns for a 2 mm thick specimen, dropping gradually to about 97% at 3.6 microns, beyond which point the cutoff was rapid and temperature-dependent. Somewhat similar results were obtained at Alfred University (1957) for the same temperature range, and about the same specimen thickness, but with different test equipment. The values ranged from about 82 to 85% emissivity at 1.0 micron to 84 to 86% at 3.6 microns, with a rapid cutoff, commencing first at the higher temperatures. Gaunt (1951) obtained emissivities of about 75% on sapphire 3 mm thick, substantially unchanged within the infrared region from 1.0 to 4.4 microns.

Lowenstein (1961) found that sapphire is highly transparent in the region 100 to 250 microns, with the transmission dropping off to zero near 110 microns.

Olson and Morris determined room-temperature emissivities on prospective commercial aircraft structural coatings: Rokide A (Norton) on molybdenum and on No. 446 stainless steel, and Norton LA-603 and RA-4213 oxides. Emissivity decreased sharply in the visual wavelengths for all specimens to less than 30% at 0.6 micron; beyond 0.6 micron the coatings on metal increased in two levels to 30% at 1.0 micron, and about 40% at 1.8 microns. The oxides dropped to even lower emissivities, less than 14% at 0.8 micron, and exhibiting a minimum at 1.4 microns.

Sully, Brandes and Waterhouse; Pattison; and Robijn and Angenot also investigated the total emissivity of the pure oxide at elevated temperatures. The total emissivity of pure dense polycrystalline alumina decreased from about 50% at 325°C to about 25% at 875°C, but increased thereafter to 30% at about 1350°C (Sully; Kingery and Norton 1955). The oxides, RA-4213 and LA-603, also showed increasing emissivity above 875°C (Pattison). The coatings of Rokide on metal also showed decreasing emissivity with increasing temperature in the range from 175 to about 1740°C.

Wade (1959) measured the total hemispheric normal emissivities of refractory oxide coatings of alumina and zirconia. The emissivity values ranged from 0.69 to 0.44 for the alumina coatings. Flame-sprayed coatings showed higher values than those applied by other methods.

10-3 PHOSPHORESCENCE, FLUORESCENCE, AND THERMOLUMINESCENCE

Pure alumina is neither phosphorescent nor fluorescent (de Ment, Kröger). Less than 0.001% of chromic oxide produced a decided fluorescence (Kröger). Thosar ascribed the bright red fluorescence of ruby, stimulated by short wavelengths of light, to Cr^{3+} replacing Al^{3+} in the lattice of alpha aluminum oxide.

Thermoluminescence, induced by irradiation with gamma rays, depends on the amount of hydration and the crystal structure. Rieke and Daniels found that the intensity increased with calcination temperature, and was the highest for synthetic sapphire. The glow curve was resolved as peaks, each of which corresponds to a separate electron-trapping level. The principal peak for tabular alumina (Alcoa T-61) fell at 164°C, with lesser peaks at 103 and 310°C. A peak at 236° was ascribed to sodium impurity. The intensity was quenched quickly in water, which suggests that the trapping centers are surface sites, and that the water is bound chemically.

Atlas and Firestone applied this method to an investigation of lattice defects in alumina ceramics, as influenced by impurities. The energy depth of each trap was estimated from the glow-peak temperature and the temperature at which the light intensity reached one-half its maximum value. Glow curves were obtained for alumina disks

constaining less than 100 ppm of impurities, prepared by dissolving 99.999% pure aluminum metal in HCl, precipitating with NH_4OH, and calcining at 1300°C. the high-purity product was doped during milling (in plastic jars with polystyrene balls), was dry-pressed and fired at 1850 to 1960°C in air. The effects of low concentrations of the oxides of silicon, titanium, iron, magnesium, and calcium on the development of peaks were determined, where possible. The impurity ions were found to contribute not only by being instrumental in creating anion and cation vacancies, but also by acting directly as trapping sites for electrons or by capture of positive holes. Trapped electrons appear to be the principal agency for thermoluminescence.

Vishnevskii et al. investigated the X-ray and thermoluminescence of pure and doped single crystals of corundum. Kelly and Laubitz examined γ-irradiated, pure, and MgO-doped alumina ceramics. Goldschmidt, Low, and Foguel determined the spectrum of corundum doped with trivalent vanadium.

Gabrysh et al. concluded that the glow curves from gamma-ray-induced thermoluminescence of sapphire have the form: $I = I_0 [b/(b + t)]^m$, which is a second order decay process. In ruby, the principal emission bands have maxima at about 150 and 240°K. Ruby, damaged by gamma-ray irradiation, shows an afterglow long after the crystal has reached room temperature.

Przibram (1960) drew attention to the blue fluorescence shown by purified alumina and other minerals excited by 0.365-micron light. The effect is caused by traces of organic matter and is destroyed on heating for a few minutes to redness.

The emission of electrons (exoelectron emission) from metallic surfaces as an after effect of mechanical working or glow discharge has been investigated by Kramer (1949); Haxel, Houtermans, and Seeger; and Seeger, among others. The phenomenon has been interpreted on the basis of removal of a chemisorbed layer of oxygen by the abrasion or discharge, followed by reformation of the adsorbed oxygen layer. The changes in the freshly formed surfaces can be correlated to the chemical nature, the adsorption properties, and the catalytic activities of the solid. Gouge and Hanle investigated the exoelectron emission from synthetic corundum and ruby. Greenberg and Wright found a very strong exoelectron emission from abraded aluminum irradiated at 0.470 micron. They suggested that metal-excess oxides may be present in the oxide oxygen-ion vacancies occupied by 2 electrons, called F' centers. The oxide film on a mechanically polished metal surface would probably contain large numbers of F' centers, and this might be expected to increase the dielectric constant of the film in the region of 0.470 micron. Ramsey measured the optical constants by the method of Drude in the visible region and found a considerable increase in the absorption index, k, and a marked lowering in the refractive index in the vicinity of 0.470 micron. Petrescu measured the intensity of exoelectron emission from both scratched metal and crystalline aluminum oxide (which does not require mechanical activation for exoelectron emission). He concluded that crystalline alumina layers of about one micron in thickness completely mask the emission of the metal below, and that there may be several forms of emission that do not verify the hypothesis of the identity of emission centers.

Irradiation of sapphire with light from a low-pressure mercury lamp caused a faint phosphorescence detected by a photomultiplier with S-11 spectral response, for times exceeding one hour. The luminous decay obeyed the expression: $I = k/t^n$, with $n = 1.664$ for t up to 4 minutes, and $n = 0.445$ thereafter (Coop and Hammond).

Alumina and its salts have entered into phosphor compositions, as for example, a mixture of alumina, lithia, and iron oxide, excited by ultraviolet radiation at 0.2537 micron, and emitting at 0.6800 micron with one-half the peak intensity at 0.6520 and 0.7320 micron (S. Jones, 1949). Hatch (1943) claimed that zeta alumina is the phase that fluoresces in this combination, because neither corundum nor $Li_2O \cdot Al_2O_3$ fluoresce when activated with Fe_2O_3. Froelich and Margolis produced a phosphor consisting of aluminum phosphate, activated with 12 to 25% Ce_2O_3 and 0.2 to 15% ThO_2.

Trofimov and Tolkachov followed the phase changes during dehydration of hydrated alumina by the phosphorescence spectra with about 1% (mole) $Eu(NO_3)_3$ present as an activator. Band spectra were observed for calcinations below 900°C, and line spectra appeared above 900°C, with wavelengths in the visible spectrum of 0.594, 0.600, 0.620, 0.628, and 0.650 micron, and in the infrared of 0.709, 0.716, and 0.718 micron.

10-4 OPTICAL SPECTRA OF ALUMINA

According to Meggers (1941) each chemical element is characterized by Z totally different atomic spectra, in which Z is the atomic number, 13 for aluminum. The wavelengths of optical spectra are measured in angstroms (10^{-10} meter), millimicrons (10^{-9} meter), or microns (10^{-6} meter). In the investigation of series spectra, the wavelengths are often converted to wavenumbers or the number of waves per centimeter (cm^{-1}). The first atomic spectrum of aluminum, designated Al I, is characteristic of neutral atoms; the second is ascribed to atoms that have lost one outer electron, and so forth. The total number and types of spectra are fixed by the electron orbits; two (1s) in the innermost shell, two (2s) and six (2p) in the intermediate shell, and two (3s) and one (3p) in the outermost shell. A classification of the different spectra in wavenumbers, including the ground levels (cm^{-1}) and ionization potentials (ev), is given by C. E. Moore (1949). The ionization potentials range from 5.984 ev for Al I at a ground level of 48,279.16 cm^{-1} to 2085.46 ev for al XII at a ground level of 16,825,000 cm^{-1}. The binding energy of an electron of given type in any state of a neutral or ionized atom may be defined as that energy required to remove it along successive terms of a spectral series to its limit. The data permit the calculation of the binding energy (Moore and Russell).

Emission spectroscopy in the visual and ultraviolet range has been applied by many investigators in the qualitative and quantitative analysis of ceramic materials, for example, by the modified excitation method proposed by Ryan and Ruh (1964). The accuracy of such methods has been shown to be sufficient for general analysis in many cases, and the sensitivity (ppm) has been adequate, except perhaps for a few applications, as in semiconductors, involving impurity levels in parts per billion.

Shreve (1952) discussed the application of infrared, ultraviolet, and Raman spectroscopy to the analysis of complex materials. Hunt et al. (1950) described a method for utilizing the infrared spectrometer in the analysis of mineral and ceramic materials. Miller and Wilkins presented a table of infrared spectra and the characteristic frequencies of 33 polyatomic ions. They found that the chief limitation of infrared analysis is the practical necessity of working with powders, which makes it difficult to place the data on a quantitative basis.

Imelik, Petitjean, and Prettre (1954) obtained infrared spectra for aluminum hydroxide gels. Frederickson applied infrared methods to the examination of bauxite ores, and showed that the four distinct alumina hydrate phases: boehmite, gibbsite, bayerite, and diaspore possess hydroxyl groups, of sufficiently different absorption patterns in the O-H stretching and deformation regions, to permit their identification. The infrared O-H patterns observed with the pure hydrates were used to identify the phase composition in a variety of bauxitic ores and diaspore clays. Petitjean (1955) investigated the absorption of gibbsite during its thermal dehydration, and Kolesova and Ryskin the absorption of gibbsite and its deuterium analog, $Al(OD)_3$ in the spectral region 2.7 to 23.8 microns. The bands at 9.9, 10.5, and 11.0 microns gradually lost intensity during dehydration and disappeared below the structure $Al_2O_3 \cdot H_2O$. The 9.35 band of boehmite showed first at $Al_2O_3 \cdot 2.5\ H_2O$, and the 8.7 band at Al_2O_3-H_2O. Upon heating to the nearly anhydrous condition (500°C), the spectrum became more transparent around 8 microns and more absorptive beyond 11 microns. The spectrum retained the characteristics of boehmite in diminishing intensity from $Al_2O_3 \cdot 0.8\ H_2O$ to $Al_2O_3 \cdot 0.2\ H_2O$, but at $Al_2O_3 \cdot 0.10\ H_2O$ the vestiges of OH bands had disappeared (Orsini and Petitjean). Kolesova and Ryskin concluded that the Al-O bond was partially covalent. The Al-O stretching vibrations for diaspore and boehmite, and the Ga-O bond in GaO·OH correspond to bands at 13.16, 13.33 to 12.82, and 15.62 microns, respectively. Additional references to the infrared absorption of diaspore include: Lecompte; Nakamoto, Margoshes, and Rundle; and Cabannes-Ott. Rundle and Parasol related the wavelength of the OH absorption to the OH-O distance, which is the length of the hydrogen bond.

Fichter measured an OH-stretch, OH-bend, and two unassigned bands for anodic Al_2O_3. L. Harris (1955) determined the infrared transmission of anodic films as thin as 250 A. The films were prepared by anodizing aluminum foil in a 3% aqueous solution of ammonium citrate at 17 to 200 volts. One anodized layer was removed by NaOH, and the metal by HCl, leaving the free aluminum oxide film. Even the thinnest films showed marked absorption beyond 11 microns. The index of reflection of all infrared wavelengths decreased with increasing film thickness. The index varied with wavelength, and had a minimum at 9 microns. The absorption coefficient did not vary with thickness but increased to a maximum at 15 microns.

Kolesova (1959) observed in the infrared spectra of alumina, aluminum silicates, and aluminates in which the atoms of aluminum are in the anion shell, a band in the region 12.84 to 13.89, which was attributed to Al-O stretch bonds.

Boreskov et al. (1964) investigated the infrared behavior of ethanol adsorbed on gamma alumina. When the ethanol was adsorbed at 20°C, bands were observed at about 9.43, 9.27, 8.93, and 7.52 microns. The first of these is eliminated by pumping at 170°C, the second and third only at 350°C. When the ethanol was adsorbed at 20°C on theta alumina or on gamma alumina, pretreated with HCl (which blocks the tetrahedral aluminum atoms), and was then pumped at 170°C, bands were observed at about 9.18 and 9.27 microns, respectively, but none at 8.93 microns. The band at 9.43 was ascribed to physically adsorbed molecules, that at 9.27 microns to AlOEt structures, octahedrally coordinated, that at 8.93 microns to AlOEt, tetrahedrally coordinated, and that at 7.52 microns to the CH_3 radicle.

Tanabe and Sugano, and Neuhaus and Richartz examined the absorption spectra of natural ruby and chromia-doped single crystals and alumina powders. Poole and Itzel, and Loh determined the reflection spectra. The Hartmann-Bethe theory of chromogenous complex ions was applied to crystals containing Cr^{3+}. The chromium enters into octahedral or quasioctahedral coordinated lattice positions. The change in color from red to violet to green in such crystals was explained as a contraction or expansion of the (CrX_6) complex, resulting from variable polarization by different adjacent central ions. Spinel and alpha alumina were successively expanded by exchange of Ga for Al, but maintaining the chromium content constant. The room-temperature ultraviolet reflectance spectrum of flux-grown ruby (about 4% Cr_2O_3) at 6 to 14 ev and at an angle of 45 degrees showed an exciton-like peak at 9.1 ev. Alumina has a spectrum very similar to that of magnesia.

Mandarino observed a strong peak in the absorption of synthetic ruby at 0.55 micron, and also measured the changes in the indices of refraction with compressive strength, applied parallel to the ç-axis. Bates and Gibbs measured the absorption coefficient of sapphire doped with various impurities in the region 0.21 to 15 microns. Color centers were found in some crystals. They could also be induced by X-rays and by heating in oxygen above 1300°C.

Krishnan determined the Raman frequencies of sapphire and the luminescence and absorption spectra of ruby. The values found for sapphire and their relative intensities are: 26.66 (8), 23.97 (10), 23.15 (4), 22.25 (2), 17.30 (3), 15.60 (6), and 13.34 (7) microns. Bands appearing in the luminescence and absorption spectra of ruby arise from the combination of the vibration frequencies of the alumina lattice with the electronic transitions of Cr^{3+} ions. The Raman spectrum was confirmed by C. V. Raman in 1951. The seven expected intensities were found for sapphire when using an argon laser source (Porto and Krishnam).

The band spectrum for the molecule AlO has been identified (Pearse). Hebert and Tyte measured the relative integrated band intensities of 29 bands in the blue-green region of AlO, the $A^2\Sigma-X^2\Sigma$ system of aluminum oxide. The bands were excited in a low-pressure arc. K. Innes (1958) investigated the conditions under which diatomic aluminum as well as the oxides, AlO, Al_2O, and AlO^+ exist and exhibit electronic spectra. Of this group, only AlO was observed previously. Becart and Mahieu used a Schuler lamp to measure the band spectra in the visible system of the AlO molecule, $X^2\Sigma^+\rightarrow^2\Sigma^+$. A 4-mm thick aluminum cathode was used in oxygen at 2.4 mm pressure. The lamp operated at 25 pulses/sec. at about 1800 volts and 90 milliamperes. After 2 hours exposure, 16 new band tips were identified, belonging either to the principal sequence $\Delta V = 0$, or to adjacent sequences. The intensity of the head of band 6-6 is affected by the presence of the aluminum line 4962.1 A. The visible limit of the spectrum of the sequence $\Delta V = -1$ is near 19,046 cm^{-1} (0.52504 micron).

Becart (1962) developed a formula connecting the frequency of the zero line of a band spectrum with the frequency of the band head, which enabled the determination of the constants of other levels. Applied to the ultraviolet bands of AlO by using the constants derived from the visible bands, the calculated dissociation energy was 5.04 ev, in comparison with 5.2 ev, determined experimentally by DeMaria et al.

Rautenberg and Johnson investigated the excitation mechanism of the continuum and the AlO spectra. The color temperature and the intensity of the continuum were limited by the boiling point of alumina. As applied to photoflash lamps, the light emission results from blackbody radiation of AlO and Al_2O_3 at temperatures of 3800°K or less.

F. Coheur and P. Coheur used the molecule AlO as an example of a spectral method for determining temperature. It was shown theoretically and experimentally that the spectral method does not necessarily measure the total spherical temperature of the source, but only the localized temperature of certain zones from which the bands of the molecules selected for measurement are emitted. AlO, burning in oxygen, had a temperature of 4225 ± 200°C; in air 4210 ± 300°C; and in a vacuum 4275 ± 300°C. When AlO was replaced by Al_2O_3, $AlCl_3$, or $Al_2(SO_4)_3$, the respective temperatures were 4175 ± 225°C, 3877 ± 250°C, and 3950 ± 300°C.

Mal'tsev and Shevel'kov (1964) have investigated the infrared absorption spectrum of Al_2O at 1500 to 1840°C, and from 5 to 435 microns wavelength. The configuration of Al_2O is linear.

Tarte mentioned potential applications of infrared investigations in the study of isomorphous replacements in complex anionic groups such as $(AlO_6)^-$, as well as in nonisomorphous silicates. Polymorphism in the Al_2O_3 and SiO_2 systems, and the investigation of analogous problems in glass constitution has disclosed important structural information, as for example, in the role of such groups as (AlO_4). Order-disorder conversions, as shown in the special case of $LiAl_5O_8$, and the feldspars may also prove to be useful applications (Laves and Hafner).

Emschwiller et al. applied infrared analysis to the hydrated calcium aluminates to show that $3CaO \cdot Al_2O_3 \cdot 6H_2O$ could not be dehydrated at room temperature. Ease of dehydration or the replacement with heavy water was found to be dependent on crystal structure. Nuclear magnetic resonance was more useful to place the nature of the water.

Trombe and Foëx (1957) applied the high-energy infrared radiation of the sun, 300 watts at 250 to 300 watts/cm^2, collected by a parabolic mirror of 2 meters diameter and 0.85 meter focal length. A filter of manganese glass absorbed visible radiation. The radiation that passed was equivalent to a black body at 2350°C. Alpha aluminum oxide and silica absorbed very little infrared radiation under these conditions, and were heated only slightly.

10-5 COLOR IN ALUMINA

Alum has been used from ancient times as a mordant for fixing native dyes and developing their brilliance. The basic aluminum salts, gelatinous aluminum hydroxide, anodized coatings on aluminum, and partially dehydrated Bayer alumina trihydrate serve as effective carriers for fixing organic dyes strongly as pigments. The intensity of dye fixation is very weak in Bayer hydrated alumina and in hydrothermal boehmite, but it reaches a maximum at activation around 350 to 400°C, and then gradually decreases and practically disappears on conversion to alpha aluminum oxide. Very stable, fade-resistant organic paint and ink pigments are formed in such structures (Harrison, 1930; Ellis, 1940). These high-surface alumina forms also adsorb and fix the soluble salts of the colorant metals (Fe, Co, Ni, Cr, Ti, V, etc.) by precipitation as hydroxides and oxides on the active surface.

Thermally stable ceramic pigments are produced by heating intimate mixtures of the colorant metal oxides with alumina at temperatures sufficient to obtain color development, apparently by solid solution. As inferred previously, various gem-stone colors are developed by adding the appropriate oxides to the feed in the Verneuil process, or to the nutrient alumina in the solution-grown processes.

The threshold of visual detection of coloration of polycrystalline alumina is quite sensitive, as determined on spectrographic beads fused in the electric arc. The sensitivity varies for different oxides, being about 0.02% for Fe_2O_3 and only 0.00035% for Cr_2O_3. The internal reflection from hollow, translucent ware is a more sensitive method of color detection. The red colorants (Cr_2O_3, Fe_2O_3) appear to augment each other's intensities, while cobalt masks their effect by producing gray tints.

Color in sintered alumina and high-alumina porcelains, picked up from specks of contamination introduced during various stages of processing, diffuses spherically from the point of contact. When diffusion is incomplete during firing, the coloration appears with circular boundaries in the ware. A reducing atmosphere at about 1500°C was effective in removing or concealing spotted brown discoloration of this type.

Spinel formation is an effective method for obtaining a range of brilliant colors from the metal oxides, for use as ceramic stains and pigments. Singer (1946) recommended that ceramic stains be prepared by combining a monoxide and a sesquioxide, either one of which may be the coloring agent. Some of the direct colors available with alumina spinels include: $CoAl_2O_4$, dark blue; $CuAl_2O_4$, light green; $FeAl_2O_4$, black; $MnAl_2O_4$, tan; $NiAl_2O_4$, blue; $ZnAl_2O_4$, violet. Partial substitution of these with spinels produced by other common bivalent and trivalent metal oxides increases the color spectrum. Among these are: $BaCr_2O_4$, dark green; $CoCr_2O_4$, blue green; $CuCr_2O_4$, green; $MgCr_2O_4$, yellow green; $NiCr_2O_4$, green; $SrCr_2O_4$, green; $ZnCr_2O_4$, green to brown; $BaFe_2O_4$, gray to brown; $CaFe_2O_4$, gray; $CoFe_2O_4$, black; $CuFe_2O_4$, brown; $MgFe_2O_4$, brown; $NiFe_2O_4$, black; $SrFe_2O_4$, gray; and $ZnFe_2O_4$, gray to brown. Evers found that the coloration of fused spinel containing from 0.1 to 15% Cr_2O_3 was homogeneously green for the low concentrations, but showed points of red ruby at the higher concentrations. Specimens containing 25% Cr_2O_3 had alternating zones of red and green, while spinels containing 50% Cr_2O_3 showed some metallic chromium. Magnesia vaporized more readily than alumina. Annealing at 950°C caused separation of alumina, generally as ruby. Absorption coefficients showed a maximum at 5893 A and a minimum at 5016 A.

Steindorff (1947) discussed in detail the various coloring matters which are added to aluminum oxide and spinel to produce different colors. Grum-Grzhimailo and Lyamina applied refractive index and absorption curves as tests of the homogeneity of alumina and spinel colors. Colors containing $3MnO-P_2O_5 \cdot 12Al_2O_3$, in which the P_2O_5 acted as a flux, were yellow when fired at 900°C; their refractive index ranged from 1.650 to 1.667 in the batch; and the absorption curve did not resemble that for Mn^{3+}. When fired at 1000°C, the refractive index ranged from 1.683 to 1.690, and at 1060°C, to 1.735 to 1.760, and resulted in the formation of anisotropic crystals of corundum; the absorption curve had the maximum characteristics of Mn^{3+}. At 1100°C the portions of material having the lower index surrounded the more highly refracting crystals and bound them into aggregates. With increasing temperature, the grain size of the corundum increased, and at 1300°C a maximum amount of lamellar corundum was formed. Colors fired at 1350°C for one hour varied in index from 1.71 to 176, regardless of the chromium content, indicating incomplete reaction, but at 1750°C, the reaction was homogeneous and complete. Similar lack of homogeneity during color formation was demonstrated for magnesium and zinc spinels. The addition of 5% boric acid to samples composed of $Cr_2O_3 \cdot Al_2O_3$ enabled the same development of color at $1350°C$ as without it would require $1750°$. The addition of boric acid to a composition (0.1 CoO, 0.9 MgO)$\cdot Al_2O_3$ stabilized the refractive index at 1.736.

Fujii et al. described the method of manufacturing "Toshiko" or manganese pink from a mixture of 2 parts manganese phosphate to 8 parts $Al_2O_3 \cdot 3H_2O$. Yamaguchi and Tomiura also investigated the manganese pinks, and Yamaguchi et al. (1954) the following systems: blue, CoO-MgO-Al_2O_3; blue, CoO-NiO-MgO-Al_2O_3; blue, $NiO \cdot Al_2O_3$; brown, $CuO \cdot Al_2O_3$; green, $CoO \cdot 3MgO \cdot 2Cr_2O_3 \cdot 2Al_2O_3$; pink, ZnO-Cr_2O_3-Al_2O_3; brown, MgO-Cr_2O_3-Al_2O_3. The brown spinel structure, $CuO \cdot Al_2O_3$, is stable only below $750°C$; blue color is obtained at $1100°C$. Yamaguchi (1954) recommended the spinel with lower valence aluminum ion, which has 0.8 equivalent reducing power per formula weight, for the preparation of ruby glass and glass transparent to ultraviolet rays.

N.S.G. Rao (1958, 1960) investigated various colors including chrome alumina pink. At a calcination temperature of $1300°C$ with 3% potassium dichromate, some successful compositions (concentrations in parts) were as follows: dark rose pink, Al_2O_3 20, ZnO 40, and Be_2O_3 40; dark rose pink, Al_2O_3 20, ZnO 40, B_2O_3 20, and SiO_2 17.23; brick red, Al_2O_3 60, CaO 20, and B_2O_3 20. Calcia, added as $CaCO_3$, is not necessary to produce pink but it lowers the firing temperature and deepens the color. A mat blue composition contained: Al_2O_3 64.5, ZnO 32.25, and CoO 3.25, fired at $1200°C$. The preparation of various other ratios, fluxed with borax, was investigated. Vecchi evaluated the use of nickel in the production of colors ranging from bright greens to steel blue.

Sono reported experiments showing the influence of alumina in altering the shade of vanadium colors in the systems ZrO_2-SiO_2-V_2O_5 and SnO_2-SiO_2-V_2O_5.

The foregoing references are indicative of the role of alumina in the development of a very large number of ceramic colors.

P. Levy (1961) determined the peak energy (E_o) and the full width (U) in ev of the color centers found in alumina both before and after gamma ray and reactor irradiation to induce defect structure. The following values were observed:

Before Irradiation		30,000 roentgen X-ray		Reactor	
E_o	U	E_o	U	E_o	U
5.45	0.60	5.45	1.25	6.02	0.60
4.84	0.54	4.28	0.70	5.35	0.40
6.2	0.4	3.08	1.50	4.85	0.54
				4.21	0.80
				3.74	0.88
				2.64	0.64
				2.00	0.44

The curves of color-center concentration vs. irradiation time for the reactor-induced bands at E_o = 6.02, 5.35, and 4.85 ev could be accurately represented by a saturation exponential plus a linear increase expression. This conforms with a simple theory which assumes that the defects present before irradiation contribute to coloring as well as the radiation-induced defects.

Mie scattering, particularly in the infrared, and absorption cross sections of micron-size alumina spheres has been investigated by Bauer and Carlson and by Plass. In both cases, the sphere diameters ranged from 0.5 to 10 microns, but Bauer and Carlson presented cross sections for absorption and for differential and total scattering for several particle size distributions with mean radii in the region of one micron. For wavelengths below 8 microns, the absorption cross section was generally less than one percent of the scattering cross section. The data were found applicable to the calculation of the emissivity of optically thick rocket plumes. Plass observed that when the particles are small compared with the wavelength, the absorption cross section varies as the radius of the particle, and the scattering cross section as the 4th power of the radius. Unless the radius is several orders of magnitude smaller than the wavelength, the scattering cross section is larger than the absorption cross section. The ratio of forward to backward scattering as a function of particle size of the alumina spheres at wavelengths of 4 and 8 microns was derived, as well as the angular distribution of the scattered radiation at 4 microns.

10-6 CHROMIA-ALUMINA SYSTEM, LASER APPLICATIONS

Besides the ceramic applications of the Cr_2O_3-Al_2O_3 system involving refractories, abrasives (Skaupy), and jewels, the use in optical lasers has recently received attention. A significant factor in this case was the development of single crystals of sufficient length and structural perfection to satisfy the requirements of this use.

Neuhaus stated that the incorporation of coloring, transition-element ions, especially of Cr^{3+}, into sapphire provides information on the structure, such as the determination of atomic positions, expansion and contraction effects, and the existence of unmixing phases. The change from red to violet to green of Cr^{3+} is caused by specific shifts in wavelength of the characteristic maxima of absorption in the Cr^{3+} spectrum. Dils, Martin, and Huggins analyzed ruby boules in a plane normal to the growth direction by the electron probe microanalysis technique. The chromium content of boules whose overall chromium content ranged from 0.03 to 0.5%, fluctuated markedly, with a general decrease from the center towards the periphery. A microscopic heterogeneity of considerable magnitude but on a very fine scale was also found. Saalfeld (1964) observed a considerable distortion of the oxygen lattice by the introduction of Cr^{3+} ions, although the structure was ordered and strain-free. Janowski and Conrad (1964) used a technique of polishing in H_3PO_4, followed by etching in $KHSO_4$ to reveal dislocations on the various crystallographic planes of crystals from four sources to be used as ruby lasers. Dislocations and subboundaries were revealed on the (0001) and (1120) planes. Less distinct pits developed on the (0111) planes, and none were observed on the (1100) planes. The average dislocation density on the basal plane was (1.5 to 3) $\times 10^6/cm^2$ and about $5 \times 10^5/cm^2$ on the (1120) plane. Wander in the c-axis for a typical sapphire crystal 1.25 inches long varied as much as 1 degree 35 minutes, as determined by Laue back-reflection. Slight etching of a mechanically polished crystal revealed numerous polishing scratches, suggesting a plastic-flow layer.

The general function of "maser" devices is molecular amplification by stimulated emission of radiation. Observations by Forrester, Parkins, and Gerjuoy (1947) on the possibility of observing beat frequencies in photoelectric emission generated by two separate light sources, and by Schawlow and Townes (1958) on the potentialities of masers foreshadowed the developments of "lasers." Makhov, Kikuchi, Lambe, and Terhune (1958) "pumped" a 0.1% chromia ruby crystal by a klystron tube at 2.4×10^{10} cps in a 4230-gauss field at 4.2°K, and it radiated at 0.9×10^{10} cps, short electromagnetic radiation.

Maiman (1960) made and described a laser, a device capable of generating and amplifying coherent infrared and visual light by stimulated emission of radiation. His laser consisted of a ruby rod placed in the axis of a helix-shaped xenon tube. The ends of the rod were plane-polished and silvered—one more heavily than the other. Light from the xenon tube entered the crystal through the sidewall, and nearly coherent flashes of concentrated light energy were emitted from the rod.

Lewis (1963) discussed the significance of coherent light, spectral purity, and laser operation. The waves of coherent light are in phase over the width of the beam and of uniform wavelength. The propagation as a parallel beam spreading only through the diffraction angle (less than one minute of arc) has suggested the use for communication through long distances, particularly in outer space. A beam to the moon is calculated to be spread less than one mile in diameter (Bell Telephone Laboratories, March 1961). At any one frequency, a laser crystal may emit billions of times as much energy as is emitted by a comparable area of the sun's surface.

More effective methods for pumping the energy have since been devised. Schawlow; Luck; Devlin, McKenna,

May, and Schawlow; Holland; Rechsteiner; Olt; Nelson and Boyle; and Humphreys contributed to the information on laser systems. Threshold pumping power was reduced and power output was increased by cladding the ruby laser in a sapphire shell, and by growth on a sapphire base. Output energies of the order of 300 joules at an efficiency around 1% have been obtained for ruby lasers.

Murray, Lamorte, and Vogel discussed such factors as crystal growth development, the preparation of scatter-free crystals, crystal annealing, the valency of the dopant, the preparation of laser rods and the reflecting surface, and laser crystal testing. Nelson and Remeika found that flux-grown ruby lasers (as from lead oxide-boron oxide flux) were superior to flame-fusion (Verneuil) crystals for laser uses. Kisliuk and Krupke investigated the biquadratic exchange energy in ruby. Birnbaum and Stocker showed that in the operation of solid-state lasers (ruby) near the threshold, many axial modes are present, even when a train of regularly spaced damped spikes appears in the output of the laser.

Hoskins (1959) observed a relaxation time of about 10^{-2} second for the inverted state of the Cr^{3+} ion in an adiabatic, rapid passage of the spin transition, $\Delta m = 1$ at about 4.1°K. Kushida and Silver (1963) induced a nuclear spin, $\Delta m = \pm 2$ transitions, using a double resonance technique in the application of a radiofrequency electrical field orthogonal to the trigonal axis of the ruby crystal. The line observed for the quadrupole transitions was more than twice as broad as the dipolar lines, and with a distinct asymmetry.

The ability to focus coherent laser beams has suggested many applications involving high concentrations of radiant energy to burn, melt or drill holes in refractory ceramic materials, including the diamond.

Laser techniques for welding both ceramic and non-ceramic materials have been developed, and the repair of retinal detachment in the eye (coagulation) has been suggested. In the latter case, coagulation by the use of conventional light optics has also been applied (Cibis; Kühn and Weinhold).

Bonem used the laser beam to evaporate aluminum oxide, tantalum metal, etc., in the preparation of thin films. No boat is required, and the film purity is claimed to equal that of the specimen before evaporation.

11 RADIATION AND ALUMINA

The effects of radiation on the properties of alumina are of interest because of the present and potential uses of ceramic materials around nuclear reactors. Some investigators who have drawn attention to these applications include: Lambertson (1951); Murray; White (1953); Warde; McCreight and Sowman; Sargent; Livey; and Wisnyi and Taylor (1960). The requirements extend from crucibles used in the laboratory processing of metal components to applications within the reactor itself, the fuels, moderators, controls, structural materials, and shielding. Aside from the conventional use of refractories in nuclear power plants, the uses of ceramics in the storage of nuclear wastes, and as pumps and container material in heat exchange systems are also of interest. One of the objectives of using ceramic materials in nuclear applications was to raise the operating temperatures into a more efficient range.

The effects of radiation on the physical properties of alumina and other refractory single oxides are given in detail in *Effects of Radiation on Materials and Components* (Editors: Kircher, J. F. and Bowman, R. E.) Reinhold Publishing Corporation, New York 1964.

Dickinson presented a classification of the radioactive isotopes, and their potential uses in industry. The types of nuclear radiation include:

Neutron (n), at 1837 X the electron mass, but without a charge, and unstable. These include fast neutrons (n_f) having a speed greater than 10^4 ev, and thermal neutrons (n_s) having a speed of 0.025 ev or less.
Protons (p), the basic nuclear charge, 4.8×10^{-10} esu, and having a mass 1835 X the electron mass.
Deuterons (d), neutron-proton units.
Alpha particles (a), helium nuclei.
Fission products (F/p), fragments of fissioned sources.
Electrons (e), beta rays (β), unit mass 9.1085×10^{-28} g.
Electromagnetic radiation (photons), including gamma rays (γ) and X-rays (about 10^{-7} cm or less).

The radiation damage in ceramics caused by each of the types of radiation is about as follows:

n_f causes scattering of ions in the crystal lattice.
n_s induces nucleon reactions (neutron lattice).
p, d, a cause nuclear reactions, lattice displacement.
F/p causes displacement and severe ionization.

β causes ionization by recoil electrons, a minor effect.
γ causes secondary ionization.

The neutron flux (nv), neutrons/cm^2 or n/cm^2, the integrated neutron flux (nvt), and neutrons/cm^2 X time (sec), are measured in terms of "barns," in which one barn = 10^{-24} cm^2/nucleus. A unit often used is megawatt days per central ton, (MWD/ct = 3×10^{17} nvt).

Nuclear radiation data reported by Way (1950) for aluminum is as follows: absorption (σ_a) 0.23 barn; scattering (σ_s), 1.5 barns; average loss in natural logarithm of energy in a collision (ξ), 0.070; slowing down power ($N\sigma_s\xi$), 0.0056; moderating ratio (σ_s/σ_a)ξ, 0.47. N = the number of nuclei per cubic centimeter.

In general, irradiation by high-energy radiation particles usually causes lattice defects in ceramics with the least damage to simple ionic crystal structures. Anisotropic structures and highly covalent crystals are particularly sensitive. Most radiation-induced defects can be annealed by heating at 600 to 1200°C. Less damage was incurred by graphite (and probably other ceramic types) at higher temperatures. Gamma-induced defects are usually annealed at 200 to 400°C.

Patrick found that thermal neutrons and fast neutrons in the range from 1.36×10^{21} to 3.17×10^{21} n/cm^2 at 425 to 700°C were quite damaging to beryllia, magnesia spinel, and yttria ceramics, but 97.5% Al_2O_3 ceramics and compositions containing from 0.5 to 2% Y_2O_3 were resistant to damage. Howie listed three types of radiation damage (1) ionization and electronic excitation, (2) transmutation, and (3) knock-on damage.

When either single-crystal or polycrystalline alpha aluminum oxide was subjected to 2×10^{20} nvt of radiation, no change was observed in the X-ray pattern or in thermal conductivity (Johnson, 1957), contrary to the observations of Stevanovic and Elston. A color change, probably an F-center phenomenon was observed.

Antal and Goland found that fast-neutron irradiation of sapphire at temperatures below 40°C resulted in little damage, but some volume expansion. The defects were only about one-fortieth the predicted number, believed to have been caused in part by Al-O vacancy pairs. Annealing at 400 to 1250°C produced no decrease in the damage, or in neutron scattering and discoloration beyond 1250°C. Annealing at 1800°C did not remove the coloring, but the density returned to its pre-irradiation value. The maximum decrease in density was at 10^{21} n/cm^2 (Hackman and Walker). Levy concluded, in the case of glasses, that each of three optical absorption bands produced by gamma radiation was associated with a specific oxide component of the glass. The effects on the absorption spectrum of additions of alumina to the glass, suggested that the aluminum atom replaces the silicon atom in the glass network.

Berman, Bleiberg, and Yeniscavich determined that alumina and zircon show little change in properties when exposed to 10^{18} to 10^{19} nvt of thermal neutrons. When, however, a source of fission fragments is provided in the form of a UO_2 dispersant, the grain boundaries and the peaks of the X-ray diffraction pattern disappear on irradiation. The alumina was observed to increase about 30% in volume concurrent with the destruction of its crystal lattice. Apparently only a small proportion of the atoms were displaced by elastic collisions or other applicable mechanisms, but more probably indirectly through anisotropic effects which distorted the lattice and made it unstable. Several ceramic materials having the cubic structure exhibited better stability against fission fragment recoil. Berman, Foster et al. (1960) measured the thermal conductivity both of single-crystal synthetic sapphire and of polycrystalline specimens at 2 to 100°K, before and after gamma ray and reactor irradiations. The reactor irradiation introduced two types of defects that influenced the thermal resistivity. The gamma radiation also induced an extra thermal resistivity at low temperatures which seemed to depend on the initial crystal imperfections.

Compton and Arnold claimed that lattice defects cannot be generated in alpha-alumina by low-energy ionization processes; fast electrons with energy less than 0.37 Mev cannot generate the lattice defects responsible for the optical absorption band appearing at 0.205 micron. Three Gaussian-shaped optical absorption bands were induced in calcium-boron-aluminum (cabal) glasses by gamma rays, with their maxima at 0.550, 0.350, and 0.250 micron (2.3, 3.5, and 5.0 ev). The intensity of the 0.350 micron band increased with increasing Al_2O_3 content in the glass (Bishay). Gabrysh et al. observed enhanced absorption along most of the optical energy range for sapphire when subjected to pressure of 30 kbars after gamma irradiation.

Dau and Davis devised a model for predicting the electrical conductivity of alumina ceramics as a function of temperature and radiation dose-rate. McChesney and Johnson observed the effect of fast-neutron irradiation on the dielectric constant (k′) and the dissipation factor (tan δ) of polycrystalline alumina (Coors AD-995) at one Mc/sec. Irradiation to 6×10^{17} and 2 to 5×10^{19} fast neutrons/cm^2 at 95°C and 47°C, respectively, caused a slight decrease in k′ with the treatment. The dissipation factor rose sharply for doses of this range, but decreased for higher doses.

C. G. Young (1962) investigated the radiation effects in alpha alumina by electron-spin resonance. Y. R. Young

(1957) measured the energy distribution of electrons penetrating thin specimens of sintered alumina. For a given bombarding energy, the average loss per unit thickness remained almost constant over the entire depth of penetration, but this loss factor decreased slowly as the bombarding energy increased.

McCreight found that a composition containing 1 part BeO to 1 part Al_2O_3 had remarkable resistance to corrosion by sodium metal at 500°C, only 0.0003 mg/cm^2/month. This was considered as a container material in reactant cooling. Gangler (1954) showed that single-component ceramic materials such as Al_2O_3 were far superior to most metals in resisting attack by such heat-transfer agents as molten lead-bismuth alloys. Hahn and VanderWall found that lead chloride ($PbCl_2$) rapidly dissolves zirconium and uranium-zirconium alloys at 510°C, allowing consideration of alumina as a container material.

Primak and associates, Kreidl and associates, and Lukesh reported the damage to glasses, silica, and silicon carbide. In some cases devitrification, metal precipitation, and expansion as great as 3.7% occurred in nuclear irradiation.

Gift (1957) described the use of a bed of 0.5-in. silvered alundum Raschig rings in a homogeneous reactor (HRT). During operation, a steam-gas mixture entering the bed was dried by a steam coil preceding the bed. The purpose of the bed was to remove iodine, a poisoner of the recombiner catalyst. During shutdown, it was necessary to remove decay heat for about 10 hours by a combination of steam flow and internal cooling of the bed by water-jacketed annuli. Richt examined by metallographic methods $Cr-Al_2O_3$ cermet specimens that had been previously irradiated to about 10^{20} nvt (thermal). The cermets showed no evidence of structural changes alone, but when fueled with UO_2 and irradiated to 2.53×10^{20} nvt at 1000°C, changes were observed.

Elleman et al. measured the height of the expansion step at the boundary between irradiated and unirradiated regions when fission fragments were recoiled into flat alumina specimens. Expansion normal to the exposed surface was independent of crystal orientation for specimens of sapphire cut along three different axes. Expansion was not a linear function of fission-fragment concentration. Radiation expansion coefficients, calculated from the expansion normal to the surface, were at least three times larger than those calculated from the bending moment of thin, fission-fragment-irradiated disks. It was concluded that fission-fragment-induced expansion is not isotropic, but that the stresses are preferentially relieved by expansion normal to the free surface.

Smalley, Riley, and Duckworth investigated the preparation of an alumina-clad uranium nuclear fuel. A special, high-purity, sinterable alumina powder was applied to unsintered uranium oxide pellets by a spraying-tumbling method followed by isostatic pressing and sintering in hydrogen. Crack-free, spheroidal pellets (1000 to 2000 microns diameter) with a sintered alumina coating (300 to 500 microns) of near-theoretical density were sufficiently impervious to prevent oxidation of the uranium for at least 100 hours at 1200°C. No measurable release of fission products was observed at 927°C in vacuum for seven days after exposure to 6.0×10^{12} nv for one hour at room temperature. Browning et al. applied dense, high-purity alumina coatings to UO_2 fuel particles by coating in a fluidized bed with hydrolyzed $AlCl_3$ vapor. The coatings were sufficiently impervious to provide good fission-gas retention on heating to 1300°C after low radiation exposure. K. Johannsen suggested the use of molten alumina as a possible matrix for the fuel in thermal nuclear reactors in which the liquid melt is retained within a thin solid cooled surface layer. Deportes (1967) reviewed the use of solid oxides in the construction of high-temperature reactors.

Whittemore (1963) claimed a method comprising surrounding a refractory metal core including a radioactive material with a ceramic alumina sheath, molded around the core, and firing the composite to an impervious integral nuclear fuel element. Oel discussed the use of ceramic alumina for breeders, moderators, reflectors, control rods, shielding, and construction elements for thermoelectric power generation.

The radioactive wastes from nuclear power generation contain such materials as salts of cesium, strontium, and ruthenium, having half-lives of 600 to 1000 years. A potential method for segregating these salts comprises reacting them in silica-alumina gels to form zeolites or preferably nonbase-exchange silicates. The waste materials have a nominal thermal energy of about 5 watts/gal. (Warde 1955).

Fleshman conducted tests to determine the leaching effect of rainwater on buried radiation-contaminated alumina. When the acid-leached alumina was treated with caustic soda before burial, the uranium contamination was converted into an isoluble diuranate which was absorbed by the alumina. Bridoux used sintered aluminum powder (SAP) as a canning material, but found (1) diffusion of fission products into the SAP, (2) the appearance of cracks from prolonged heating, and (3) poor ductility of the SAP, particularly in tubes that receive no support from the fuel. At 400 to 500°C, however, the creep strength of SAP is comparable to that of beryllium and is hence acceptable.

Dalmai et al. found that the catalytic activity of activated gibbsite (as measured by the decomposition of formic acid) is increased when the gibbsite has been irradiated with gamma rays or neutrons. Tsitsishvili and Sidamonidze similarly found that fast neutron and gamma radiation increased the catalytic effect of aluminum hydroxide and gamma alumina in the adsorption of water and benzene at 25°C, and in the dehydrogenation of isopropyl alcohol. The phase composition is unaffected.

Schwab and Conrad found that irradiation of alumina with reactor neutrons affected the rate of conversion of para hydrogen. Bogoyavlenskii and Dobrotvorskii stated that the incorporation of $Na_2\,^{185}WO_4$ into the anodic film on aluminum (duralumin) in the course of formation caused the specific surface radioactivity, i, of the alumina film formed to increase proportional to the duration of anodization t, with the anodic current density, D_a and with the specific radioactivity of the solution, X. Empirically, $i = a \log t + b$, and $\log i = -1.85 - 0.67/D_a$. When the temperature was increased from 20 to 25°C, i decreased very rapidly, and finally, asymtotically, to zero at 35°C.

Gorodishcher and Mashneva removed ^{32}P from drinking water by adsorption in aluminum hydroxide gel from concentrations from 7×10^{-8} to 8×10^{-7} curies/liter. The activity of the clarified water was reduced to the maximum permissible dose of $(0.82 \text{ to } 2) \times 10^{-8}$ curies/liter.

12 CHEMICAL PROPERTIES OF ALUMINA

The chemical properties of alumina may be separated arbitrarily into two general classifications (1) thermal and (2) hydrothermal. These classifications are exemplified by *Phase Diagrams for Ceramists*, Levin, Robbins, and McMurdie (1964), which contains 132 phase diagrams involving Al_2O_3 as a component, of which 17 also contain water. Early exhaustive compilations of the chemical properties include: Mellor, *Comprehensive Treatise on Inorganic and Theoretical Chemistry*, (1924), and Gmelin, *Aluminum*, Vol. II (1934). More recent accounts are found in *Chemical Background of the Aluminum Industry*, J. G. Pearson (1955); *Encyclopedia of Chemical Technology*, Kirk and Othmer (1947, 1957); *Encyclopedia of Chemistry*, Clark and Hawley (1957); *Progress in Inorganic Chemistry*, Vol. I (1959); and *Handbook of High Temperature Materials: No. 2, Properties Index*, Samsonov (1964).

Chemical reactions of alumina of general ceramic interest include the resistance to attack of sintered alumina by various reagents, particularly at high temperatures. High-temperature chemistry includes those chemical phenomena which occur above 1000°C. Such temperatures are attained by combustion, by electrical heating, or by chemical explosions and nuclear reactions (Margrave, 1962). The high-alumina regions of the phase diagrams are of particular interest in interpreting the influence of additives on the sintering behavior and refractory properties of polycrystalline alumina. Equilibrium is probably more nearly realized in these cases than is usually expected in the practical application of the phase diagrams in general (W. R. Foster, 1951). Of the ten large-scale phase diagrams selected for publication by the American Ceramic Society and the Edward Orton, Jr., Ceramic Foundation (1960), five include Al_2O_3 as a component, namely: $CaO-Al_2O_3-SiO_2$, $MgO-Al_2O_3-SiO_2$, $Na_2O-Al_2O_3-SiO_2$, $K_2O-Al_2O_3-SiO_2$, and $FeO-Al_2O_3-SiO_2$. Among the many contributors to the literature on these systems are included: Day, Shepherd and Wright; Bowen and Greig; Toropov and Galakhov (1953); Aramaki and Roy; Osborn and Muan (1960); Schairer and Bowen (1955, 1956); and Richards and White (1954).

Wygant and Kingery (1952) discussed the stability of ceramic oxides in terms of the equilibrium oxygen pressure, in the case of a-Al_2O_3 about 10^{-184} atm. The stability at 298°K (the free energy of formation per gram atom of oxygen) = -125,600 cal. The order of increasing stability for some refractory oxides on this basis is as follows: 1/2 ZrO_2, 1/3 Al_2O_3, BaO, MgO, BeO, CaO, 1/3 Y_2O_3, 1/3 La_2O_3, and 1/2 ThO_2. From heat content, entropy, and heat capacity data, they concluded, for example, that sintered alumina would not be a suitable container material for molten calcium metal at 1200°K. Contrariwise, Jaeger and Krasemann (1952) found by actual experiment that calcium, strontium, and barium are inert in contact with sintered alumina to their boiling points. Tripp and King (1955) prepared graphs of the variation in free energy of the various ceramic oxides of interest, in order to simplify the use of such information. Fiegel, Mohanty, and Healy (1962) have investigated the equilibrium of the refractory metal oxides. Gulbransen and Wysong (1949) measured the rate of formation of amorphous alumina from purest aluminum and oxygen at about 8 micrograms/cm²hr at 500°C, or a thickness increase of 400 A/hr. The activation energy was 22.8 kcal/mole, and the entropy of activation was -26 cal/mole deg.

12-1 WET CHEMICAL REACTIONS OF SINTERED ALUMINA

Impermeable polycrystalline alpha alumina has marked resistance to wet chemical corrosion. As a rule, the lower phases of alumina and the hydrous aluminas show increasing chemical reactivity as they decrease in density. Early experiments on the resistance of sintered alumina to attack were intended to demonstrate its suitability as a container, crucibles, etc., for thermal reactions (Winzer, 1932). Concentrated H_2SO_4, HCl, HNO_3, H_3PO_4, and 20% NaOH dissolved no more than 0.02% of a 30 × 35 mm crucible within six hours, as indicated by loss in weight of the crucible. This is not necessarily indicative of chemical inertness, as for example, phosphoric acid readily reacts even with coarsely crystalline tabular alumina to form slowly soluble phosphate bonds at temperatures below the boiling point of the acid. Dawihl and Klingler (1967) state that sintered alumina containing 3% silicates is far more resistant to corrosion by HCl, HNO_3, and H_2SO_4 in concentrations from 10 to 95% acid and up to 100°C than are titanium, cast silicon, and Cr-Ni steels.

Finely divided alpha alumina is rapidly dissolved by HF, hot concentrated H_2SO_4, mixtures of these acids, ammonium fluoride, molten alkali bisulfates or pyrosulfates, and by concentrated HCl, especially when under pressure. All these reagents have been used to dissolve alumina in analytical procedures. Sintered alumina dissolves in concentrated H_2SO_4 faster than some high-alumina porcelains containing clay binders (85% Al_2O_3). Karpacheva and Rozen found that the densest sintered alumina reacts with water, even at temperatures as low as 200°C. With heavy water, $H_2^{18}O$, the following reaction rates, as percent reaction within 80 minutes, were observed:

Temp °C	200	400	600	900
Rate (%/80 min)	20	30	50	97

Hot-water solutions of the free alkali hydroxides and carbonates cause perceptible reaction, the rate being correspondingly faster at higher temperatures and under pressure.

Jaeger and Krasemann (1952) extended the earlier work of Winzer on the resistance to corrosion of sintered alumina crucibles. In many instances, the weight loss estimates were based on multiple tests. Crucibles lost little weight on successive heating, but in many cases after treatment with acid, especially with 1:1 hydrochloric acid, they shattered on subsequent heating as a result of intercrystalline attack. In general, attack by the alkali oxides was very aggressive, but sodium oxide could be contained in alumina at 1350°C. The order of decreasing attack for alkaline melts was Na_2O, $Na_2S_2O_7$, Na_2O_2+KOH, $Na_2CO_3+KNO_3$, Na_2CO_3+S, NaOH, Na_2CO_3, K_2CO_3, and $NaKCO_3$. Ashed precipitates showing an adherence to sintered alumina include SiO_2, Al_2O_3, $BaSO_4$, CaO, BeO, SnO_2, ZnO, and graphite. Little adherence was shown by Fe_2O_3, Cr_2O_3, or $CaSO_4$. The sulfides, MnS and CuS were strongly fluxed, and also $Mg_2P_2O_7$.

12-2 REACTION OF THE CHEMICAL ELEMENTS WITH ALUMINA

Ryshkewitch (1960, p 177) discussed the chemical properties of sintered alumina, particularly in its applications as a high-temperature container material for various chemical reactions. Norton and Kingery (1952), Norton, Kingery, Economos, and Humenik (1953), and Reed (1954) have investigated high-temperature metal-ceramic reactions. Lyon (1954) assembled the available data on the corrosion of alumina and other prospective materials by low-boiling metals and alloys that might be suitable for high-temperature heat exchange systems. Gangler determined the corrosion resistance of 18 ceramic compositions and cermets to molten lead-bismuth alloys. Colligan et al. related the corrosion by some molten alloys to foreign components in sintered alumina used in crucible and mold refractories, such as the oxides of chromium, iron, and manganese.

Some reactions of interest between the chemical elements and alumina are shown in the following references. Reactions of molten alumina with some refractory metals and alloys are included. These are presented in alphabetical order, in general, for convenience of reference.

Alkali Metals

Ryshkewitch (1960) stated that molten lithium attacks alumina aggressively, potassium to a lesser degree, and sodium the least. Reed claimed that the corrosion resistance of sintered alumina to both liquid and vaporized sodium is good at 900°C. A commercial sintered alumina article disintegrated within 168 hours at 940°C, however. A synthetic sapphire cylinder (1/4 in. diameter by 1 in. length) lost about 1% in weight in 168 hours at 900°C but remained clear. Kelman, Wilkinson, and Yaggee found that sodium-potassium did not attack below 500°, but caused corrosion at 600°C. Kolosova et al. found a weight loss of only 0.01 to 0.06% in 35 hours at 400°C for sintered alumina immersed in molten alkali (79%K, 21% Na).

The behavior of vaporized alkali metals on oxides and other dielectric materials has been of interest for thermoelectric converters. Wagner and Coriell (1959) tested Al_2O_3, BN, ZrO_2, MgO, HfO_2, ThO_2, CaO, and NbC to exposure to cesium vapor at temperatures as high as 1475°C. Only fused alumina remained unaffected by the test. It was concluded that cesium probably reacts with impurities in the contacting material, but the effect can be observed only when the surface areas available for reaction are relatively large, as is the case with sintered specimens. Higgins attributed pitting of single crystals by cesium vapor at 600°C to silicon and barium impurities; neutron irradiation increased the attack. C. E. Adams (1959) observed that rubidium effectively condensed at high temperatures only on oxides with which it could form stable complex compounds. Cowan and Stoddard (1964) stated that glasses could not be used in thermionic converter seals because of alkali metal (cesium) corrosion.

Alkaline Earth Metals

Jaeger and Krasemann (1952) observed no reaction of calcium, barium, or strontium to their boiling points (about 1150°C). This is likely, since BaO and SrO can be reduced to the metals (thermite reaction) by metallic aluminum at 1100°C in a vacuum (Gvelisiani and Pazukhin). Magnesium also shows no attack to its boiling point (1110°C).

Aluminum

Aluminum reacts with alumina at 1100°C to form Al_2O, and at 1600°C to form AlO (Hoch and Johnston).

Antimony, Arsenic

Jaeger and Krasemann claimed no attack of sintered alumina.

Beryllium

Beryllium (m.p. 1500°C) shows no attack below 1600°C. A slight darkening of the alumina, caused by formation of an interfacial layer of chrysoberyl occurred at 1800°C (Economos and Kingery). Chrysoberyl, $BeO \cdot Al_2O_3$ (m.p. 1870°C), and $BeO \cdot 3Al_2O_3$ (m.p. 1910°C) are identified compounds (Lang, Fillmore, and Maxwell, 1952; Galakhov, 1957).

Bismuth

The corrosion resistance is good to 1400°C (Reed). The reaction product is $Bi_2O_3 \cdot 2Al_2O_3$ (Levin and Roth, 1964).

Carbon

Carbon reduces alumina to the normal carbide Al_4C_3 (Prescott and Hincke) at above 1700°C (Jaeger and Krasemann), initiating at as low as 1310°C (Komarek et al.), but requiring about 2400°C for complete conversion (Kohlmeyer and Lundquist). Two oxycarbides also exist at about 2030°C, Al_4O_4C and Al_2OC (Foster, Long, and Hunter; Cox and Pidgeon). The presence of Fe_2O_3, SiO_2, TiO_2, and V_2O_5 as impurities in sintered alumina might induce deterioration at lower temperatures (Stroup), as low as 1380°C (Kroll and Schlechten). The reduction to metallic aluminum occurs at about 2000°C (Miller, Foster, and Baker). Graphite does not wet molten alumina, but is severely pitted in contact with it in water-free, inert atmosphere or vacuum (Bartlett and Hall).

Cerium

Ceric oxide, CeO, forms no compounds with alumina (Wartenberg and Eckhardt). Cerous oxide, Ce_2O_3, forms $Ce_2O_3 \cdot 11Al_2O_3$ and $Ce_2O_3 \cdot Al_2O_3$, both of which decompose in air to form Al_2O_3 and CeO_2 at temperatures above 800°C (Leonov and Keler).

Chlorine

Chlorine does not attack, except in the presence of carbon (Singer and Thurnauer).

Chromium

Chromium wets alumina at 1650°C in a reducing atmosphere (Blackburn, Shevlin, and Lowers).

Cobalt

Cobalt neither wets nor reacts with sintered alumina to above its melting point (1480°C) in a reducing atmosphere (Sieverts and Moritz; Tumanov et al.).

Copper

Copper reacts with the transition aluminas at 800°C in air to form $CuAl_2O_4$, which is stable to 1000°C. It then converts to $CuAlO_2$, which is stable in air to about 1260°C (Hahn and de Lorent, 1955; Misra and Chaklader, 1963).

Fluorine

Winzer claimed attack of sintered alumina at 1700°C by dry fluorine.

Gallium

Sintered alumina is inert to gallium to 1000°C (Kelman et al.). Wartenberg and Reusch (1932) found solid solutions above 810°C with Ga_2O_3.

Hydrogen

Jaeger and Krasemann observed no reduction of alpha alumina in hydrogen up to the melting point, only a surface darkening. L. J. Trostel, Jr. (1965) noted that hydrogen attacks alumina refractories below 1600°C, however, in the presence of water vapor.

Iron

Iron can be melted in sintered alumina under reducing conditions, but wets at about 1600°C (Blackburn et al.). The spinel, $FeO \cdot Al_2O_3$, dissolves to about 6% in alumina at 1750°C (McIntosh, Rait, and Hay; Fischer and Hoffman). Muan and Gee (1956), and Muan (1958) found limited solubility for $Fe_2O_3 \cdot Al_2O_3$ in corundum. Richards and White; Atlas and Sumida; and Turnock and Lindsley investigated the spinel reactions; $Fe_2O_3 \cdot Al_2O_3$ has a structure similar to kappa alumina, and requires above 1320°C to prepare.

Lead

No reaction occurs with alumina at the melting point of lead (327°C). In lead-bismuth eutectic alloy (44.5% Pb), Gangler found 0.000 mils/year loss at 1090°C. Lead aluminate is unstable above 970°C (Geller and Bunting).

Lithium (see Alkali Metals)

Manganese

Manganese does not attack sintered alumina to above the melting point (1260°C) in a reducing atmosphere. Although it is more active than iron, cobalt, or nickel, it can be distilled in sintered alumina to give a spectroscopically pure product (Sieverts and Moritz).

Mercury

Kelman et al., and Hahn, Frank, et al. found no reaction with alumina at 300°C.

Molybdenum

Alumina is not reduced by molybdenum even above the melting point of the alumina. Discoloration may occur at 2100°C in a dry, inert atmosphere (He).

Nickel

No attack occurs in dry inert atmosphere. Nickel can be melted in sintered alumina in hydrogen atmosphere (Economos and Kingery). Wetting occurs at 1800°C.

Niobium

Mass spectrometric and thermogravimetric analysis at 1800 to 2200°C indicates the principal reaction is: $Al_2O_3 + 3Nb = 2Al(g) + 3NbO(s)$. Secondary reactions under neutral conditions are: $Al_2O_3 + Nb = Al_2O(g) + NbO_2(g)$, and $Al_2O_3 + 2Nb + \frac{1}{2}O_2 = Al_2O(g) + NbO_2(g) + NbO(s)$ (Grossman, 1966).

Nitrogen

Nitrogen does not attack (Jaeger and Krasemann).

Palladium, Platinum

Both metals can be handled in the molten condition in sintered alumina (Jaeger and Krasemann).

Phosphorus

No attack was observed at moderate temperature (Jaeger).

Silver

No compounds could be prepared with silver and alumina (Hahn, Frank et al.).

Sulfur, Selenium, Tellurium

These elements do not attack alumina.

Tantalum

Tantalum reacts only slightly with molten alumina in water-free inert atmosphere, H_2, N_2, CO, or in vacuum (Bartlett and Hall). Alloy 90Ta-10W shows slight reaction. Tantalum carbide and $4TaC \cdot ZrC$ react only slightly (Bartlett and Hall).

Tin

No reaction occurred in molten tin at $1000°C$.

Titanium, Zirconium

No reaction occurred in an inert atmosphere below $1800°C$, at which temperature black discoloration of the grains and corrosion occurred (Economos and Kingery). Titanium nitride wets molten alumina with negligible corrosion in water-free inert atmosphere (Bartlett and Hall).

Titanium + Aluminum-Hardened Nickel-Base Alloy

Decker, Rowe, and Freeman found that trace amounts of zirconium or boron picked up from zirconia or magnesia crucibles reduced cracking of the hardened nickel-base alloys during hot-working and increased their rupture strength and ductility. It was desirable to compensate for this effect when making heats in sintered alumina crucibles.

Tungsten

Jaeger and Krasemann observed no reaction between tungsten and sintered alumina. Wallace et al. (1961) investigated the reactions in a Knudsen cell-oven operable at $2500°C$ in conjunction with a Nier-type mass spectrometer. Jaccodine (1960) observed a growth of nodules or hillocks of alumina in the investigation of heater-cathode breakdown of alumina-coated heaters operated at $1200°C$ and 180 volts dc. Tungsten wets molten alumina with negligible corrosion in water-free inert atmosphere (Bartlett and Hall).

Uranium

Jaeger and Krasemann found no reaction between uranium and sintered alumina to $1200°C$. Dykstra (1960) found no solid solution between Al_2O_3 and any uranium oxide.

Vanadium

Burdese obtained reaction between V_2O_5 and gamma alumina at $500°C$ to form $Al_2O_3 \cdot V_2O_5$.

Zirconium Boride (ZrB_2) and Zirconium Carbide (ZrC)

Molten alumina wets and reacts moderately in water-free, inert atmosphere (Bartlett and Hall).

12-3 SLAGGING EFFECTS

An important application of the phase diagrams for ceramists is in the interpretation of slag reactions and the corrosion of refractory linings. In many cases, alumina is a major component involved either in the slag or the refractory.

Early investigators were concerned about the melting and softening of refractories as a viscosity problem (Hartmann, 1938), and the influence of fluidity and solvent action of slags on the destruction of refractories at high temperatures (Endell, Fehling, and Kley, 1939). McGill and McDowell determined the chemical reactions and their temperatures for different refractory materials in contact at high temperatures. The chemical reactions contributed by atmospheres, as for example, the destruction of refractory brick by carbon monoxide disintegration (Hogberg and Heden) have been recurring problems. The significance of wetting as a factor in corrosion, and its appraisal by various methods, as for example, the sessile drop method, has been demonstrated (Dinescu, 1962). Marked wettability was shown by Al_2O_3, and BeO in comparison with V_2O_5 and ZrO_2 for molten lead glass. Physical effects, such as volume changes in slagging reactions, have shown that high-alumina refractories expand markedly in contact with portland cement (Hyslop, 1959). Reich and Panda characterized the slagging resistance of mixtures of sintered alumina and K_2SO_4 and other slagging combinations by the use of a hot-stage microscope. High liquidus temperatures could be associated with low values for the softening point, the "hemisphere" point, and the flow point of a small pellet observed on the hot stage.

12-3-1 Ash Slags

The combination of phases affecting the fusibility of coal ash slags was investigated by Schairer (1942). Coal ash and petroleum ash, particularly from some South American oils of high vanadium content, are destructive to aluminosilicate refractories. While the ash content of petroleum is extremely small, varying from less than 0.001 to about 0.05% in the crude product, this component is nonvolatile and concentrates in the residual fractions of heavy fuel oil. Jones and Hardy found that alkaline earth and alkali oxides (particularly sodium oxide) lower the fusion point of refractories, but are not as destructive as vanadium. McLaren and Richardson (1959) described experiments in which cone deformations showed that compositions containing alumina in the mullite concentration were unaffected by vanadium pentoxide up to 1700°C.

Foster, Leipold, and Shevlin applied a simple phase equilibrium approach to oil-ash corrosion by devising a compatibility diagram for the system Na_2O-SO_3-V_2O_5. Volatilization experiments and phase identification by microscopic and X-ray methods confirmed the validity of the diagram. The diagram was found to be consistent with most of the previous recorded experience. It was noted that Na_2SO_4 and V_2O_5 are not compatible with each other in either the molten or the crystalline state. They react upon heating to form $NaVO_3$, as well as several complex vanadates, but they do not form Na_3VO_4 or $Na_4V_2O_7$. The complex vanadates cause the most severe attack of any of the possible corrodents. Vanadium pentoxide, previously credited with a major role in oil-ash attack, is absent from all but a few fuel-oil ashes, leaving only $NaVO_3$ as the simple compound which might cause slagging. Sulfur trioxide is a potentially severe corrosion factor.

One method that has been suggested to counteract the effect of ash slagging of refractory linings in power-plant installations is to incorporate nonvolatile additives to the oil, as for example, finely divided alumina, which change the nature of the ash to less fusible, powdery consistency.

12-3-2 Slags Containing Sulfates

Ellingham (1944) considered the reducibility of oxides and sulfides and the value of the slags produced. Gow devised experimental tests on the reaction of sodium sulfate, either vaporized or in the liquid state, on aluminous refractories. The progress of the reaction was followed by the change in weight, the determination of SO_3 by precipitation, and the determination of Na_2O by the flame photometer. In comparing the effect of sodium sulfate on a commercial-grade, high-alumina brick (88% Al_2O_3, 11% SiO_2) and fused crystalline alumina (99.5% Al_2O_3), it was found that the dissociation of the sodium sulfate was accelerated more by the presence of SiO_2, probably as the compound mullite, than by the crystalline corundum. It was deduced that the ultimate product in the commercial brick was nephelite, $NaAlSiO_4$, but in the fused alumina, sodium aluminate, $NaAlO_2$. The rate of deposition of sodium sulfate in the brick was doubled by exposure in a vacuum (150 microns Hg) at 1200°C, and was increased by one-third in the fused alumina, as compared with the rate in air at atmospheric pressure. Lambertson (1952) measured the loss in weight, checked by the water extraction of unreacted salt, and showed that liquid sodium sulfate reacts with mullite more rapidly than with alumina. Nephelite was the principal reaction product. The resistance of high-alumina firebrick to soda attack could be improved by a refractory coating, particularly by one composed of alumina, nephelite, and sodium aluminate. On the other hand, Steinhoff claimed that sillimanite rings made from kyanite

with very little free Al_2O_3 proved to be most resistant to attack by sodium sulfate vapor in laboratory ring tests simulating the normal operating conditions of recuperator tubes. The degree of attack increased rapidly with increasing alumina content. In 99% Al_2O_3 rings, a small amount of alkali induced the formation of beta alumina and cracking of the forms. The formation of feldspathic compounds could be avoided by glazing the inside of the tubes or by using acid refractories.

Bovensiepen, Wolf, and Schwarz claimed that actual service behavior of one year in glass-tank operation and laboratory slagging tests reveals the mechanism of corrosion of refractory brick by sodium silicate-sodium sulfate melts. The sulfate is reduced to sulfide, which reacts with part of the remaining sulfate to form sulfite, SO_2 and free Na_2O. The Na_2O and sulfides diffuse into the refractory and decompose mullite and silica to form $Na_8Al_6Si_6O_{24}\cdot SO_4$ and SiS. The resistance to corrosion of a refractory therefore increases with an increase in the Al_2O_3 content and with a decrease in the SiO_2 content. Hyslop and McLeod found that the loss in weight resulting from reacting a mixture of sodium sulfite and a metal oxide for one hour at 1050°C depended upon the acidity of the oxide. Plotting the weight losses against the atomic numbers of the elements involved gives curves that agree with known chemical properties. In the case of the oxides SiO_2, Al_2O_3, ZnO, Cr_2O_3, Fe_2O_3, and CaO, they agree with the order of mobility or diffusion of these oxides in solid-state reactions.

12-3-3 Steel Furnace Slags

Hay et al. (1934, 1936) investigated ferrous slagging systems. Blast furnace slags vary in basicity from about 0.9 to 1.2, in which basicity means the weight ratio of percentages of CaO and MgO in the slag to the SiO_2 and Al_2O_3. Chemical reactions between refractories and these slags are usually interpreted in terms of their acidity or basicity. Certain physical factors are also important. Hartmann and Schultz (1937) stressed the importance of the viscosity of the slags from steel production. Röntgen, Winterhager, and Kammel measured the viscosities of synthetic slags in the systems FeO-Al_2O_3-SiO_2 and FeO-SiO_2. The constitutions of the slags are similar in both the solid and the liquid state, and the solid-stage bonding forces are still present and active in the liquid state. Mikiashvili et al. obtained the viscosity of molten slags of the system Mn-Al_2O_3-SiO_2 corresponding in composition to the products of deoxidation of steel in the regions of the lowest melting temperature. The viscosity of homogeneous liquid slags increased with increasing SiO_2 content. The viscosity of heterogeneous liquid slags at 1400 to 1590°C increased with decreasing SiO_2 content, and was dependent on the proportion of solid phase present. The lowest viscosity (0.5 to 10 poises) was observed for melts with MnO:Al_2O_3 = 6, and 20 to 30% SiO_2. With decreasing Mn:Al_2O_3 ratio, the viscosity of the melts increased. The most fluid melts contained 22% Al_2O_3.

Muan (1958) applied phase equilibrium data for the system $FeO-Fe_2O_3-Al_2O_3-SiO_2$ in evaluating the reactions between the iron oxides and aluminosilica refractories under various idealized conditions which approach those occurring in actual practice. The extent of attack and the nature of the reactions changed with different levels of oxygen pressure. At the highest oxygen level considered (1 atm), much iron oxide was present in the Al_2O_3-containing crystalline phases because of the similarity in size of Fe^{3+} and Al^{3+}. Under strongly reducing conditions, the amount of this substitution was trivial. Three different brick compositions were considered for evaluation: fireclay (46 wt % Al_2O_3, 54 SiO_2), mullite (70 Al_2O_3, 30 SiO_2), and high-alumina (90 Al_2O_3, 10 SiO_2). From isothermal planes taken at temperature levels of the phase diagram between 1200 and 1470°C, it was possible to evaluate the performance of the refractories as a function of $Al_2O_3:SiO_2$ ratios of the brick and oxygen partial pressures. The temperature at which a liquid phase developed in the refractory upon reaction with iron oxide decreased as the partial pressure of oxygen was reduced. At temperatures above 1390°C, one atm oxygen, and 1380°C in air, iron oxide, absorbed in excess of 7 wt % Fe_2O_3, caused a liquid phase to develop in the fireclay brick. In mullite brick, about 60% iron oxide absorption was tolerated without liquid formation at 1390 to 1440°C. In the high-alumina brick, no liquid appeared below 1470°C at one atm oxygen, and 1460°C in air.

Under strongly reducing conditions, the refractoriness was much lower than under oxidizing conditions. In the range 1088 to 1205°C, a liquid phase developed in all three brick compositions after 35 to 25% wt % iron oxide had been absorbed. Above 1380°C, in the strongly reducing atmosphere, all three refractories failed rapidly by liquid formation.

Hyslop (1953) found the following order of decreasing stability against reduction, iron oxide-chromite growth reaction, and slagging, for heating effects to 1600°C in the presence of carbon: alpha alumina, magnesia spinel, magnesite, magnesia-chromite, chrome-magnesite, and silica. Minowa and associates found that the reaction between molten steel and high-alumina refractory crucibles under various oxygen pressures (produced by changing the ratio of hydrogen and water vapor) caused the following changes: at oxygen pressure 1.9×10^{-9} atm, the iron spinel, hercynite ($FeAl_2O_4$), was found, and it increased with increasing oxygen pressure; at oxygen pressure 3.74×10^{-9} atm, limonite ($Fe_2O_3 \cdot xH_2O$) was present, and its rate of production increased with increasing H_2O in mixed gas; when the iron was melted in air, hematite (Fe_2O_3) increased and limonite decreased. When molten iron and a refractory oxide contact each other, carbon in the iron reacts with the oxide to liberate carbon monoxide. The apparent activation energy of the reaction was calculated as follows: 50.8 kcal/mole for SiO_2-C, 59.0 kcal/mole for Al_2O_3 (zero porosity)-C, and 48.0 kcal/mole for Al_2O_3 (30% porosity)-C. When a mixed powder of graphite and either SiO_2 or Al_2O_3 was heated in a high vacuum, an increase in pressure for the carbon monoxide reaction occurred at 1170°C for the graphite-SiO_2 couple, and at 1300°C for the graphite-Al_2O_3 couple.

Bachman and Eusner (1959) claimed an improved high-alumina refractory, especially resistant to ferruginous slags, with lower porosity and greater resistance to spalling, characterized as an aggregate of separate particles of calcined fireclay dispersed in a matrix of mullite prepared from either high-purity alumina or high-grade bauxite. The brick, fired for 4 hours at 1510°C (2750°F), had a net Al_2O_3 content of 58%, an apparent porosity of 12.1%, a bulk density of 2.51 g/ml, a 0.0% load deformation under a load of 25 psi for 1.5 hours at 1510°C, 0.0% spalling loss, and a modulus of rupture of 3000 psi. Slagging tests showed only 0.50 in. as compared with 1.0 in. for a 70% Al_2O_3 brick.

McCune, Greaney et al. (1957) found in laboratory tests that $Al_2O_3-SiO_2$ linings in the low-temperature region of a blast furnace, where 3 to 10% K_2O was the principal contaminant, failed by peeling. Superduty fireclay brick and 70% alumina brick treated with $KCl-K_2CO_3$ mixtures resisted the effect at 870°C. The probable cause of peeling was the formation of leucite ($K_2O \cdot Al_2O_3 \cdot 4SiO_2$).

Changes in steel making since 1957 involve the introduction of oxygen through lances to blast furnaces to speed up the process (from one heat every eight hours to one in 45 minutes) and the present shift to basic oxygen furnaces (BOF). Silica brick and aluminosilica brick, formerly adequate for these applications temperaturewise, are being replaced by more refractory, wear-resistant types. Tar-bonded magnesite and dolomite, and alumina (about 25,000 tons in 1964) have supplied replacements in the steel industry (Chemical Week, May 15, 1965, page 116). Slagging tests by Miller and Shott on six commercial refractories, including fusion-cast types, indicated that slag resistance is directly related to their alumina contents.

Bron, Savkevich and Mil'shenko obtained a slag in the production of metallic chromium containing 80 to 86 wt % Al_2O_3, 9 to 12% Cr_2O_3, and 2 to 4% Na_2O (mainly beta alumina), which they suggested for use in open-hearth checkers, cupolas of hot-blast stoves, ladles, and roofs of steel-melting furnaces.

12-3-4 Glass Furnace Reactions

Stanworth and Turner (1937) found that up to 5% calcined Bayer alumina added to a glass-batch mixture of the molecular composition Na_2CO_3, $CaCO_3$, and $6SiO_2$ has no measurable effect on the rate of decomposition at 700°C, or on the amount of reacted silica formed. The alumina reacts with sodium carbonate at this temperature but not with calcium carbonate or silica, even at 800°C. Abou-El-Azm arranged the oxides in their order of decreasing rate of reaction with silica as follows: K_2O, Na_2O, Li_2O, PbO, B_2O_3, BaO, CaO, ZnO, MgO, TiO_2, Al_2O_3, and ZrO_2. The anionic element or group was also found influential, the order of decreasing reactivity being: fluorides, nitrates, carbonates, chlorides, sulfates.

Baudewyns (1938) discussed the attack of refractory materials used in glassmaking. Hyslop et al. (1947) observed in bench tests that the order of decreasing acceptance of refractories for soda-lime glass is: chrome-magnesite, magnesite, alumina, 42% alumina firebrick, kyanite, chrome, forsterite, 76% alumina firebrick, fusion-cast mullite, 95% silica, zircon, and clay tank block; the order for lead glass is: alumina, 42% alumina firebrick, chrome-magnesite, magnesite, kyanite, forsterite, zircon, fusion-cast mullite, 25% silica, 76% alumina firebrick, clay tank block, and chrome; and for borosilicate glass: alumina, zircon, 76% alumina firebrick, fusion-cast mullite, kyanite, clay tank block, 95% silica, chrome-magnesite, magnesite, chrome, and forsterite.

Day and Ambrosone compared the solubility of single-crystal sapphire with Corhart ZAC refractory in contact with common lime glass at 1400 to 1550°C. The corrosion resistance of the sapphire at the metal line was found to be less than that of the commercial refractory. Busby and Eccles (1962) observed a wide variation in the rate of corrosion of single-crystal corundum immersed in four commercial glass melts (soda-lime-silica, barium crown, borosilicate, and lead) at 1400 and 1500°C. The location above or below the glass level affected the rate. At least two corroding methods were involved: (1) a surface tension effect at the glass line, and (2) a density-diffusion effect below the glass level. McCallum and Barrett found that with lime-alumina-silica slags, an increase in temperature of 10°C had a much more pronounced effect on the dissolution rate of single-crystal corundum than had an increase in the lime content of the slag from 35 to 55%. The variation in the rate of dissolution with basic oxide content was greater with sodium silicate slags. The dissolution of corundum by these slags is probably an anion-controlled diffusion process. Solomin and Galdina found the material most resistant to lime-soda glasses to be Bakor (14.3% ZrO_2, 12.9% SiO_2, 72.1% Al_2O_3), followed by fused high-alumina refractories. W. F. Ford applied phase equilibrium principles to the corrosion of acid refractories. Solubility data were used to assess the comparative resistance to glass of fused tank blocks.

Kunugi presented a survey of the corrosion of fused-cast tank blocks by soda-silicate melts. Two fusion-cast beta alumina refractory blocks lasted about 16 months as superstructure for sheet glass production. The decomposition products were corundum and a glassy phase (K. Kato). The addition of small amounts of ZrO_2 or Al_2O_3 to some base glasses ($R_2O-BaO-SiO_2$) was found to reduce corrosion of $Al_2O_3-ZrO_2-SiO_2$ glass tank blocks, ZrO_2 being more effective (Schlotzheuer and Hutchins).

Comeforo and Hursh stressed the effect of wetting as a factor in the penetration of molten glass into porous refractories. A break was observed in the relationship between the rate of penetration and the pore size of 99% alumina bodies with two liquids of markedly different properties. This break was ascribed to a change in pore volume with a change in pore radius.

Siliceous glass-furnace lining is made more resistant to alkalies by coating with a mixture of MgO and Al_2O_3; $3Al_2O_3$ and $2SiO_2$; or $2Al_2O_3$ and SiO_2 in an amount up to 10%, thus increasing the life of a furnace 3 to 4 times by this means (Parodi, 1957). Busby and Eccles (1961) observed that both fusion-cast zirconia-corundum and fusion-cast mullite refractories exude a glassy phase at above 1200°C, which leads to considerable sticking to, and subsequent reaction with, other types of refractories except those containing MgO. Very high-Al_2O_3 fusion-cast refractories were the most stable materials of those tested.

Steinhoff noted that fireclay brick (43% Al_2O_3) had a preferred position in the durability of checker brick for glass-melting furnaces, because it forms a nepheline, high viscosity coating with the soda in the alkali atmosphere. Corundum and zircon brick take soda from the alkali atmosphere and form beta alumina or sodium zirconate, which cause cracking because of the resultant volume expansion. Yamanouchi and Kato (1946) found, however, that beta alumina is superior to alpha alumina, diaspore, or pyrophyllite for glass melting crucibles, and ascribed this to the formation of a thin surface layer of nephelite ($Na_2O \cdot Al_2O_3 \cdot 2SiO_2$).

12-3-5 Calcium Aluminate Slags

Some slags produced in metallurgical operations fall in the $CaO-Al_2O_3$ and $CaO-Al_2O_3-SiO_2$ systems, and may have economic value because of their ability to develop hydraulic bonding power or because the alumina may be recovered. Nagai and Katayama found that the composition $3CaO \cdot 5Al_2O_3$ is a high quality refractory cement in combination with aluminous grogs in resisting corrosion by glass fluxes, slags, and ashes. The slags in the system $CaO-Al_2O_3-SiO_2$ show an exothermic peak at 750 to 1100°C by DTA, at which temperature crystallization occurs and the latent hydraulic property decreases.

Samaddar, Kingery, and Cooper (1964) found that the corrosion rate of single-crystal and polycrystalline corundum was the same in a 40 wt % CaO, 20% Al_2O_3, 40% SiO_2 anorthite slag, indicating that the presence of grain boundaries and crystal orientation does not have a significant effect on the corrosion rate of dense (97.2% theoretical) alumina in this slag. At 1350°C alumina is by far the most corrosion-resistant in comparison with mullite (92.5% theoretical density), anorthite (85% theoretical density), and vitreous silica; at 1500°C alumina is still superior, but the differences are less pronounced. Nagai, Suzuki, and Ota found that the mullite component of brick is eroded by molten lime cement; and gehlenite and glassy substances form. Diaspore is less resistant to molten cement than corundum. DeKeyser and Wollast measured the initial reaction temperature between lime and aluminosilicate refractories at 900°C. Carter and MacFarlane determined the sulfur equilibrium between $CaO-Al_2O_3$ slags and $CO-CO_2-SO_2$ gas mixtures at 1500°C, from which the activity of lime and alumina in these melts could be deduced. The approximate values for the free energies of formation of $3CaO \cdot Al_2O_3$, $12CaO \cdot 7Al_2O_3$, $CaO \cdot Al_2O_3$ from their constituent oxides were derived.

Sharma and Richardson measured the sulfide capacities and limiting solubilities of calcium sulfide in lime-alumina melts at 1500°C, using a gas and slag technique. The activities of lime and alumina were obtained over the entire liquid range.

Bertrand applied the ternary eutectic CaO-Al_2O_3-SiO_2 in a eutectic range suitable for making a slag for the reduction of lead and antimony in order to obtain these metals directly from the ores in the crude state without roasting or the use of costly flotation equipment. Toropov and Volkonskii synthesized slags from a mixture of $CaCO_3$, Al_2O_3, and SiO_2 by fusion in a Tammann-type furnace. The slags were rapidly cooled by water granulation. The vitreous products were crystallized, at 850°C for small crystals, chiefly spherulites; at 1300°C the crystals were much coarser. The crystalline slags were more active hydraulically than the vitreous. The slags heated at 850 to 1300°C developed higher strength than the original slag, attributed to the development of the crystalline phases $12CaO \cdot 7Al_2O_3$ and $CaO \cdot Al_2O_3$.

Volchek developed hydraulic slags ($12CaO \cdot 7Al_2O_3$, and $CaO \cdot Al_2O_3$), and Mori carbonating hydrated products from $6CaO \cdot Al_2O_3$ compositions in the system CaO-Al_2O_3-SiO_2. Ermolaeva investigated the viscosity of melts in the system. Samaddar and Lahiri concluded from tests on synthetic slags containing fixed ratios of SiO_2:Al_2O_3 but varying CaO content, that the liberation of lime during hydration is the mechanism of activation of the slags.

12-3-6 Aluminum Slag Reactions

Schurecht (1939) prepared refractory coatings with good adherence on firebrick by spraying a slip composed of clay and aluminum powder on the brick and firing. Similar coatings were produced by spraying molten aluminum on the firebrick. Hard, abrasion-resistant surfaces, obtained by this method, and the tight bond were ascribed to heat generated by aluminothermic reactions starting at 750 to 930°C, but capable of increasing the temperature of the coatings to 1500°C within five minutes. The coatings increased the resistance of the fireclay refractories to attack from basic slags, but reduced the resistance to acid slags.

Suzuki and Fujita stated that corrosion of chrome spinel and magnesia spinel by aluminum was negligible in comparison with that by cryolite. Brondyke (1953) found that all commercial aluminosilica refractories could be wetted and subsequently penetrated on exposure to molten aluminum. The penetration and associated pickup of silicon, produced by an aluminothermic reaction, was independent of the porosity, grade of refractory, and the source of the refractory material. The rate was controlled by the diffusion of aluminum and silicon. The reaction caused mechanical disruption of the refractories because of volume expansion. The rate of buildup and adherence to the aluminosilica refractories depended on the silica content.

A. F. Johnson (1950) claimed that a refractory composed of a fused mixture of aluminum nitride and alumina is particularly effective in resisting the attack of fused cryolite in the electrolytic reduction of alumina to aluminum. Aluminum nitride is produced by heating aluminum powder and fine charcoal to about 1000°C in air. The nitride grain is mixed with 50 to 85% alumina and fused in a graphite crucible at 2000 to 2300°C.

12-3-7 Miscellaneous Reactions

West and Gray (1958) found that the greatest reactivity in mixtures of extremely pure silica and alumina is encountered with stoichiometric ratios to form mullite, but not with the formula $3Al_2O_3 \cdot 2SiO_2$, rather with the formula $2Al_2O_3 \cdot SiO_2$. The reactivity also depends on the crystalline modification of the alumina used. A sharp exothermic differential peak at 980°C was ascribed to three simultaneous reactions dependent on the silica-alumina ratio of the mixture: (1) crystallization of gamma alumina, (2) crystallization of a hydrogen spinel HAl_5O_8, and (3) reaction of silica with the spinel to form mullite.

Eitel (1955) found none of the common refractory materials, including the aluminosilica and high-alumina compositions, satisfactory against corrosion of a fluoride-silica melt used in the manufacture of synthetic fluorine-mica. Ebner placed the order of decreasing resistance to ablation (surface loss by heating effects) of six refractory materials in fluorine-rich hydrogen-fluorine diffusion flames as follows: graphite, silicon carbide, alumina (Remmey AD 99), zirconia, and magnesia. Graphite and silicon carbide were the recommended types, particularly for flame-resistant deflectors to direct the flame away from the rocket motor base. Alumina and magnesia form liquids which flow from the impingement area.

Norton and Hooper found that commercial grades of firebrick containing 60% or more Al_2O_3 in contact with powdered magnesia showed no reaction up to 1540°C (2800°F). Brisbane and Segnit examined the slagging effects in rotary portland cement kilns. The interstitial material in both high-alumina and magnesite-chrome refractories was attacked. Alkali penetrated the brick, causing an increase in thermal expansion (beta alumina?), which in some cases induced spalling. Kao found that high-alumina bricks (80 to 85% Al_2O_3) provided good corrosion resistance over alkali electrofurnace roofs.

Slags produced in iron-titanium metallurgical processes average about 20% TiO_2 and 60 to 70% Al_2O_3, with small amounts of SiO_2, Fe_2O_3, CaO, MgO, and Na_2O. They are composed chiefly of corundum with TiO_2 in solid solution as colorless crystals in oxidizing atmospheres above 1450°C (Dolkart and Gul'ko).

Chaklader found that traces of alumina (0.22 and 0.44 wt%) catalyzed the transformation of quartz to cristobalite.

13 COLLOIDAL PROPERTIES OF ALUMINA

13-1 PLASTICITY

Alumina generally has been classed as a nonplastic material, without qualification or degree, in referring to its wet-forming behavior. This is too arbitrary. All the wet-forming operations, that are applicable to the plastic clays, can be successfully applied to calcined Bayer alumina, although perhaps not with the same facility. To the extent that crystal form and particle size contribute to plasticity, the crystals of Bayer alumina are a fair match of the clays. The form factor is lost substantially in alumina ground from fused or sintered alumina, however. The addition of as little as 3 to 5% of various ingredients to Bayer alumina compositions often provides adequate plasticity in many forming operations.

Schwerin (1910) and O. Ruff (1924) found it necessary to grind fused and Bayer aluminas with an acid to stabilize the suspensions for slip-casting—a forming process which depends little on plastic properties. The slow release of alkalinity from the Bayer alumina was a significant factor affecting its stability.

Green (1949), Ventriglia, and others have attempted to improve on the inadequately defined concept of plasticity developed by Bingham. Roller treated the water suspensions as solids rather than liquids. Under a compressive force, the dispersions undergo a permanent deformation which does not commence until a certain stress, the Bingham yield point, has been exceeded, a characteristic of the solid state.

Gruner described the fundamental mechanism of plasticity as it applied to amorphous and polycrystalline solid-liquid systems. Clay-water systems were considered too complex for a simple fundamental examination. In the simpler oxide-water systems, chemical forces of hydration of the crystal surface and of ion adsorption were believed to predominate in determining the plasticity. Three types of plasticity were considered: (1) true plasticity caused by reaction between the liquid and solid phases, in which particle size is of secondary importance, and materials with a larger lattice structure have the highest plasticity; (2) limited plasticity, which is caused by chemical reaction such as the formation of oxychloride on the particle surface (cation adsorption) or the adsorption of hydrophilic colloids (humic acid substances); and (3) false plasticity, or the surrounding of nonplastic particles with a highly viscous or sticky liquid or gel.

Haase (1957) concluded that a body has plastic workability if (1) the stress/strain relation shows a threshold value of stress, and (2) the stress necessary to produce a desired deformation is less than the cohesive strength of the body. On this basis the workability was defined as being proportional to strength/viscosity. The workability may be increased either by increasing the strength (organic binders) or by reducing the viscosity.

13-2 SURFACE CHARGE AND ZETA POTENTIAL OF ALUMINA

The surface charge which a powdered oxide forms in liquid suspension is thought to determine the extent of flocculation or dispersion and the viscosity of the suspension. These properties affect the value of suspensions in forming processes such as slip-casting. The mixing of different phases, as for example, alumina and clay in water suspension may be affected by a difference in the charge.

Schmäh observed that the surface of very pure alumina gel is alkaline. By electrophoretic methods Fricke and Keefer (1949) determined the isoelectric point or zero point charge (zpc) of gamma alumina to be at pH 9.0, that of amorphous $Al(OH)_3$ at pH 9.4, that of gibbsite at pH 9.20, and that of boehmite at pH 9.40 to 9.45. The potential-determining ions were considered to be H^+ and OH^-, which enter into electrochemical reaction at the surface in the case of aluminum oxide. The essential part of the surface reaction schematically is as follows:

$$\begin{array}{c} H \\ {>}O \\ H \end{array} (+) + H_2O \underset{H_3O^+}{\overset{}{\rightleftharpoons}} {>}OH \underset{}{\overset{OH^-}{\rightleftharpoons}} {>}O(^-) + H_2O$$

| Positive surface | Uncharged surface | Negative surface |

Tewari and Ghosh found values for the isoelectric point of hydrated alumina, precipitated and aged under the conditions specified, as follows:

Precipitation	Aging Conditions		
	Fresh preparation	Rapid aging by heating	Suspensions aged 4 months
Excess alkali	5.08	6.79	5.78
Equivalent alkali	6.63	7.28	7.06
Deficient alkali	7.29	7.43	7.32

The zeta-potential is considered by some to be a significant factor affecting the behavior of colloidal suspensions. It is the electrokinetic potential which represents the difference in potential between the immovable liquid layer attached to the surface of the solid phase and the movable part of the diffuse layer in the body of the liquid. O'Connor et al. (1956) found that the surface of natural corundum has a positive zeta potential by the streaming potential method in water at 17 to 20°C, but it changes to negative on heating to 1000°C. They suggested that the formation of a surface charge was probably caused by dissociation of a surface hydroxide, but the type of hydroxide formed depended on the reactivity of the surface; severe surface distortion, caused by grinding, might induce complete hydroxylation to $Al(OH)_3$, but less distorted particles might hydroxylate only to $AlO \cdot OH$. In any case, the hydrated aluminas are more stable in contact with water at room temperature than alpha alumina. They also suggested that heating the samples restored the corundum surface and annealed out the strains, hence $AlO \cdot OH$ was the main surface phase dissociating to produce a negative surface.

The zpc of corundum and the hydrated aluminas has been investigated also by Johansen and Buchanan; Modi and Fuerstenau; Gayer, Thompson, and Zajicek; Pike and Hubbard; Dobias, Spurney, and Freudlova; Robinson, Pask, and Fuerstenau; and Yopps and Fuerstenau. The reported values for zpc of corundum vary from pH 6.7 to pH 9.4; the values for the hydroxides vary even wider, from pH 5.1 to pH 9.4. This may result from the different methods of determination used: electroosmosis, microelectrophoresis, pH of minumum solubility, etc.; or perhaps impurities may be a factor. From the Helmholtz-Smoluchowski equation, in aqueous systems at 25°C, the zeta potential = 9.69×10^4 (E/P) λ (millivolts), in which E is the streaming potential, P is the pressure drop in the cell, and λ is the specific conductance of the solution.

Anderson and Murray correlated the characteristic minimum that occurs in the apparent viscosity pH curves for oxide slips to the zeta potential of the dispersed oxide. Bakardiev stated that the surface OH groups are distributed inhomogeneously on the surface of gamma alumina. Particles with greater pores (smaller curvature) have a higher OH group concentration than those with smaller pores, Koz'mina and Dobrynina found that the pH of calcined bayerite varied from 8.9 for material calcined at 1100°C to 3.2 for that calcined at 1600°C. The zeta potential of alpha alumina was markedly dependent on the calcination history.

13-3 FLOCCULATION AND DEFLOCCULATION EFFECTS

Much early investigative work was done to show the relative power of different ions in flocculation and in optimal thixotropic gelation of different types of colloidal suspensions. Comparative flocculation values for a positively charged alumina sol in millimoles per liter are as follows: NaCl, 43.5; KCl, 46; KNO_3, 60; K_2SO_4, 0.30; $K_2Cr_2O_7$, 0.63; $K_3Fe(CN)_6$, 0.080 (Ishizaka, 1913). The ion in the electrolyte of opposite charge to that on the alumina sol is the effective ion. The increased potency of the effective ion with increasing valence is similarly reflected in the relative values for thixotropic gelation in millimoles per liter for an alumina sol: KCl, 330; Na_2SO_4, 3.5; $K_3Fe(CN)_6$, 1.5; $K_4Fe(CN)_6$, 1.0 (Aschenbrenner, 1927).

Tar et al. concluded that aluminum hydroxide obeys the Freundlich absorption equation for dye absorption: $x/m = ac^{1/n}$, in which x is the weight of substance adsorbed by a weight m of adsorbent from a solution whose volume-concentration at equilibrium is c, and a and n are constants. In the presence of the dye and an adsorption peptizer the solubility of the gels increased in the order: base-gel, gel treated for 6 to 12 hours, gel treated for 1 to 3 hours.

Raychaudhuri and Hussain compared the buffer curves of freshly prepared gels of alumina, silica, and mixed aluminosilicates of varying ratio of SiO_2/Al_2O_3 with similar gels that had been aged for one year, as well as with some native clay minerals, bauxite, halloysite, kaolin, limonite, and montmorillonite. The freshly prepared materials possessed much less buffer capacity, relatively. Their buffer capacity passed through a maximum with increasing SiO_2/Al_2O_3 ratio, but the aged gels' buffering capacity continued to increase, attaining a maximum value with pure silica gel. Krleza found that the mutual flocculation of $Al(OH)_3$, $Fe(OH)_3$, and silicic acid sols, when their constituents were present in about the proportions found in bauxite, differed little between nascent and aged conditions. Decreasing concentration caused a decrease in turbidity, and increased the pH required for flocculation. At maximum turbidity, negative and neutral potentials were observed in the different sols; some were acid, others neutral or basic. The flocculated material was acid, but all three ingredients took part in the flocculation.

Howard and Roberts (1953) reported a continued increase in pH with aging of suspensions of commercial Bayer alumina that had been adjusted with hydrochloric acid to the pH range 0.4 to 8.6. The time required for final equilibrium and the final pH depends on the ratio of Al_2O_3/acid. The change on aging was ascribed to the gradual diffusion of sodium aluminate from the interior of the Bayer grain. This effect has a marked influence on the stability of aqueous alumina suspensions for slip casting or extrusion.

Shulz classed alumina as an acid oxide with SiO_2 and B_2O_3. Voitsekhovskii and Vovnenko found that the potential (E_o) of Al_2O_3 suspended in water or in 4.5N NaCl,

4.5N LiCl, or 1M Na_2SO_4 solutions became positive with respect to the potential of the liquid, as the pH of the suspension was raised. The curves going in both directions of pH formed a hysteresis loop, which met only at the points of origin of both low and high pH.

Shiraki investigated the slip-casting properties of calcined aluminum oxide. An addition of gypsum to the suspension increased the viscosity, but additions of $AlCl_3$, $MgCl_2$, $CaCl_2$, HCl, H_2SO_4, and HNO_3 yielded the maximum viscosity. Casting in an acid medium was favorable, but became difficult at pH values below 3.0, and also in the alkaline range. At pH 7 to 8, cracks occurred in the body and the body stuck to the mold. Above pH 9.2 the strength of the cast body suddenly decreased. The addition of 5% alumina, which had been heated in concentrated HCl, to nontreated alumina gave a body of high strength. In casting alumina slips, HCl and HNO_3 were considered more suitable peptizers than H_2SO_4 or an alkali. Alumina, calcined at 1300°C and having hygroscopic properties, was found to make a good casting slip when treated with hot HCl. Barium compounds were detrimental as binders or mineralizers, but mixtures of talc with 3% ZnO, ZrO_2, $MgCl_2$, $SnCl_2$, or $AlCl_3$ were considered necessary to good sintering.

Hauth (1953) stated that acid-leached, ground, fused alumina can be deflocculated in either an acid or a basic medium. The viscosity, dilatancy, yield value, and sedimentation volume of both types of suspension were related to the concepts of diffuse double layers and zeta potentials. Hauser (1952) has drawn attention to the shortcomings of the Helmholtz double-layer theory and of the Gouy-Freundlich diffuse double-layer theory. A new concept of solid double layers caused by dipole formation in the surface layers of colloidal matter was presented.

Howard and Roberts (1951) investigated the effect of various electrolytes on the stability of a commercially pure alumina suspension by observing the change in concentration with time at a fixed distance below the surface. In water alone, the suspensions flocculated with time. Neutral salts flocculated the suspension, but no further deflocculation took place on further additions. Flocculation and deflocculation, following the progressive addition of an acid or acidic salt, was explained by a reversal of the charge on the particles, a theory which was supported by electrophoretic measurements from which the mobilities of the alumina particles could be calculated.

Ludvigsen and Andsager mentioned that ceramic processes require rather high concentrations of body or glaze slips to prevent unmixing of the plastic and nonplastic components. In peptized suspensions for casting porcelain, a concentration of 65 to 70% is necessary, but it is often possible to prevent unmixing by a 15% concentration in flocculated suspensions. Barium chloride was found to have a marked flocculating effect on thin suspensions of kaolin and feldspar, so that these suspensions can settle without losing the original composition in the upper and lower layers. Similar considerations are applicable in the case of alumina slips.

Van Olphen (1964) discussed the nature of the flocculation behavior of hydrous clay suspensions in relation to the anisometric shape of the particles and the double layer structure. The interior net charge on the crystallites is compensated by positive counter ions adsorbed on the flat face surfaces which carry a negative electric double layer of constant charge. At the edges of the crystal plates, a broken bond surface is exposed. Under neutral or acidic conditions, the double layer on the edges is positive. The formation of voluminous flocs or gels is ascribed to crystal orientations edge-to face, but this formation can be prevented by the adsorption of suitable anions at the edge surfaces. Presumably, a similar situation holds for alumina crystals in suspension, but with the charges on the particles which apply to this system.

G. W. Phelps (1961) described the effects of soluble sulfates on the casting properties of ball clays.

The sensitivity of colloidal suspensions of alumina to very small amounts of electrolytes indicates the importance of the quality of the water used in preparing wet suspensions for forming operations. Phelps and Maguire (1956) discussed the effect of the usual soluble-salt impurities in water on the plastic clay systems. Durum discussed the quality and availability of suitable water for ceramic purposes. Morrison (1941) and W. E. Coombs stressed the need for softening or de-ionizing ceramic-processing water.

Nirmala and Srivastava observed that stable emulsions of aluminum hydroxide could be prepared in nonaqueous liquids such as kerosene-aniline. This was not possible with other dry solid oxides such as Cr_2O_3. In this case the aniline was the continuous phase. Colloidal suspensions are often stabilized in hydrophobic liquids by the addition of small amounts of polar compounds, such as oleic acid.

Iler (1964) stated that the mutual absorption of colloidal silica on alumina and of colloidal alumina on silica and silicate materials occurs at about pH 4. The film is essentially a monoparticle layer. Positively charged alumina particles acquire a negative charge from the adsorbed silica. Since the coated particle has the same charge as the surrounding silica colloidal system, no flocculation occurs. This well-known protective principle can be applied to prevent the thickening or flocculation of alumina when mixed with clays or other negatively charged ceramic components. On the other hand, Okura et al. (1963) stated that the presence of silicic acid or of SO_3 tends to shift the isoelectric point of aluminum hydroxide toward its acid side, increasing its negative charge, which aids flocculation.

Kukolev and Karaulov investigated the rheological properties of acqueous suspensions of fine-ground, technical-grade, calcined alumina over a wide range of pH and concentration of water suspensions. The alumina had been precalcined at 1450°C. Two pH ranges were found in which maximum zeta potential values were reached, corresponding to the minimum viscosity and also to the viscosity of the limiting structural breakdown. Applied to casting slips, higher densities were attainable with acid pH, preferably under de-airing conditions. Heating the slips to 30 to 40°C reduced the viscosity and increased the casting rate. Dextrine was considered the best bond for increasing the wet strength.

Graham and Thomas found that hydrous alumina reacts in at least two steps with acid, the initial step being attributed to a rapid conversion of surface hydroxo groups into aquo groups. A subsequent slower reaction is deolation of the oxide structure. The reactivity decreases with increasing calcination temperature of the alumina, the effect being ascribed to increasing oxolation of the alumina structure. With increasing temperature, the number of surface hydroxo groups passes through a maximum and becomes zero for fully calcined alumina. The rate of neutralization of various 0.2N acids by hydrous alumina at 25°C was most rapid for HF, and decreased in the order: H_3PO_4, HC_2O_4, H_2SO_4, HNO_3, HCl, CH_3COOH, H_2NSO_3H, and $HCLO_4$, which agrees reasonably with the tendency for the negative ion to form stable aluminum complexes. Graham and Horning found that the addition of hydrous alumina at 25°C to 0.1M or 0.05M solutions of the potassium salts at pH 7.0 increased the pH to the following values: fluoride 9.6, oxalate 8.6, sulfate 8.0, nitrate 7.3, chloride 7.3, perchlorate 7.1, and phosphate 7.0. The increase was attributed to the displacement of OH groups from the surface of the alumina by the anions. Marion and Thomas claimed that the pH of maximum yield of aluminum hydroxide resulting from the addition of hydroxyl ion to solutions of aluminum ion is not a constant, but varies with the nature and concentration of the anion present. Fluorides or oxalates raise the pH of maximum precipitation; sulfate, lactate, and chloride gradually lower it; while citrate and tartrate cause a sharp drop. These investigators advanced the theory that the basicity of the anion, the binding affinity of the anion for aluminum, and the resistance of the bound anion to displacement by added hydroxyl ion account for the change in pH.

Fricke and Schmäh said that the solubility of very pure aluminum hydroxide in water is less than 10^{-6} mole/liter. According to Szabo et al. (1955), the solubility products for aluminum hydroxide are as follows:

	20°C	30°C
$(Al^{3+}) \times (OH^-)^3 =$	1.25×10^{-33}	1.92×10^{-32}
$(AlO_2^-) \times (H^+) =$	1.80×10^{-13}	1.34×10^{-13}

Ashley and Bruni prepared colloidal alumina sols by reacting weak aliphatic acids with fine aluminum pellets. Ziese et al. prepared them by reacting precipitated aluminum hydroxide containing 0.1 to 5.0% of the anions of strong inorganic acids to steam at 5 to 12 atmospheres for from one to 10 hours. Gregg (1960) discussed the preparation of highly dispersed colloidal solids by various methods including fine grinding, calcination of a decomposable salt, sublimation, precipitation, and removal of one constituent by dissolution or evaporation. Bugosh (1956) prepared a stable colloidal, amorphous, hydrated alumina sol by reacting a water-soluble aluminate with a monobasic acid

having a dissociation constant greater than 0.1, and an aluminum salt of the acid within the pH range 3.5 to 5.5. Conrad and Lenne prepared peptizable sols by milling bayerite, $Al(OH)_3$, until only a trace of the bayerite X-ray pattern was detectable. Yuille prepared a stable thixotropic aqueous dispersion by adding a water-soluble aluminum salt to an aqueous solution of an alkali metal base until a pH of 3 to 5 was reached, then by adjusting the pH to 6 to 9 with ammonium hydroxide, followed by dispersing the precipitate by means of a water-soluble phosphate, and by milling the dispersion until the particles averaged one micron or less.

Weiser, Milligan, and Purcell (1941) observed that the alumina floc precipitated in water purification practice at pH 5.5 to 8.5 is boehmite, free of basic aluminum sulfate. Flocs precipitated from aluminum sulfate below pH 5.5, particularly after aging for 24 hours show the X-ray pattern for basic aluminum sulfate, $Al_2O_3 \cdot SO_3 \cdot 1.5 H_2O$. Gels prepared from aluminum nitrate and aluminum chloride solutions at a pH around 4.0 conform to boehmite. The formation of basic aluminum sulfate has been observed to occur in hot, concentrated, highly acid solutions of aluminum sulfate on dilution, if the water is not mixed in rapidly.

13-4 ADDITIVES

Several classifications of organic binders for ceramic forming have been presented, as for example, that of McNamara and Comeforo (1945). The natural and synthetic resins and other temporary ceramic binders, applied in adjusting the rheological properties of clay ceramics, are also applicable to alumina.

Luks (1956) classified the types of binders for alumina used as aids in improving the unfired strength and to facilitate machining operations in three general groups. These include: (1) the water-soluble types (dextrine, gum arabic, polyvinyl alcohol), (2) the water-swelling types (gum tragacanth, carboxymethyl cellulose, starches), and (3) emulsion types (wax and resin emulsions). As a class, the water-soluble binders afford the greatest increase in strength, but they tend to migrate, causing difficulty with prefired machining, and in addition promoting the formation of hard-shell, soft-center forms. The water-swelling types create plasticity, bond strength, and generally uniform distribution. Difficulty is sometimes encountered with slow drying rates, resulting in fractured ware. The emulsion types contribute appreciably to internal lubrication, especially advantageous in dry or nearly dry forming operations.

Brown and Coffin claimed a marked freedom from migration for polyvinyl alcohol in partly or completely hydrolyzed condition. Popil'skii and Nemets preferred oleic acid to paraffin as a plasticizer for highly dispersed alumina mixes, but the amount used (18%) was excessive by present standards. Lawson and Keilen mentioned the water-soluble forms of pine-wood lignin as deflocculants in casting operations. Bartos described a method for evaluating the gel-forming strength of bentonite-water suspensions in forming operations. Krannick (1961) stressed the use of ethyl silicate and its hydrolysis to polysilicic acids with subsequent gelling and dehydration to solid SiO_2. Hydraulic bonding is provided up to 230°C, and ceramic binding at higher temperatures. The binder is often used for quick set of dipped coatings.

Some organic binders have high ash contents (Ref. Organic Binders, 1962). Alginates contain 25% ash, as well as requiring a bactericide. Methyl cellulose and polyvinyl alcohol have not over 1% ash, but the latter is precipitated by many sodium salts (Na_2CO_3, $Na_2B_4O_7$, etc.), it forms thermoplastic films, has a tendency to foam, and decomposes at 180°C. Methyl cellulose is soluble in cold water, is not ionized, does not react with metal salts, and decomposes at 230°C.

Budnikov, Marakueva, and Tresvyatskii claimed that alumina wax-stearin-oleic acid suspensions exhibited thixotropic and dilatometric properties.

Isostatic pressing requires less than 4% wax emulsion lubricants for satisfactory molding of small forms.

14 GRINDING CERAMIC ALUMINA

High mechanical strength in sintered alumina requires fine grain size of the heat-matured body. Since some grain growth occurs during sintering, the raw alumina must be still finer. Fine grain size, achieved by grinding massive particles to size, requires the expenditure of much energy. Fine grain size can also be obtained by producing the raw alumina in the desired crystal size, by controlled calcination of Bayer hydrated alumina or an aluminum salt. In this case the separation is easier because of a weaker particle bond. Owing to the platy nature of the Bayer crystals, the ground alumina possesses more physical anisotropy than massive-ground particles, and might be expected to behave differently during forming and firing of ceramic articles.

The use of different types of grinding equipment alters the properties of the ground alumina. This may be observed as a change in the length of the ground particle size distribution or as a difference in surface area. Vydrik stated that the sintering density of the body at a given firing temperature is related primarily to the inverse square of the mean surface radius of the particles. Ease of flow and of compaction of the ground material are influenced by the length of the size distribution, as is also the pelletizing ability.

Avgustinik observed that hexagonal alpha aluminum oxide is more brittle than stabilized zirconia. The grinding rate constant shows good correlation with impact strength. Materials with higher coefficients of thermal expansion have lesser bond strengths and higher values of grinding rate constant. The mean square thermal displacement values showed good correlation with microbrittleness, while higher values of the grinding rate constant corresponded to higher values of shear modulus. No correlation was found between microhardness and the grinding rate constant. As the grinding rate constant is proportional to the effective power of the mill, mills operating in the brittle rupture range are the most effective.

Bond (1963) has discussed recent advances in grinding theory and practice. Three principles of particle breakdown were defined and compared with the Rittinger and Kick theories. Equations were given for calculating product sizes from surface area measurements, and for calculating power requirements and mill capacities. Forchheimer discussed the hardness scales and the factors that affect the toughness of alumina (crystal size, chemical composition, porosity, grit size, and grain shape).

Several different types of grinding equipment have become available during the last two decades for preparing ceramic materials in the finer sizes for forming operations. These include, among others, various types of fluid energy mills, vibrating ball mills (Sweco), and the Attritor. Improvements in conventional ball milling have also been realized with the development of high-alumina mill liners and grinding balls, rubber liners, air-swept mills, Hydroclone classifiers, and other classifying means for removing the ground product from the sphere of action efficiently. The Stag mill is a ball mill provided with peripheral slots to allow removal of the completely ground material from the mill. It should be advantageous in grinding massive alumina. Novel grinding methods include igniting an explosive mixture and allowing the waves to pass in disintegrating contact with the solid material, and continuously repeating the operation (Haltmeier, 1958), or generating compressional shock waves of ultrasonic frequency, adapted to produce alternate compressive and expansive forces in a small area of the material (Nilsson, Nilsson, and Hagelin, 1961).

Each of these grinding devices has its advantages and disadvantages. The air classification systems are limited in their ability to separate finer particle sizes. The fluid energy mills generally produce narrow, less plastic size distributions, devoid of extreme fines, and have poor efficiency in breaking up massive alumina such as the fused and tabular grades. The ground products have markedly different surface properties than the usual ball-milled products, as indicated by standard oil titrations. The Attritor may accomplish in a few minutes the same extent of grinding that requires hours of ball milling, but wear and contamination from the equipment, and power consumption may be high. It is also necessary to operate in a liquid suspension.

Bergmann and Barrington claimed an increase in surface area for alumina of at least 350%, on the basis of X-ray line broadening, or 29% by actual measurement, of explosively shocked powder. Heckel and Youngblood calculated the stored energy (V) resulting from the microstrain to be 0.37 cal/g for alumina, using the formula: $V = 2.81 \, Ee^2$, in which E is Young's modulus (55×10^6 psi). Lewis and

Lindley estimated strains in alumina of the order of 7 cal/g by ball-milling.

Ceramic-grade alumina is usually wet-ground in conventional ball mills in present commercial operations. Tabular alumina for spark plug manufacture is ground with steel balls, which requires that the product be leached with mineral acid to remove the iron pickup. This operation entails careful handling of the iron solution and of washes to prevent hydrolysis and adsorption of the iron salts on the alumina surface, and to avoid loss of the finer fractions of alumina.

Surface active agents such as Darvan may be added during ball milling to reduce changes in the viscosity of the slurry, and to allow use of the minimum amount of water. Pickup of contamination from mill and ball wear in bench-scale tests were found to be about ten times higher in wet ball-milling than in the equivalent dry ball-milling in grinding reactive alumina. Slightly higher reactivity, found for the wet-milled product, is attributed to the pickup of finely divided impurities. When using a 30/1 ratio of grinding balls to alumina, required for supergrinding reactive alumina, pickup from the balls and mill (two-thirds from the balls) amounted to 0.88%/hr in the ground alumina. This indicates the desirability of approximately 99.5% Al_2O_3 balls for grinding high-purity alumina. Kitaigorodskii and Gurevich recommended changing the ratio of water to alumina from 1 to 0.75, at the same time adding to the mill an adsorbing agent, such as sugar (0.1% by weight of the water), in order to reduce the grinding time by two-thirds.

Dry ball-milling involves phenomena that are not significant in wet-milling, such as mill packing and agglomeration of the fines. Some aluminas, particularly those containing small amounts of alkalies, are prone to pack when the average particle size approaches one to two microns. The formation of a coating on the balls and mill lining marks the practical limit of size reduction. Bond and Agthe ascribed this effect to (a) static electricity, (b) adsorption, and (c) mechanical impact. They recommended the addition of 0.75% water and 0.13% of a patented lignin compound to prevent the effect.

Heat was found to be an effective preventive of mill packing, and means for supplying it to mills have been devised (Russell and Hardinge). Grinding in a vacuum has practically no effect on the grinding efficiency of ball mills (Viro). The idea of grinding aids developed from the addition of wetting agents. Byalkovskii and Kudinov observed that the hardness of a specific material (a ceramic color) varied as a function of the medium in which it was ground; when dry-ground in air the hardness index was 1910 (sclerometric), whereas in oleic acid it was 420. Berry and Kamack recommended naphthenic acid as a grinding aid, added in small increments to keep pace with the rate of new surface produced.

A. V. Somers (AC Spark Plug Company) contributed significantly to the technical grinding of reactive alumina by demonstrating the effectiveness of a seemingly high ratio of grinding ball weight to alumina weight, 32:1. A charge of 125 grams of alumina of normal Bayer particle size distribution "supergrinds" within 5 hours at 50 rpm when using 4 kilograms of one-inch balls (85% Al_2O_3) in a 1.3-gallon ceramic mill. Efficient grinding ensues because substantially all the alumina being ground coats the grinding balls early in the course of grinding, and the material remains continuously in the sphere of action. An effective grinding procedure at a ratio of about 20:1 has been developed, and has been translated to the plant scale under conditions that enable the clean separation of the superground alumina from the balls and the mill.

Rose and Sullivan developed a general equation to explain the factors involved in dry ball-milling, indicating that the number of impacts per piece of grinding medium per mill revolution is a constant. This, plus the assumption that the energy per impact is linearly related to the mill diameter, enabled Hart and Hudson to derive an expression: $K = RNDt$, in which K is a constant that yields equivalent properties (e.g., equal fired density), R represents the ratio of grinding medium to alumina being ground, by weight, D is the mill diameter in feet, t the hours of grinding, and N the revolutions per hour. K was found to have a value of about 230,000 for low-soda reactive alumina, applicable for mill diameters ranging from about 1.5 to 4 feet or more, and for ball sizes of about 0.75 to one inch (85% Al_2O_3). The constant K conforms closer than stated if the expression is corrected for the "effective" mill diameter by deducting the ball diameter in each case. A grinding aid (0.3% naphthenic acid), and a compaction aid are generally used. Hydrophilic and hydrophobic grinding aids have been developed. Green densities exceeding 3 g/ml at only 5000 psi forming pressure are attainable, making it possible to sinter an all-alumina composition to a density of 3.88 to 3.90 g/ml within one hour at 1700°C at less than 9% shrinkage.

Rose and Trbojeric showed that the charge in mills at 1.5 times the critical speeds oscillates in a pendulum-like manner; the leading edge of the charge ascends to an altitude of about 45 degrees above the horizontal, and then, owing to deceleration of the charge, because of slippage along the shell, the leading part collapses and falls as a coherent mass. This produces a heavy blow which can cause more grinding than merely the sliding of the charge on the shell.

McCreight (1954) correlated the ceramic properties (compressive strength) of alumina and an alumina porcelain with ball-milling time, granulation, and dry pressing. Minima in the curves of the single-component oxide were found between 30,000 and 50,000 revolutions of the ball mill. Granulation by slugging, that is, by forcing pressings at 6000 psi through a 10-mesh sieve, provided 30% greater compressive strength, 1.5% less shrinkage, and 1% less absorption in the specimens, as compared with specimens formed from ungranulated material. High-pressure hydrostatic forming was preferred.

Poluboyarinov and Kirshenbaum claimed that dry grinding of Bayer alumina with small steel balls is faster than grinding with porcelain balls. It requires only four to six hours with steel balls, while 10 to 15 hours are required with porcelain. The amount of iron pickup is small and does not affect the quality of the production for many purposes. Actually, the Bayer structure of the calcined aluminas can be substantially separated into unit crystallites within one-half hour in a laboratory-scale one-gallon ball mill, rotating at 70 rpm, and charged with 12 kilograms of 1/4-inch steel balls and one kilogram of alumina. With this ball size, attrition of the crystals is relatively very slow. The product is suitable for appraising crystal size distribution, but is too discolored for fine ceramic ware.

Poluboyarinov and Ershova (1949) found that alumina, ground with porcelain balls, does not attain the proper sintered condition until 1750°C has been reached. Alumina ground with steel balls (HCl-leached) attains the same sintered condition at 1600°C. Alumina prefired at 1450°C sinters much better than unfired alumina. Rybnikov suggested that Bayer trihydrate (gibbsite) can replace calcined ceramic alumina, and claimed that it grinds more easily in ball mills, needs no precalcination, and sinters more readily with clays. This was refuted by Arandarenko and Poluboyarinov. Bayer trihydrate is usually quite difficult to grind because of its gummy nature. It has been added in small amounts (about 5%) to alumina during grinding to improve the isostatic molding properties. It has an adverse effect on the reactivity and green density of reactive aluminas, however.

Berry and Kamack investigated the surface activity during dry ball milling, and claimed that there would be no limit to attainable particle fineness except for the tendency to reagglomerate. The practical limits in their dry-grinding tests on ilmenite, limestone and silica, however, were about 1.0 to 1.3 microns, unless a grinding aid was used to prevent packing. They found an increased grinding rate with decreasing ball size, but the minumum effective ball size depends on the mill diameter. With balls between 1/2 and 1/16 inch diameter, the ball size and the mill size are both significant factors affecting the grinding efficiency.

Reagglomeration during dry grinding is a very advantageous factor affecting the reactivity of alumina in dry forming processes. The reagglomerated particles contribute to better free-flow in transport to molds, and are relatively denser than agglomerates formed by wet methods, thus providing higher compacted density. Crystal orientation is likely to be more randomized. The addition of lubricants and binders can be restricted to the surface requirements of the pelletized particles, thus reducing the amount of burnout necessary in the subsequent firing operation. Papadakis summarized research on the ability of a material to break down and to reagglomerate, and derived equations for the optimum use of ball mills.

The agglomerated nuclei formed during ball-milling act as seed for dry pelletizing operations. Actually, the pelletizing equipment may consist of a small horizontal cylinder, rotating at a speed sufficient to cause cascading of the powder. Lubricants, compaction aids, e.g., stearic acid, wetting agents, and temporary binders can be added during the grinding and pelletizing operations. The pelletizing operation may be performed in heated apparatus to enable the use of high-melting waxes or to remove moisture. Most isostatic compositions require removal of the free water below about 0.2%. Also, because of the high compaction densities attainable with the modern reactive aluminas, it may be necessary to preheat pressed forms at temperatures just sufficient to volatilize organic additives prior to firing.

Alternatively, the agglomeration and pelletization effects may be destroyed, often with some difficulty, by aqueous suspension. If hydrophobic additives were used in the grinding or pelletizing operation, their resistance to wetting can be circumvented by the addition of soaps or detergents to the alumina. The main advantage of completely dry forming operations, namely high compaction density, is substantially lost by converting to wet operations.

Okuda et al. described the effect of rubber linings about one-third of an inch thick on the efficiency of wet milling alumina and other hard materials. Rubber linings withstand hard wear relatively better than ceramic linings, but ball wear is greater with the rubber lining.

Sintered alumina grinding balls (99.5% pure), formed by isostatic molding and fired at temperatures above 1550°C, may have densities exceeding 3.88 g/ml. A common appraisal for resistance to wear of alumina grinding balls consists of charging 6 kilograms of the balls (1-inch diameter) with 200 g tabular alumina, minus 325-mesh, and 150 ml water, to a 1.3-gallon mill and rotating at 70 rpm for a 2-hour period. The ball charge is washed, dried at 110°C for 15 hours, and weighed. The test is repeated until the weight loss remains constant, renewing the charge of tabular alumina for each 2-hour period. Wear losses not in excess of 0.1%/hr are considered satisfactory. Ball wear appears to be proportional to the surface crystal size, suggesting that impacts during ball-milling cause the loss of particles in crystal units rather than by wear.

The advantage of alumina in comparison with steel grinding balls lies in its freedom from introducing objectionable discoloration, as in the grinding of pigments, glazes and colors. The advantage of alumina in comparison with porcelain or native ceramic materials is its superior hardness and relatively higher density. Clay discussed the use of high-density grinding media particularly for milling porcelain enamels.

15 FORMING ALUMINA CERAMICS

The early forming methods for alumina ceramics were taken directly from methods in vogue for clay ceramics (B. Moore, 1942; Kiley, 1946). The poor plastic properties of the available alumina types and lack of knowledge of their physical properties restricted the scope of forming methods. Slip casting, dry and semidry cold pressing, extrusion, isostatic forming, and the Verneuil process were among the available methods. Ball rolling was soon developed (Derfler). More recent developments such as hot pressing, hot rolling, flame spraying, and fusion casting appear to have developed with aluminum oxide as the principal agent.

The physical properties of alumina particles on which the forming and firing behavior depends, have been discussed to some extent in other sections. The particle form or habit can range from completely anhedral crystal fragments to substantially well-developed crystalline platelets. The ability to control particle size distributions in the subsieve range, so as to obtain higher packing densities during forming has been mentioned. The adverse effect of porosity in incompletely ground Bayer particles on the fired density is a factor. Methods for improving the flow of dry particles in the subsieve range by pelletization and the incorporation of lubricants and binders in the pellets have also been considered.

Various forming operations have been discussed in *Critical Compilation of Ceramic Forming Methods*, A. G. Pincus, project coordinator, January 1964. It contains an extensive bibliography on "Characterization of Particles and Assemblies of Particles" by Mular and Mee, covering the period 1958 to July 1963, and another on "Forming of Particles" by Pincus, Rose, and Mee for the period 1955 to July 1963. Many of these latter references apply specifically to the formation of alumina, but in many, forming is incidental to other properties, particularly fired properties, which are not indicated by the classification method used. In the cited survey the role of particle preparation as an important factor in forming operations was emphasized, and different methods of producing fine particle size were mentioned, including: comminution, precipitation, atomization, thermal decomposition, vapor deposition, and thermal quenching, all of which can be applied to alumina particle production.

In the cited survey the forming processes were arbitrarily divided into the classification: cold forming, hot forming,

melt forming, and miscellaneous. Cold forming includes such operations as: rigid die forming; flexible die forming (isostatic, polymeric die liners); extrusion; vibratory compaction; plastic forming (organic aids); slip-casting; jiggering; impacting; ramming; cement gunning and casting; surface coating from suspensions (enameling and dip coating); spraying and slush coating; electrostatic spraying; and possibly other distinct modifications.

Hot forming includes many of the preceding dry operations performed at high temperatures and pressures and combines forming and sintering operations in many cases. Isostatic hot pressing has been applied to specimens in a volume 1.5 by 3 inches at 2200°C and 40,000 psi. Larger installations capable of forming specimens 13 inches by 4 ft, operating at 1550°C and 15,000 psi helium gas pressure have been constructed. Hot rolling of ceramic plates and sheets has been performed on isostatically formed bodies, preheated to 1520°C with metal rolls having ceramic shells of sintered alumina.

Melt-forming processes include operations most generally applicable to glass, but also applicable to alumina or to high-alumina compositions, as in fiber drawing (Fiberfrax, high-alumina glasses); fusion casting (abrasive and refractory fused alumina, and high-alumina cast refractories); flame sprayed coatings; and crystallization of glasses, self-nucleated; sprayed forms (free standing shapes); single-crystal-growing by flame fusion (Verneuil), temperature gradient in molten alumina or in saturated molten lead salt, or by vertical pull (Czochralski).

The miscellaneous forming methods in the cited reference include: forming from vapors; forming with fibers; electrophoretic-forming; impregnation; reaction sintering (particularly applicable to alumina as one final component of the thermitic reaction); high-energy rate forming; and machining and grinding.

15-1 COLD FORMING OF ALUMINA

Hofmann, Scharrer et al. (1962) discussed the fundamentals of dry pressing. Bodies consisting of clays or oxides, when dry pressed, contain some water. The main cause of the cohesion of these moist particles is the Coulomb force between the ions adsorbed on the surface of the particles. The cohesion is greatest with platelike

121

particles as is the case with Bayer alumina. The cohesion also increases with increasing valency (in the order Na, Ca, Ba, La, etc.), which is the basis of the Hardy-Schulze flocculation rule. An unexpected result is that the greater the cohesive strength of the moist framework, the less the shrinkage during drying and the lower the compaction density. They found that lanthanum ions provided very high strength with decreased plasticity. An increase in dry strength resulted when the calcium ions of natural clays were replaced by sodium ions.

Haase (1960) investigated the mechanism of dry pressing. The facility with which the pressure was distributed through the dry powder in a steel die was measured as the ratio of the pressure transmitted to the bottom punch to that applied by the top punch (Pb/Pt), the difference representing the friction between the particles and each other and the wall of the die. A pressure-calibrated, mercury-filled steel chamber measured the compression. For dry samples Pb/Pt decreased exponentially as the height of the pressed cylinder increased and was approximately 0.5 for height/diameter = 0.5. As the moisture content in the sample increased (up to 8%), Pb/Pt increased. Unsaturated long-chain fatty acids (especially oleic acid) increased Pb/Pt, but lubricants such as machine oil were no more effective than water. Oleic acid also formed a more or less permanent film on the die, which greatly increased the transmission of pressure. Entress and Steiner formulated an organic binder for compression-molding alumina, consisting of polyvinyl acetate, glycerol, oil, and cellulose, which requires no heat treatment to set or decompose the binder. Bruch (1967) discussed the problems involved in die-pressing submicron-size alumina powder.

Cooper and Goodnow (1962) found that radiographing of lead grids introduced into a powder provides a simple means for inferring the density distribution after compaction. The technique was used on alumina dry-pressed in one direction, and showed qualitatively that wall friction is the major source of density variation. W. C. Allen (1961) reviewed the development of the fundamental mathematical relations between compositional, pressing, and firing factors for ceramic materials. It was believed that apparent fired density is mathematically predictable from the volume shrinkage, pressed density, and ignition loss. The temperature required to meet a desired density is affected by the amount of additives, particle size, forming pressure, and the time of holding the maturing temperature. The ratio of the linear shrinkage values parallel to and normal to the pressing direction (HS/LS ratio) is a function of the particle orientation, and it allows the estimation of linear shrinkage from the volume shrinkage. The HS/LS ratio is affected by the type of additive, granule density, granule size distribution, and forming pressure. Linear shrinkage can be controlled by the use of controlled firing techniques which result in constant density and loss on ignition. Meerson (1962) concluded that the compressive force consumed by external friction may be several times higher than the extrusion force for brittle, hard materials. Friction between

powder particles was not considered a cause for non-uniform density in the compact, and the assumption that the walls exert a direct compacting effect was rejected. External friction is the main cause of density gradient in specific directions in the compact.

Croskey (1949) developed a production method for casting slip relying on controlled weight and casting rate. Hauth described the mechanics of slip casting using fine-ground fused alumina fines. Deflocculation could be attained by the addition of the proper amount and type of either an acid or base, following ball milling and acid treatment to form a suspension in water. Both drain and solid castings were possible from acidic or basic slips, through a range of pH from 3 to 12.5, and of density from 2.60 to 3.00. Shiraki observed the effect of grinding time on the particle size distribution, the effect of concentration and particle size on the viscosity of the slip, and the effect of temperature on the casting operation. He found that the viscosity of Bayer alumina slip changes during aging, and that the most suitable condition for casting is attained after more than a week's aging. Casting in the acidic range was recommended. Yamauchi and Kondo (1949) found that one-half the alkali in calcined alumina can be removed by washing and heating to $1300°C$, and two-thirds is removed by the addition of 20% HCl (1N). The slip viscosity was maximum at pH 5 to 8, but was most suitable for casting at pH 4 to 5, obtained by adding about 1% $AlCl_3$ to the alumina. Drained alumina slip recovered after casting contained dissolved gypsum from the mold, and showed increased viscosity, but this high viscosity could be decreased to the original value by the addition of barium chloride. Shiraki found that the addition of $AlCl_3$, $MgCl_2$, $CaCl_2$, gypsum, HCl, H_2SO_4, or HNO_3 caused an increase in viscosity of the slip. Casting became difficult at pH less than 3 and in alkaline media. At pH 7 to 8, ware cracks and sticking to the gypsum mold ensued; at pH above 9.2, casting strength disappeared.

Ramsay (1950) recommended the use of flocculant that is insoluble in cold water but soluble in hot water (about 0.1% of either hydrolyzed polyvinyl alcohol or phenylacetic acid) to control deflocculation obtained with sodium phosphate. Blaha (1950) added albumin or gelatine to the slip at a temperature at which the cohering substance is soluble. Intricate slip-castings could then be prepared by adjusting the temperature in the mold to cause the cohering substance to gel. Maxwell, Gurnick, and Francisco developed a freeze-casting method which allowed the slip-cast form to dry by sublimation. Smoot and Ryan found that the addition of a small amount of a silicone oil emulsion (0.01 to 0.03% of a 3 to 50% dilution) produces a cast shape without a central cavity defect. E. D. Miller, U.S. 3,235,923 included 2.5 to 5% hydrated alumina in the dry mix to avoid folds and depressions.

Blackburn and Steele invented the method of embedding a fluid pressure supply in the plaster mold so that after the water of plasticity had been absorbed from the ware by the mold, a uniform pressure (50 to 80 psi) could be applied to release the ware from the mold (Ram method). Suction could be applied during the initial casting period to remove water.

Chatterjee and Chakravarty noted that the alumina slip remained flocculated during aging. It acquired a net positive charge in acid media. Dietzel and Mostetzky investigated the mechanism of dewatering of a ceramic slip by the plaster mold, and found that the diffusion coefficient is satisfied by the expression: $Dp = (x/\sqrt{t})^2$, in which x is the distance to which the water has penetrated in time t, for values of D ranging from 8.5 to 17.0×10^{-2} cm^2/sec. On the basis of this theory, the casting growth is parabolical. Nies and Lambe (1956) considered the absorption capacity of the gypsum mold from its crystal size and pore spaces, and evaluated the drying rates of gypsum mold structures under typical commercial conditions. Data were developed to show the maximum rates of water movement into and out of the molds under practical conditions. Salmang (1956) stressed the importance of surface tension on ware cracking in slip casting. Rempes, Weber, and Schwartz developed a "film technique" for slip casting a variety of metals, cermets, and ceramics, employing a thin layer of sodium alginate on the inside surface of the mold so as to serve as a tenacious porous membrane (0.2% Keltex solution). Schifferli (1959) has given a general review of slip casting showing some of its advantages and limitations when used with refractory oxides, cermets, and powdered metals. Recent investigations are concerned with better control of flow properties (Worrall, 1963; Hamilton, 1959; Kukolev, 1963; Buehler, 1963; and Hermann and Cutler, 1962).

Alumina compositions are slip cast in sizes as large as glass tank blocks. Popil'skii and Galkina (1960) describe the vacuum treatment of slips and the repeated vibration of the plaster molds during the casting of high-grog-containing refractories. Eldon Miller (1964) developed a slip-cast alumina refractory containing from 1 to 10 wt % of volatilized silica and from 0.1 to 0.3% dispersant such as sodium phosphate, which could be vibrated during setting without surface cracking.

Schwartz, White and Curtis compiled slip casting fabrication data for various types of crucibles.

Hydrostatic or isostatic pressing, in its primitive form, consisted of immersing the powder compact encased in a rubber bag in a pressure vessel containing water or oil that could be subjected to about 4000 to 5000 psi.

Vassiliou prepared crucibles, combustion boats, etc., by a simple hydrostatic pressing operation, using rubber molds to distribute the pressure uniformly. The isostatic process was a pioneer development in spark plug manufacture by Champion Spark Plug Company (B. A. Jeffery, 1932; H. W. Daubenmeyer), using rubber molds. From spark plugs the method has been applied to electron tube elements, and in the fabrication of ceramic radomes, and grinding balls, among many increasing applications (Anderson, 1958; Schaefer and Stoia, 1962). Alumina radomes have been prepared by the optional methods of forming on a rigid, smooth-surfaced, nonabsorptive mandrel (Anderson, Brandt, LeClercq, and Fargo) or within such a mandrel.

As isostatic forming is substantially a dry-forming process (moisture less than 0.2%), it was necessary to create free-flow of the particles for easy transport to the mold, and this was done by spray drying to form spherical particles in the nominal range of about 20 to 70 mesh. The usual pressures applied in isostatic forming have been around 4000 to 5000 psi. G. D. Kelly (1961) investigated the effect of higher hydrostatic pressures (up to 100,000 psi) on the body properties. Powder-packed unfired bulk densities increased from 2.03 to 2.59 g/ml, and fired densities from 3.75 to 3.89, for a change in forming pressure from 3000 to 100,000 psi. Fired strength improved only slightly at low pressures and actually declined at the higher pressures, suggesting that fired strength (and pore structure) is independent of forming pressure, and that optimum strength is attained only under a particular set of firing conditions. (It is possible that as the compacted density approaches the fired density, the removal of small amounts of volatiles causes strains and cracking defects.) It was concluded that the grains attained beyond 6000 psi hardly justify the higher forming pressures. No significant difference was claimed between the pure powder-packed bodies and specimens prepared by spray drying a composition containing combustible binders (Cumar P-10 resin and collodion), although it might be expected that the removal of the organic matter would cause loss in density and in strength.

Bell, Dillender, Lominac, and Manning investigated vibratory compactions as a forming method. Three "nonplastic" powders, pure alumina, an alumina-chromium cermet, and a titanium carbide-nickel cermet were subjected to mechanical, electromechanical, pneumatic, magnetostrictive, and exponential horn driver units as vibratory sources. In the alumina compositions the size distribution of particles was varied by mixing AWIF 38900, 38600, and 38500 (Norton Company) with T-60 tabular alumina (Alcoa). For comparison, the same compositions were formed by dry pressing and were re-pressed hydrostatically at 47,500 psi. The vibratory energy was applied at 2 impacts per cycle, one in each axial direction, at about 70 cycles per second. The alumina samples were dampened with 0 to 8% water prior to vibrating. The conclusions derived from the investigation were that specimens formed by low-frequency vibration have properties which are about the same as specimens formed by dry-pressing followed by hydrostatic re-pressing. The vibratory method has these advantages over hydrostatic forming (a) no binder is required, (b) the method is considerably faster (on a laboratory scale), (c) less equipment is required, and (d) intricate shapes can be formed in one operation. Different powder compositions and particle size distributions may require different vibratory compaction procedures, but all compositions investigated were successfully formed by vibration. Particle size distribution has a considerable effect on the compacted density.

Quayle et al. (1945) described a method of making a ceramic body, which comprises, mixing the ceramic material with a thermoplastic polymerized bonding material, shaping, and heating the form in two steps. One involves the depolymerization of the bond to form a volatile material which leaves the body uniformly without decomposition; the other involves the sintering of the ceramic material.

Schwartzwalder (1949) described injection molding of single and multiple oxide spark plug compositions. In this process an intimate mixture of the alumina composition, thermoplastic resin, and a plasticizer is heated to about 150°C and injected into a cold die (water-cooled). The press opens automatically and the set pattern is ejected. Howatt (1948) used as the thermoplastic material, ethyl cellulose, polystyrene, or wax, and a plasticizer such as pine oil or solvent resin. The amount of resin required for this process was large and the shrinkage during firing was unusually high, making the technique less attractive than other forming methods for spark plug manufacture.

Moteki applied injection molding to the production of high-alumina porcelain and found that the physical, mechanical, and electrical properties were practically the same as those obtained by dry pressing or by casting. The method was considered satisfactory for making cutting tools and insulators. Randolph (1961) pointed out that careful matching of the thermoplastic resins and oils with the particle size and density of the ceramic is necessary. The oils must be baked out at a slow rate, leaving the resin to provide sufficient green strength for machining operations prior to final firing. Attainable tolerances are ± 1% of the fired dimension, and the thickness ranges from 0.02 to 0.25 inch.

Ryshkewitch (1960) summarized the prior information on the extrusion of alumina. Cold extrusion processes tax the ability of the "nonplastic" single oxides to acquire plastic properties sufficient to flow and assume a continuous formed shape. The addition of small amounts of various organic plasticizers or of the plastic clay materials to ground Bayer alumina in the proper consistency with water is sufficient, however, to allow auger extrusion.

White and Clavel (1963) reexamined the extrusion properties of non-clay oxides, using alumina powder of less than 5-micron particle size. The extrusions were forced through a 1/8 in. diameter die from a 1-in. diameter extrusion cylinder. From tests in which 23 wt % of hydroxymethyl cellulose solution (20 wt % organic) acted as the plasticizer, a pressure of 4400 psi for a flow rate of 0.1 in./sec, and 4650 psi for a flow rate of 0.5 in./sec were measured. It was concluded that the plastic dispersion was a thixotropic type. The alumina was not thixotropic in rubber-xylene plasticizer, however.

Rogers and Mooney (1955) molded an 88% Al_2O_3 composition containing 10% $3CaO \cdot 5Al_2O_3$ and 0.52 part polystyrene, 0.11 part hydrogenated peanut oil, and 0.27 part of a more volatile oil (50 to 150°C). Cornelius, Milliken, and Mills (1957) extruded attrition-resistant alu-

mina trihydrate particles impregnated with an aqueous solution containing nitric acid sufficient to form a minor amount of aluminum nitrate in a moist compact. Brennan and Field (1960) kneaded finely divided calcined alumina containing 1 to 18% combined water with 0.2 to 1.1 volumes of aqueous ammonium hydroxide per volume of the calcined alumina, extruded the composition, dried, and calcined it.

Richter (1962) plasticized alpha alumina for extrusion with gel-like aluminum hydroxide of nearly nonthixotropic properties (5 to 10% Al_2O_3, ratio oxide:sulfate, or chloride 10:1). Zet (1963) discussed various organic plasticizers for extruding alpha alumina shapes; end-closing techniques, and methods for firing ceramic pipe in a tunnel kiln at 1700°C.

Howatt et al. (1947) prepared thin ceramic sheets of titanates for capacitor dielectrics by extrusion of heavily loaded plastic pastes onto a moving metal belt. Siegrist also extruded thin ceramic plates, having a fired thickness of about 10 mils, through an orifice having the width and thickness of the unfired extrusion. The methods apply to alumina.

Thompson (1963) mentioned two methods for forming thin ceramics for mounting devices and vacuum tube spacers in a thickness range from 0.005 to 0.100 inch (1) extrusion (J. L. Park, 1960) and (2) casting or knife coating (J. L. Park, 1961). The extrusion process uses a water-soluble polymeric organic binder (polyvinyl alcohol) and a water-soluble plasticizer (propylene glycol, glycerol), both of which volatilize above 100°C. The moisture content is adjusted to form a heavy paste which is deaired and extruded at moderate pressure through a rectangular die to form a continuous strip or tape. The tape is carried through a dryer in which the moisture is reduced to less than 0.5%.

Young, Wilkins et al. (1960) described nonglaze-containing, screen-printed, conducting patterns comprising silver-platinum alloys as two-dimensional electrodes. Bulk alloy ultimate tensile strengths for the silver-platinum alloy system at firing temperatures of 1000 to 1500°C on two ceramic substrates were given. Alumina substrates for silver glazes were discussed.

Young, Owen et al. (1960) described the production of screen-printed electrode films of alloyed silver and platinum. Powders containing no glaze have good adhesion to refractory substrates and acceptable conductivity. Extensive physical and electrical property data were presented. The use of evaporation-annealing, or vacuum deposition on a hot substrate, as a substitute for cycle-annealing of gold-palladium thin-film resistors was discussed. The effect of electron bombardment and radiation damage was discussed. Radiation damage varied from none for the resistors to a permanent effect for a film capacitor. The useful operating temperature for screenprinted $MoSi_2$ resistors was about 650°C.

The mass production at RCA of miniaturized alumina and titanate wafers for the "micromodule," a small block-form electronic circuit was described (Ref. Anon. "Ceramic Wafers"). Held (1963) described a capacitor element comprised of a ceramic wafer composed of dense nonporous aluminum oxide, characterized by a relatively high and uniform dielectric strength and a nonvarying dielectric constant up to 540°C, and having conductors of gold, silver, or platinum, or their alloys on each side of the ceramic wafer. A thin layer of glass was fused to and covered the exterior portions of the wafer, bonding the conductors to it, and increasing the resistive path around the wafer.

The knife coating process casts a coating onto a nonporous carrier material. The ceramic powder is mixed in a ball mill with an organic volatile solvent, a wetting agent, and an organic plasticized binder. The viscosity is adjusted to 500 to 1000 cpoise sec. One of the best water-soluble binders is methylcellulose.

15-2 HOT PRESSING

Ridgway and Bailey (Norton Company) in 1937 described a method of making a self-bonded granular material of a metal oxide of the group Al, Mg, Zr, Ti, Th, Ce, and Cr, which comprised the steps of enclosing the granular refractory metal oxide in a mold space, heating the material while subjecting to pressure within the mold to a temperature close to its melting point, at which it is plastic, and compacting it into an integral mass, and as soon as it has been compacted, cooling to a point materially below its melting point. Comstock (Norton Company) in 1952 described the method of preparing molded articles of alumina having high compressive strength, high density, and substantially no porosity by hot pressing in a graphite mold at a temperature of 1650 to 1800°C and at least 500 psi pressure, a mixture consisting of 98.7 wt % alumina, 0.4 magnesia, and 0.9 cobalt oxide. The alumina used should have a purity better than 99%, and have microcrystallinity (98% of its particles under 2.5 microns, with 50% under 1 micron). The formed product has a density of at least 3.94 g/ml, and compressive strength above 450,000 psi.

Watson (1950) patented a graphite mold for hot pressing refractory crucibles, and Ballard (1951) patented the process for hot-pressing any oxide having a melting point above 1850°C by heating at 1650°C and pressing at 500 psi. Livey et al. (1964) have patented a method which comprises hot pressing at 600 to 1000°C and not less than 10,000 psi, a stoichiometric refractory metal oxide powder having an average crystallite diameter of less than 0.1 micron, or a nonstoichiometric powder having insufficient excess oxygen in the oxide to alter the crystalline form of the corresponding stoichiometric compound and having an average crystallite diameter of less than 0.1 to 0.25 micron. Murray, Rogers, and Williams (1954) discussed the theoretical aspects of hot pressing in terms of the Shuttleworth-Mackenzie theory relating the end-point density with the applied pressure, surface tension, yield stress, viscosity, and the number of closed pores per unit volume. The predictions of the equations agreed fairly with the experimental results.

Riskin and Goncharov (1957) claimed that hot pressing is more suitable for obtaining dense bodies (65% Al_2O_3)

than recrystallization by sintering at 600°C and pressing at 7100 psi. Pointud and Caillat (1957) found that a temperature of at least 1550°C was necessary to attain a density of 3.96 g/ml. Mangsen, Lambertson, and Best (1960) developed a rapid method for evaluating aluminous materials for their hot-pressing characteristics. A rate equation proposed by Murray et al. was found to hold for two alumina grades, enabling the prediction of the effects of changes in the hot-pressing conditions.

Francis, Swica et al. (1960) determined the relations between time, temperature, pressure, and grain growth in hot pressing with alumina grain to determine the technique necessary for the production of dense polycrystalline alumina having a grain size in the 4 to 5 micron range for use in thermal conductivity studies. Alumina having a density of about 95% theoretical and grain size in the desired range could be hot pressed rapidly at the maximum temperature. Jackson 1961, Accary and Caillat, Scholz (1963) investigated new theories regarding hot pressing. Densification cannot be expressed by a linear plot of log porosity vs time, and various modified expressions have been suggested. Vasilos and Spriggs (1963) observed that apparent diffusion constants calculated for the densification process are orders of magnitude greater than for pressureless sintering, which might be explained by enhanced diffusion under stress. Alternatively, however, lower calculated coefficients, which are more in agreement with pressureless-sintering results are obtained when pressure correction terms, modified by porosity, are applied to the existing relations. It was concluded that in the pressure range covered, densification beyond the initial stages is a diffusion-controlled process.

Hamano and associates found that enhanced grain growth and orientation occurred in alumina, hot-pressed at 1700 to 1900°C. The grain growth was associated with the Na_2O content of the alumina. A temperature increase under high pressure caused a layered structure; the c-direction of the alumina crystallites tended to be parallel to the hot-pressing direction. McClelland and Zehms (1963) examined the asymptotic end-point density of hot-pressed aluminum oxide over the range 1100 to 1500°C and 500 to 6000 psi. The pressure dependence was explained on the basis of a plastic-flow model which exhibits a critical shear stress. The temperature dependence suggested that the critical shear stress is a thermally activated process having an activation energy of 39 kcal/mole. Spriggs, Brissette, Rossetti, and Vasilos (1963) hot-pressed in high-density alumina dies. The contamination and reducing atmosphere of conventional graphite die-pressing was eliminated; it was possible to attain higher pressures and lower temperatures; and fabrication of otherwise easily reduced oxides was achieved.

Rossi; and Rossi and Fulrath determined the kinetics of vacuum hot-pressing. At 1150 to 1350°C and 2000 to 6000 psi, the final stage densification conformed with the Nabarro-Herring diffusional creep model, at an activation energy of 115 kcal/mole. Densification occurs by particle rearrangement. Entrapped gases in pores cause residual porosity. A method claimed by them for reducing the residual porosity of the hot-pressed alumina compacts was to replace the residual water in the submicron-size powder by a nonpolar solvent such as acetone, 2-propanol, 1-propanol, or butanol, but not methanol or ethanol. Stett and Fulrath described a hot-pressed aluminum oxide-glass composite with nepheline ($Na_2O \cdot Al_2O_3 \cdot 2SiO_2$) nucleating at the surface of the alumina and growing into the glassy matrix.

Budworth, Roberts, and Scott (1963) devised a technique for joining alumina components by hot pressing. Polycrystalline pieces have been joined to each other and to single-crystal alumina.

Rhodes, Sellers et al.; and Sellers et al. investigated hot-forging (high-pressure, high-temperature deformation and shaping). Crystallographic and mechanical texturing of polycrystalline alumina was observed, as well as an improvement in strength of as much as 200% of that expected from the existing grain size and other factors. Additions of MgO increased the high-temperature ductility, and lowered the brittle-ductile transition temperature by about 50°C. Large single crystals could be produced, presumably by strain-annealing or secondary recrystallization, in conventional, graphite, die-pressure, sintering equipment.

Reed (1961) used the induction plasma torch to grow crystals of sapphire at higher temperatures than is possible with flames and in inert or reactive atmospheres. Meyer (1964) found that the fusibility of various refractory powders, including alumina, is not related to the melting point or to the energy of melting. The pores which are found in plasma-fused spheres of alumina (and zirconia) are due to the formation within the plasma of nitrides which are later reoxidized in contact with air, with the liberation of nitrogen. A noble gas plasma is recommended for oxides of this type.

15-3 MISCELLANEOUS FORMING METHODS

Hallse (1963) applied a high-energy unit (Dynapak, General Dynamics Corp.) to alumina forming, able to compress a ram with energies as high as 2 million in.-lb at velocities approaching 850 in./sec. He concluded that it probably would be necessary to evacuate the whole die assembly, to heat the ceramic powder, and to use a shock absorber to obtain sound compacts.

Some factors which influence forming, or are influenced by it, are shown in the following references.

According to the U.S. Atomic Energy Commission patent Brit. 832,309 (1960), finely divided ceramic alumina powder is granulated homogeneously in order to obtain uniform density in pressed and sintered shapes. The powder is first mixed with two liquids of widely different volatility. The more highly volatile liquid, which predominates initially, facilitates mixing to a plastic mass, but is volatilized during granulation, leaving the desired amount of moisture in the granules. In the case of alumina and magnesia a sudden increase in strength occurs at temperatures which correspond apparently to the destruction of the absorbed water film on the surface of the particles and to the replacement of hydrogen or hydroxyl bonding between particles by ionic bonds. This occurs at temperatures well below those at which sintering proper commences (Clark et al., 1953). Mong and Donoghue found it advantageous to include a portion of a compound of aluminum that decomposes to alumina upon heating in the preparation of alumina refractories. Brandes (1957) claimed the method of including not less than 20% gamma alumina by weight for compositions containing alpha alumina or other principal components. Ryshkewitch, Strott, and Utz claimed the process of forming by cold-pressing, hot-pressing, extrusion, or slip-casting refractory oxide compacts that sinter to over 99% theoretical density if about 0.03 to 0.06% of 5 to 25 micron powder of the same metal as the oxide is added before compaction.

Viola and McQuarrie investigated the effect of different treatments of raw alumina (Alcoa A-14) on the relationship between the unfired and fired densities of alumina compacts. They obtained an expression:

$$K = \frac{[(1/\sqrt[3]{D_u - D_o}) - (1/\sqrt[3]{D_f - D_o})]}{[(1/\sqrt[3]{D_u - D_o}) - (1/\sqrt[3]{D_t - D_o})]} ,$$

in which D_u, D_f, and D_t are respectively the unfired, fired and true (3.99 g/ml) densities.

16 SINTERING

16-1 INTRODUCTION

Sintering may be defined as the consolidation, densification, recrystallization, and bonding obtained by heating agglomerated powders during or following compaction, at temperatures below the melting point of the principal component.

Alumina has remarkable properties for a variety of applications when it has been prepared in massive form by the process of sintering. It may appear as a single-oxide component or in combination with various ceramic oxides or other refractory materials, including metals. The composition may be substantially crystalline or may contain vitreous phases. The composition may vary throughout a wide range of texture, porosity, transparence, and other properties. The versatility of alumina in these applications is dependent in large part upon its favorable refractoriness, hardness, chemical inertness, high strength, high electrical resistance, good thermal shock resistance, and availability in cheap, high-purity forms.

Only a brief review of the theoretical aspects of sintering is possible within the scope of this review. Our present knowledge of the process has been largely aided by investigations of the sintering of alumina, however. Several theoretical models have been proposed for the sintering process, representing conditions in which no vitreous phase is present as well as conditions involving vitreous phases. These models are probably not faithful representations of the actual mechanisms of sintering (Wilder), because they postulate uniformly sized spherical particles and other idealized conditions. In some cases the observations are valid only for the initial stages of sintering. Surface tension has generally been considered a driving force of sintering, whether the sinter consists of pure crystalline phases or includes a liquid phase. Three stages of sintering are usually considered: an initial period of neck growth between adjacent particles, a stage of material transport or densification, and a final stage of grain growth with elimination of isolated voids. These are overlapping phenomena.

Kuczinski in 1949 developed an equation to distinguish between four transport mechanisms for inducing neck growth during the first stage of sintering, having the form: $x^n/a^m = f(T)t$, in which $F(T)$ is a function of temperature, t is the sintering time, and n and m characterize the mechanism responsible for the formation and growth of the neck of radius a. Transport by viscous flow fits x^2, by sublimation x^3, by volume (lattice) diffusion x^5, and by surface diffusion x^7. Herring derived linear scaling equations analogous to those of Kuczinski for determining the time, t_1, required to attain the same degree of sintering for particles or clusters of different sizes. The equations are: $t_1 = \lambda t_2$ for plastic flow; $t_1 = \lambda^2 t_2$ for sublimation; $t_1 = \lambda^3 t_2$ for volume diffusion; and $t_1 = \lambda^4 t_2$ for surface diffusion; where $\lambda = a_1/a_2$, the ratio of particle diameters. The rate of each transport mechanism increases in proportion to the ratio of particle sizes raised to the power representative of the particular mechanism. Ultrafine particle size is particularly effective for those stages of sintering dependent upon volume and surface diffusion.

Rhines proposed yet another concept of sintering, based on the topology of closed surfaces and which ignores dimensions, area, contained volume, and difference in direction. The parameters are: P, the number of particles in the powder; N, the number of separate surfaces required to define the shape of the material at any stage; C, the number of contacting points between particles; and G, the topological genus. The genus $G = C-P+1$. During the first stage of sintering, the genus increases somewhat, to complicate the exposition. During the second stage, the genus decreases to zero as the pores become isolated. In the third stage, the genus remains at zero and the pores disappear. Mackenzie and Shuttleworth, and Clark and White derived a sintering model based on plastic-flow transport, however, which has served to account for the phenomena of hot-pressing in the hands of Murray, Livey, and Williams. Felten refuted this for low-temperature hot-pressing of alumina, and suggested boundary sliding as the process responsible during the initial increase in compact density. Coble in 1963 observed that neck areas in hot-pressed sintering at $1530°C$ are greater than in pressureless sintering, and are constant for all periods between 10 and 480 minutes. For alumina, plastic flow contributes only slightly to densification at the pressures normally used (less than 10,000 psi). It was concluded that the final stage of densification occurs by increased diffusion under the influence of stress.

Wilder and Fitzsimmons used the relative rates of sintering fused alumina in a wide range of particle sizes, in an attempt to confirm the plastic flow transport mechanism. They found that even loose aluminum oxide powders exhibit sintering if the particle size is small (less than 20 microns). The values of n ranged from less than one to greater than 5. It is difficult to conclude from these experiments that the transport is by plastic flow. Navias in 1956 performed sintering tests on 40-mil sapphire spheres to verify the plastic-flow theory of neck development. Contact lenses were observed by heating the spheres by induction in a vacuum or in hydrogen at about 1900°C. The spheres eventually developed the hexagonal shape of cut gem stones. The tendency for fragmentary crystals to perfect their crystal form suggests this effect as a factor in sintering. Ground Bayer aluminum oxide is likely to be predominantly well-developed single crystals, while ground fused alumina or tabular alumina is substantially anhedral.

Kuczinski, Kingery and Berg; Coble; Johnson and Cutler; and others find that the sintering of alumina is a bulk diffusion process.

Burke investigated the role of grain boundaries in sintering. Grain growth changes the configuration of grain boundaries relative to pores, and thus may affect the shrinkage rate. He concluded that the grain boundaries act either as sinks or as diffusion paths for lattice vacancies. Additives increase the sintering rate if they increase the diffusion rate or impede grain-boundary movement. Secondary recrystallization, that is, discontinuous grain growth of a few large grains that are nucleated and grow at the expense of a fine-grained matrix, is the general mechanism of sintering. Grain growth involves grain disappearance. Secondary recrystallization engulfs the pores. Jorgensen and Westbrook observed that high-purity materials exhibit the most pronounced grain growth; there are less secondary phases. Kingery estimated the energy change corresponding to a change in crystal size from about 0.1 micron to 1 cm at 0.1 to 0.5 cal/g. The difference in free energy on the two sides of a grain boundary is the driving force that makes the boundary move toward its center of curvature, thus decreasing its curvature. If all grain boundaries are equal in energy, they will meet to form angles of 120°. Hamano and Kinoshita found the relation: $dD/dt = K(1/D-2f/d)$ held to explain the effect of pores on grain growth, in which D is the average grain size; d is the average pore diameter; f is the volume fraction of pores; and K is a constant. The addition of the growth retarder MgO changed the relation to: $dD/dt = K(1/D-2f/d)-K'K$. Patrick and Cutler, on the basis of experiments with seeded coarse crystals of alumina in fine distributions, concluded that the abnormal growth is not caused by large grains growing unimpeded through a matrix of fine grains in which normal growth has been impeded by pores or inclusions.

McHugh et al., and Whalen et al. used finely divided molybdenum as a dispersion in alumina to serve as inclusions for inhibition of grain growth and secondary recrystallization during sintering. Specimens sintered in a vacuum had densities approaching 98% of theoretical at an average grain size of 2 microns or less, and an average transverse strength of 80 kpsi. The molybdenum could be introduced as fine-ground powdered molybdenum or as

MoO_3 which could be reduced by hydrogen prior to sintering.

Coble in 1958 obtained experimental measurements of the rate of shrinkage of Linde A aluminum oxide compacts, the neck growth between single-crystal sapphire spheres and plates, and the effect of particle size on neck growth. The temperature dependence of the rate of shrinkage and neck growth in aluminum oxide is characterized by an activation energy of 165 kcal/mole. The observed data are mutually consistent with a bulk diffusion sintering model.

In 1961 Coble presented photomicrographs of pore and grain boundary structures that provided a qualitative description of the course of densification. During sintering the density increases linearly with the logarithm of time, and the grain size increases with the one-third power of time. Incorporation of the time dependence of grain size growth into late-stage bulk diffusion sintering models leads to corrected models by which a semilogarithmic behavior can be predicted. The sintering process is then controlled by bulk diffusion of aluminum ions; the oxygen transports along grain boundaries.

Shackelford and Scott (1968) determined the relative grain boundary energy and surface diffusion coefficient of aluminum oxide from thermal grooving behavior of bi-crystals with symmetric tilt angles. From agreement of their vacuum-etch data with air-etch values, they concluded that the oxygen ion was not rate-controlling.

In summation, it appears likely that each of the cited theoretical models may contribute to the sintering operation under specific conditions.

The distribution of pores and their location with respect to the grain boundaries influence the time required to attain fired density. Normal grain growth in the final stage reveals the pores generally at 4-grain corners. An alternative condition, however, is that in which the pores collect in the centers of crystals often with a clear margin around the boundaries (Burke), resulting from secondary recrystallization. The conditions required to obtain these effects, about 12 hours at 1800°C or 2 hours at 1950°C, are well beyond those normally used in sintering alumina, both in duration and in temperature. Schatt and Schulze observed that the free surface (that of the original particles which have not lost their identity for complete intersintering with adjacent particles) shows a polyhedral terraced structure, considered to be the natural result of grain growth by solid diffusion.

Dillingham investigated the microstructure of sintered high-alumina surfaces, using a diamond-polishing method, a technique also described by Angelides; Elyard; and Swan. Etching in solution to render microstructural details visual has been described by McVickers, Ford, and Dugdale; Taylor; Bassi and Camona; Alphord and Stephens; Tighe; Schloemer and Mueller; and Soden and Montforte. Molten salt etching techniques were described by Factor et al.; and A. G. King. Bierlein, Newkirk, and Mastel etched by ion bombardment, and Beauchamp flash etched by flame-polishing. This can be applied in vacuum, or in air, oxygen, or hydrogen (Heuer and Roberts; Savitskii et al.). Krokhina et al. etched by ionic bombardment in an inert atmosphere.

Entress and Skatulla used electron micrographic methods for studying crystallization, pore dislocation, grain boundary intergrowths, etc., in sintering high purity commercial alumina for sapphire production (99.5% Al_2O_3) and industrial alpha aluminum oxide (98% Al_2O_3). They claim that all of the phenomena observed in sintering can be explained by an elementary diffusion theory. The one-component oxide systems are separated into two classes: (1) "active oxides" in which the sintering process is caused by structural defects (below 10^{-5} mm) cooperating with heat radiation effects, namely, collective crystallization and secondary recrystallization, and (2) "passive oxides" which are either fused products or naturally grown minerals that do not undergo any crystalline conversion or thermal disintegration. Secondary recrystallization confirms the diffusion theory and cannot be reconciled with an explanation based on surface tension effects or plastic flow.

Tresvyatskii found that the rate of the process and densification are greatly retarded if large numbers of closed pores are formed within the crystals.

Pavlushkin (1963) stated that the shrinkage of corundum compacts commences at 1100°C (0.5%), reaching 10.5% at 1450°C. Actual sintering begins at 1400°C, and crystal growth is observed at 1600°C. The sintering temperature decreases with decreasing particle size down to 0.5 micron average diameter, then increases again. Hijikata and Miyake claimed that the temperature of incipient sintering (1000 to 1050°C) does not depend upon the particle size and the sintering atmosphere. The finer the particles, however, the faster and greater is the shrinkage. Bruch found that the densification rate is markedly dependent upon both the initial density and the sintering temperature. The rate of grain growth is independent of the initial density. Bruch recognized that the rate of sintering is very strongly controlled by the size of the pores in the initial compact. His empirical equation relating densification to initial porosity has the form: $\log P = 10 \log P_0 - 0.434 nQ/RT + n \log t - 23.43$, in which P_0 and P are the initial and final percent porosity, n and Q are constants having the values, -0.4 and 150 kcal/mole, respectively for their particular sample of Linde alumina, t is the sintering time in minutes, R is the gas constant, and T is the sintering temperature (°K). His equation relating grain growth to sintering temperature has the form: $\log D = \log K - 0.434 mQ/RT + m \log t$, in which D is the grain diameter in microns, K is a constant having a value of 5.66, and m has a value of 1/3. The other terms have the same values as in the preceding equation.

Shrinkage during sintering of alumina is proportional to the deformation only at low temperatures; at high temperatures around 1400 to 1500°C, deformation increases rapidly (Kainarskii, Orlova, and Degtyareva).

Dawihl and Doerre attempted to clarify the functional dependence between texture and deformation of polycrystalline alumina (<0.2% impurity) below 1000°C, the range of significance for interpreting the lifespan of alumina cutting tools. Change in surface area during sintering, initially about 11.6 m^2/g, was used to measure the diffusion of alumina in the specimens. Incipient diffusion was detected at 850°C, and was marked at 1100°C (2.3 m^2/g surface area). The pressed specimens were claimed to have less than 3 microns grain size when sintered between 1800 and 1900°C. Strength and hardness tests at room temperature, 500°C, and 1000°C showed that porosity was the major factor decreasing compressive strength until the grain size exceeded 3 microns. Initial grain growth began at 1500°C, accompanied by screw dislocations and etch pits.

Bruch assumed that the large pores forming in "subnormal" sintering result from coalescence of pores rather than from the presence of relatively large void spaces in the original compact. Bolling emphasized that the ratio of the time required for bubble coalescence to the time required for void space to disappear is a measure of the probability that the void space will disappear faster than it can form larger pores. If the ratio is small, large pores should form, and sintering will be slowed; if the ratio is large, the pores which form should be small and isolated, and rapid sintering rates will be maintained. He concluded that the exponents n and m in the densification and grain-growth processes (Bruch's equations) are related as follows: n is about equal to -m.

Vergnon, Juillet, Elston, and Teichner found that rapid heating of alumina, whether alpha or delta phase, is more effective for shrinking the form. Both the ultimate percent compression and the rate of compression increase with the temperature. The rate is proportional to the initial specific surface, but is much greater for delta alumina than for alpha alumina. The activation energy for delta alumina is 145 kcal/mole, and for alpha alumina 55 kcal/mole. Pearson found that the apparent activation energy for intermediate-stage densification in combustion-gas atmosphere increases with the initial crystallite size.

Poluboyarinov and Vydrik found that sintering is dependent upon the prior firing and dispersion of the alumina. With increasing firing temperature, the activity falls off, making it necessary to grind finer to obtain a completely sintered body. Volosevich and Poluboyarinov showed that the density of pure corundum ceramics increases with increasing firing temperature (T_f) and is affected by the particle size of the raw material; for instance the density of the body formed from corundum with a particle size of one micron increased from about 3.0 g/ml at 1400°C to a maximum of 3.80 at 1600°C, and was not affected by a further increase in T_f up to 1850°C. The density of a body prepared from corundum with a particle size of 5.6 microns increased from 2.5 to 3.6 g/ml as T_f was increased from 1400 to 1850°C. If the raw material had a particle size less than 2 microns, the temperature of the preliminary heat treatment had little effect on the sintering. The rate of crystal growth became significant when the density of the fired body reached 3.5 to 3.6 g/ml. In the firing of pure corundum (99.8% Al_2O_3) with a particle size of one micron, sintering was completed at 1750°C, and the average crystal size was as coarse as 120 microns. The maximum bending strength, 36,000 psi, was obtained on the product fired at 1650°C. These data should not be considered optimum, because other factors of significance relating to the conditioning of the raw material were not considered. Improved sintering is illustrated by the performance of the reactive aluminas.

Kitaigorodskii and Gurevich in 1959 found that the strength of sintered alumina passed through a maximum with increasing temperature, in their tests at about 1800°C.

Sintering in the presence of a liquid phase has been considered to be an expedient process for obtaining the desirable properties of high-alumina porcelains at reduced maturing temperatures. Bulavin; Kukolev and Leve; and Kingery investigated the mechanism of this process. The surface tension of the liquid phase is the major driving force. The viscosity of the liquid phase is a significant variable. Three separate densification processes were believed to occur: rearrangement of solid particles to give increased density; solution precipitation in which material is dissolved away from contact points, allowing the centers of the particles to approach each other; and finally, coalescence of the solid particles. Bulavin assumed that there was no liquid formation at the expense of partial solution of the crystalline phase. In this case, the first stage of sintering involves redistribution of the primary fusion and neck formation. A second stage involves the filling of the intercrystalline spaces with the liquid.

Lay investigated grain growth in the system UO_2-Al_2O_3, in which Al_2O_3 was the minor phase (only to 0.5%) and was liquid at sintering temperatures of 1960 to 2200°C. The grain diameter satisfied the expression: $d^3 = 0.106t \exp(-93000/RT)$. This conforms to recent theory that with liquid phase present, $d^3 = kt$, whereas if it is not present, $d^2 = kt$. The absolute growth rate was found to increase as the amount of liquid decreased.

It is generally observed that the strength and elasticity of a porcelain is increased by additions such as alumina, having high elasticity. The effect is often ascribed to changes in the matrix, as for example, increased mechanical strength of the vitreous phase by partial solution of the alumina, localized around the alumina grains. Binns investigated the changes in fracture strength and Young's modulus of hot-pressed mixtures of glass and dense alpha alumina grain, within the concentration range, 0 to 40% alumina. The mixtures attained maximum density at 600 to 700°C. It was concluded from this work that elasticity and strength of the composite can be increased markedly beyond that of the glass. The rate of increase in elasticity is independent of the grain size of the alumina, and indicates that the composite solid behaves partly as a constant-strain system. The resulting uneven stress distribution is believed to be responsible for the observed increase in strength. Differences in thermal expansion of the glass and alumina always lowers the strength and elasticity (for coarse crystalline grains), the reduction being most pronounced when the thermal expansion of the glass is higher than that of the

crystalline inclusion. This is explained by the observed crack system formed during cooling. When the thermal expansion of the crystalline grains is higher, the cracks tend to follow a line parallel to the glass-grain interface, and to enclose individual grains; when the thermal expansion of the grains is lower, cracks tend to run normally to the interface and to pass from grain to grain, thus producing more disruption of the glass. With finer crystalline inclusions (around 10 microns), the strength reduction and tendency to crack formation is less.

Allison, Brock, and White (1959) applied the rheological principles derived from aqueous suspensions to the behavior of close-packed ceramic refractories bonded by a Newtonian liquid at high temperatures. Such systems are dilatant, that is, the particles assume a less dense packing when the body is sheared. With liquids that wet the solid particles, the magnitude of the cohesive force between the particles decreases progressively with increased liquid content. The yield point is usually considered the compressive strength. At loads above the yield point, three stages of flow can be detected, (1) decelerating, (2) constant, and (3) accelerating; the curves are similar in form to creep curves of a metal. The accelerating flow is associated with dilatancy. A distinction is made between highly vitreous bodies, whose behavior under stress approximates that of a viscous glass, and bodies which consist primarily of solid crystals with minor amounts of liquid phase. A further distinction is made between bodies in which the liquid phase forms continuous films around the solid particles and those in which the liquid phase forms discrete globules, so that solid-solid bonding occurs. The magnitude of the compressive and tensile strengths observed near the hot-strength peak, however, is apparently too great to be accounted for entirely on the basis of liquid bonding; partial solid bonding is probably present.

16-2 SINTERING ATMOSPHERE

The nature of the atmosphere during the sintering of alumina may influence the rate of sintering or the residual porosity. The oxygen deficiency observed by Bevan et al. when alumina is heated in water vapor might explain differences in sintering rate. Walker found that the rate of sintering of aluminum oxide at 1650°C is faster in hydrogen or argon than in air. Water vapor retards the rate in comparison with air. Oishi and Hashimoto found that the rate of grain growth of aluminum oxide at 1600 to 1960°C is in the same order, that is, dry hydrogen was most effective, followed by carbon monoxide, argon, and air. There was little difference between firing in air and in a vacuum. Vines et al. found that the internal porosity of dental porcelains can be reduced by firing in a vacuum, or in a diffusible gas (helium, hydrogen or water vapor), and by finish-firing and cooling under pressure. They showed that gas diffusion controls the shrinkage rate during the final stage of firing. Low final densities resulted from firing in nitrogen or in carbon dioxide at one atmosphere. Theoretical densities were achieved in vacuum or in helium. In this case, however, the pores were trapped in a continuous glass structure. The mechanism of pore elimination from porcelains may differ somewhat from that of crystalline structures.

Coble in 1962 found no observable effect on the sintering rate of alpha alumina containing 0.25% MgO in hydrogen or oxygen atmospheres introduced at a dewpoint of $-70°F$. The hydrogen reacted with the alumina furnace wall to form H_2O and aluminum suboxide. If the hydrogen was too dry when it met the specimen, the specimen slowly evaporated. Complete elimination of porosity was attained in hydrogen, oxygen, or vacuum, if 0.25% magnesium oxide was used as a grain-growth inhibitor. The pores generally became sealed off at about 95% of the theoretical density (about 3.98 g/cc).

Kuczinski, Abernethy, and Allan reported no action of oxygen pressure on the rate of neck growth between single-crystal sapphire spheres except in dry hydrogen mixtures, in which a faster rate was observed. Paladino and Coble have since found that the oxygen ion diffusion coefficients in single-crystal aluminum oxide are several orders of magnitude less than aluminum ion diffusion coefficients in polycrystalline aluminum oxide. In polycrystalline alumina, oxygen ion diffusion is enhanced by the presence of grain boundaries. As the slower rate controls the process, this would indicate that below a certain grain size, about 20 microns, the aluminum diffusion rate controls, while above this size the oxygen rate controls. Ringel investigated the effect of firing atmospheres on the sintering of alumina. Warman and Budworth pointed out that even when sintering in a vacuum, bloating may occur when the outer skin of the compact sinters to an impermeable state before the interior pores are freed of gases, presumably mainly atmospheric nitrogen. The use of vacuum to sinter alumina to theoretical density is feasible.

Budnikov and Kharitonov stated that high-pressure water vapor increases the open porosity, with a corresponding decrease in strength, depending on the amount of glassy phase present. Hydration products found on the surface include boehmite, gibbsite, and rarely, diaspore.

16-3 SINTERING ADDITIVES

Additives other than temporary binders have been used in ceramic alumina compositions for several purposes. Among these purposes might be cited: crystal growth repression, crystal growth acceleration, acceleration of sintering or shrinkage rate, reduction in maturing temperature, alteration in porosity, changes in physical or chemical properties, and removal of impurities. Examples of these are shown in the various applications and preparation methods for ceramic aluminas.

Crystal growth in sintered alumina was early recognized as an unfavorable factor affecting strength and thermal shock resistance. Although relatively pure fine-crystal aluminas were available, it was found that mere fineness of the starting alumina is not sufficient to avoid excessive crystal growth. Ryshkewitch ascribed excessive crystal growth on the surface of sintered alumina to water vapor, since

hot-pressed specimens do not show this effect. He observed that giant crystal growth is favored by small amounts (above 0.2%) of TiO_2 or ZrO_2 at sintering temperatures around 1800°C. Crystal growth was repressed by SiO_2, Cr_2O_3, and particularly MgO, if the sintering temperature was held below the point at which eutectics form with the alumina. He explained the action of growth repressors as the result of physical separation of the alumina particles. Magnesia, first suggested by Ryshkewitch, was considered the ideal additive (about 2.5%) for sintering aluminum oxide because its eutectic point is at 1925°C, well above the practical sintering point of pure alumina. Usually, as little as 0.02% magnesium ion is effective, but the temperature of inhibition is somewhat lower than the accepted eutectic point.

Dawihl and Doerre state that as little as 0.25% MgO added to alumina causes spinel formation at 700°C. The spinel is not soluble in alumina, but dissolves up to 50% (mole) Al_2O_3 in solid solution at 1700°C. Roy, Roy, and Osborn observed that Al_2O_3 should precipitate to about 5% (mole) in cooling below 800°C. The spinel is observed at grain boundaries. Microscopic examination shows that amorphous substances exist on the edges of fused corundum crystals, that unite the grains; they appear to be silicates of metals (Miyabe and Nishigaya).

Kharitonov (1966) found the linear growth rates of the grains in MgO-doped alumina compositions at the temperatures 1650, 1700, 1750, and 1800°C were 2.80, 2.50, 1.40, and 3.00 cm $\times 10^{-8}$/sec, respectively; and in the undoped compositions at the same temperatures the rates were 10.0, 13.8, 13.4, and 12.5 cm $\times 10^{-8}$/sec. The activation energies were 87.5 and 120.0 kcal/mole, respectively. It was observed that the degree of crystal growth for the doped samples rose sharply above 1750°C, but the rate for the undoped samples was about an order higher and fairly constant.

Silica and silicates, particularly the native clays, have been common additives to sintered alumina. Reactions between SiO_2 or kaolinite and Al_2O_3 at about 1600°C may involve a glassy phase formed by diffusion of alumina into a quartz zone surrounding cristobalite particles, and additionally, mullite crystals may form in the alumina (de Keyser, 1965).

Much of the early investigation of the effect of additives during sintering was undertaken to obtain suitable spark-plug porcelains. Reichmann in 1931 found that all-alumina sinters prepared from 2-micron fused alumina contained air pores and were grossly crystallized. Better results were obtained by calcining slip-cast forms of a mixture of gamma alumina and alpha alumina obtained from the Bayer process. When AC Spark Plug Division of General Motors Corporation commercialized the process, it was found more practical to develop other forming methods, as for example, injection molding (Schwartzwalder, 1949).

A search was made for additives that would reduce the firing temperature and obtain more compact and pore-free crystalline porcelains. McDougal; Fessler; and Schwartzwalder and their coworkers obtained patents on the addition of oxides of barium, strontium, cerium, manganese, nickel, cobalt, tantalum, or thorium; clay-lithia-magnesia, zirconia-silica; calcia-magnesia-silica; zirconia-silica-alkaline earth oxide; zircon-kyanite; and silica-strontia-magnesia. Monazite sand was also claimed as a suitable additive. The effective compositions usually ranged between 85 and 99% Al_2O_3.

Jeffery in 1942, and Riddle developed compositions containing up to 90% sintered alumina and the fluxes: calcia-magnesia, alkaline earth oxide-aluminum silicate, or beryllia-magnesia-calcia. Feichter reduced the sintering temperature to about 1650°C at 92.2% Al_2O_3 by adding a mixture of talc, fluorspar, clay, and chromic oxide. Bonnet prepared compositions containing about 80% Al_2O_3 and calcium or magnesium phosphate and hydrous silicates of magnesium or aluminum (bentonite or kaolin). Products having zero porosity were claimed upon firing at 1440 to 1540°C. Luks added manganese oxide as the principal flux. Austin and Rogers added CaO in the form of tricalcium penta-aluminate. Early German patents assigned to Robert Bosch specify the addition of natural or synthetic zeolite in which the alkali oxide has been replaced, wholly or in part, by an alkaline earth oxide, and all or part of the silica has been replaced by titania, to enable firing at 1500°C. Plastic talc-kaolin-feldspar binders were used with ground fused alumina. Eitel described the status of the German spark plug compositions prior to 1946.

Nakai and Fukami reported increasing strength and crystal growth in the following order for 2.5% additions of the following to alumina: $Na_2B_4O_7$, CuO, Li_2O, and Fe_2O_3.

Bron found that sintering of alumina is retarded by alkalies and alkaline earth oxides. At 1400 to 1500°C, about 2 to 5% TiO_2, added as titanium dioxide, ilmenite or magnesium titanates forms solid solutions which induce sintering to densities around 3.84 g/ml. The solid solution is of the subtraction type for additions of TiO_2, but of the substitution type for additions of magnesium orthotitanate. The former is more effective because it stabilizes the lattice defects in corundum at different temperatures. Defect structure favors the transport.

Zachariesen had pointed out that Ti_2O_3 and a-Al_2O_3 are isomorphous. The solubility of Ti_2O_3 in Al_2O_3 varies from 1.0 to 2.5 mole% between 1400 and 1600°C by X-ray diffraction, and with no evidence of TiO_2 in solution under reducing conditions (McKee and Aleshin). Horibe and Kuwabara (1964) found 1.6, 2.7, and 1.7 mole% Ti_2O_3 soluble at 1600, 1710, and 1840°C, respectively. Winkler, Sarver, and Cutler found both Ti_2O_3 and TiO_2 in samples air-fired at 1300°C, TiO_2 amounting to as much as 0.25 mole%, as determined by diffuse reflectance spectra and cathodoluminescence. S. K. Roy and Coble (1968) confirmed that the solubility of both TiO_2 and MgO in alumina, heated in air, is too small to detect by X-ray lattice shifts. The effect of an additive may not be the same throughout the effective sintering range; Gerasimov and Kovachev claimed

that MgO affected sintering unfavorably at 1600°C but promoted it at 1700°C.

Robinson used 0.1 to 10% H_3PO_4 or 0.05 to 2% metal fluoride (preferably magnesium fluoride) to control grain growth. Locsei (1964) claimed that up to 10% AlF_3 accelerated sintering and improved strength.

Smothers and Reynolds, using Alcoa Alumina A-11 (0.20% Na_2O, 2 microns median crystal size), investigated the densification and grain growth of pressed compacts at each of four temperatures, 1300, 1400, 1500, and 1700°C. They found that one percent additions of TiO_2, Ti_2O_3 or Nb_2O_5 caused marked shrinkage accompanied by some grain growth at 1300°C. Other additives, effective at 1500°C, include MnO, Cu_2O, CuO, or GeO_2. Additives which did not affect the sintering or grain growth include Ga_2O_3, Y_2O_3, P_2O_5, Fe_2O_3, ThO_2, CeO_2, ZrO_2, and Co_3O_4. Additives which repressed sintering and grain growth include NaF, MgF_2, Sb_2O_5, CaF_2, KI, KCL, KBr, K_2CO_3, $NaNO_3$, Na_2CO_3, $CaCO_3$, MgO, $BaCO_3 \cdot SrCO_3$, SnO_2, Cr_2O_3, La_2O_3, V_2O_5, and SiO_2. They found that grain growth did not commence until an apparent porosity of 32 to 36% had been reached, regardless of the additive used. They agreed with Bron that additives which form solid solution with alumina (TiO_2) favor sintering, but recognized that there are others which repress sintering (Ga_2O_3 and Fe_2O_3). Kovatschev and Serbezova found that Nb_2O_5 MnO_2, and ZrO_2 accelerated sintering, but Na_3AlF_6 retarded it. Hauttmann found that finely divided iron oxide (0.05 to 0.5 micron) accelerated sintering. Cahoon and Christensen obtained faster sintering by additions of oxides of iron, manganese, copper, or titanium provided the amounts and firing temperature were within certain bounds, but crystal growth was accelerated by the iron and manganese. Pavlushkin in 1956 studied the effect of nearly all the elements in the periodic table as additions to alumina. He classified such additions into three groups: the crystal growth repressors Mg, Ni, Co, Th, W, Cr, U, Ta, Si, Ca, Zr, Mn, Tl, Y, Pr, Te, Li, Be, Na, Bi, Fe, Ti, Nb, Mo, Cs, Sr, and V; the growth accelerators Ag, Pb, Rb, Sb, F, Cu, Hg, K, Zn, Sn, Cl, I, S, Br, and P; and the intermediate agents Au, Pt, Pd, Ge, Ba, Cd, As, Ce, Se, C, and B. This is an oversimplification. The concentration of additive, different combinations of anion and cation, and sintering temperature alter the effectiveness. For example, $Na_2B_4O_7$ does not act the same as H_3BO_3. Fluorides are accelerators, but not magnesium fluoride. Pavlushkin pointed out that CaO and Na_2O should be used in amounts not over 0.02% and 0.05%, respectively, to avoid excessive formation of the corresponding beta aluminas.

Bartuska classified the effectiveness of the elements on sintering into somewhat different groups: (positive effect) Cu, Ti, V, Mn, and Fe; (less positive) Li, Be, Mg, Ca, Zn, Zr, Pb, Br, W, and Co; (no effect) Na, Sr, Sn, Sb, and Ni; and (negative effect) K, Ba, Si, and Cr. Reaction with alumina can proceed by formation of an isomorphous solid solution, by simple substitution of aluminum ions by ions of lower or higher valency, by metathesis, by formation of a eutectic melt or a single component melt, and by formation of an inhibitor to sintering.

Pavlushkin in 1962 investigated additives in amounts from 0.001 to 0.030 mole per mole Al_2O_3. Additions of Hg and Mg salts accelerated sintering; additions of Be, Zn, and Cd compounds accelerated sintering in the range 1580 to 1750°C. The alkaline earths retarded sintering. The strength of sintered corundum fired at 1710°C was higher with additions of MgO and BaO. Kukolev and Mikhailova also investigated the effect of impurities on the sintering of alumina.

Cutler, Bradshaw et al. selected additives from metal oxide combinations having low-temperature eutectics in the systems MnO-Al_2O_3-TiO_2 and Cu_2O-Al_2O_3-TiO_2. Bodies having small grain size, and bulk densities above 3.80 g/cc were obtained at maturing temperatures of 1300 to 1400°C. Evidence of a "bleeding" liquid that drained away from the sintered pellets justified the conclusion that eutectic liquids formed. It seems probable that such compositions would not have stable high-temperature properties. Kovaschev claimed that the addition of 3 to 4% of an equiweight mixture of MnO_2 and TiO_2 to commercial alumina (99.3% pure) gave a sintered compact having a density of 3.71 to 3.75 g/ml at a maturing temperature of only 1250°C.

Bagley investigated the reaction rate between alumina and TiO_2. A rapid contraction was observed when TiO_2-doped samples were heated in air after first having been heated in hydrogen. Alumina sinters containing both Ti_3O_5 and TiO_2 were in solid solution. The sintering rate increased up to a certain concentration, beyond which the rate decreased. When TiO_2 was added to alumina of less than one micron size, the sintering mechanism was by bulk diffusion; for a crystal size of one to two microns the sintering was by a combination of bulk diffusion and grain boundary diffusion. Addition of soda in amounts from 10 to 10,000 ppm to alumina (Linde A-5175) eliminated exaggerated crystal growth. Samples pressed at high pressures had more exaggerated growth, probably because of trapped soda. A soda atmosphere also stimulated exaggerated growth.

Volosevich and Poluboyarinov in 1957 found that 0.5 to 1.0% additions of MgO, MgF_2, CaO, ZnO, SrO, ZrO, or Cr_2O_3 influence both crystal size and mechanical strength when using "pure" alumina containing only 0.05% Na_2O and of 2-micron maximum grain size, when sintered at 1550 to 1800°C. The initial powder size controls the pore diameter. Reduced mechanical strength and crystal growth were obtained with TiO_2, ZnO, MnO, Cr_2O_3 and Fe_2O_3. In reducing atmospheres the TiO_2 reduces to lower oxides, soluble in the alumina. Glasses in eutectics in the systems CaO-Al_2O_3-SiO_2, MgO-Al_2O_3-SiO_2, and CaO-$B_2O_3$$SiO_2$ lower the sintering temperatures to 1500 to 1550°C, when added in about 5% (wt) concentrations. Kiyoura and Sata claimed that oxides having a high softening point in their systems with alumina are effective in improving high-temperature strength (1000°C); especially effective are CoO, NiO, MgO, and Cr_2O_3, about 2 to 5%.

Degtyareva and associates found that the shrinkage of alumina containing small amounts of TiO_2, $MgTiO_3$, Al_2TiO_5, and $ZrTiO_2$ is proportional to the TiO_2 content, with an activation energy for sintering of 58–87 kcal/mole at 1200–1500°C. The sintering rates in alumina containing small amounts of MgO and SiO_2 increase with the SiO_2 content, with an activation energy estimated to be 46 to 65 kcal/mole.

Kainarskii, Orlova, et al. found that alumina containing $MgTiO_3$ sintered at 1200 to 1500°C shows a nonlinear relationship between deformation and shrinkage, due to the development of a constant-rate deformation, but a preliminary heat-treatment followed by a subsequent heat-treatment at higher temperature converts the relationship to a linear rate. Orlova, Kainarskii, and Prokopenko claimed that $MgTiO_3$ affected the fired properties, particularly flexural strength, very little, but titanates of aluminum and zirconium, as well as TiO_2, ZrO_2, and SiO_2 lowered the strength. Magnesium silicates, presumably by formation of spinels, increased flexural strength. Regardless of the type of additive and its amount, whether TiO_2, MgO, $MgTiO_3$, Mg_2TiO_4, $ZrTiO_4$, or Al_2TiO_5, the growth of corundum crystals starts when the initial porosity (35 to 40%) decreases to 8 to 10%. At about 4% porosity, the unit crystal size (3 microns) increases only to 4 to 4.5 microns; gross crystallization starts at 0.5 to 1.0% porosity, at which point the size depends on the additive, especially TiO_2) (Degtyareva, 1965).

Jones, Maitra, and Cutler observed increased rates of material transport in the sintering of alumina containing additions of titanium, chromium, manganese, iron, or zirconium oxides in atmospheres of oxygen, nitrogen, and hydrogen. Pampuch reported that additives having lower electronegativity of the metal-oxygen bond than the basic oxide, in general about 1.6 to 2.0, improve sintering. The electronegativity (Pauling) measures the atom's ability to attract electrons, and is proportional to the sum of electron affinity (energy required to add an electron) and the ionization potential (energy required to remove an electron). The most effective oxide additions found were those that form solid solutions with the basic oxide, and in which the cation of the added oxide does not differ greatly in size from that of the basic oxide. Alumina of 4-micron size, containing 1 to 2% additive could be sintered at 1200°C. Kato, Okuda, Iga, and Okawara, in discussing the effects of 2 to 5% (wt) additions of $Mg(OH)_2$, Co_3O_4, Fe_2O_3, Cr_2O_3, TiO_2, or MnO_2, concluded as follows: sintering is promoted by the addition of MnO_2 but retarded by Cr_2O_3; the other oxides promote it slightly. The retarding effect of Cr_2O_3 was confirmed by Gerasimov et al. (1965). Dils investigated the cation interdiffusion in Cr_2O_3-Al_2O_3. X-ray analysis shows spinel structures when $Mg(OH)_2$, Co_3O_4, or MnO_2 is added. Since the lattice constant of the manganese spinel increases with the content of MnO_2, this is regarded as indicative of solid solution. Alumina and Fe_2O_3 or Cr_2O_3 form perfect solid solutions, but Fe_2O_3 may form a spinel, depending on the sintering atmosphere. Both MnO_2 and Fe_2O_3 accelerate grain growth remarkably.

133

Keski; and Keski and Cutler found that the sintering rate at 1550°C of alumina containing small amounts of MnO increases to a maximum between 0.2 and 0.4% MnO, and then decreases with further additions. They concluded that, although pure alumina sinters by grain boundary diffusion, the addition of even 0.1% MnO changes the method to bulk diffusion. Isothermal shrinkage measurements at 1450 to 1650°C indicated oxygen-ion diffusion as the rate-limiting step during densification, on the assumption of the bulk diffusion mechanism.

Britsch's criteria for sintering corundum are as follows: the raw alumina should be as pure as possible and contain less than 2% Na_2O; the particle size should be uniform. An extremely fine fraction favors irregular crystal growth. Kaolin is the recommended sintering additive. Popil'skii et al. state that without magnesia, sintering in vacuum, hydrogen, helium, or combustion gases in the temperature range of 1600 to 1900°C produces corundum with a pore volume of not less than 3.5%. With not more than 0.2% MgO, pore volume can be reduced to 0.2% (99% of theoretical density) only in vacuum or in hydrogen. Tuleff also discussed the development of pores as a function of their firing temperature.

Jorgensen and Westbrook concluded that the addition of magnesia or nickel oxide inhibits grain growth during the sintering of alumina by maintaining a high diffusion flux of vacancies from the pores to the grain boundaries, causing a decrease in grain boundary mobility. The segregation was inferred from microhardness tests and autoradiographic tests with ^{63}Ni.

The main purpose of a growth repressor is to allow substantially complete sintering, approaching theoretical density before secondary or discontinuous crystallization has reached excessive limits. If secondary recrystallization has occurred and pores become trapped, the theoretical density cannot be obtained within a reasonable sintering time and temperature.

From a practical standpoint, the factors which determine the relationship of pore removal to secondary recrystallization are those factors that determine the reactivity of the raw alumina, the attainable degree of compaction, and the grossness of the pores in the raw compact. Little attention has been given to these factors until recently. A better relationship is likely to be obtained with the "reactive" aluminas which have higher compaction densities than have been attainable with conventional alumina powders in the past. These newer types also have good flow properties for mold filling in the dry condition, and require only low amounts of lubricants and growth repressors.

17 ALUMINA IN REFRACTORIES

17-1 GENERAL

A study of the phase diagrams of which Al_2O_3 is a component indicates the refractoriness that alumina contributes to compositions. General technical information on refractories is provided by such texts as F. H. Norton, *Refractories* (1950); J. R. Coxey, *Refractories* (1950); and more recently by Harder and Kienow, *Refractories* (1960); Singer and Singer, *Industrial Ceramics* (1963); and Salmang, *Physical and Chemical Fundamentals of Ceramics* (1961); among others. *Refractories Bibliography* 1928–1947, and 1947–1956; *Modern Refractory Practice* (Harbison-Walker Refractories Company), McDowell, Scott, and Clark (1962); Mamykin (1958); Whittemore (1959); Jaffee and Maykuth (1960); Pechman (1953); among many others, serve to establish the position of alumina among the refractory oxides.

Alumina is classed as a neutral or amphoteric refractory. Trombe (1949) listed the "superrefractory oxides" as those with melting points between 2050 and 3000°C (Al_2O_3 to ThO_2). According to Navias (1960) the order of increasing refractoriness is: Al_2O_3, Cr_2O_3, La_2O_3, Y_2O_3, SrO, BeO, CaO, CeO_2, ZrO_2, MgO, HfO_2, UO_2 and ThO_2. While the melting point of alumina is at the low end of the superrefractory scale, it is one of the most stable oxides in both reducing and oxidizing atmospheres, and is one of the cheapest and most readily available in very pure form.

Konopicky (1959) confirmed the correctness of classifying refractory fireclay brick by their alumina content. Birch evaluated the refractories of the future (in 1945) as being those with melting points above 1800°C, of which the mainstays are magnesia (2800°C), alumina (2050°C), and zircon (2430°C). In 1964 the most promising future refractories were magnesia and alumina, with zirconia for limited demand, based on such factors as melting point, susceptibility to hydration, reducibility, toxicity, cost, and availability. These oxides and carbon are available in high purity. Berezhnoi (1963) pointed out the likelihood of new refractories is from binary mixtures. Brandt (1949) considered the use of oxygen as a means of increasing the flame temperature in the open-hearth steel-melting furnace. An analogy was drawn between this type of furnace and the glass-tank furnace. In glass furnaces the maximum permissible roof temperature is a limiting factor. Thielke (1950),

Hiester et al. (1956), and White (1963) discussed the changing pattern in the technology of new refractories. The advantage in increased production rate by substituting oxygen for air in the production of steel is one factor. Kirby (1951) recognized alumina, magnesia, beryllia, zirconia, ceria, and thoria as the pure refractory oxides capable of being used at abnormally high commercial temperatures (above 1800°C). Smutny and Tomshu (1958), and Bray (1951) believed that high-alumina refractories were favored in steel production from consideration both of cost and properties.

Another factor influencing the selection of refractories is abrasion resistance, as pointed out by Roe and Schroeder (1952) and Lesar and McGee (1956). Venable measured the erosion resistance of refractory linings in fluid catalytic cracking units in petroleum refining. Phosphate-bonded tabular alumina was about as resistant as silicon carbide shapes. Ramming mixes and refractory concretes had decreasing resistance.

Refractories made from calcined bauxite or diaspore may be classed as high-alumina refractories, to distinguish them from extra high-alumina refractories made from dense sintered alumina (tabular alumina) or fused alumina grain. Alumina refractories containing 80% Al_2O_3 and approaching 90% Al_2O_3 can be made from bauxite. It would seem to be uneconomic to debase bauxite or Bayer alumina in refractories of significantly lower Al_2O_3 content than mullite, in view of the excellent properties attained with limited fluxes in high-alumina refractories.

17-2 HIGH-ALUMINA REFRACTORIES

White (1938) noted that fused alumina refractories are preferably used where high temperature and high bearing loads are combined (piers and supports). Corundum shapes have high resistance to all common slags (Schaeffer, 1939). Fitzgerald (1943) fired a molded mixture of hydrated alumina with 7 to 30% uncalcined clay containing 60 to 75% SiO_2 at a temperature above cone 10 (1260°C), and stopped the firing while the refractory was still capable of further expansion on reheating (approximately 95% Al_2O_3). Schroeder and Logan (1951) added sufficient clay to contribute from 2 to 8% SiO_2 to the fired body, composed of fused alumina grain. The flexural strength was

135

much higher than that for brick made from pure fused alumina grain.

Poluboyarinov and Balkevich (1949) sintered mixtures of electrofused corundum (98 to 99% Al_2O_3) with mixed clays and/or kaolin not to exceed 15%. The clays contained 38 to 43% Al_2O_3. Specimens containing kaolin deformed at temperatures 40 to 90°C below those containing the clays (Latna type). The sintering process continued to 1700 to 1800°C. Reheat shrinkage amounted to 3 to 4% at 1600 to 1750°C. The maximum temperature of initial deformation was about 1650°C, and for two-side service, 1750°C (with substantially no load). Various other bonds were used, including (1) pure technical (Bayer) alumina, (2) alumina and alum, (3) alumina and clay, and (4) alumina, clay and lime, in total amount of bond from 15 to 45%. The apparent porosity ranged from about 23 to 26% and decreased with increasing bond content. Shrinkage and strength increased with increasing bond strength. The initial deformation occurred at 1830 to 1840°C. The admixtures did not affect the results; it was concluded to use only the technical alumina as the binder, and to form by semidry pressing (6 to 7% moisture). Firing at 1700°C for 3 to 5 hours should give practically constant volume, and an allowable service temperature to 1850 to 1900°C if the load is very small. The same investigators prepared high-density grog (3.80 g/ml) by heating pressed briquets prepared from ground technical alumina (99% Al_2O_3) at 1600 to 1710°C. The crushed briquets were then bonded with wet-ground technical alumina, with the addition of 1 to 2% sulfite cellulose liquor (tall oil) or aluminum sulfate to increase green strength. Mixes fired at 1710°C showed increased shrinkage and bulk density and decreased porosity with increasing bond content. The characteristics of these refractories were superior to those of the fused alumina with a clay bond. The initial temperature of deformation was 1850°C at 28.4 psi. Volume stability was high up to 1900°C. Visible cracks appeared after 15 to 16 heat-shock cycles. The compressive strength was as high as 420,000 psi.

Whittemore (1949) placed the effective operating temperature of pure alumina at 1900°C, in comparison with magnesia and stabilized zirconia at about 2300°C.

Lepp and Slyh (1951, 1952) calcined mixtures of alumina and aluminum in almost any proportion, but preferably with 35 to 90% Al_2O_3, at 815 to 1700°C enclosed in graphite in an oxidizing atmosphere. The product contained alumina, aluminum nitride, aluminum, and carbon. The products had excellent strength and thermal shock resistance.

Morgan Crucible Company (1953) prepared refractory alumina bodies by forming from fine-ground alumina, either with or without a combustible binder. The calcined alumina was ground until 95% passed a 300-mesh sieve. Iron contamination was removed by magnetic separation and by acid leaching. Zhikharevich et al. (1958) presented data on the use of high-alumina refractories of sillimanite, mullite, and mullite-corundum for blast-furnace linings.

These linings had low porosity and gas penetration and were volume-stable at 1550 to 1600°C.

Karklit and Gruzdeva (1950) prepared shapes weighing up to 8.8 pounds (97% Al_2O_3) and good for service at 1700 to 1850°C by dry-pressing Bayer alumina with 5% melted paraffin at about 14,000 psi, and by firing at 1710°C for 5 to 6 hours. The total shrinkage was 17.9%, apparent porosity 0.00%, and density 3.86 g/l ml, which did not change on reheat. The average crushing strength was 95,000 psi. Cracks appeared after 2 to 3 cycles of heat shock (heating to 850°C and cooling in water), and complete destruction after 6 to 8 cycles.

Riddle (1945) mixed from 11 to 29 parts silicon carbide with 100 parts alumina to obtain a refractory having high strength at 1455°C. E. D. Miller (1964) prepared a readily castable high-alumina refractory by the addition of 1 to 10% pure volatilized silica and 1 to 5% clay, the remainder being coarse-ground dense alumina or bauxite. The product was characterized by coarse-textured alumina particles rigidly held in a dense matrix of submicron mullite crystals. Pieper (1956) patented a composition containing 17.5 to 25% by weight of silicon carbide with about 75 to 82.5% aluminum oxide of specific size distributions, formed with water containing lignin, and fired at 1650°C.

Saunders, and Somer and Brady described silicon carbide refractories coated with alumina or iron oxide and alumina protective layers. Kappmeyer and Manning (1963) evaluated about 140 brands of high-alumina brick ranging from 60 to 99% Al_2O_3 including mullite brick. The results showed that many brands of high-alumina brick were not as good as some brands of fireclay brick currently available. The alumina content alone is not sufficient to determine suitability for specific steel-plant use. Kappmeyer, Lamont, and Manning (1964) found, similarly, that high-alumina plastics and ramming mixes in the steel industry must be selected for their individual properties rather than their overall composition. Powers and Kappmeyer (1965) concluded that none of the newly developed refractories to date (borides, carbides, nitrides, beryllia, and thoria) provide appreciably better performance in contact with molten steel than alumina, magnesia, or zircon.

Khemelvskii and Minakov (1957) found high-alumina refractories satisfactory for the baffles in glass-melting tanks. Koldaev et al. (1957) developed a refined technique for preparing mullite-bonded high-alumina blocks (50 to 96% Al_2O_3) which were claimed to be superior to electrofused blocks. The improved technique involved finer grinding of all constituents, the use of TiO_2 as a mineralizer, plastic and pneumatic tamping to shape the blocks, and a 16-hour soak at 1620°C. C. L. Norton (1964) developed a sintered composition for contact with molten glass, having high thermal- shock- and corrosion-resistance, formed from a mixture of 10 to 50% alumina-silica grain (80 to 95% Al_2O_3) in graded sizes to produce low porosity, 4 to 12% bonding clay, and the remainder corundum grain in graded sizes.

17-3 FUSED CAST ALUMINA REFRACTORIES

The terms: fusion-cast, fused cast, and electrocast are equivalent, meaning melting in an electric furnace and casting the homogeneous liquid into a monolithic refractory shape in molds. Though trials had been made earlier with a fused cast refractory "siemensite" by Haglund (German Patent 539,682, June 1927), the commercial manufacture started in 1925 when Corning Glass Works filed a patent application for a mullite block cast from bauxite (Fulcher, U.S. Patent 1,615,750, July 1925). Knuth (1935) appraised this "Corhart Standard" as approximately mullite with free corundum and a vitreous phase. A French subsidiary began operations in Modane, France, in 1929 (Bortaud and Rocco, 1964). The composition was melted at about 2200°C and poured into sand molds.

Benner and Baumann (the Carborundum Company) in 1935 patented a refractory for glass tank use, made by fusing 94% bauxite with 6% soda ash by weight, and containing about 85 to 90% Al_2O_3, most of which was present as beta alumina. Other variations were tried, as for example, incorporating in a bath of molten alumina a mineralizer composed of a halide of an alkaline earth metal, in amount about 1% (Benner and Baumann, 1939).

The Carborundum Company developed three early types of fused glass-tank blocks, having about the following properties:

	Beta Al_2O_3	Beta-alpha Al_2O_3	Alumina-Chrome
PCE	39-40	39-40	38-39
Wt (9 X 4-1/2 X 2-1.2)	10.2 lb	11.4 lb	11.6 lb
Thermal cond. at 620°C cal/cm°C	30	30	30
Coeff. thermal expansion 25 to 1400°C (°C)	7.4 X 10^{-6}	7.4	9.3

Benner and Easter (1940) developed a glass tank refractory by fusing over 5% iron oxide in the alumina. Zhilin and Ignat'eva cast blocks from blast furnace slags (maximum CaO 39%). Refractories Committee C-8 of the American Society for Testing and Materials included high-alumina block and brick and molten cast blocks in recommended refractory types in the report on refractory service in glass furnaces in 1939.

The fusion casting process avoids the usual bonding of refractories by developing interlocking crystal growth in high-density forms which withstand deformation at high temperature and offer exceptional resistance to corrosion, particularly in glass tanks. The blocks are presently cast in graphite demountable molds, which are removed as soon as possible after pouring. The casting is blanketed with insulation to minimize thermal stresses. After cooling, the casting is cleaned and subjected to finishing operations

which may involve cutting or grinding with diamond abrasives. Because of the great change in expansion upon cooling, a central cavity forms, which is difficult to fill completely by repeated pourings. Various treatments were recommended. Corning Glass patents (1940) mention filling the cavity with crushed, prefused material of the same composition, having enough glassy phase to induce sintering at the lower cooling temperature at which it is introduced, or the cavity is made to take a central position in the block by rotating the block during casting. Field and Smyth (1945) filled the "pipe" with additional molten refractory of the same type, mixed with a sulfate capable of dissociating into sulfur trioxide at temperatures between 400 and 1100°C. Alternatively (1952), they placed a gasifier in a container within the mold prior to pouring; the bottom of the mold was covered before the container disintegrated, the delay allowing the pipe or void to be redistributed into many small closed pores throughout the casting.

Field, Fulcher and Field, and Fulcher, early in 1942 disclosed a cast refractory consisting essentially of crystalline zirconia and corundum in a siliceous noncrystalline matrix (around 20% SiO_2 by weight). The silica could be substituted by alkali oxide in the ratio with contained alumina of less than 1:19. A third variant consisted essentially of zirconia (above 60°) and alumina (not over 25%) substantially free from iron oxide and titania. These led to a low-silica refractory (U.S. Patent 2,424,082, Field, 1947) containing 45 to 92% Al_2O_3, 1.9 to 40% ZrO_2, and only 1.5 to 5% SiO_2, distributed as crystals in an amorphous matrix. These ZAC cast refractories were developments of Corhart Refractories Company and their French associates Electro-Refractaire. ZED blocks contain about 50% Al_2O_3, 33% ZrO_2, 11 to 13% SiO_2, and 1 to 2% Na_2O. The soda (added as soda ash) provides electrical conductivity and aids recrystallization (Moore, 1950). Lamy (1952), and Hartwig (1956) discussed their favorable behavior in glass tanks even under unfavorable conditions in borosilicate glass melting.

Seki (1937) examined melts prepared in the Al_2O_3-SiO_2, Al_2O_3-MgO, and Al_2O_3-MgO-Cr_2O_3 systems to determine their resistance to basic open-hearth slags, acid open-hearth slags, borosilicate glass, and soda-lime glass. Refractories of the Al_2O_3-MgO-Cr_2O_3 system were found resistant to the glasses but were believed to be not suitable for this use because of glass discoloration from the chromium. The refractories containing high alumina were suitable for glass tanks and acidic open-hearth steel furnaces; refractories of the Al_2O_3-MgO and Al_2O_3-MgO-Cr_2O_3 systems were found excellent for basic open-hearth furnaces.

The widespread use of basic refractories in the steel industry of the United States following World War II led to the success of electrocast magnesia-chrome-alumina refractories (Field, U.S. Patent 2,408,305, 1946); and magnesia-alumina-silica-zirconia basic cast refractory (Field, U.S. Patent 2,409,844, 1946). The latter contains not less than 15% ZrO_2 and not less than 11% SiO_2, and the moles of

MgO are about equal to the moles of Al_2O_3 plus twice the moles of SiO_2. A heat-cast refractory designed for use in open-hearth roofs, consisted of about 15% of a crystalline solid solution of FeO in MgO, a spinel phase consisting of FeO, MgO, Al_2O_3, and Cr_2O_3, and a siliceous matrix. The composition contained 12 to 50% chromia, and 5 to 25% alumina (Corhart, British Patent 665,209, 1951).

The high-alumina fused cast blocks have received further attention. Galdina and Deri (1962) described a fused refractory 'Korvishit', containing 99.2% Al_2O_3 and only 0.15% Fe_2O_3, SiO_2, and alkali. It has a bulk density of 3 g/ml, is refractory to 1930°C, has no open porosity (apparent porosity 10 to 17%), coefficient of expansion (20 to 1200°C) 8.5×10^{-6}, and thermal conductivity 4.5 $kcal/hr^{-1}°C^{-1}m^{-2}m$. Corhart (1964) claimed that fused alumina refractories substantially crack-free are obtained if they contain 0.36 to 16.5 mol % of an alkaline earth oxide and 0.06 to 2.63 mol % fluorine, preferably calcium fluoride. Castings $1 \times 1 \times 3$ in. withstood 11 to 13 cycles of heating for 10 minutes at 1650°C and cooling to room temperature before spalling loss occurred. The bending strength was 7000 psi. Another Corhart disclosure (1964) revealed the influence of very small amounts of calcia (0.5 to 4%) on the resistance to thermal shock of substantially all-alumina castings, by altering the crystal structure from long straight crystals to a random distribution of fine interlocking Al_2O_3 and $CaO \cdot 6Al_2O_3$ crystals. The raw material must not introduce more than 1.5% MgO or 1% total of other impurities (SiO_2 and Na_2O). A pouring temperature of 2030 to 2050°C was specified. Specimens withstood the previously cited spalling test for 44 cycles before spalling, whereas the pure alumina castings broke after one cycle; compositions with 0.67 to 4% MgO broke after 2 to 9 cycles; and a composition with 4% BaO broke after 7 cycles. These blocks were recommended for arc furnaces and roofs of steel and glass melting furnaces. Alper and McNally (1964) found that the resistance of substantially all-alumina castings to thermal shock was considerably increased by addition of 0.2 to 1.7% SiO_2. In this case, the content of Fe_2O_3 and TiO_2 must be kept below 0.1%, and of Na_2O below 0.6%. A block specimen failed the thermal shock test after 20 cycles of heating. Pure alumina failed on the first cycle, and one containing 0.2% SiO_2 and 0.1% Fe_2O_3 failed in 12 cycles.

Sandmeyer and Miller (1965) showed the following table for typical fused cast alumina refractories (Harbison-Carborundum).

The thermal conductivity of the alpha alumina type is about twice that of the beta type, with the alpha-beta type falling intermediate.

Fused cast spinel refractories containing beta alumina were developed for use in contact with molten magnesium metal (Field, 1949). The refractories were prepared from metal-grade alumina, soda ash, and calcined magnesite low in silica. The required alkali was about six times the amount of silica present. Sekiguchi (1959) found that the spinel refractories resisted molten Fe_2O_3 strongly in open-hearth

	Alpha Alumina	Alpha-Beta 40:60	Beta Alumina
Al_2O_3	99.34	94.84	94.46
Fe_2O_3	0.06	0.06	0.07
SiO_2	0.08	1.09	0.12
Alkali oxides	0.39	3.58	5.17
Alkaline earth oxides	0.13	0.43	0.18
Bulk density g/ml	3.45	3.17	2.89
Apparent density	3.76	3.41	3.06
True density	3.93	3.54	3.26
Cold abrasion resistance Abrasion loss (cm^3/cm)	0.20	0.55	—
Thermal expansion (0 to 1200°C)*	0.98	0.95	0.79

*Approximated from curve

furnaces and against cement clinker in rotary kilns. Small amounts of Fe_2O_3 prevented peeling or bursting of the refractory. Ribbe and Alper (1964) found that rapid chilling of mixtures in this system show an enrichment of the low-melting spinel in the outer skin, called inverse segregation.

Fused cast mullite refractories, produced from the iron oxide-containing bauxites have been used in certain zones of furnaces containing glass-corroding baths. A suitable composition contains 70 to 80% mullite, 7 to 12% corundum, and 13 to 18% glass (Rustambekyan, 1964).

Chrome-alumina and chrome-magnesia-alumina fused refractories were prepared by mixing chromite ore (preheated at 800°C to develop crystallinity) with fused alumina, and optionally some magnesia, a preferred composition containing 75 wt % fused alumina to 25% Philippine chrome ore (Sandmeyer, 1959). Alper and McNally (U.S. 3,140,955, 7/14/64) developed a refractory of this type containing 56 to 85% MgO, 13 to 40% Al_2O_3, 0.4 to 4.5 Cr_2O_3, 0 to 3.5% FeO, and minor amounts of CaO, SiO_2, and B_2O_3.

Much interest has been shown in fused cast refractories in the Al_2O_3-SiO_2-ZrO_2 system. Litvakovskii, Busby and Partridge, Solomin and Galdina, Solomin, Schlotzhauer and Wood, and Ono and Shibata have described the preparation from zircon and technical-grade alumina (sometimes bauxite), to keep the iron content low. A highly refractory and corrosion-resistant glass tank body was prepared substantially to the stoichiometric proportions of the equation: $2ZrSiO_4 + 3Al_2O_3 = 2ZrO_2 + Al_6Si_2O_{13}$.

Steimke (U.S. 2,919,994) developed the composition: 40 to 60 wt % Al_2O_3, 12 to 22% SiO_2, and 25 to 45% ZrO_2, bonded with a borosilicate glass (0.5 to 1% B_2O_3, but not over 1% Na_2O, K_2O, and/or Li_2O, and substantially free from TiO_2 and Fe_2O_3. Nagashima, Miyake et al. (1960) prepared crack-free castings at 1900 to 2100°C with mixtures of zircon, alumina, and ores containing rare earth elements. The ZrO_2 constituted 20 to 40%, the rare earth oxides 2 to 50% of the ZrO_2, the Al_2O_3 30 to 50%, SiO_2 10 to 20%, Na_2O 0.1 to 5%, and CaO + BaO 2%. Ono, Fendo, and Onaka (1962) modified the formula to obtain practical freedom from alkaline earths, to obtain corrosion-proof glass refractories.

Alper, Begley, Londeree, and McNally (1964) claimed that the castings crack less in the molds if they are composed of 15 to 60% ZrO_2, 20% SiO_2, 0.5 to 2.5% Na_2O, 0.05 to 1.5% halide, and the remainer Al_2O_3. The preferred halides are calcium and aluminum fluorides.

Baque (1950), Carrier (1950), and Fabianic (1950) discussed the use of fused cast and other types of refractories for glass-furnace construction extant at that time. Roy Brown (1962) concluded that beta alumina, fused cast refractory is alumina saturated with soda, a single phase completely free of glass. It has excellent thermal shock resistance, refractoriness, and alkali resistance, and low thermal conductivity. It has the disadvantage of low abrasion resistance, poor resistance to molten batch carry-over, and is costly for certain shapes. Its use was recommended in port necks, feeder entrance covers, and non-glass-contact positions beyond the first and second port areas. Frischbutter and Schroeder (1966) evaluated the applications of beta alumina in electrocast refractories.

Survey of Refractories Used in Glass Tank Furnaces, J. Soc. Glass Technol. *42*, (209) 63-99P (1958) describes the qualities and properties of the various types of refractory ingredients (brick, cements, jointing materials, etc.), and their application in tank furnaces.

17-4 CLAY-BONDED ALUMINA REFRACTORIES, MULLITE REFRACTORIES

Alumina-silica refractories increase in refractoriness with increasing alumina content. Harvey and Birch (1944) pointed out, however, that a superduty silica refractory should not contain significant amounts of alumina (total Al_2O_3, TiO_2, and alkali oxides not over 0.5%), a specification for this grade (Lynam et al., 1952). These impurities cause a rapid decrease in melting point as the high-silica eutectic is approached (Mackenzie, 1952). Keltz (1951) claimed, however, that 0.1 to 0.3% fused alumina provides strength to lime-bonded silica brick.

High-alumina, clay-bonded refractories have been prepared from diaspore clays, but with depletion of the Missouri deposits, it became necessary to resort to calcined bauxite and Bayer plant alumina, calcined, sintered, and fused types, in order to produce refractories of comparable grade (Day). The following references are to refractories mainly from these sources.

Impure mullite grog has been prepared by rotary-kiln calcination of native Indian and African kyanites. Pure, fine-crystal mullite is prepared by sintering the stoichiometric ratio, $3Al_2O_3$ to $2SiO_2$, or preferably with a slight excess of Al_2O_3, using Bayer calcined alumina and kaolin of suitable purity. Welch (1960) claimed, however, that the

primary phase in the mullite region is approximately $2Al_2O_3 \cdot SiO_2$. Coarse-crystalline, pure mullite for special refractory purposes is prepared by electrofusion (Hutchins, 1941). Pevzner (1946) described the production of "thermite" mullite by the oxidation of aluminum. Low titania and alkali content is claimed for this product.

In the synthesis of mullite from Bayer alumina and clay the usefulness of various prospective mineralizers in promoting the reaction has been investigated. Hawkes (1962) described methods based either on Bayer alumina or on bauxite. Kitaigorodskii and Keshishyan used 3% MgO to promote the reaction of 72% Al_2O_3 to 25% SiO_2, during a heating at 1500°C for 6 hours and then at 1700° for 0.5 hour. The product was a highly resistant refractory, negligibly attacked by molten glass at 1300°C. Fenstermacher and Hummel (1961) attempted to sinter relatively pure $3Al_2O_3 \cdot 2SiO_2$ and $2Al_2O_3 \cdot SiO_2$ compositions to low porosities at 1710 and 1650°C, respectively, with the addition of 1% magnesia in each case to facilitate the reaction. The $3Al_2O_3 \cdot 2SiO_2$ body sintered to a porosity of 7.1% and was substantially all mullite; the $2Al_2O_3 \cdot SiO_2$ body sintered to 10.9% porosity and was composed of mullite and corundum. This would tend to refute the claim of Barta and Barta (1956) and others to the existence of many mullites between sillimanite ($Al_2O_3 \cdot SiO_2$) and pragite ($2Al_2O_3 \cdot SiO_2$). Especially in a reducing medium, SiO_2 was found to evaporate from $3Al_2O_3 \cdot 2SiO_2$ until virtually only corundum was left (Toropov and Galakhov, 1951). They reported in 1958, no evidence of solid solutions of $3Al_2O_3 \cdot 2SiO_2$ and $Al_2O_3 \cdot 2SiO_2$ at 1400 to 2100°C. Tromel et al. (1957) claimed that the very pure mixtures of Al_2O_3 and SiO_2, when held just above 1810°C for 2 hours, contained only mullite or mullite and glass, but no corundum. In the presence of 0.2% Na_2O (a common impurity in alumina samples), the mullite decomposed with the formation of corundum. The pure oxide mixtures required many hours at 1650°C to attain equilibrium.

Mineralizers were found to be of little value, contrary to the claims of Reinhart (1955) for B_2O_3, Kirillova (1958) for CaF_2 or $2MgO \cdot B_2O_3 \cdot H_2O$, Locsei (1964) for AlF_3, and others. Goncharov et al. found the maximum quantity of mullite was formed for a ratio of Al_2O_3:SiO_2 of 1.5:1. The composition of the mullite depends on the proportions of alumina to silica as well as on their calcination temperatures. De Keyser (1963) noted four zones of reaction between the interface of quartz and corundum particles: corundum; mullite; a mixed phase of cristobalite, corundum, mullite, and glass; and cristobalite.

Weill (1963) investigated the relative stability of the crystalline phases in the Al_2O_3-SiO_2 system. The morphology of mullite was shown by the electron microscope to be the characteristic needle-like crystals if sufficient glassy phase was present in the melt (Lohre and Urbain). Halm (1948) observed that acicular mullite, formed from the near-eutectic fusion, is very prone to coalesce into globular mullite by alternate dissolution and recrystallization with fluctuating temperature.

Poluboyarinov and Popil'skii (1947–1955) attempted to prepare mullite-containing refractories from technical Bayer alumina and several types of Russian clays (Latnenski, Chasov-Yar, and kaolin). Wet-ground charges containing 51% alumina and fired at 1400°C had high apparent porosities (38 to 44%). Firing at 1500°C reduced the porosity only to about 30%. No sintering was observed at 1600°C with charges containing 82% alumina. Mullite refractories containing about 60% Al_2O_3 were prepared by wet-grinding technical alumina (7.96% SiO_2, 88.86% Al_2O_3), mixing a clay slip on the base of 58.5% clay with 41.5% alumina, forming briquets and firing at 1570 to 1580°C. The reground composition was formed into bricks with 15% clay (6% as a slip, 9% as a dry powder). The pressed brick, fired at 1450°C for 1.5 hr had 0.7% shrinkage, the reheat shrinkage at 1600°C was 0.4%, and at 1650 to 1680°C, 0.8%; the bulk density was 2.31 g/ml; the apparent porosity was 17.8%; and the initial deformation under load occurred at 1480°C. During the firing of mullite-corundum refractories, dimensional changes were observed. Rybnikov (1950) reduced the porosity to about 1.7% by the addition of 1% sodium silicate in the grog product and by increasing the firing temperature to 1620°C. Compressive strength was about 47,000 psi. Glass tank blocks were fabricated from mixes containing 75% grog, 25% of the original mullite mixture, and 9 to 10% water, and fired at 1650°C for 8 hours. The blocks ranged in porosity from about 16 to 20%.

C. Jones (1954) prepared synthetic mullite by calcining an intimate mixture of Bayer alumina, clay and ethyl silicate binder at a temperature between 1500 and 1800°C, the initial ratio of Al_2O_3:SiO_2 falling between 72 and 78% Al_2O_3 and between 28 and 22% SiO_2 by weight. Vinogradova et al., and Stavorko (1960) reported on trial batches of thermally stable mullite-corundum brick. Brick suitable for blast furnace stoves contained only 45% Al_2O_3. Brick containing 77% Al_2O_3 plus TiO_2 in the walls of a laboratory furnace, working at 1700 to 1750°C with rapid heating and cooling, showed high stability. After 50 firings (21 hours at 1600°C or higher) the linings were in excellent condition. Heeley and Moore prepared refractory mixes from calcined alumina and specially fine china clay in the proportions corresponding to sillimanite and mullite and containing up to 0.5% Na_2O, added as sodium silicate, in the sillimanite, and up to 3% in the mullite mixtures. The soda addition reduced the true porosity of the sillimanite at firing temperatures of 1400 to 1450°C, the effect diminishing with increasing temperature. The corrosion by molten glass increased as the firing temperature approached 1590°C for both compositions, but was more severe in the case of mullite.

Richardson et al. (1947, 1948) found that high-alumina brick and aluminous firebrick were attacked more severely than silica brick by ferrous oxide. Mullite was quickly decomposed by ferrous oxide to form hercynite; fayalite did not form until the mullite and corundum disappeared. Mullite appeared to be attacked more readily than corundum in high-alumina brick by mixtures of calcium and ferrous oxides. CaO was apparently more reactive than FeO. If the SiO_2 and Al_2O_3 were insufficient to satisfy the CaO, it combined with FeO rather than remain in the free state.

Refractory-grade bauxite used in the production of aluminosilica refractories has been subject to National Stockpile Material Purchase Specifications P-50 (December 5, 1951), conforming to the following percentages by weight: Al_2O_3 minimum 85.00; SiO_2 maximum 7.00; Fe_2O_3 maximum 3.75; TiO_2 maximum 3.75; and L.O.I. maximum 0.50. A stringent specification requires that the bauxite be well calcined to a bulk density of 3.10 g/ml.

Notwithstanding the hard-burned densification of the bauxite (or of pure alumina types) the reaction in aluminosilica compositions to form mullitized refractories causes the phenomenon called secondary expansion. J. L. Hall (1941) ascribed the secondary expansion of high-alumina grogs in clay binders to the formation of the relatively less dense mullite. McGee and Dodd (1961) claimed that the expansion results from crystallization of secondary mullite by the action of impurities in aiding the diffusion of alumina and silica. Secondary mullitization and growth began at 1400°C and caused as much as 23% expansion at 1600°C. Small additions of sodium fluoride were effective in controlling the expansion. Patzak (1964) observed the presence of tialite (Al_2TiO_5) in high-temperature calcination of bauxite, and emphasized the effect of the additive aluminum fluoride on mullite formation. Sedalia attributed the higher expansion resulting from bauxite to the formation of beta alumina by the reaction of the alkalies from the bonding clay with the alpha alumina in the bauxite. The linear expansions measured by Hall and Sedalia were less than 4% after 5 hours at 1600°C. Heilich (1962) found that the volume stability of compositions in the mullite range, prepared from either Suriname or Demerara bauxite, could be controlled by adjusting the particle size distribution of the components. Both the strength and the expansion in the desired low range were achieved by increased fineness of the bauxite, at least partially, in the 200 to 325 mesh range.

Leduc (1956) investigated the effect of the bonds: (a) 5% aluminous clay, (b) 5% hydrated lime, (c) 5% alumina cement, and (d) 15% alumina cement on the properties of bauxite bricks. Specimens a, b, and c deformed at 1420°C under a load of 28 psi; specimen d deformed at 1200°C.

Mullite Refractories (1940) presented data showing typical applications of mullite refractories in electric induction and arc furnaces, industrial boilers, open flame and crucible furnaces, glass tanks, ceramic kilns, iron and steel furnaces, etc., in the 3300°F (1820°C) range of melting. Mullite is claimed to have a high softening point, a low coefficient of expansion, good resistance to thermal shock and slag erosion, negligible tendency to vitrify, high load-bearing capacity, and chemical neutrality. Remmey (1948) described the properties of sintered mullite structures bonded with mullite. Partridge and Busby (1955) described slip cast mullite and zircon tank blocks. Knauft, Smith, Thomas, and Pittman (1957) described bonded sintered mullite and zircon refractories for use in glass contact areas, superstructures, and feeder forehearths of furnaces melting borosilicate and dense opal glass.

Rutman (1959) recommended the bonding of mullite brick used in rotary kilns with a mortar to convert the refractory into a monolithic liner. The mortar was composed of 90% ground grog (88% alumina, 10% clay), 10% raw clay and 5% sodium silicate (density 1.45 g/ml). Wicken and Birch (1963) reviewed the properties of high-alumina refractories, among others, used in modern rotary kilns. Good refractories of the unfortified type range from 82 to 93%, averaging about 85% mullite. Fortified types (addition of pure alumina) contain 70% or more mullite (20% alpha alumina). The fortified mullites have increased load bearing resistance, spalling resistance, and resistance to slagging.

17-5 SPINEL AND ALUMINA-CHROMITE

Alumina reacts with magnesia to form $MgAl_2O_4$, or $MgO \cdot Al_2O_3$, the end product of a family of analogous compounds, called spinels, many of which occur in nature. As some of these spinels are produced commercially by synthesis using Bayer process alumina, they constitute a significant application of alumina.

Rankin and Merwin (1916) melted "normal" spinel at 2135°C, redetermined by Alper, McNally, Ribbe, and Doman (1962) at 2105°C, using spinel synthesized from Bayer alumina (Alcoa A-14). Because of the high melting point, a major use of spinel is in refractories, but other ceramic uses for members of the family are found, as for example, ceramic pigments.

The unit cell structure of normal spinel is described in Section 4-1 (eta alumina). The magnesium ion can be replaced by similarly sized bivalent ions such as Zn, Co, or Ni, without changing the lattice appreciably, and the aluminum can be replaced by trivalent ions. Univalent, tetravalent, and hexavalent substitutions are also possible for the cations.

Examples of the types of spinels are as follows, assuming A, B, C, D, and E to represent metal ions of valence equal to 1, 2, 3, 4, and 6, respectively, and O to signify oxygen or its substitution.

BC_2O_4 $MgAl_2O_4$, $FeO \cdot Fe_2O_3$
DB_2O_4 $TiMg_2O_4$
$(AC)C_4O_8$ $(LiAl)Al_4O_8 = LiAlO_2 \cdot 2Al_2O_3$
EA_2O_4 $MoLi_2O_4$
C_2O_3 γ-Fe_2O_3

Spinels composed of a single metal in two states of oxidation are designated "inner" spinels, as for example, magnetite, $FeO \cdot Fe_2O_3$, and the claimed $AlO \cdot Al_2O_3$ of Filonenko, Lavrov, Andreeva, and Pevzner. The bivalent cations may be replaced by ions of different valence, in which case the spinel is designated "inverse" to distinguish from the "normal" spinels. Inverse spinels include the

magnetic iron spinels, the nickel, and manganese (partially). The cobalt spinel is almost completely normal (Greenwald, Pickart, and Grannis), as are the nonmagnetic iron spinels. The complexity of the spinels increases because of their ability to take in solid solution an excess of either the bivalent or trivalent ion; $MgO \cdot Al_2O_3$ picks up as much as 60% Al_2O_3, according to Winchell (1940), but to $MgO \cdot 4Al_2O_3$, according to Biltz and Lemke (1930). Magnesia is hardly soluble in $MgO \cdot Al_2O_3$, but dissolves in other types (to 1.5 $MgO \cdot Cr_2O_3$).

Jagodinski and Saalfeld (1958) using the Fourier analysis of X-ray data, and Brun (1960), by measurement of paramagnetic nuclear resonance, found that natural spinel crystals show almost completely the "normal" spinel structure with the aluminum ions in the octahedral positions. Synthetic spinels in $MgO:Al_2O_3$ ratios of 1:1, 1:2, and 1:3.5 show more disordered distribution from normal. Natural spinels became disordered by heating at 800 to 900°C, but the process could not be observed in the reverse direction, that is, ordering in synthetic spinels. Since all the crystals investigated did not correspond strictly to the space group $Fd3m(O_h^7)$, it was concluded that spinels are composed of twinned crystallites of lower symmetry. Arnold (1960) interpreted the visible acicular inhomogeneities observed in synthetic spinels as voids caused by a decrease in internal strains during annealing. Stoichiometric spinel ($MgO \cdot Al_2O_3$) showed decreased absorption in the 3-micron band and a shift of the transmission limit to longer wavelengths. It also had a minimum index of refraction with increases on both the alumina-rich and the magnesia-rich ($MgO \cdot 0.8\ Al_2O_3$) sides of the stoichiometry (Wickersheim and Lefever). Rasch developed two-dimensional schematic diagrams of the spinels and corundum.

Rinne (1928) determined the index of refraction of Mg-Al spinel to be 1.718 to 1.728 (n_D). The change in density with change in Al_2O_3 content was as follows:

Al_2O_3 (%)	71.7	81.5	89.4	91.0	92.7
g/ml	3.578	3.604	3.621	3.625	3.624

The molecular weight is 142.27, and the ideal alumina concentration 71.67%.

Ryshkewitch (1941, 1942, 1948, 1960) recognized the importance of spinel as a pure oxide ceramic material, and drew attention to its chemical stability and refractory properties. Early attempts to realize these properties, particularly in refractory applications, were marred by variability in the properties, coarse grain, and incomplete densification. The structural properties of spinel are largely dependent on the microstructure, which is influenced by the raw materials and the processing conditions (Palmour 1964).

Normal spinel ($MgAl_2O_4$) has been successfully fabricated by the usual ceramic processes, such as casting, extrusion, or cold pressing, followed by sintering at about 1650 to 1850°C. Well-crystallized, pure spinel is produced by electrofusion. McCreight and Birch (1957) prepared spinel by mixing magnesium hydroxide and aluminum hydroxide in water sufficient to obtain good mixing, prior to firing. Noguchi (1948) prepared the normal spinel by heating cylindrical pressed specimens of the equimolecular mixture of oxides at 1535°C for 30 minutes with 0.01, 0.03, or 0.05% of various mineralizers. Compounds of Li, B, Ti, and V were effective, but not those of Ca, Sr, or Ba. Toropov and Sirozhiddinov found that Y_2O_3 had the greatest mineralizing action of the rare earths, La_2O_3, Y_2O_3, and Nd_2O_3 on spinel synthesis. The formation was almost complete (97%) after 5 hours at 1300°C for 1 and 3% additions. The additions of 1 and 5% Nd_2O_3 and of 1% La_2O_3 required 5 hours at 1500°C, or an effectiveness equivalent to that of H_3BO_3. Spinel production occurred in two steps (1) an intense reaction at the boundaries of the particles up to 1300°C, in which a layer of spinel formed at the surface of separation; (2) above 1300°C a decreasing reaction rate with time and temperature regulated by the denser spinel layer, which decreased the rate of diffusion of the Mg^{2+} and Al^{3+} ions. Diesperova and Bron found that the rate of synthesis of MgO-Al_2O_3 commences below 900°C, and finishes at 1500°C when catalyzed by Y_2O_3 or B_2O_3.

Hedvall and Loeffler found that the reaction between CoO and Al_2O_3 occurred at a faster rate when gamma alumina was used instead of dead-burned alumina, and Co_2O_3 instead of CoO. An increased velocity (Hedvall effect) is generally observed when one or more participating components undergo a transformation during the reaction.

The preparation of normal spinel by electrofusion and casting was described by Alais in 1944. The product was characterized by high refractoriness and exceptional dielectric properties. Navias (1961, 1963) described the preparation and some properties of spinel made by the vapor transport of MgO from periclase in hydrogen atmosphere at 1500 to 1900°C. The products diffused into alumina (sapphire) to form a uniform outer layer of spinel on all surfaces. At 1900°C, the spinel layer was 20 mils thick in one hour and 48 mils in 16 hours; at 1600°C, the rate was about 2.6 mils in one hour and 6.1 mils in 16 hours. The calculated activation energy was 100 kcal/mole. The lattice constant, Vickers hardness, and refractive index of the spinel layer varied with depth in a uniform manner, indicating a continuous change in composition from MgO·Al_2O_3 to MgO·2 to $3Al_2O_3$. In conversion to spinel, a volume increase of 47% over that of the original sapphire occurred. Clear shapes of spinel were produced from clear sapphire and from translucent polycrystalline alumina. Vapor of alumina diffused into fused periclase blocks to form spinel. In an oxidizing atmosphere, spinel was formed only on the surface of sapphire directly in contact with periclase, in the range 1500 to 1900°C. Altman (1963) observed that gaseous MgO is of little importance in the vapor phase at 2000°K. A negative heat of formation from the oxides is probable, from Knudsen experiments. Rossi and Fulrath (1963) examined the solid-state reaction between MgO and Al_2O_3 in air at 1560°C. Two spinel layers were observed, one thick layer grown contiguous with the sapphire crystal, and a thinner layer grown

contiguous with the sapphire crystal, and a thinner layer grown from the periclase-spinel interface. Often a fracture appeared within the sapphire about parallel to the sapphire-spinel interface. From the ratio of thicknesses of the two layers, the mechanism of counterdiffusion of cations through the oxygen framework was verified. The crystallographic orientation of the parent crystals had no effect on the rate of spinel growth. The orientation of spinel grown from sapphire was always determined from the orientation of the sapphire, as was also the case when epitaxy was observed in periclase-grown spinel. The orientation of spinel grown from periclase appeared to depend on the mechanism of transport across the periclase-spinel interface.

Bruch (1964) claimed that spinel bodies produced by treating compacted polycrystalline alumina with magnesia vapor carried in hydrogen, or as a directly contacted hot solid body, are free from an internal zone of weakness that splits readily, if all but one side of the alumina body is kept from contact with the MgO, as by masking. Crystal faults, where growths from opposite directions meet, are prevented. At 1900°C, the activation energy is 92 kcal/mole.

Carter (1961) ascribed the solid-state reaction forming $MgAl_2O_4$ in single-crystal and polycrystalline alumina to the counterdiffusion of the Mg^{2+} and Al^{3+} ions through the relatively rigid oxygen lattice. Inert markers were used in the experiments, but Carter demonstrated that pores could serve equally well as markers.

Budnikov and Zlochevskaya (1958) found that the use of salt solutions instead of magnesite, and fine dispersions of alumina (0.9 to 2 m^2/g) gave uniform production of spinel at temperatures above 1400°C. At lower temperatures, spinel formation practically ceased within two hours. Rybnikov (1956) obtained bulk densities of 2.98 g/ml for normal spinel prepared from magnesite and aluminum hydroxide, but only 2.84 g/ml when it was prepared from calcined alumina and magnesite. The initial softening under load of 28.5 psi at 1640°C was obtained for spinel containing 30% MgO (ideal, 28.3%).

Spinel refractories are credited with good resistance against molten iron oxide in open-hearth furnaces and against cement-clinker slag in rotary kilns for portland cement (Sekiguchi, 1962). Niwa found that after 361 heats in open-hearth furnaces, spinel stood up better than magnesia-chrome brick in roofs. The main cause of failure was spalling, but the spalled pieces were thinner than with magnesite-chrome brick. Suzuki found that spinel refractories are sufficiently matured at about 1485°C (cone 18) if boric and silicic acids are used as mineralizers.

Kriegel, Palmour, and Choi (1964) investigated the effect of intimacy of mixing on the reactivity and mechanical properties of synthetic normal spinel. Samples prepared from wet-milled mixtures of calcined alumina (Alcoa A-14) and magnesium carbonate, fired for 2 hours at 1450 to 1470°C, had ultimate crystal size below 2 microns and marked recrystallization occurred at 1800 to 1850°C. Samples prepared by the stoichiometric coprecipitation of the corresponding hydroxides by neutralization of the chlorides with NH_4OH, had an ultimate particle size below 0.04 micron after converting at only 850 to 860°C for 24 hours heating. Choi, Palmour, and Kriegel (1962) found that specimens, hot-pressed at 1227 to 1750°C at 1000 to 3000 psi, required fine grain size (2 to 15 microns) to attain maximum strength (34,000 psi in transverse strength, and 3.5×10^7 psi in modulus of elasticity). Palmour, Choi, Barnes et al. added from 0.001 to 0.1% graphite to improve strength. Chung, Terwilliger, et al. (1964) measured the elastic moduli of magnesium aluminate spinels varying in molar ratios. Smoke (1954) measured the low-loss dielectric properties of magnesian spinel compositions.

Yamaguchi (1953) observed for various mixtures of MgO, CoO, NiO, CuO, ZnO, and MnO with Al_2O_3, heated between 900 and 1700°C, that spinel was most readily formed in the $NiO-Al_2O_3$ system, was formed with difficulty in the $MnO-Al_2O_3$ system, and was not obtained in the $CdO-Al_2O_3$ or $CaO-Al_2O_3$ systems. MgO, CoO, and ZnO were intermediate in reaction. Pattison, Keely, and Maynor obtained synthesis of $NiO \cdot Al_2O_3$ and $CoO \cdot Al_2O_3$ in solid-state reactions when calcining mixtures of the hydrated oxides at 1200°C, but did not achieve conversion when alpha aluminum oxide was used. The spinel $MnAl_2O_4$ is stable in the region 1520 to 1560°C (Hay, White, and McIntosh, 1934–35). Compositions on both sides of the spinel $ZnAl_2O_4$, extending from 20 to 60 mole% melt at about 1950°C (Bunting, 1932).

Unicrystalline precious stones or shapes, such as rods, filaments, and filament guides, prepared by the flame-fusion process from spinel, can be deformed and bent to shape, after heating until the form becomes plastic (Linde, 1948).

Mixtures of Bayer alumina and native chromite are also used as refractories. Alumina-chromite linings containing only 10 to 12% Cr_2O_3 developed the best refractory properties (Chatterjee and Panti, 1963). Chromite ores containing over 10% SiO_2 were not suitable for glass furnace use. Rammed chromite-alumina linings had twice the stability of chromite brick linings, owing to a somewhat better homogeneity of the material and the absence of joints. These linings applied over a calcium aluminate base in the sintering zone of rotary kilns for cement manufacture provided more stability than was attained with talc linings (Khvostenkov and Tararin, 1939).

17-6 REFRACTORY EQUIPMENT

Alumina enters into the composition of refractory container and handling equipment such as ladles, stoppers, crucibles, saggers, molds, troughs, thermocouple protection tubes and insulators, etc. It is also used in heat-exchange in the form of powder, pebbles, balls and checkerwork. Examples of some applications in these categories are given.

Ryshkewitch (1960), Stott (1938), Gurr (1939), Thompson and Mallet (1939), Pointud and Roger (1957), and Avgustinik and Kozloskii (1959) described the produc-

tion of impervious, all-alumina or high-alumina sintered ware, including laboratory crucibles, combustion boats, and pyrometer tube porcelains. In most cases the small-size ware is made by slip-casting or extrusion, and fired at temperatures below 1700°C.

Meister coated porous alumina (Alundum) crucibles with thoria, beryllia, or zirconia to impregnate the pores effectively. Rutman et al. (1957) prepared thermally stable high-alumina ladle brick and stoppers of mullite-corundum composition (Al_2O_3 75.9%, SiO_2 21.1%, Fe_2O_3 1.0, Na_2O 0.6). The refractory failure was at 1850°C, bulk density 2.62 to 2.75 g/ml, porosity 15.7 to 18.6%, cold crushing strength 10,000 to 14,000 psi, and the initial deformation under load of 28.5 psi was at 1540 to 1560°C. Slag resistance was high, and thermal shock resistance good (18 melts life in comparison with 11.8 melts for fireclay brick). From an assessment of 1990 individual ladle linings ranging from 25.8 to 43.0% Al_2O_3, G. Workman found that in the tapping of steel slags of high Fe_2O_3 content, ladle life increased with decreasing porosity and with increasing alumina content in the range 32 to 43% Al_2O_3. Grebenyuk and Zhuravleva added small amounts of TiO_2 and ZrO_2 to alumina crucible slip-casting compositions to reduce the maturing temperature. Similar rammed compositions were used to line vacuum induction furnaces. The crucibles were good for 18 melts or more at a metal temperature of 1600°C. Karklit and Timofeev (1959) recommended high alumina ladle linings (15 to 25% corundum, 55 to 65% mullite, and 17 to 22% glass). Aldred (1962) claimed a nonglassy composition consisting of sintered crystals of alumina, bauxite, uncalcined and calcined kyanite, fused silica, mullite, or stabilized zirconia, finer than 100 mesh (Brit.S.S.) and embedded in a coarser matrix (10 to 50 vol. %) of a different additive from the same group of refractory materials. Bron (1963) selected refractories in the system $Al_2O_3-Cr_2O_3-SiO_2$ because of the higher viscosity of their liquid phase, which adds resistance to deformation and clinkering.

Lifshits (1950) recommended high-alumina brick (60 to 70% Al_2O_3) and a porosity of 18 to 20% for the upper row of nozzles of glass melting furnaces. Urban (1955) reported on trial installations of checker brick at the Sparrows Point Plant of Bethlehem Steel Company. When installed in the top eight or ten courses of checkers (which received the most severe treatment), brick having an alumina content of 55 to 60% gave better service than the regular fireclay quality; two brands of high-alumina brick with 60 to 70% Al_2O_3 appeared very satisfactory. Makarychev, Rybnikov et al., and Ivanov et al. (1958) were in agreement, that high-alumina brick give the longest service in open-hearth checkers and glass furnace regenerators. Some explosive spalling resulting from expansion of iron oxides at 1600°C was observed.

Hicks (1955) recommended basic refractories, specifically a 90% periclase, for checker brick in glass tank regenerators "because they remain dry," and because high-alumina brick fail by formation of nephelite.

Ceramic balls or pebbles were developed as tower packing for the high-temperature heat-exchange in various operations, such as steam generation, shale oil recovery, the cracking of crude oil, etc. The problem involved not only suitable forming methods for the ball, but also the design of the equipment for maturing the balls and equipment for effecting the heat exchange. The ball form was preferred in order to flow the heat-exchange medium into the reaction chamber and out, so as to make the process as continuous as possible. The requisite properties of the exchange medium; good heat capacity, high refractoriness, chemical inertness, and good resistance to heat shock, impact, and attrition, favored sintered alumina as a suitable medium. Several methods for producing free-flowing ceramic shapes were developed, including the tabular alumina ball process of Alcoa and extrusion processes in which the extrudate was cut into slugs which were then compacted into spheres. S. P. Robinson (Phillips Petroleum Company) prepared extruded pebbles from alumina and magnesium aluminate spinel, among others. Bearer combined 1.8% methyl silicone, with the alumina so that during the maturing of the bond at 1590 to 1760°C the silicone oxidizes to form free silica, which reacts with the alumina to form a mullite bond. Kistler prepared pebbles (1) consisting of an outer shell separated from an inner solid sphere or (2) an outer shell and an inner hollow sphere, for which good resistance to thermal shock was claimed. Renkey and Reardon (1963) discussed the preparation of high-purity alumina spheres for heavy-duty ceramics in pebble-heater beds, which are serviceable to about 1930°C. Isostatic forming was the preferred forming method. C. L. Norton (1946) designed a pebble heater in which refractory pebbles were moved continuously from a firing chamber to a heat interchanger chamber, for the purpose of preheating air to 1260°C or steam to 980°C.

Yuzhaninov (1964) measured the loss in alumina as dust from fluidized bed furnaces. The losses depend on the size of the charge, the chemical composition, the velocity of the gas stream, and the depth of the separation zone. The proper depth of separation zone was found to be double the depth of the fluidized bed. Equations were derived experimentally.

Unground, free-flowing calcined Bayer alumina has been used in the placement (bedding) of pottery in saggers, as a substitution for flint, a requirement in British practice (Elliott, 1937). This change was made by all the fine china firms and many others by 1938, not only because of improvement in the quality of the ware but in the health of the ceramic workers (Meikeljohn and Pozner, 1957). In this respect, a voluminous literature on silicosis and its treatment has been generated, of which the following is typical. Cunningham (1940) stated that silicosis in industry is not known to occur from silica combined as silicates, except in the case of asbestos and talc. Jephcott, Johnston, and Finlay (1948) observed that the fumes from the electric furnaces for fused alumina production from bauxite consisted chiefly of amorphous particles ranging in diameter from a few hundredths up to nearly 0.5 micron, and that

from 85 to 90% of the fume was alumina and silica. Hannon (1958) described the method for producing a finely divided aluminum oxide powder formed from substantially pure metallic aluminum and less than about 1.2 microns particle size, to be used in the prevention or mitigation of silicosis by inhalation.

Alumina is used in the preparation of saggers for supporting or enclosing ceramic ware during firing. Schaefer and Schwartzwalder (1959) supported ceramic bodies containing at least 90% Al_2O_3 and which require a firing temperature of at least 1700°C, on a sintered mass consisting essentially of 80 to 90% alumina and the remainder zirconia. The grain size of the alumina prior to sintering was not over 100 mesh. Caton (Champion Spark Plug Company) disclosed a sagger composition in 1951 composed of at least 7 parts alumina and from 1 to 3 parts silica to which is added about 2% calcium aluminate ($3CaO \cdot Al_2O_3 \cdot 6H_2O$). The alumina content is at least 85%. Ortman (AC Spark Plug) in 1959 described a sagger composition prepared from either of two different size distributions of tabular alumina but in amount 72 parts, 20 parts of synthetic mullite, and 4 parts each of No. 1 ware clay and Florida kaolin. The composition was mixed with sufficient water and 0.2% sodium phosphate deflocculant for casting in plaster molds. The molds were agitated during casting, and fired at cones 14 to 30, for example at 1410°C.

Zeemann (1957) prepared a monolithic mullite refractory plate to be used as kiln furniture by mixing 49.83% bauxite (64% Al_2O_3, 2% Fe_2O_3) and 50.17% pure kaolin. The mixture was milled to 200 mesh, formed by slip casting with 30% water, and fired at 1500°C to 23% porosity.

Alumina is used as the mold material in different types of casting of metals. Neiman (1944) described the precision casting process by investment molding. Principal materials used by the dentist, jeweler, and precision caster were given. Feagin (1951), Janssen (1957), and McIntire (1957), presented data on the use of alumina for high-temperature precision casting of metals.

Combustion tubes for small metal-resistor furnaces are prepared from high-alumina porcelains and sintered pure aluminum oxide. Impervious tubes have been prepared by slip casting, dry pressing and extrusion. Combustion tubes are often used in dry hydrogen ambients in order to protect the metal resistor windings from oxidation. The reduction of alpha alumina to suboxides has been mentioned previously (Section 3-9). Reduction of the oxides involved may occur according to the following reactions:

$$Al_2O_3 + 2H_2 = Al_2O + 2H_2O$$
$$SiO_2 + H_2 = SiO + H_2O$$
$$Na_2O + H_2 = 2Na + H_2O$$

Trostel (1964) measured the temperatures at which the rates of volatilization of reduced oxides became appreciable from an alumina specimen containing 0.5% SiO_2, 0.25% Na_2O, and 0.25% MgO, and found that when flowing dry hydrogen (-57°C dew point), the loss of Na_2O was noted at 1100°C, SiO_2 at 1500°C, MgO at 1700°C, and Al_2O_3 at 1900°C. During a two-hour heating period at 1500°C, about 50% of the Na_2O was lost; in the same period at 1900°C, virtually all the Na_2O, SiO_2, and MgO were lost. A mullite tube, rated nominally at 1650° service in air, might deteriorate rapidly in dry hydrogen at the same temperature.

Amerikov and Pirogov (1960) added about 1% TiO_2 to technical grade alumina (98.8% Al_2O_3) for use in preparing small-diameter tubes about 4 to 6.5 ft in length. Aluminum phosphate binders were formed by addition of 20 to 23% H_3PO_4. The green tubes (38% porosity) were fired at 1600°C. Kassel (1964) described the manufacture of large alumina cylinders, 48 in. in length, 20 in. outside diam., and 1 in. thick for use as the "magnetic bottle" included in the Stellerator, a device for creating thermonuclear energy through the process of heating a plasma of ionized gas to (0.5 to 1) million degrees K. At this point fusion is expected with the release of marked thermonuclear energy. The maximum heat would be contained well within the walls of the tube. A high alumina composition (Almanox 4462, Frenchtown Porcelain Company) was selected for forming operations. Core casting and joining of segments were the forming methods used. Core casting was successful, and isostatic pressing at 30,000 psi, without the use of a binder was successful.

Monolithic furnace linings of tabular alumina are often prepared by ramming tabular alumina moistened with aluminum sulfate, phosphoric acid, or mixtures of the two. Pevzner (1954) mentioned some applications for corundum refractories including burner ports in a furnace melting sulfate-fluoride charges. Keith and Whittemore lined a high-temperature laboratory kiln, capable of being fired to 1950°C, with alundum. Kupffer (1962) used 99% Al_2O_3 brick for ceramic skid rails in a pusher furnace. Weltz (1956) described a refractory composition containing 93 to 98 parts of alumina to 7 to 2 parts of boron nitride, for use as journals and bearings.

17-7 REFRACTORIES FOR ALUMINUM AND OTHER NONFERROUS USES

McDowell (1939) recommended high-alumina brick (50% Al_2O_3) for nonferrous metallurgical furnaces. Alumina brick were used in roaster hearths; alumina brick (60% Al_2O_3) were used in the side walls of nickel anode furnaces, and 70% alumina brick were used in different types of lead furnaces.

Burrows (1940) discussed the effect of molten aluminum on various types of refractory brick. Brondyke (1953) found that all commercial alumina-silica refractories were wetted and subsequently penetrated on exposure to molten aluminum. Penetration and silicon pickup were independent of the porosity, grade of refractory, and source. Stock and Dolph have recently reviewed the mode of attack of refractories in aluminum melting furnaces.

The melting point of aluminum is about 660°C, but destruction of the refractory ensues by wetting and penetration by molten aluminum, together with oxidation and thermitic reaction with easily reduced oxides such as silica. Knauft (1943) recommended zircon refractories for aluminum melting furnaces. Zircon and special, bonded silicon have excellent resistance to attack by molten aluminum (Caprio, 1965). Nitride-bonded silicon carbide (Carborundum's Refrax and Refrax-50) or oxynitride-bonded silicon carbide (Norton's Crystolon-63) are types that appear promising in remelt furnaces, metal transport, trough tile, pyrometer tubes, and tube assemblies for induction furnaces (Ref. Norton introduces . . . . 1960). Peskin (1961) found no penetration of the nitride-bonded silicon carbide by molten aluminum in 200 hours.

Brown and Landback discussed the specific properties of ceramic and nitride-bonded silicon carbide, fused cast alumina, and high temperature, ceramic-fiber products in applications in the aluminum industry. Ceramic fiber (about 50% alumina—50% silica) has good resistance to metal attack and is used in many molten-metal applications.

The refractory linings used in the critical areas (hearth, charging ramp, lower walls) by the primary aluminum producers are mainly high-alumina brick in the 75 to 95% Al_2O_3 class. The alumina grogs selected for these bricks are usually tabular alumina, fused alumina, or calcined bauxite. The most satisfactory bonds for resisting metal attack are aluminum phosphate, frit, calcium aluminate, and combinations of the latter two (Caprio, 1965). A satisfactory frit material consists of 5 to 15% of a preformed vitreous product consisting of 15 to 80% B_2O_3, 5 to 50% CaO, and 2 to 60% Al_2O_3. The remainder contains not over 15% of one or more oxides, MgO, BaO, BeO, ZrO_2, ZnO, V_2O_3, Cr_2O_3, or Mo_2O_3 (McDonald and Dore, 1961). Dolph (1963) described a somewhat similar refractory using alumina or bauxite as the main ingredient. It was claimed that the brick is free of glassy phase. Roudabush (1958) clad high-alumina brick with aluminum, but with bare ends. Oxidation of the cladding at the joints was expected to prevent corrosive penetration. Dewey (1963) used a rammed mixture of from 20 to 75% alumina in cryolite as the lining for the reduction cell to replace the conventional carbon lining.

Monolithic linings, in general tabular alumina bonded with calcium aluminate cement, are gaining favor for furnace walls and ladles. These structures are inferior in abrasion resistance to the conventional brick, however. Singer (1946) attempted to protect conventional refractories against attack by molten aluminum by applying surface coatings of (a) calcined aluminosilicates (kyanite, sillimanite, or andalusite), (b) carbonates or sulfates of metals of the second periodic group of the elements, and (c) sodium silicate, all applied as an aqueous slurry.

17-8 LIGHTWEIGHT ALUMINA REFRACTORIES

A. J. Metzger compiled a bibliography of articles on lightweight ceramics (1956) and a bibliography on the relevant patents (1956). In the case of those lightweight ceramics containing alumina, the application is generally as a refractory.

Callis (1946) prepared a heat-insulating structure composed of finely divided monohydrated alumina, which is reactive with hydrated lime at room temperature, a fibrous material such as asbestos fiber, and water to form a paste. The material was partially dehydrated. Langrod (1949) mixed wet ball-milled Bayer alumina with 40 to 80% dry unground alumina (Alcoa A-1) to form a thick slurry. The dried cake acquires from 20 to 50% external porosity after calcining at 1300 to 1850°C. Morgan Crucible Company (1951) patented the method of ball milling a metal with iron balls to disperse it in alumina. An acid was added to generate hydrogen and to obtain frothed slip, which is poured into a plaster mold to set.

Lesar and Glen (1958) prepared a lightweight insulating firebrick from a composition containing 18.5% expanded perlite (10 mesh), 55.5% calcined bauxite (48 mesh), 5% raw kyanite (35 mesh), 5% raw kyanite (100 mesh), and 3% each of monoaluminum orthophosphate and dialuminum orthophosphate. After firing the wet-pressed brick at 1260 to 1400°C, the finished brick had a density of 64 to 70 lb/ft^3, a modulus of rupture of about 300 psi, no shrinkage on a 1650°C panel preheat test for 24 hours, and no loss on a 1400°C panel spalling test.

Misra and Puri (1961) used powdered aluminum in bauxite, with either sodium chloride or hydrochloric acid to generate hydrogen. Booth and Hess (1960) combined fluxed and expanded volcanic glass with from 1 to 10 wt% alumina, and refired at 760 to 1090°C. Zimmerman and Haeckl (1960) obtained fired bulk densities of 62 to 115 lb/ft^3 for a cellular alumina ceramic prepared by mixing Bayer alumina and phosphoric acid with about 0.1% fine metal powder and 1 to 2% water-absorbing foam stabilizer. The foamed product was heated slowly and ultimately to about 315°C to develop 750 psi compressive strength. Wheeler and Olivitor (1961) used orthophosphoric acid and a powdered gaser, calcium silicate as a form stabilizer, and silica sol as the bond.

Guzman and Poluboyarinov (1959) used a combustible (petroleum coke) as the gas generator, and 1 to 2% TiO_2 to lower the firing temperature from 1700°C to 1550°C. The best results were obtained with alumina that had been precalcined at 1450 to 1600°C. By this method, Trigler et al. (1962) obtained insulating liners for furnaces operating at 1550°C. Products containing about 82% Al_2O_3 and fired at 1200°C, had a softening point of 1855°C, an initial load deformation temperature of 1370°C (14.2 psi), no contraction at 1500°C, an apparent porosity of 59%, a compressive strength of above 470 psi, and a thermal conductivity of 0.5 to 0.6 kcal/hr °C.

Powers (1962) tested alumina foams to determine their suitability at high temperatures (1930°C). The foams had a nominal density of 37 lb/ft^3. Kainarskii and Gaodu, and Gaodu and Kainarskii (1963) prepared lightweight insulators, good to 1900°C, with mixtures of alumina, gypsum, magnesium hydroxide, and powdered aluminum. Petroleum coke and phosphoric acid were also used as the foam generator.

Patented variations of lightweight alumina compositions have been described by Kawashima et al. (1962) Griffith, Olsen, and Rechter (1962), Holland (polystyrene and polyethylene foamers, 1963), Konrad and Stafford (hydraulic setting calcium aluminate, nodulated mineral wool), Powder, 1964 (addition of an inorganic gas generator to the stream of molten aluminosilicate), Pechniney-St. Gobain (1962), Dreyling and Dreyling (1964), and Perotte. Eubanks and Hunkeler prepared a low-temperature foamed castable with the following ingredients: 31 g H_3PO_4, 30 g $Al(OH)_3$, 6.5 g $Al(PO_4)$, 1.2 g SiO_2 (10 to 20 millimicrons size), 1 g bentonite, 0.15 g aluminum. The phosphoric acid reacts with the aluminum to generate hydrogen, and with the aluminum hydroxide to produce an aluminum phosphate gel. The remaining ingredients strengthen the foam. After cold-setting for about 0.5 hour, the lightweight ceramic is heated at 85°C for about 2 hours to cure. The compressive strength ranges from about 120 psi for a density of 20 lb/ft^3 to 460 psi for density of 37.5 lb/ft^3.

Nowak and Conti (1962) developed lightweight ceramic honeycomb structures. Alumina-silica paper, impregnated with an alumina-silica slurry was used as a refractory model system for developing fabrication techniques. The paper originally acts as a carrier, but finally serves as an integral part of the fired refractory. A simple mechanical method was developed for continuous fabrication of ceramic corrugated ribbon or sheet. A literature survey of unclassified data on ceramic foam and ceramic honeycomb for the years 1956 to 1962 was published (Ref. Ceramic Foam . . .).

A porous refractory is claimed for a mixture in which one component consists of metallic aluminum particles that have been fluxed and slowly oxidized optionally with a filler to form hollow particles of Al_2O_3 (Talsma).

17-9 BINDERS FOR ALUMINA REFRACTORIES

Pole and Beinlich (1943) claimed that pure alumina is less satisfactory than low-alumina compositions in the CaO-Cr_2O_3 and CaO-Cr_2O_3-ZrO_2 systems as a refractory in contact with molten rock phosphates. Alumina is very reactive with the phosphate ion. Greger (1950) found that aluminum hydrogen phosphates are excellent bonding materials for ceramics. The amorphous, glassy, water-dispersible phosphates containing 1.5 to 1.8 moles of Al_2O_3 to 3 moles of P_2O_5 were considered the most useful binders. When dehydrated above 260°C the aluminum phosphates become insoluble in water. Kingery (1950, 1952) reviewed the previous literature on phosphate bonding, and found three major methods: (1) phosphoric acid with siliceous materials, (2) phosphoric acid with oxides, and (3) the direct addition or formation of acid phosphates. Except for dental cements, no data on bond mechanisms were found. Stone, Egan, and Lehr (1956); and St. Pierre investigated the phases in native calcium aluminum phosphates as sources of bonds for alumina refractories.

Beck (1949) showed the analogy between the crystallographic inversions of the aluminum orthophosphate polymorphs and those of silica. A crystalline form of anhydrous

$Al_2O_3 \cdot P_2O_5$, probably the β-tridymite form, resulted when $Al_2O_3 \cdot P_2O_5 \cdot 4H_2O$ was heated to about 90°C. The α-tridymite form was obtained at 300°C, and the cristobalite form at about 1000°C. $Al_2O_3 \cdot P_2O_5$ apparently lost some P_2O_5 during 40 hours at 1450°C, and as much as 0.7%/hr during 5 hours at 1600°C.

Gitzen, Hart, and MacZura (1956) prepared refractory castables composed of tabular alumina grog bonded with phosphoric acid. Both heat-setting and cold-setting compositions were developed, which were characterized by high bond strength, developed at 340°C, and remarkable resistance to erosion over a wide temperature range. Excellent serviceability of these castables in the range to 1870°C was indicated. Formation of the bond in place by use of concentrated H_3PO_4 gave much stronger bonds than were achieved by aluminum phosphate additions. Hansen and King (1958) reacted phosphoric acid with mixtures of "inert" refractory aggregate, with one member selected from the group consisting of hydrated alumina, raw bauxite, raw diaspore and raw clays.

Bechtel and Ploss (1960, 1963) claimed excellent strengths at low and medium temperature (600 to 1000°C), provided by $Al(H_2PO_4)_3$ without any adverse effect on the high-temperature properties of the refractories (fireclay, bauxite, chrome, and silica). The bond $Al_2O_3 \cdot 3P_2O_5$ began to sinter at 1150°C, and to soften at 1340°C, but a retaining skin of $AlPO_4$ formed which was crystalline and solid at 1600°C. The bond $Al_2O_3 \cdot P_2O_5$ began to soften at 1260°C. Preusser (1961) obtained compressive strengths in excess of 150 kg/cm^2 (2130 psi) for corundum or chamotte bodies with diluted phosphoric acid (2 to 15% H_3PO_4) at reaction temperatures of 100 to 400°C. A reaction period of 4 hours at 300°C was equivalent to 6 hours at 200°C. The compressive strengths were much lower than those attainable with 85% H_3PO_4. Margulis and Kamenetskii (1964) found that a bond with the molecular ratio of $Al_2O_3 : P_2O_5$ of 1:3.5, at a density of 1.57 g/ml was most suitable as a bonding agent in the manufacture of articles and coatings. $AlPO_4$ raised the thermal stability of corundum materials and at the same time increased their strength.

W. O. Lake (1936) reported that alumina refractories containing from 70 to 80% Al_2O_3 were employed in oil-fired basic furnaces, bonded with high-alumina cement (Lumnite-type cement). One of the earliest applications was as a cast liner for furnace doors. Anderson (1938) divulged the method of preparing articles, as for example, brick, by mixing 4 parts of ground sintered aluminum oxide (natural corundum, emery, etc.), one part of high-alumina cement, and water to provide a satisfactory casting consistency.

H. N. Clark (1946) specified a furnace lining made from a refractory concrete consisting of an intimate mixture of an inert filler with an aluminous cement binder comprising about 60 to 75% alumina, about 20 to 40% calcia, and less than 2% iron oxides. The recognition of the necessity for increased purity of both the cement and specific single-oxide aggregates in hydraulically bonded refractories marked an advance in the development of refractory linings. S. F. Walton (1941 developed a 50:50 batch mixture by weight for hydraulically forming a refractory body containing fused alumina and a high-alumina cement. The cement formed a refractory bond with the alumina having a higher melting point than that of the original cement. Herold and Hoffman (1941) also showed the value of various refractory grogs bonded with Lumnite cement. Gitzen, Hart, and MacZura (1957) appraised a pure calcium aluminate cement (empirical molar formula $CaO \cdot 2.5Al_2O_3$) in a grog which was substantially tabular alumina and determined an optimum balance between refractoriness and bond strength in high-temperature castables. Additional references to the production and use of calcium aluminate cement in other classes of refractories are found in Section 20-1.

L. Jacobs (1960) developed a laminated plastic refractory which consisted essentially of a primary layer stable up to 1870°C, and consisting of at least 60% tabular alumina and a supporting layer of plastic refractory which converted to a glass above 1200°C, and which consisted of a homogeneous mixture of about 60% calcined flint clay (3 to 10 mesh), 10% kyanite (20 to 50 mesh), 10% plastic kaolinite fireclay, and 10% plastic ball clay, the remainder being water.

Scott and Emblem (1951) described the application of ethyl silicate in refractories. This binder has been used in specific applications of alumina in which its remarkable ability to undergo hydrolysis and condensation to yield an alkali-free silica gel is useful. Quick-setting dip coatings for refractory forms is one example.

Eldon Miller (1965) claimed that most properties of refractories containing over 50% Al_2O_3 are improved by additions of 3 to 7% of volatilized, amorphous silica, as deposited from vapors. G. Bayer claimed the addition of from 0.5 to 5.0% or more of Ta_2O_5 or Nb_2O_5 to alumina refractories and firing at only 1400 to 1500°C. Such compositions are claimed to serve as well as pure alumina fired at 1700 to 1800°C.

Hoepli and Klasse claimed the preparation of sintered, high-purity alumina refractories by using aqueous acidified suspensions of $Al(OH)_3$ with virtually colloidal alumina, and flocculating with ZnO. The filtered precipitate in the form of briquets, is fired for 6 hours at 1600°C.

18 ALUMINA AS AN ABRASIVE MATERIAL

18-1 INTRODUCTION

The hardness and abrasiveness of alpha alumina from general sources were considered in Section 5-10. These qualities in native corundum had been utilized from antiquity in the sharpening of tools and in the grinding and polishing of metals and ceramic materials, such as jewels of lesser hardness. The Bayer alumina grades, calcined and sintered, have applications as polishes and abrasives. Bauxite and Bayer alumina are both used to prepare electrofused alumina for abrasives and other products associated with the fusion process and fusion casting. Parché has given a description of the industry. The historical development is mentioned in "History of Abrasive Grain." Grinding and Finishing 9 (2) 22-7 (1963). A. Schneider (1959) provided chemical and mechanical data on electrocorundum and silicon carbide, and a survey of the technology of their production.

Initially, domestic bauxite sources (Arkansas) were sufficiently pure for fused abrasive purposes, but with depletion of the pure grades, it has become necessary to draw on foreign sources. Calcined South American bauxites (Suriname, Demerara) are sufficiently pure to meet the requirements of about 80% Al_2O_3, and not more than about 8% each of SiO_2 and Fe_2O_3, and 3.5% TiO_2. The fusion process relies on the formation of magnetic ferrosilicon (<15% Si) by addition of iron, if necessary, to provide purification of the product. The heavier ferrosilicon can be run off while fluid, or removed by chipping after the melt has solidified.

Many references relate to methods that have been considered for purification of the melt, or for obtaining the ingot in a more advantageous condition for crushing and sizing. Glezin (1937) mentions the addition of 5% TiO_2 to increase abrasive properties, and 5% Cr_2O_3 to increase the tenacity. Baumann and Benner (1944) crystallized the alumina with at least one metal of the group Fe_2O_3 and Mn_2O_3 in solid solution, and also in a fused matrix of a manganese, cobalt, or nickel spinel (1947). Baumann and Wooddell (1945) prepared an abrasive material comprising at least 90% crystalline aluminum oxide and a matrix of beryl or zircon. Benner and Baumann (1943) claimed an abrasive containing alumina and a minor amount of monazite. Schrewelius (1947) separated the silicon and

silica slag as upper layer of the melt from mixtures of chamotte and a reducing agent. By addition of a member of the sulfur group (selenium), aluminum selenide was formed, which could be broken up by hydrolytic action with water to liberate the corundum grain (1948). Ridgway (1939, 1943) purified the fused bauxite product by immersion in fused alkali metal cyanide to dissolve a considerable portion of the slagging constituents still clinging to the grain. Klein (1942) coated the grain with an alkaline earth metal sulfate or with cryolite. Yamaguchi and Tanabe (1954) synthesized artificial emery by firing powdered bauxite at 1350 to 1550°C, and found a slightly poorer abrasive efficiency than that of the native product. Klein and Ridout (1949) described the properties and applications of 32 Alundum, a grade of fused alumina (99.6% Al_2O_3), crystallized by cooling a molten bath of bauxite containing a metal sulfide. The sulfide in the solidified mass is water-soluble, releasing the alumina grains of controlled size and rough surface.

Frost (1958) prepared a fused abrasive consisting essentially of about 10 to 50% by weight of fine crystals of titanium carbide dispersed in crystalline alpha alumina. Ueltz (1963) sintered (to zero porosity) preformed grains of natural bauxite, characterized by having a skin produced by the sintering process and a large proportion of the crystals randomly oriented in the size range 5 to 30 microns and composed of alpha alumina and mullite, with glass present as an interstitial component.

Pevzner (1949) considered the production of corundum abrasives without the use of furnaces but by heat generated by the thermitic power of aluminum in the reduction of some less stable oxide than Al_2O_3. A product containing 95.5% Al_2O_3, 1.24 MnO, 0.95 CaO, 0.56 TiO_2, 1.08 Fe_2O_3, and 0.64 SiO_2 was claimed to be equivalent to normal electrocorundum from bauxite. It could hardly be justified from an economic standpoint for large-scale operation. Nesin (1964) described the process of making crystalline alumina lapping powder from coarsely crystalline Bayer sources by ball milling and elutriation. Polishing grades of Bayer alumina have been described previously (Section 5-10). Funabashi (1952) stated that aluminum oxide prepared by calcining ammonium alum was the most suitable polishing grade; when calcined at 1250°C the particle size was 0.1 to 0.2 micron, at 1400°C, 0.4 to 0.6

micron. Ordinary alum, calcined at 1400°C and separated by sedimentation provides a particle size of 0.7 to 0.8 micron. Wagner (1956) prepared an abrasive with porous grain by decomposing hydrogen peroxide at 60 to 70°C in a moist cake of ground alumina containing a surface active agent and a manganese compound. A higher degree of uniformity of pore distribution, shape, and size was claimed by this method than is attained by use of NH_4HCO_3 or other CO_2-forming agents. The porous cake was only partially sintered. When the crushed grain was used in making grinding wheels, the porosity was first blocked with gelatine or glue before addition of the cementing compounds.

The Bayer alumina used in producing fused alumina abrasive is a special type selected for its current-carrying ability, moderate to low soda content, and suitability for producing a white product. No additives are required during the fusion process. The fused product is more porous (induced by the volatilization of soda) and is more friable than the bauxite grain, a desirable property for cool cutting.

Silicon carbide and corundum are often confused because of their similar properties and uses, not to mention the close similarity in some of the trade names for the artificial products.

The abrasive applications of alumina are legion, but they generally fall into categories using loose grain, coated grain (cloth, paper, fiber), and wheels and tumbling shapes. The major applications of the finer loose-grain sizes are in optical polishing, lapping, and fine grinding and polishing of plate glass. Coarser grit sizes are used in quarry sawing of granite and marble, lithoplate graining, pressure blasting, tumbling operations, grinding and buffing compounds, in nonskid floor and road compositions, and in set-up wheels.

18-2 LOOSE GRAIN ABRASIVE

Jacquet (1939) described the techniques of surface treatment of metals using files, emery papers, polishing cloths and powders. Four historical periods include (1) before 1900, empirical, (2) 1900 to 1926, scientific methods and the metallurgical microscope, (3) 1927 to 1933, electron diffraction, and (4) 1934 to 1939, electrolytic polishing.

Both Bayer alumina trihydrate and kappa alumina have been claimed as dental polishes (Broge, Hedley). Gamal, a gamma alumina structure, and Microid (Griffin and Tatlock) are finely divided metallographic polishes. Manuilov and Zimbal recommended a 2-hr calcination of Bayer alumina at 1300°C in the alpha alumina range for metallographic polishes. Sira, a coarsely crystalline sharp alpha alumina, produced from Bayer hydrate, relies on the relatively soft bond of the Bayer grain to produce fine scratching in the polishing of glass (Carlisle). McAleer, Powell, and Klein discussed the properties of polishing grades prepared from Bayer alumina and fused alumina.

The Bayer polishing aluminas have been applied to finishing chrome plate and stainless steel. The ware is buffed

on rapidly rotating wheels of sewed canvas layers to which the polishing compound has been applied. The compounds consists of proprietary mixtures of the ground alumina in stearic acid and other greases in the proper consistency for holding on the wheel, but with minimum coating of the ware with a wax film. Polishes of different grades are designated "cutting" for fast surface removal, and "coloring" for high polish development. The abrasive action is thought to depend upon a combination of partial disintegration of Bayer agglomerates having the proper crystal size and the development of a suitable Bayer grain bond for the specific operation. The Bayer grain provides backing for producing scratches, but the Bayer particle bond is weak, and disrupts easily under pressure.

Rollason et al. (1949) claimed that platy crystals are most suitable for polishing mild steel. Gamma alumina at $0.5 \ m^2/g$ and alpha alumina at $1.5 \ m^2/g$ were considered to be suitable. The lapping method, introduced by Chrysler, was described by Mauzin (1949). Silicon carbide was used for cast iron, aluminum, brass, bronze, and brake bands; aluminum oxide for alloy steels, rapid tool steels, malleable iron, and hard bronze. Grain sizes were 180 to 320 for roughing, 320 to 500 for finishing, and 600 for the finest surfaces. Schulz (1952) claimed that fire polishing and acid polishing (etching) processes are suitable only for wellground glass shapes which do not have rigorous requirements of sharp detail. For optical glass, only mechanical polishing is permissible. Burkart (1952) discussed the fields of application of alumina among the various commercial abrasives in different categories: (A) Coarse abrasives competitive with alumina in some applications include quartz sand, Italian pumice, silicon carbide, and the relatively more expensive boron carbide; (B) Finer surface-treatment abrasives include tripoli, diatomaceous earth; (C) Polishing materials include vienna chalk (a mixture of CaO and MgO finer than 1 micron which hydrolyzes and carbonates undesirably), chalk ($CaCO_3$), rouge, and chromium oxide. Stead, Korelova, and Ingalls discussed abrasives used in grinding and polishing glass, especially optical glass. Cerium oxide and a mixture of cerium oxide with other rare-earth oxides (barnesite) is faster than rouge or alumina in optical polishing.

The fused grain is crushed and sized to standard sizings adopted in 1930 but revised at later dates (Simplified Practice Recommendations R118-36, National Bureau of Standards, Sept. 1, 1936). Various treatments have been applied to the grain to improve its stability or the final shape by impact crushing, air blasting, or pan mulling, and roasting, washing, or chemical treatment with acid, alkali, or salt baths. The coarser sizes, 6 to 240 grit, are obtained by sieve sizing. If insufficient fines are produced in coarse crushing operations, they may be generated by ball or rod milling. Classification of the finer sizes is by water sedimentation methods, hydraulic flotation, or air classification (Eigeles, 1936; Benedicks and Wretblad; Puppe, 1938). Dumas (1960) described a battery of centrifugal classifiers that accurately produces six standard fine sizes ranging from 22 to 5 microns average grain size. The improved classification system assures consistency of surface removal rate (scratch free) of optical glass and electronic components. Kaempfe recommended that fused grain should be tested for gain on oxidation, titanium content (not over 2%), and the thermal expansion of the unfired material. Terminasov and Kharson used X-rays to detect internal defects in corundum abrasives.

Funabashi compared aluminum oxide with magnesia and silica sand in the polishing of synthetic resins and celluloid.

Wills (1943, 1944) discussed the factors governing fine surface quality obtained by grinding, and pointed out the necessity for a more scientific classification of surface finishes.

Funabashi and Terada (1962) appraised the polishing ability of corundum and mullite barrel-finishing media. There are three classes of media for barrel finishing: metallic, natural mineral, and synthetic aluminum oxide. With the addition of silica, fused alumina, emery, and synthetic alumina, the amount of metal removed increased. Silicon carbide or alumina removed the most metal; iron oxide or chrome oxide gave the best finished surface. The rate of metal removal increased with increasing crystal size, and the surface of the pieces became rougher.

Ryshkewitch (1960) p. 208, stated that polished, sintered alumina shows no surface structural details unless it has been etched. Suitable "structureless" surface is attained by diamond polishing, using methods described by Dillingham (1956), Angelides (1961), Elyard (1962), and Swan (1963).

Rea and Ripple (1957) and Finnigan (1963) described the methods used for precision grinding of sintered alumina products to tolerances of 0.0001 in. or better. Grinding provides improved surface finishes, the quality of which depends on the abrasive used, the body being ground, and the grinding practices. The use of diamond wheels, slitting tools, band saws, and powders makes the machining of ceramics easy. Royal Worcester Industrial Ceramics, Ltd (1962) surface grind alumina ceramic parts (Regalox) for industrial use to tolerances of 0.0002 in. Even with shrinkages of 14 to 15% when fired at 1500°C, ware dimensions have been held to 2%. Three types of diamond tools have been developed (1) impregnated diamond wheels (sintered metallic or resinoid bond), (2) electroplated wheels having a single layer of natural diamond grit on the grinding face, held in electrolytically deposited nickel. and (3) micron paste, used for lapping surfaces for ceramic seals and other parts that must have an extremely fine finish. Rishel, Infield, and Kirchner claimed that sintered alumina forms, preferably about 96% pure (intergranular material substantially magnesia), are machinable with carbide tools, if they are leached first with aqueous or gaseous HF and subsequently refired at about 1690°C. Some loss in strength and increase in surface roughness results.

Baab and Kraner devised a sand-blasting test for determining the relative abrasion resistance of refractory forms. Blasting with uniformly sized abrasive at supersonic speeds is used for precise cutting and machining of hard brittle

materials. The "Airbrasive" tool, a product of the S. S. White Company, has eliminated shattering and flaking (Ref. Cutting Hard Brittle Ceramics . . . .). Aluminum oxide is the most widely used abrasive; the best cutting is obtained with 30 to 60 mesh size. Air pressure, grit size, and blast angle all affect the surface finish. Boron carbide is suggested for the blasting nozzles (Ref. Use of Abrasive Grain . . . (1963).

Rushmer and Elsey (1964) described the production of intaglio stones for finishing rotogravure cylinders to 1 to 2 microinches. The stones are made by molding graded alumina particles with synthetic resin. Sintered alumina balls have been used for conditioning aluminum engraving plates to make them receptive to inks. Many types of abrasive are used in the blast cleaning of surfaces for painting and for maintenance repainting.

The applications of coated abrasives and the equipment in which they are used are well covered in the book *Coated Abrasives—Modern Tool of Industry*, McGraw-Hill Book Company, New York, 1958, 426 pp. Spencer (1959) described the main types of abrasives (flint or quartz, Turkish or American emery, garnet, aluminum oxide, and silicon carbide) and the main factors influencing their operating efficiency. The methods of manufacture of coatings have been described in "Use of Abrasive Grain in the Manufacture of Coated Abrasives" (Anon., Grinding Finishing 9 (7) 33-6 (1963).

Coated abrasive articles, not necessarily restricted to alumina abrasives, are represented by the following disclosures. Hanford (1959) described a coated abrasive article comprising a backing having a layer of abrasive grain attached to a surface by animal glue and phenolformaldehyde, ureaformaldehyde, or melamineformaldehyde, and a substantial amount of wollastonite (80% minus 325 mesh). Quinan and Sprague (1962) patented the disclosure of a flexible coated abrasive sheet, comprising a woven cloth backing precoated with an elastomer which is a terpolymer of butadiene, styrene, and acrylonitrile containing at least 5 wt % of combined styrene, a film-forming, strength-imparting, thermosetting resin compatible with the elastomer, consisting of phenolaldehyde and urea-aldehyde resins, a resinous bonding coat, and abrasive grains bonded by the bonding coat.

Coated abrasives are prepared in the form of rolls, sheets, belts, and disks, coated in different grit sizes, and with several different alumina types. A preferential orientation of the grit is also possible, obtained by electrostatic methods.

18-3 GRINDING WHEELS

Abrasive wheels are produced from a wide variety of fused and Bayer alumina materials. There are five types of wheel bonds in general use (Parché): vitrified or ceramic, rubber, resinoid, shellac, and sodium silicate. The wheels are made in an extensive series and degrees of abrasiveness and porosities. Vitrified wheels, hones, and rubs are used in tool grinding, cylindrical grinding (center-type and centerless), form grinding, ball grinding, and heavy-duty snagging.

Rubber-bonded wheels are used in cut-off and snagging wheels where impacts are high. Rubber-bonded wheels provide finer finishes relatively, hence are suited for finishing ball bearing races, drill flutes, cutlery, and in centerless grinding. Resinoid wheels are used in cut-off and snagging wheels, in thread grinding, intaglio stones (engraving finishes) and many others. Shellac-bonded wheels are used for softer abrasive action, as in hemming wheels for cutlery, grinding, and in roll grinding. The sodium silicate-bonded wheels also provide a soft abrasive action suitable for hemming wheels. Knife hones may be silicate-bonded.

Forrester (1956) described the production of abrasive aluminum oxide for vitrified grinding wheels; Buchner (1956) described the manufacture of the wheels. Aluminum oxide as the basic grain in bonded abrasives has been described (anonymously) in Grinding Finishing 9 (5) 36-9 (1963). Kingery, Sidhwa, and Waugh (1963) devised a simple model for the microstructure of a vitrified bonded abrasive wheel based on a coordination number of 6 for neck growth of the vitrified bond on assumed spherical particles having a packed solid (abrasive grain) content of about 50%. McKee (1961) presented a table of comparative designations of abrasive grain types used in grinding wheel manufacture. American standard markings for identifying grinding wheels and other bonded abrasives include a letter symbol for the type of abrasive (A, aluminum oxide; C, silicon carbide), a number indicating grain size, a letter indicating grade (A, soft to Z, hard), a number indicating structure (0, dense to 15, open), and a letter indicating the bond or manufacturing process. Both grade and structure are complex characteristics difficult to define precisely. Grade defines the tenacity of the bond for the grain. The grade of wheel has no generally accepted quantitative definition, but is based on the proportion of bond to grain in the original mix.

Schlechtweg derived a simple formula for fracture of rotating brittle disks, showing the fracture velocity as a function of the ratio of the outer diameter to the inner diameter of the abrasive wheel. McKee (1963) presented the results of a panel discussion on the merits of using aluminum oxide and silicon carbide bonded together or cemented on a backing. Kistler and Barnes (1940) claimed a method of making an abrasive body consisting of pouring a polymerizable compound which is fluid at room temperature onto the abrasive grain in the mold, heating and without pressure thereby polymerizing the compound around the abrasive grains to a hard condition, and finally stripping the abrasive body from the mold.

Barnes et al. (1942) patented a method for bonding, comprising mixing fused alumina with rubber latex to form a molded wheel. Abbey (1951) and Taylor (1953) described rubber-bonded grinding wheels strengthened by the inclusion of fibrous material, and combining soft-cutting and polishing abilities.

Hessel and Rust (1951) described bonds of resinous condensation products, Ahrens and Lappe (1951) sintered metaloxide bonds, especially Al_2O_3, at firing temperatures above 1700°C. Maziliouskas (1958) described formulas for frits, and Houchins (1956) vitrified borosilicate bonds. Foot (1960) prepared a new abrasive wheel using alumina bubbles (steamblown aluminum oxide from the poured melt) with a resinoid bond, which operates at 9500 sfpm for grinding rubber, leather, and cork without loading.

Zimmermann and Burton-Banning showed that illitic and montmorillonitic clays give higher strength than clays composed mainly of kaolinite, fireclay, or halloysite in abrasive wheel bonds. Bibbins (1962) patented an abrasive wheel with a metal bond consisting of 86% iron and ferrite-strengthening metals, 0.4 to 5 phosphorus, a trace to 0.8 carbon, and 0 to 7% sulfur. The bond melts above 750°C, and after sintering at 750 to 1100°C, grain weakening is induced by the phosphorus.

Horibe and Kuwabara (1961) showed by X-ray examination of the unit cell spacing, that a considerable part of the Ti^{3+} dissolved in alpha alumina abrasive grain becomes oxidized to Ti^{4+} during the firing of the vitreous bond.

Marshall (1964) designed a surface-grinding wheel having the correct degree of strength and impact resistance for surface grinding, without too rapid or insufficient wear of the abrasive face, by crushing fused masses of conventional alumina abrasive material with 0.25 to 1.0% Li_2CO_3. This produced an abrasive grain containing about 0.1 to 0.4% Li_2O in the form of 1.6 to 7.2% zeta alumina phase ($Li_2O \cdot 5Al_2O_3$). Grain containing no zeta alumina was found to be too tough, and that containing about 14% was too brittle.

18-4 CERAMIC TOOLS

One of the earliest patent applications for a sintered alumina body was British Patent No. 4887, February 27, 1913, granted to Thomson-Houston Company, Ltd., for a wear-resistant body suitable for use as tools, dies, rolls, bearings, drills and the like. Very finely divided aluminum oxide was mixed with about 10% of a suitable binder (gum tragacanth) and the mixture was compressed in the form of a flat cylinder, e.g. 0.25 in. by 0.75 in. in diam., and was baked at 1300 to 1400°C in an inert environment (tungsten or molybdenum wound resistance furnace). After baking to the consistency of chalk, the piece could be machined or bored easily with diamond dust. After machining, the specimens were refired at 1800 to 2000°C in an inert environment to a "dense condition free from crystalline structure." This may be compared, for example, with a more recent disclosure, German Patent 1,098,427, January 26, 1961, to Trent and Comins, "Process for Preparing Hard Aluminum Oxide Sinter Bodies," which claims (1) A process for preparing hard aluminum oxide base sinter material, characterized by the fact that a homogeneous powdered mixture which consists of 87-96% Al_2O_3 and the remainder, in addition to impurities, of at least 2% each of titanium nitride and titanium oxide, is compressed in the usual manner into a briquet, this briquet being sintered at a temperature of between 1600 and 1750°C.

Ryshkewitch obtained U.S. Patent 2,270,607, January 20, 1942, for a ceramic cutting tool consisting of a molded self-bonded sintered aluminum oxide and containing chromium oxide, the body of the cutting tool having a fine crystalline structure throughout. Westmoreland-White (1946) disclosed the British advances in ceramic lathe tools as being represented by a composition containing a well-ground mixture of 94% alumina, 2.5% calcium fluoride, and 3.5% bentonite. The dried particles (minus 200-mesh) were dry-pressed at 20,000 psi and sintered at about 1500°C. Singer (1949), and Graham and Kennicott (1950) felt that ceramic oxide tools did not meet the requirements for cutting tools, and the toughness of titanium carbide and tungsten carbide relative to the pure oxides was cited as a prime advantage. Isaev et al. (1952) stated that red-hot hardness is a main quality in cutting tools. The alumina ceramic tools retained hardness even at the melting point of steel, but certain defects necessitated special attention to tool shape, machining conditions (speeds, etc), the attachment of the tip, and grinding of the tip. Russian ceramic tools were claimed to exceed the hard-metal tools in the quality of the surface finish, and the cost of the ceramic was only about one-tenth. Speeds 60 to 80% higher than with the hard-metal tools were possible. Both detachable and nondetachable tips were used. Kitaigorodskii and Pavlushkin (1953, 1955) described the properties of "Mikrolit," a superstrong corundum structure containing 99.0 to 99.2% Al_2O_3, a crystal size of 1 to 3 microns, a bulk fired density of 3.96 g/ml, Rockwell hardness of 92 to 93 (scale A), flexural strength up to 64,000 psi and crushing strength up to 710,000 psi. The tools were probably made by hot pressing. Because of their brittleness they were used only in finishing and semifinishing operations that did not involve impact loads. Crucibles made from the composition had extremely high chemical resistance. Trippe (1957) described a technique developed by the Russians featuring the metalizing of the ceramic tips in a vacuum in preparation for brazing to the tool shank.

Rea (1956); Richter and Kammerich; Blanplain; and Tangerman examined the likely applications of alumina tools and some of the precautions necessary to their use particularly for cutting steel. This culminated in the Rodman Laboratory report PB 111757, which was concerned with the possible substitution of ceramic materials for strategic alloys, a possible cost reduction by using the ceramic tools, and the use of the tools to perform machining not possible with conventional hard metal types.

E. and N. Labusca (1957) described the properties of tools prepared from Romanian sources, mixtures of tungsten carbide and cobalt with more than 90% alpha alumina, designated ENC. These were claimed to excel tungsten carbide in the machining of hard materials. Cutting ability was retained to 1200°C at speeds to 720 ft/min; tungsten carbide was found satisfactory only to 850°C. Gion and Perrin gave full details on "Ceroc" ceramic tips, a French development comprising 95 to 100% alpha aluminum oxide of crystal size within 1 to 10 microns. Strength in flexure was 60,000 to 70,000 psi, and in compression 280,000 to 350,000 psi.

During a ceramic tool symposium, held in Houston, Texas, March 1957, Zlatin discussed the progress of ceramic machining operations at Wright Aeronautical Division, Curtis Wright Corporation; D. R. Kibbey and W. T. Morris analyzed the variables in ceramic tool cutting; Haeme and Hook mentioned potential industrial applications; W. B. Kennedy described his experimental machining tests with ceramic tools and the types of applicable ceramic compositions; and J. F. Allen appraised the performance of ceramic cutting tools in production jobs.

McAuliffe, and Siekmann and Sowinski discussed the optimum tool geometry for cutting high-speed steels, super-alloys, and titanium. Rake, edge-cutting angles, and nose radii dimensions were investigated. Ryshkewitch (1957) cautioned the importance of preventing chatter in both the work and the tool tip. F. Singer and S. Singer stated that the desired properties of cutting tools include: high compressive strength, toughness, resistance to abrasion and heat, and unwettability by metals. Matthijsen thought cermets could provide sufficient elasticity and hardness to lathe tools.

Many investigators have lately investigated the performance of ceramic tools. Dawihl and Klingler claimed that impact resistance could be improved by heating the bit (by medium frequency current during the cutting operation). They found that the maximum resistance to vibration lies between 400 and 700°C, confirming the theory that some recovery processes occur in cutting tools at high speeds. Mii et al., and Yamada et al., tested the tools on high speed lathes (to 5000 ft/min) at cutting depths of 0.04 inch. At feed rates of 0.004 in./revolution, the ceramic tools withstood the conditions well, but carbide tools wore exceptionally fast. Wear of the ceramic tools was rapid, however, for tools having chamfers 0.002 inch in width and 45 and 30 degree angles. Marked cratering appeared on the tip surface. Of the three components of cutting resistance, the horizontal component increased the most. Roughness of the machined surface increased, tending to generate burrs, and the surface layer of the work became deformed. Although the results were ascribed to excessive chamfering, the blunt cutting edge was probably unsuitable for high-speed cutting. Nakayama et al. (1961) concluded that the use of ceramic tools without chamfer and at very high speeds is desirable. Ceramic tools having tool geometry of -5, -7, 5, 7, 15, and 1.5, and negative land, 30 degrees $\times$ 0.04 inch, were used successfully for high-speed turning of hardened alloy steels at about 650 ft/min, and at a depth of cut ranging from 0.008 to 0.016 in. and a feed of 0.0016 in./rev. The hardness of the ceramic tools was found to decrease markedly between room temperature and 400°C, but this tendency decreased above 400°C, and above 700°C, the hardness exceeded that of carbide tools.

Haidt (1960) noted the lack of toughness of alumina tools in comparison with the carbide tools. Kolbl (1958) cited the advantages of high cutting speeds at high temperatures, Puehler and Wagner the economy and better surface finishes obtained in comparison with diamond. Shaw and Smith (1958) found that the compatibility of the ceramic tools for various metals ranged from excellent to poor in the following order: nickel, cobalt, gray cast iron, steels and inoculated cast iron, lead, silver, aluminum, beryllium, and titanium.

Fleck and Gappish examined the grinding surface of ceramic tools with the electron microscope and also by means of the Triafol-SiO-pressure method and found that the edge-life depends largely on the grain size of the sharpening abrasive. Only diamond-impregnated grinding wheels can be used to grind and polish ceramic tools, and owing to the brittleness, great care must be exercised. Brewer (1959) discussed such factors as porosity and grain size, tool life, power requirements, and surface roughness. Gill, and Gill and Spence (1958) reviewed the advantages of aluminum oxide tools, and the physical properties of the different commercial compositions.

Staudinger (1960), Gion (1960), Widmann (1964), and Droscha (1964) have reviewed the literature on recent ceramic alumina tool developments. Oishi, Hamano, Kinoshita, and Takizawa (1963) attempted to manufacture alumina cutting tools by hot-pressing, adding oxides of Mg, Zr, Cr, or Co. Oxide cutting tools of long life were manufactured on a commercial scale by hot-pressing Bayer alumina with the addition of magnesia at 1800°C at a constant pressing temperature.

Following are some tool compositions. Goliber (1959) claimed one composition containing oxygen-deficient titanium oxide in solid solution in alpha alumina in an atomic ratio of Al:Ti from 0.8 to 20. The claimed hardness was 94 Rockwell A, the bending strength 90,000 to 95,000 psi; the product was electrically conductive; another composition contained oxygen-deficient titania in solid solution in alumina, and up to 60% by weight of reduced chromium (preferably 25 to 50%). The Rockwell A hardness was 87.5, and the flexural strength at room temperature was 110,000 psi. Both compositions were cold pressed at 20,000 psi and sintered at 1600°C.

Gill (1959) pressed a mixture of 40 to 93 wt % alpha aluminum oxide (particle size <10 microns), 20 to 59% Cr_2O_3 (<10 microns), and 1 to 54% Cr, W, Mo, Fe, Mn, Co, Ni either alone or mixed together (<15 microns). The mixture was hot-pressed in impervious graphite molds at 1000 to 5000 psi and at 1400 to 700°C. An addition of 1 to 15% pure metallic chromium was found especially suitable. The hardness of the finished pieces was 93 to 94 Rockwell A. Gill (1960) hot-pressed at at least 1400°C and at least 1000 psi an extremely fine powdered gamma alumina having a surface area of at least 200 m^2/g, for such uses as cutting tools, thread guides, nozzles, and wire-drawing dies. White (1960) coated fine 5 to 10-micron sintered alumina particles with ammonium molybdate solution (about 10 wt % Mo) and ignited the mix at 850°C for 15 minutes prior to calcining at 1100°C for one hour. The sieved powder was cold-pressed at 20,000 psi with wax lubricant to give small bars. The bars were sintered in hydrogen at 1800°C for 2 hours. The product had an even texture, a cold bending strength of 25,000 psi, a density of

4.11 g/ml, and an open porosity of only 0.7%. Evans et al. (1967) used a composition of pure alumina containing 2 to 3% nickel to retard grain growth. Sintering at 1600 to 1730°C for 5 to 60 minutes gave the best cutting performance.

Csordas et al. (1960) claimed a process comprising the sintering at 1450 to 1500°C of pure aluminum oxide (99.99% Al_2O_3), the particles being all less than 3 microns, and 90% less than one micron. The sintering time was held for such a short time that 90% of the crystal size was below 10 microns. Ryshkewitch and Taylor (1961) developed a dense, nonporous, and wear-resistant tool material consisting of an intimate mixture of fine-grained silicon carbide particles (8 to 50% by volume) in fine alumina. Schmacher (1963) claimed sintered mixtures of titanium oxide and alumina, obtained by mixing from 1 to 50% titanium hydride with the alumina and sintering the molded form in a mixture of nitrogen and hydrogen at 1800 to 2000°C. Whittemore et al. (1963) prepared molded alumina cutting tools and wire-drawing dies by hot-pressing powder mixtures finer than 325 mesh and containing at least 95% Al_2O_3, at 1400 to 1800°C. Densities of over 3.95 g/ml with crystals averaging not coarser than 7 microns were achieved. As much as 2% Cr_2O_3 or Ti_2O_3 was found to be permissible, and 0.015 to 1% MgO was beneficial in preventing grain-growth. The abrasion resistance increased with decreasing crystal size below 4 microns, and maximum toughness was found with 6 to 7-micron average crystal size. Both crystal size and density increased with increasing time and temperature in pressing; with no MgO, one minute at 1700°C gave high density but the crystal size was too coarse; about 20 minutes at 1450°C gave an equal density with only 2-micron crystals. The crystal size was only 5 microns after 9 minutes of heating at 1750°C when magnesia retarder was used.

Bugosh (1964) hot-pressed fibrous boehmite having 200 to 400 m^2/g surface area, and 25 to 1500 millimicrons average length, and containing from 0.5 to 5% of a grain-growth inhibitor such as magnesium fluoride or an oxide of nickel, cobalt, chromium, or magnesium. Pressing at 1600 to 1800°C and 1000 to 6000 psi resulted in bulk densities as high as 3.97 g/ml, 64,000 psi cooled modulus of rupture, and 92.5 Rockwell A hardness.

Bradt (1964) summarized the comparison of performance of present ceramic tools with those of 1956 as shown in the following tabulation.

Tool Properties	1956	1964*
Rockwell A	86–90	93–94
Transverse Rupture	40–60 (K psi)	90–110 (K psi)
Impact fracture	6–8 in. lb	11–13 in. lb
Porosity	0.5–2.0%	0.1%
Compression	300–350 (K psi)	425–475 (K psi)
Elast. modulus	42–46 (KK psi)	56–60 (KK psi)

*Conservative, especially strengths, which may exceed by 25%.

Material Machinable with VR-97	Tool Performance	Best Finish RMS	Usual Speeds SFPM
Steel-Soft to RC30	Good	35	600–1500
RC30–RC45	Excellent	25	400–1200
Hard RC45–RC65	Excellent	20	100–800
Cast Iron	Excellent	35	400–2500
300 S Stainless	Poor	35	400–1000
400 S Stainless	Good	35	600–1500
Carbon-Graphite	Excellent		200–1000
Tungsten	Excellent		150–800
Stellite	Excellent		60–600
Tantung	Excellent	–	60–140
Molybdenum	Fair	–	500–800
Titanium	Poor	–	Not recommended
Alum. Bronze	Good	–	1500–2500
Hard Rubber	Excellent	–	To 1000
High temp. alloys:			
Iron base	Poor	–	Not recommended
Nickel base	Poor	–	Not recommended
Cobalt base	Good	–	60–600

Hardness of Steel	Cut Depth	Maximum Feeds	Recommended Speeds
RC-40 to 48	To 1/32″	.030	300–1200 SF
	1/32 to 1/8	.025	300–1000
	1/8 up	.021	300–800
RC-48 to 55	To 1/8″	.015	200–800
	1/8 up	.010	200–600
RC-55 to 65	To 1/32″	.010	120–450
	1/32 up	.008	75–350

155

19 ELECTRICAL APPLICATIONS

19-1 SPARK PLUG INSULATORS

The early development of spark-plug insulators by Gerdian and Reichmann (1927) from substantially pure alumina required forming by slip casting in highly acid solution to avoid difficulty with the poor plasticity of contemporary alumina. It was also soon found expedient to resort to the addition of fluxes to avoid the higher than conventional temperatures required to mature the ceramic forms. These compositions often contained from 10 to 20% vitreous additions, although compositions, containing as high as 95 to 97% Al_2O_3 had been developed, which could be fired at feasible temperatures of 1620 to 1700°C.

Haldy, Schofield, and Sullivan investigated high-alumina porcelains in the concentration range from about 80 to 99% Al_2O_3. The flux consisted of 2.5 parts dolomite, 2.5 parts flint, and 15.0 parts kaolin clay. The specimens were dry-pressed at 7500 to 10,000 psi. The maturing temperature applied to obtain apparent porosity of not over 0.1%, varied from 1540°C for the 80% Al_2O_3 compositions to 1810°C for the 99.0% Al_2O_3 compositions. Flexural strengths ranged from 30,000 to 40,000 psi. The room-temperature resistivities, measured with a capacitance bridge, were as high as 10^{17} ohm-cm. The T_e values were affected by the residual soda contents in the alumina, being higher with decreasing alkali. The power and loss factors were quite low, falling in the classification grade of excellent insulators for radio and other communication equipment. The dielectric constant, calculated from capacitance measurements, increased with increasing alumina content of the specimens.

Kato and Okuda (1960, 1962) discussed the physical properties of alumina porcelains prepared from three different alumina sources, namely, calcined ammonium alum, Bayer process, and electrically fused alumina fines. When using 2% MnO_2 and TiO_2 as additives, the microscopic examination of specimens fired above 1400°C revealed a mosaic structure for the calcined alum and the fused alumina, but an intergranular structure for the Bayer alumina. The effects of the addition of 2 to 5% (wt) of various oxides including MgO, Co_3O_4, Fe_2O_3, Cr_2O_3, TiO_2, or MnO_2 were investigated. The electrical properties of specimens containing MnO_2, such as insulating resistance and dielectric loss at 1 megacycle, were superior to those of

the others (see Luks, 1942). The electrical properties were poor for specimens containing Fe_2O_3 or TiO_2, and intermediate for those with MgO or Co_3O_4. Kainarskii, Degtyareva, and Alekseenko found that small additions of titanates of aluminum or zirconium did not affect the dielectric properties, although only 0.5% $MgO \cdot 2TiO_2$ lowered the breakdown voltage.

Most commercial electrical insulator compositions in the high-alumina class have been based on the use of Bayer process alumina of reduced alkalinity (below 0.2% Na_2O). The compositions may contain from 80 to 95% aluminum oxide. Aluminum silicates and alkaline earth oxides usually comprise the major portion of the remaining constituent but many metal oxides have been claimed to provide additional favorable improvements in the properties.

Spark plug insulators, as presently produced in the United States, contain from about 85 to 92% Al_2O_3. Low-soda calcined aluminas and wet-ground, acid-leached tabular alumina are used by the major producers. Sintering additives (Section 16-3) and other components may be introduced during wet or dry grinding. The spark plug blanks are formed generally by isostatic molding from a spray-dried suspension containing the ground alumina, wax lubricants, binders, etc., but the compositions are also extruded, cut, and profiled to shape in the leather-hard condition. Oversize ware may be diamond-tooled. The saggered ware is fired in tunnel kilns at about 1700°C. Completely dry processes have been proposed. The use of more reactive alumina types would extend the process substantially to 100% Al_2O_3 bodies at about the same firing conditions.

Skunda (1955) described the process of making a spark plug insulator comprising the steps of forming a mixture of ceramic material with a heat-softening resin binder, forming an elongated blank from the mixture, rotating the blank on its longitudinal axis, heating the surface of the rotated blank to above the softening temperature of the resin binder, forming the heated surface to an insulator contour, and firing to remove the binder and to sinter the insulator.

Kato and Okuda (1960) said that insulators prepared from Bayer alumina, the fusion process, and ammonium alum decomposition, with MnO_2 and TiO_2 mineralizers, were excellent high-frequency insulators. Lungu and Nagy,

Poluboyarinov (1958), Budnikov, and Smoke et al. (1959) examined various compositions for ultra-low loss characteristics. Rigterink (1958) predicted that it was unlikely that an inorganic insulating material appreciably better than sapphire would be found, unless it had covalent bonding rather than ionic. Improved structures would have no glass phase but very fine crystallites; and 98 to 100% of the theoretical density and lower firing temperatures might be realized by a more sophisticated use of organic additives, more homogeneously distributed. Some advances have been made in this direction.

Schurecht (1957) found that zinc oxide reduces the specific electrical resistance below 1 megohm at about $730^{\circ}C$ when mixed in amounts above 25 mol. % with alumina. Promising compositions were found in the system $BeO\text{-}MgO\text{-}ZnO\text{-}Al_2O_3$, but the toxicity of beryllia might restrict its use in spark plug manufacture. Schurecht (1959) found compositions in the $Al_2O_3\text{-}SnO_2$ system in a wide range of concentration of SnO_2 (19 to 81%) satisfactory for the firing end of a spark plug insulator.

Meredith and Schwartzwalder devised a process for forming a low-tension spark plug which included preforming a semiconductor composition around the center electrode, assembling an outer electrode about the preform, positioning the electrodes for a predetermined gap between their ends. The assembly is heated and pressure applied to compress the semiconducting preform so as to cause good adherence to the electrodes and to form a good heat-conducting path between them. Schurecht (1962) produced a semiconducting body by firing a mixture of 5 to 80% copper metal powder and 95 to 20% alumina in an oxidizing atmosphere at 1290 to $1485^{\circ}C$ for a time equivalent to cone 10 to cone 18. Yenni and Runck (1961) sprayed consecutive layers of a refractory electrode metal, a refractory insulating material, and a refractory electrode metal onto the end region of a base member. An end portion of the coatings was ground away to provide electrode surfaces.

Norin and Magneli (1960) showed the presence of new phases in addition to $AlNbO_4$ in the system Nb-O, having the approximate formula $NbO_{2.40}$ to $NbO_{2.48}$, and in the monoclinic system. Somers (1963) claimed an insulator body prepared from alumina containing from 0.9 to 1.5% Nb_2O_5 with up to 1% by weight of other fluxing oxides (MgO, kaolin, and talc), having such a smooth surface that glazing the insulator could be circumvented. A 2% wax emulsion in water provided a temporary binder for bisque firing at about $1100^{\circ}C$, with final firing at about 1430 to $1590^{\circ}C$ for 2 hours.

Sifre (1963) prepared an insulator body by adding from 0.1 to 1.0% lanthana to alumina (99% pure) by pressure sintering at 1600 to $1700^{\circ}C$ to a density of about 3.85 g/ml. The mechanical and electrical properties are the same as for a lanthana-free alumina body, but a sintering temperature above $1700^{\circ}C$ would otherwise be required.

Zaionts (1958) described the preparation of spark plug insulators for automobile engines in Czechoslovakia from the ingredients: fused corundum, Bayer alumina, kaolin, flint, limestone, magnesia, sillimanite, and prepared glazes. The body contains 92% nonplastics and 8% kaolin as the bond. The materials are wet ball-milled, filter-pressed, dried, sieved, remoistened (12 to 13% H_2O), and aged for 14 days before forming. The formed insulators are fired at about $1715^{\circ}C$ on a 14-hr schedule. The glazing is done at $1300^{\circ}C$, and the insulators are tested at 12 to 15 kv. The tensile strength averages 10,700 psi, and effective breakdown voltage about 18.7 kv/mm. The specific resistance at $600^{\circ}C$ was claimed to be (1.37 to 1.75) $\times 10^7$ ohm-cm.

Solomin (1959) described two types of spark plug compositions used in the Soviet Union: ground mixtures of magnesium and iron silicates or of alumina and alkaline earth oxides. From 5 to 10%, free or chemically combined iron oxides, is added to the starting mixture.

Jeffery (1940) developed a glaze spray suitable for applying to isostatically formed alumina insulator bodies. Fisher and Twells (1957) described the use of feldspathic glazes with the alkaline earth group composed of alkalies from high potash feldspars, augmented, if needed, by small amounts of potash frits. The area of best glazes was centered around 1.0 RO, 1.5 Al_2O_3, and 15.0 SiO_2 and 1.0 RO, 1.75 Al_2O_3, and 15.0 SiO_2 for one-fire and two-fire glazing, respectively. RO represents alkali oxide.

19-2 ELECTRON TUBE ELEMENTS, HIGH-FREQUENCY INSULATION

A review of ceramic materials for high-frequency insulation in 1937 showed no data up to that time on high-alumina compositions (Thurnauer). With recognition of the superior high-frequency properties of polycrystalline alumina, the availability of suitable grades of Bayer alumina, and the development of practical, economic base compositions and suitable glazes, a rapid increase in the use of alumina in this application has taken place in recent years.

Henry (1958) pointed out the differences between electronic ceramics and ordinary ceramics with reference to vitrification, texture, polarization, purity, etc. Spencer and Turner (1958) drew attention to the electrical properties and applications of high-alumina ceramics in industry. Dressler (1964) described the development from electrical whiteware containing alkalies, through steatite and forsterite to alumina and beryllia. The better mechanical and thermal characteristics are particularly important for microwave engineering and for sealing to metal parts with suitable coefficients of expansion.

An early electronic use of alumina was as the cathode insulation in ac electron tubes. This component serves to insulate the tungsten or tungsten alloy resistance heater from the tube circuits, stabilizes the electron emission, and eliminates the ac hum. Leached ground Bayer alumina, or preferably fused alumina, was found satisfactory (Radio Corporation of America, 1941). The wire, coated with a suspension of alumina (maximum impurities 0.5%) in an organic binder (nitrocellulose) was heated at about $1800^{\circ}C$ in an atmosphere of hydrogen containing water vapor (saturated at 50 to 55°C). LaRocque described a similar treatment in 1959. Coppola et al. (North American Phillips) prepared a thermionic dispenser cathode which comprised mixing about 90% of a powdered fused mixture of at least 60% alkaline earth oxide, and the remainder aluminum oxide. The mixture was fired at 1650 to 1750°C. Rettner et al. presented data on the emission and evaporation characteristics of a porous tungsten cathode impregnated with $5BaO \cdot 2Al_2O_3$. Active barium was generated by the reaction: $2/3 Ba_3Al_2O_6 + 1/3 W = 1/3 BaWO_4 + 2/3 BaAl_2O_3 + Ba$. The barium is transported through partly clogged pores. Yamamoto et al. (1953) described methods for refining fused alumina of the soda content present as beta alumina. Four methods of removal were tried: (1) dissolution of the alkali by acid, (2) decomposition by heating at high temperature, (3) separation by heavy liquid, and (4) separation by crushing to fine grains. The last method was the most effective. To increase the heat radiation from the cathode coating, Masuch added 1 to 2% Cr_2O_3, TiO_2, or V_2O_5, or 0.01 to 0.1% Fe_2O_3 or Co_2O_3 to the alumina powder. The insulating layer can be applied by dipping or spraying, or by electrophoresis from suspensions (1 to 35 microns particle size). The dielectric strength of fused alumina cathodes was independent of temperature below 1050°C, but markedly influenced at higher temperatures. Shalabutov measured the volt-ampere characteristics of the leakage currents of an ac voltage, applied across the heater and its cathode, to determine that the coatings were ionic in nature. The temperature of the heater controlled the polarization of the coating and modified the dynamic effects of the leakage current.

Michaelson (1953) showed the suitability of sintered alumina for high-temperature vacuum service in terms of its dissociation pressure. German (Telefunken) alumina electron-tube envelopes were in service during World War II, and had been critically analyzed in this country in a number of study contracts issued by the Armed Services (Kohl and Rice, 1958). The main advantages that the ceramic envelope provided was increased ruggedness, resistance to vibration, the ability to outgas at 600 to 700°C, rather than at 350 to 450°C for glass envelopes, and a longer vacuum life.

Coykendall (1954) discussed the development of an all-ceramic annular radio tube by the Machlett Laboratories. The tube was capable of operating through a frequency range previously covered by several narrower range magnetrons. The tube was designed as a UHF power triode with a tunable frequency range of 30 to 2000 Mc and a useful power output of 200 watts at the top frequency and up to 700 watts at the lower. Brand et al. (1954) investigated zirconium silicate and magnesium silicate as ceramic sealing materials, but Mn-Mo, MoO_3, and TiH_2 were the favored methods (Navias, 1954).

Sorg (Eitel-McCullough, Incorporated) in 1956 obtained patents on an electron tube constructed of a cylindrical sintered alumina envelope joined to the end members by metallic bonds which also provided the lead-in conductors for the anode, grids and cathode, which were arrayed on

disk-shaped members appropriately interposed in the correct arrangement. McCullough and Williams (1959) obtained a patent on a stacked ceramic type. Mayer (1962) described cold cathodes for vacuum tubes consisting of the separate oxides of Al, Be, B, and mixtures of them. Bickford and Smith (1959) proposed a semicrystalline composition for electron tube envelopes operating above 700°C composed of 90% Al_2O_3 (-500 mesh, maximum Na_2O 0.05%) and 10% glass frit (60% SiO_2, 14% Al_2O_3, 16% CaO, and 10% BaO), fired at 1625°C. Claimed properties were: coef. of expansion 6.44×10^{-6}, flexural strength 42,000 psi, dielectric constant at 25°C 7.5 and at 500°C 8.2, loss tangent at 25°C 0.0003 and at 500°C 0.006, and loss factor at 25°C 0.002 and at 500°C 0.046.

Gallet (1956, 1957) tabulated the characteristics of 50%, 66%, and 95% Al_2O_3 compositions suitable for electron tube construction. Porous ceramic parts were found to be suitable for hot electrode supports because of their low thermal conductivity and ease of release of gases during manufacture. R. Williams (1954) of American Lava Corporation described the production of ceramic components for electron tubes. The majority of insulators used as vacuum tube envelopes are cylindrical; many designs can be machined from extruded stock. Diameters up to 8.5 inches maximum were readily available, with wall thicknesses about 0.1 the outside diameter for the most economical production. Washers, disks, and intricate designs (not exceeding 1.5 inches in thickness are produced normally by wet or dry pressing. Isostatic forming, except for spark plugs, had not received much attention by ceramic manufacturers previously.

By 1958, electron tubes had developed to a state of operation at temperatures as high as 500°C, and at high levels of shock and vibration. J. D. Cronin (1960) investigated the application of alumina tube envelopes at high-temperature service.

Of the various electronic effects in tubes, the photoelectric generation of secondary electrons and the voltage breakdown are influenced by the type of ceramic. The ratio of secondary electrons generated to primary exciting electrons (d) is a maximum of 1.5 to 4.8 at a primary electron energy of 350 to 1300 ev. The voltage breakdown falls into three classes: (1) across the vacuum gap, (2) within the dielectric material, and (3) along the surface. Kohl stated that the breadown across the gap was independent of temperature to about 1200°C. The volume resistivity at 500°C of sintered alumina (99%) was about 10^{10} to 10^{11}, and of sintered alumina (75%) about 10^9. Additional references of significance include: Whitehead (1951), Pomerantz (1946), Navias (1954), Jacobs (1946), F. J. Norton (1957), and P. D. Johnson (1950).

Wisely (1962), and Moulson and Popper (1963) sought to extend the range of materials available for electron tube envelopes and windows of tubes. Beryllia was considered best for windows, but manufacturing costs were high. Alumina is widely used, and the higher the Al_2O_3 content, the more generally suitable are the properties. In 1961,

159

Varian Associates used mainly high-alumina ceramic parts in microwave tubes; 70% of the high-temperature tubes used ceramic windows and 95% of the cathode types were ceramic (Ref. Anon. Trip to Varian).

Hursh (1949) described the preparation of a ceramic vacuum tube for use in a betatron. The tube was toroidal in form with two tubular arms to which glass connections were fused for the sealed-in electrical leads under high vacuum.

19-3 ALUMINA PORCELAIN INSULATION

From a practical standpoint, ceramic electrical insulation has several advantages in comparison with competitive organic (plastic) dielectrics in general and power transmission service. Among these are: superior electric properties, high mechanical strength, absence of deformation or creep at ordinary temperatures, a greater resistance to changes in the surrounding environment, and impervious seals with metals used as conducting connectors or supports. The flashing of plastic insulators is also likely to cause highly conductive carbonaceous leakage paths. Although the addition of alumina to low-alumina porcelains of this type has not been found to contribute markedly to the electrical properties, it does contribute advantageously to the mechanical strength. This enables improvement in the design of the insulators, as for example, by reduction in equivalent size.

Austin, Schofield, and Haldy (1946) investigated electrical porcelain compositions containing as little as 23% Bayer alumina, present mainly as undissolved fine particles in the glassy matrix. When substituted for the usual flint additions in standard dry-process compositions and under normal firing conditions, the alumina provided certain advantages that could not be ascribed merely to its finely divided condition. Although increasing amounts of the alumina increased the over-all thermal expansion (30 to 1000°C), the alpha-to-beta quartz inversion in the thermal-expansion curve decreased with increasing alumina content. The inversion was not evident beyond 20% alumina addition, even though the raw body still contained a relatively large amount of potter's flint, thus allowing more rapid firing schedules. The use of alumina lengthens the firing range because the glass phase is stiffer at a given temperature than in a conventional body. The tendency for pyroplastic flow is decreased decidedly, but not necessarily the warpage. Translucency is reduced, which is to be expected, because of the fine particle size and high index of refraction of the alumina. Strength was increased as much as 200% when 40% alumina was added. Modulus of rupture above 22,000 psi was obtained, in comparison with about 4000 psi for unglazed, dry-pressed ware, or 12,000 psi for unglazed extruded ware of the common feldspathic types. Zircon porcelains range in flexural strength from 10,000 to 20,000 psi. It was particularly advantageous on strength to substitute alumina completely for flint, and nepheline syenite for feldspar, in order to remove as completely as possible sources of free quartz. Selsing claimed strengths of 26,000 psi in flexure for such bodies containing in excess of one

percent quartz to which a flux is added containing sufficient sodium silicates ($Na_2O:SiO_2$ at least 1:10) to reduce the quartz content on firing to no more than one percent.

Blodgett (1961) prepared high-strength alumina porcelains (average flexural strength 31,000 psi) using 65% Alumina A-2, minus 325 mesh, 10% nepheline syenite, 2% each of lithium manganite and whiting, and 21% ball clay. A significant factor was the wet grinding of all components together for 14 hours. This is presently not a conventional practice in preparing electrical porcelains. Commercial porcelains are prepared for forming by blunging the ingredients in water suspension and by screening out the coarse oversize, including organic carbonaceous material.

Low-alumina electrical porcelains (20 to 60% Al_2O_3), in which calcined Bayer alumina has been substituted for flint in compositions containing clays and feldspar, or preferably nepheline syenite, can be fired under the same conditions as for cone-12, feldspathic, electrical porcelains. Bayer alumina additions in concentrations as high as 40 to 60% of the compositions are contemplated for high-tension insulation in the long-distance transmission of electric power. The range of alumina additions extends beyond 85% Al_2O_3, however.

Many applications exist, as for example, bushings for transformers, and cable spacers (PLM Unimold, Diamonite). These latter serve to hold the geometry of high-tension wires rigidly, and to support the cable from the messenger, for line loads to 34.5 kv.

All such applications of alumina conform with the older conventional insulators in carrying a very good glaze, adapted to the alumina composition.

19-4 RESISTORS AND SEMICONDUCTORS

Alumina is used as the nonconducting base or support for semiconducting coatings, or as a homogeneously mixed composition with conducting materials to allow various degrees of resistivity to the flow of electricity. The theoretical exposition of the subject has been treated by Hannay (1959). Resistors of the first type consisting of cylinders of sintered alumina with a layer of semiconducting carbon or metal deposited in several ways, and having low resistance terminals on each end, have served as noninductive resistors in electronic and other applications. In the second type, the alumina may or may not contribute to the electrical conductivity (at ordinary temperatures) depending on whether it has reacted to form new compounds. Otherwise the conductivity is by bridging of the conductive component.

Ridgway (1941) described an electrical resistor formed as a shaped, self-bonded, coherent, dense, continuous, monolithic, prefused, crystalline structure, having alumina as the major phase intergrown with a phase of crystals of essentially titanium dioxide and a lower oxide, which formed an electrically conductive refractory body. Palumbo

(1949) impregnated a porous refractory alumina composition with carbon, deposited from a hydrocarbon gas at temperatures between 800 and 1400°C. Verwey and van Bruggen prepared a resistor having a negative coefficient of electrical resistance, by sintering a mass of mixed crystals of FeO, Fe_2O_3, MgO, and Al_2O_3 in the ratio to form spinel, with less than 50 mol. % of iron oxide in spinel form (Fe_3O_4), the sintered mass having a uniform spinel phase which is less than supersaturated with a second phase at 500°C. Heath (1962) produced a resistor by mixing 50 to 80% silicon carbide, 5 to 35% aluminum phosphate, aluminum oxide and reactive P_2O_5, and 0.5 to 10% flux (stannic oxide), pressing to a form, and firing it at 1200 to 1300°C. Navias (1958) dispersed tungsten, molybdenum, or mixtures of the two as a powder in the alumina, and fired in a hydrogen atmosphere, to obtain a resistor with a positive coefficient of resistivity. Huffadine and Sanders mixed molybdenum disilicide with the alumina, but relied on 0.5 to 3.0% elemental carbon to provide the electrical conductivity by bridging the molybdenum disilicide particles.

Challande, Hantel, and Montgomery (1962) observed the phenomenon of the acquisition of electrical charge on crystals of alumina by friction with a metal. Wisely (1957), Anon (ref. Oxide Thermoelectric Materials), and W. A. Miller (1956) investigated thermally sensitive conductors, such as $(Al_2O_3) \times (ZnO)_{100-x}$, alkali-aluminum silicates, and porous sintered alumina containing aluminum metal diffused through the pores at a sintering temperature above 1450°C in a nonoxidizing atmosphere.

Methods of developing an electric spark by impact on certain crystalline compositions of piezoelectric characteristics, or by discharge of a capacitor through a low resistance, attracted attention for several applications, including the use as a substitute for present sparking equipment in combustion engines. Sheheen and Mooney (1958) invented such a device, using aluminum oxide impregnated with a semiconducting oxide. A first coating on the body is made up of 86% CuO and 14% CaO, fired at about 1430°C, oxidizing. A second coat containing 86% CuO and 14% $CaCrO_4$ is fired at 1120 to 1290°C. Counts et al. (General Motors) patented semiconductor compositions containing TiO_2, SnO_2, Ta_2O_5 and MoO_3 in which alumina is a component. Harris (Bendix Aviation Corporation) prepared a spark-gap semiconductor containing alumina, tricalcium pentaluminate and the oxides of cobalt and manganese.

Terry (1957) discussed the use of ceramic materials in electric motors and transformers. These applications include coatings, castable refractories, cements, ceramic encapsulants, and sheet insulation. Aluminum oxide anodic coating on aluminum wire, aluminum oxide filler in transformers, and filler in encapsulating plastic are examples of these applications (Walton and Harris, 1958; Fischer, Stone, and McGowan, 1963).

Folweiler (1962) showed that individual ceramic parts for electrical insulation purposes can be joined by filling the surfaces to be joined together with a powder of the same composition and baking under pressure for about 10 minutes at above 1600°C, or for about 100 minutes at 1700°C for a vacuum-tight joint, preferably heated in an inert gas atmosphere.

Stringfellow (1958) composed a ceramic interrupter arc chute having several ceramic arcing plates of low coefficient of thermal expansion and retaining high thermal shock resistance to repeated circuit interruptions. The plates are supported in an interleaved, spaced-apart relation, so that an arc must pass through a stack of the plates. The composition is essentially 5 to 13 wt % MgO, 37 to 57% Al_2O_3 (Alcoa Alumina A-10), and 38 to 50% SiO_2 and contains more than 50% cordierite crystals. The firing temperature is in the range 1260 to 1480°C.

Brunetti and Khouri (1946) described the printed electronic circuits for generator-powered, radio proximity fuses for the U.S. Army's mortar shells. The wiring and resistor components were printed on a steatite surface. Printed circuits on alumina wafers have been developed. Hannahs and Stein (1952) described the types of printed circuits that use ceramic and plastic bases. Svec (1966) reviewed the uses of alumina as a substrate for deposition of resistor, capacitor, and conductor networks in a modern, commercial computer-assembly plant. Favorable points are the high mechanical strength, chemical inertness, smooth surface, good thermal conductivity, and electrical insulation.

20 CEMENT

20-1 CALCIUM ALUMINATE CEMENT

Alumina is present in portland cements in relatively low amount. It is a major constituent in the calcium aluminate cements, however, and both the preparation of hydraulic calcium aluminate cements and their applications are of ceramic interest. They are designated as aluminous cement, high-alumina cement, and calcium aluminate cement, the last probably being the most acceptable designation. The presence of calcium aluminate in slags, and in the Pedersen process for alumina, and the hydration reactions have been mentioned in Section 3-2 and 12-3-5.

The major production of calcium aluminate cement is an impure product, generally prepared from bauxite, mainly European grades, and limestone, selected to hold the silica content in the fired clinker within about 6%. High iron oxide content, about 20 to 30% of European bauxites, contributes little to the hydraulic strength but acts as a flux during fusion and modifies the setting properties. It also provides metallic iron as a by-product, with consequent lowering of the content in the cement in reduction processes. Titanium compounds, usually not over 3.5%, contribute no cementing power, and magnesia, the remaining impurity contributed by the limestone in appreciable amounts, causes a marked decline in strength of the cement, when present beyond 1%.

A relatively small, but increasing tonnage of very high-purity cement is made from Bayer alumina and high-grade limestone, and containing from 70 to 80% Al_2O_3, less than 0.5% Fe_2O_3, and lesser amounts of SiO_2 and TiO_2. It is intended mainly for refractory applications to as high as 1800°C under some conditions, but is also receiving attention in other applications where refractoriness is incidental.

Typical commercial cements are identified as follows:

Lumnite (Universal Atlas Division, U.S. Steel Corporation). The raw materials are high-iron, high-silica bauxite and limestone, melted in a rotary kiln and quenched in water to cool and granulate. This is ground with small amounts of gypsum, and a retarder/plasticizer is added.

LaFarge Ciment Fondu (England). French or Greek bauxite and limestone are melted in an L-shaped reverberatory furnace, cooled on a pan conveyor, and ground without further addition.

Trade names, not well known in the United States, include Fundido Electroland (Spain), Alcement Lafarge (Scandinavia), Fondu Lafarge (France), and Lightning (England).

In an intermediate range of purity are classed the following:

Refcon (Universal Atlas), total Fe_2O_3 about 1.5%, SiO_2 5.5%.

Rolandshutte (Metallhüttenwerke Lübeck AG, Germany). Bauxite, limestone, coke and iron scrap are melted under reducing conditions in metallurgical blast furnaces, the molsten slag is poured from ladles into a shallow pit to cool in sheets, and is then ground. Iron is also a tapped product from the operation. The iron content of the cement is low (about 1.5%), but it is mainly present as ferrous oxide or as relatively coarse iron particles, a significant factor with respect to carbon monoxide disintegration.

High-purity white cements, available commercially, are represented by Secar 250 (Soc. Anon. SECAR, Paris) and Alcoa CA-25 (Aluminum Company of America Product Data, Calcium Aluminate Cement CA-25, January 1963). Babcock and Wilcox produces complete castables, called Kaocast (Norton and Duplin, 1950). Several refractories manufacturers prepare calcium aluminate cements sporadically.

Sainte-Claire Deville probably was the first to be aware of the hydraulic properties of calcium aluminate and its use as a binder for refractories, through his use of equal amounts of powdered alumina and limestone which he wet-mixed to form crucibles for high-temperature investigations. A subsequent search for a hydraulic cement which would not disintegrate in sea water and in contact with sulfates led to the development of the Lafarge "Ciment Fondu."

Modifications of the raw materials entering into high alumina cement production have been made primarily to increase the strength in the range of weakness of castables around 800°C, in which the hydraulic bond is largely destroyed and a ceramic bond has not developed strongly.

An example is the addition of topaz to the raw mix (Kocher, Atlas Lumnite).

Some data obtained on typical commercial samples of calcium aluminate cement are shown in Table 14.

Schneider and Mong (1958) determined the strengths of castables of six groups of laboratory-prepared castables using domestic calcium aluminate cements. In general, the modulus of elasticity dropped remarkably for the initial heating at 200°C, indicating a substantial loss in hydraulic bonding strength. Castables initially heated at 1050°C, retained about the same strength throughout the range of temperature as the minimum measured at 1050°C. Specimens heated at 1300°C, however, recovered a substantial amount of the strength lost at 200 to 300°C, indicating the development of a ceramic bond. The strength was substantially retained when measured at 1000°C.

Lea and Desch (1956), Filonenko and Lavrov (1950), and Wisnyi (1954), among others, have examined the high-alumina portion of the system. Wisnyi synthesized homogeneous single crystals of $CaO \cdot 6Al_2O_3$ (Filonenko's phase). Buessem and Eitel (1936) claimed that the previously described phase $5CaO \cdot Al_2O_3$ is actually $12CaO \cdot 7Al_2O_3$. Auriol, Hauser, and Wurm (1961) suggested $3CaO \cdot 16Al_2O_3$ as the formula of the alumina-richest phase. Ampian (1964) reviewed the published optical, microscopic, and X-ray data for the principal phases.

Brcic et al. (1962) found that only $5CaO \cdot 3Al_2O_3$ is formed at 400 to 600°C when $a\text{-}Al_2O_3$, boehmite, or bayerite is reacted with calcium compounds. Uchikawa et al. stated that $3CaO \cdot Al_2O_3$ is the first compound formed regardless of the proportions of the mix, when $CaCO_3$ and $a\text{-}Al_2O_3$ are heated together at 800 to 1000°C. Andouze claimed that an orthorhombic form of $5CaO \cdot Al_2O_3$ exists (stable above 950°C). Except for the phase $2CaO \cdot Al_2O_3$, all the calcium aluminate phases can be prepared by reaction of CaO and Al_2O_3 at 900 to 1600°C.

Roth and Wolf obtained the following heats of formation for two phases: -11.1 kcal/mole at 20°C for $12CaO \cdot 7Al_2O_3$, and 3.9 kcal/mole for $3CaO \cdot Al_2O_3$. Babushkin and Mchedlov-Petrosyan prepared equations for determining variations in the free energy as a function of temperature for ten possible reactions in the system.

Solacolu placed the most favorable compositions on the line from $CaO \cdot Al_2O_3$ to the invariant point $2CaO \cdot Al_2O_3 \cdot SiO_2\text{-}CaO \cdot Al_2O_3\text{-}2CaO \cdot SiO_2$ on the triaxial diagram. From the standpoint of refractoriness, silica-free compositions

Table 14

Properties of Calcium Aluminate Cements

Analysis	High-Purity Cements		Intermediate-Purity Cements		Low-Purity Cement
Total Al_2O_3	80.0	74.0	53.2	66.6	41.0
Total CaO	17.8	24.0	35.0	23.0	35.2
Total Fe_2O_3	0.4	0.3	1.5	0.7	10.0
Total SiO_2	0.1	0.2	5.7	4.1	7.9
Total MgO	0.2	0.2	0.5	0.5	1.6
Sp.G. g/ml	3.35	3.00	3.00	3.04	3.11
PCE	34-35	26	—	15-16	14

Properties in Alumina-Silica Castable (19.4% cement)

Casting water (%)	9.5	12.0	12.0	12.0	12.0
Flexural strength (psi)					
Cured	970	800	770	600	300
Dried	1190	1100	1020	750	350
Fired	1200	350	550	300	200

Properties in Sintered Alumina Castable (15% cement)

Casting water (%)	10.0	12.8	12.7	15.5	12.7
Flexural strength (psi)					
Cured	830	445	315	220	495
Dried	940	545	595	255	570
Fired	1700	395	395	230	375

would be preferred. The commercial cements are mainly composed of monocalcium aluminate ($CaO \cdot Al_2O_3$), but $12CaO \cdot 7Al_2O_3$ and $CaO \cdot 2Al_2O_3$ are usually present. The cement composition acquires faster setting properties with increasing CaO content beyond $CaO \cdot Al_2O_3$, and slower setting properties with increasing Al_2O_3 content of the cement. This is not indicated by the cement formulas, however, because the Al_2O_3 is not always combined as calcium aluminate. The refractoriness of the cement increases with increasing alumina content. Pole and Moore (1946) fused compositions conforming with $3CaO \cdot 5Al_2O_3$ and obtained good hydraulic bonding and a refractoriness of 1495°C for the neat cement. A mortar prepared with one part cement to 3 parts crushed firebrick bent at cone 30 (1650°C). Gitzen, Hart, and MacZura (1957) prepared cements from high-purity Bayer sources and obtained refractoriness to cone 34 (1760°C) for a balanced composition conforming to the empirical formula $CaO \cdot 2.5Al_2O_3$, in which the predominating bonding phases were $CaO \cdot Al_2O_3$ and $12CaO \cdot 7Al_2O_3$. The resistance to attack by sulfur acids, a normal, controllable setting time, the rapid hardening of the hydraulic bond (substantially complete within a few days), and the retention of high strength after heating at elevated temperatures are the main properties of the cement.

Lehmann and Mitusch (1959, 54 references) and Robson (1962, 589 references) have provided excellent coverage of the manufacture, properties, applications, and analysis of the refractory cements and the mortars and concretes prepared from them.

The scope of ceramic uses of calcium aluminate cements is so diverse that only general uses and a few examples can be cited in the following list. Some uses have been mentioned elsewhere (Sections 3-2-3, 12-3-5, and 17-9).

1. Neat cement and mortars

The neat cements are used to seal the suspension hardware of power line insulators and in sealing the center wire of sparkplugs. The porcelain components of the insulators are assembled with mortars of the cement and ground porcelain. The cement is used to seal the sidewalls of diamond-drill borings and oil holes. Castings of the neat cement and mortars, with colorant or texturing ingredients, against a polished surface imparts gloss to the surface. This has been used to simulate various effects in architectural panels.

2. Nonrefractory applications of mortars and concretes

These applications are dependent upon the ability of the installation to be ready for service within 24 hours of placing, which is the main consideration for their use instead of portland cement. They cover such applications as bridge repairs, paving or repair of busy industrial, business, or military locations, the pouring of piers for machinery, the construction or repair of sea equipment between tides, the sealing of caissons, concreting in mines (Robson). Where forms are used to prepare precast sections, the possibility of faster output per mold enables the more economical use of mold equipment or time.

Precast or dirven piles, in diameters seldom greater than 2 ft, were recommended for supporting loads in sulfated ground. They were driven after aging for periods from one day to several years, the essential requirement being to strip the molds within a few hours after casting and to keep the piles wet to prevent overheating. Prestressed beams, lintels, and purlins can be fabricated easier than when using rapid-hardening portland cements.

Calcium aluminate cement is used to make concrete pipes for sewers in acid environments or for carrying acid effluents, in brick jointing in railway tunnels, low-temperature brick flues and stacks. It has been suggested for radiation shielding at high temperatures. The use of the hydrated additive colemanite ($Ca_2B_6O_{11} \cdot 5H_2O$), in the castable for fast-neutron shielding was discussed by Henrie.

Calcium aluminate cement can be pelletized, nodulized, or cast into different granular sizes. Its surface readily fixes metal salts, as for example, silver, of use in catalyzing organic oxidations.

3. Refractory calcium aluminate concretes

Concretes in a wide range of refractoriness are prepared both from the standard calcium aluminate cements and the high-purity white calcium aluminate cements, Secar and Alcoa CA-25. Various types of aggregate are available. Refractory concrete is being increasingly used for kilns, furnaces, coke quenching floors, retort settings, charge-hole blocks, flues, furnace doors, dampers, and general repairs (A. E. Williams, 1951). The concretes prepared from the Lumnite-type cement with chrome ore, magnesia, or dense alumina are stable to at least 1600°C, and those with more common aggregates to about 1300°C. The advantages are good refractoriness, a low thermal conductivity, readiness for use within 24 hours, the facility of casting in any size and shape, trivial shrinkage upon drying, and no necessity for prefiring before placing in service, generally. The same advantages apply to the white pure cement, and in the higher range of refractoriness. Shapland and Livovich (1964) showed a PCE cone of about 13 (1350°C) for the lowest purity commercial cement, about PCE 15 to 16 (1410 to 1450°C) for the intermediate purity type, and cone 34 (1760°C) for the pure calcium aluminate. Robson (page 196) listed the maximum hot-face temperatures allowable for concretes containing different cements and aggregates. They varied from only 250 to 300°C for silica gravel and ordinary calcium aluminate cement to about 1650°C for high-purity cement with firebrick grog (60% Al_2O_3), and to about 1800°C with high-purity cement and fused alumina or tabular alumina. Other aggregates used with the cement include: traprock, pyrophyllite, pumice, magnesite, chromite-magnesite, zirconia, silicon carbide, and the foamed or exfoliated grogs such as vermiculite, perlite, celite, etc. In general, the ordinary cement concretes have poorer refractoriness than the high-purity cement concretes by about 100 to 300°C with increasing temperature. The refractoriness of the combination of cement and aggregate may exceed that of the cement. For example, the overall purity of a castable containing 15% cement and 85% tabular alumina may be over 96% Al_2O_3, thus raising the refractory level by 50 to 70°C into a region above 1800°C.

The concretes may be poured, rammed, or as is increasingly the vogue, gunned (Velikin, 1959; Cook, Cook, and King, 1963, 1964). Gunned specimens set hotter, up to 90°C, and faster than their cast counterparts. Gunned linings showed no decreased strength in the temperature region 800 to 1100°C (as had been previously reported). Changes in the gunning procedure markedly affected the properties of the linings. This technique, applied to calcium aluminate concretes, finds increased application in steel-industry and other metallurgical furnace operations. The method is adapted to repair of deteriorated linings.

Livovich (1961) compared portland cement with calcium aluminate cements in cyclic heating tests on castables. Only the specimens prepared with calcium aluminate cements (Lumnite and Refcon) were stable. Cracking and expansion of the portland cement types was attributed to hydration of lime formed in their thermal decomposition. Rengade claimed that mixtures of portland cement and calcium aluminate cement provide quick set, are nearly as strong as portland cement, and are not disintegrated by sea water (contrary to Dorsch).

Lafuma (1954) added expansive cement (calcium sulfoaluminate) to portland cement to counteract its drying shrinkage. Okorokov et al. prepared this form by reaction between gypsum and calcium aluminate at 1100 to 1300°C. Tanaka and Watanabe determined the chemical composition and physical properties of a commercial expansive slag. The cement consisted of 65 parts portland cement and 25 parts sulfoaluminate clinker. Armstrong and Whitehurst mixed calcium aluminate slags containing up to 0.4% fluorine with calcined calcium sulfate (anhydrite) to obtain sulfoaluminate cement.

McGrue (1937) recommended high-alumina cements as binders for concretes used in coke ovens and blast furnaces for steel. Recommended aggregates for insulating concretes included slag, Haydite, Sil-O-Cel and vermiculite. Shapland and Livovich (1964) found that castables suitable for service at 1320 to 1540°C in steel production had poor resistance to attack by alkalies, and that castables prepared from only two of five cements tested were resistant to carbon monoxide disintegration. Calcium aluminate castables follow about the same pattern of behavior as firebrick with respect to carbon monoxide attack except that castables are more likely to have free iron particles present, either as a result of the cement production method (reduction process) or from pickup during grinding of the clinker. The pickup from fused clinker is likely to be much higher than from the sintered high-purity clinkers. Castables prepared from high-purity cements and aggregate have resistance comparable with superduty fireclay brick (Gitzen, Heilich, and Rohr, 1964).

Wygant and Bulkley (1954) found that suitable concretes for lining fluid catalytic cracking vessels contained about 60 wt % of coarse aggregate (8 to 14 mesh) and 15% fine aggregate (about 50 to 100 mesh). Crowley (1964) claimed that the addition of $CaHPO_4 \cdot nH_2O$ improved high-alumina, high-purity calcium aluminate concretes. Compressive strengths ranging from 14,000 to 20,000 psi were attained on specimens containing 18 wt % cement and 82% tabular alumina, damp-cured for 24 hours and heated at about 540°C.

Feagin (1959) used 3 to 35 wt % calcium aluminate binder in refractory aggregates of periclase, zircon, silica, and kyanite in proportions to provide controlled expansion in the precision casting of metals (about 5.53% linear expansion from room temperature to 1150°C). Kuntzsch and Rabe described the use of "fire concrete" as a substitute for fireclay brick. Cobaugh (1958) described the use of castable refractories in the production of ceramics and glass, Robson (1952) the shaping of supports for electric resistance wires. Refractory concrete was claimed to have higher electrical resistance than the materials of which it was made, its conductivity being about 75% of fireclay brick. Braun (1953) recommended monolithic liners for furnace doors according to a design that eliminates the necessity of providing studs for anchoring or cooling. Tatousek and Peters evaluated the cast refractory for hot-top linings, and claimed that the cast liners were more difficult and costly to install, but their life was longer. Erosion of the cast tops caused excessive metal losses, when lined with "high aluminasilica" aggregate (48% Al_2O_3, 39% SiO_2, 10% CaO). Schwarz and Hempel (1960) successfully replaced chamotte brick walls of tunnel kiln cars with refractory concrete containing about 14 to 17% cement and chamotte granules. Williams (1958) described general metallurgical furnacing equipment in which concretes can be used. The details of methods for support and reinforcing are too voluminous to report in this survey.

Ricker claimed that up to 2.25 wt % of one alkaline earth carbonate added to the cement used in preparing castables increased the concrete strength.

The methods used for curing calcium aluminate concretes have an important effect on the strength, since the type and binding strength of the phases produced depend on the time and temperature required for hydration. Wells, Clarke, and McMurdie claimed that $3CaO \cdot Al_2O_3$ is not a normal calcination product, but the cubic, hydrated form $3CaO \cdot Al_2O_3 \cdot 6H_2O$ is the only stable hydration phase between 21 and 90°C. Near 1°C, the condition apparently reverses, and metastable $3CaO \cdot Al_2O_3 \cdot 6H_2O$ transforms to one or other of the hexagonal hydrates, $CaO \cdot Al_2O_3 \cdot 10H_2O$, $2\,CaO \cdot Al_2O_3 \cdot 8H_2O$, or $4CaO \cdot Al_2O_3 \cdot 13H_2O$ (Cirilli; Carlson, 1957; Buttler et al.). The hydrate $3CaO \cdot Al_2O_3 \cdot 6H_2O$ is stable to 215°C, and the phase $4CaO \cdot 3Al_2O_3 \cdot 3H_2O$ from 215 to 250°C (Peppler and Wells, 1954). Neville (1958) found that loss of strength always occurs in high-alumina cement concretes that have been moist-cured continuously or intermittently at 25 to 40°C. The loss was associated with the inversion of metastable hexagonal $CaO \cdot Al_2O_3 \cdot 10H_2O$ to cubic $3CaO \cdot Al_2O_3 \cdot 6H_2O$. Budnikov and Kravchenko claimed that these hydration products still may have higher strength than structures composed mainly of the hexagonal calcium aluminates. Schneider (1959) found by X-ray diffraction and DTA that $CaO \cdot Al_2O_3 \cdot 10H_2O$ was the principal phase of hydration at room temperature; $3CaO \cdot Al_2O_3 \cdot 6H_2O$ and $Al_2O_3 \cdot 3H_2O$ (gibbsite) were formed as by-products at 50°C, the principal phase being $CaO \cdot Al_2O_3 \cdot xH_2O$, in which x was less than $10H_2O$. At 500 to 600°C, the principal phase dehydrated to $CaO \cdot Al_2O_3$; the by-product $3CaO \cdot Al_2O_3 \cdot 6H_2O$ transformed to CaO and $12CaO \cdot 7Al_2O_3$. Longuet and Tournadre claimed that both $CaO \cdot Al_2O_3$ and $CaO \cdot 3Al_2O_3$ form $3CaO \cdot Al_2O_3 \cdot 6H_2O$ at about 30°C.

Gitzen and Hart (1961) observed that the tendency to explosive spalling by calcium aluminate castables on initial heating could be circumvented by decreasing the cement concentration and increasing the particle coarseness in order to increase the permeability. Control of the ambient curing temperature was a more desirable method, however, that did not sacrifice castable properties. King and Renkey claimed that the addition of up to 1.5% boric acid to the refractory castable produced an explosion-resistant castable.

Tseung and Carruthers (1963), on the basis of an arbitrary selection of size distribution of aggregate in a castable formulation for which a low water/cement ratio (0.5) was found to be optimum, concluded that curing at 75°C for 6 hours immediately after casting gave appreciably stronger castings at all temperatures than 24-hr curing at room temperature. It appears, however, that the water/cement ratio is a critical factor affecting the optimum curing temperature, particularly when the ratio does not provide sufficient water for complete hydration.

Some confusion between the results of different investigators for the mechanical properties of the castables stems from differences in the early stages of firing the castable. The prolonged duration of steam in the castable in the temperature range between 100 and about 250°C appears to be a deteriorating factor. From this, it would seem to be more effective to dry the castable completely below 70°C (if practical) before firing.

Wygant and Crowley (1964) found that in actual large-scale installations it is necessary to prevent moisture from escaping from the lining during curing. Internal temperature rise did not impair the strength. The application of impermeable membranes (organic resin or asphalt coatings) had been used successfully for 8 years in installing liners.

Tulovskaya and Segalova concluded that at 5°C, $CaO \cdot Al_2O_3$ hydrates to aqueous $CaO \cdot Al_2O_3$ and aqueous $2CaO \cdot Al_2O_3$. The aqueous $CaO \cdot Al_2O_3$ then changes to aqueous $2CaO \cdot Al_2O_3$. At higher temperatures both hydrates transform to $3CaO \cdot Al_2O_3 \cdot 6H_2O$.

Onofrio (1963) subjected castables to a high-frequency electric field (radio frequency) during curing. Crepaz and Raccanelli (1963), on the basis of DTA tests, determined that the products of hydration of calcium aluminate cement are $Al_2O_3 \cdot 3H_2O$, $3CaO \cdot Al_2O_3 \cdot 6H_2O$, $4CaO \cdot Al_2O_3 \cdot 19H_2O$, $CaO \cdot Al_2O_3 \cdot 10H_2O$, and $2CaO \cdot Al_2O_3 \cdot 8H_2O$.

Treffner and Williams (1963) observed that fine-crystal, low-burned, calcium aluminate cements develop fast heat of hydration. Increasing the Al_2O_3 content of the cement to the limit, $CaO \cdot 2Al_2O_3$, decreased the hydration rate for the same equivalent crystal size. The rate of hydration decreased with increasing impurity of the cement. The optimum castable strength was obtained with crystal sizes not less than 20 microns.

Kimpel (1962) claimed a cement composition containing a mixture of 30 to 40 wt % plaster of paris, 15 to 30% aluminum oxide, and 5 to 10% diatomaceous earth, 1 to 4% ball clay and 20 to 36% water. Weltz (1960) claimed a low-density insulating castable containing 50 wt % fused alumina bubbles, 25% dense alumina (14F and 48F sizes), and 25% calcium aluminate. Mikhailov and Litvinov (1963) found that most glass fibers corrode in lime and cement. A glass of good durability consists of 61.9% SiO_2, 6.1% Al_2O_3, 13.5% CaO, 7.9% Na_2O, 1.0% K_2O, and 8.7% other oxides.

Talaber (1961) claimed that the high alumina cements reach maximum strength after two years, then slowly decline. Resistance in compression is lost faster than in tension. Iron does not affect strength or durability. The underwater durability is superior to that in air. The final products are hydrated alumina and calcium carbonate. Mehta stated that unless this retrogression in the hydraulic strength can be circumvented, the future of calcium aluminate cements in structures intended for long duration is uncertain.

20-2 BARIUM ALUMINATE

Barium-containing cements have the advantage over common commercial types of superior radiation-stopping power. Two types, silicate and aluminate, have been prepared. As the aluminate type generally requires pure alumina or bauxite in its synthesis, this is a potential application of alumina. Purt (1960) included among the solid aluminates: $BaO \cdot 6Al_2O_3$ (m.p. 1915°C), $BaO \cdot Al_2O_3$ (m.p. 1815°C), and $3BaO \cdot Al_2O_3$ (m.p. 1425°C). Wallmark and Westgren found that the first two are hexagonal crystals, the phase $BaO \cdot 6Al_2O_3$ being ismorphous with other beta alumina types. Avgustinik and Mchedlov-Petrosyan, and Smirnov et al. observed that the conversion of a 1:1 mixture of $BaSO_4$ and Al_2O_3 proceeds between 1200 and 1400°C, and is promoted by the addition of about 5% CaF_2. Peppler and Newman (1951) determined the heats of formation of various barium aluminates from their heats of solution in HCl. Calvet, Thibon, and Dozoul (1964) claimed to have produced $2BaO \cdot Al_2O_3$ by precipitating stoichiometric proportions of the two chlorides in ammoniacal $(NH_4)_2CO_3$ solution, and by heating at 1200 to 1300°C. Massazza (1963) found the neutron absorption cross-section of hydraulic cements containing BaO to be 3 times that of CaO, 6 times that of Al_2O_3, and 9 times that of SiO_2.

Barium aluminate prepared by sintering mixtures of pure alumina and barium carbonate is soluble in water, but more so in dilute alkaline solution (Akiyama et al., 1938). The presence of silica, however, influences the solubility. Carlson, Chaconas, and Wells investigated the hydration of the barium aluminates. All the barium aluminate compositions, including $BaO \cdot Al_2O_3 \cdot 4H_2O$, $2BaO \cdot Al_2O_3 \cdot 5H_2O$, $BaO \cdot Al_2O_3 \cdot 7H_2O$, and $7BaO \cdot 6Al_2O_3 \cdot 36H_2O$, hydrolize in water to release gibbsite. The probably stable solid phases are $Ba(OH)_2 \cdot 8H_2O$ and $2BaO \cdot Al_2O_3 \cdot 5H_2O$. Carlson and Wells prepared $BaO \cdot Al_2O_3 \cdot 2H_2O$ and $BaO \cdot Al_2O_3 \cdot H_2O$ hydrothermally.

Braniski, during the period 1957 to 1961, investigated the possibility of partial or complete replacement of the lime in calcium cements (both silicate and aluminate types). The tests showed the existence of $3SrO \cdot SiO_2$ and $3BaO \cdot SiO_2$. Barium blast-furnace cements, barium slag cements, and barium cements were found to have exceptional resistance to attack by sea water. Barium aluminous cement melted at $1730^{\circ}C$. Owing to the high density of the barium aluminate cements, good protection against X-rays and gamma rays was provided, though not to the same extent as with the barium silicate cements. When BaO is substituted for CaO in aluminous cements, the products are not hydraulic but merely water-soluble, air-hardening cements. Braniski and Ionescu (1960) found that replacing part of the lime by about 3% BaO in raw mixes for barium aluminate cement promoted the sintering of the clinker and the strength increased by about 10%. Refractory concretes containing various grogs were exceptionally resistant to spalling. The refractoriness with corundum was 1865 to $1900^{\circ}C$, with bauxite 1750 to $1790^{\circ}C$, with clay 1670 to 1690, with chrome-magnesite 1880 to $1960^{\circ}C$, and with magnesia 1920 to $1980^{\circ}C$. The refractoriness of $BaO \cdot Al_2O_3$ is much higher than that of $CaO \cdot 2Al_2O_3$, and the early strength is extraordinarily high.

Braniski (1961) claimed that the hardening of barium aluminate cement involves the dehydration of the alumina gel released by hydrolysis, and also the dehydration of the barium aluminate and silicate present. Crystallization of $3BaO \cdot Al_2O_3 \cdot 6H_2O$ also caused hardening. After heating at $800^{\circ}C$ for six hours, the cold strength of $BaO \cdot Al_2O_3$ cement was reduced by 40%. It was concluded that barium aluminate cement is the most suitable binder for basic refractory concretes.

21 ALUMINA IN GLASS

21-1 INTRODUCTION

Alumina enters glass compositions in a variety of ways. R. W. Hopkins mentioned the following sources, mostly native: perlite, sand, aluminum hydrate, bauxite, clay, kyanite group minerals, topaz, cryolite, anorthosite (a feldspathic rock), and blast-furnace slag. Hydrated alumina probably represents the largest volume of Bayer-source material used in the glass industry, but actually several grades of calcined Bayer alumina and ground tabular alumina also enter into glass manufacture, not to mention the alumina used in making the refractories essential to the production of glass. Since the purified Bayer aluminas are valued somewhat higher than the native alumina sources, it is evident that the principal reason for their use is their freedom from certain impurities, as for example, iron oxide. The substitution of Bayer hydrate for feldspar in plate glass batches reduced the iron content from 0.15 to 0.09%, caused the greenish tint to almost disappear, and increased the light by 1.5% to a total transmission of 87.8%/cm thickness (Zaporozhtseva and Gorbai).

The major groups in the glass industry are container glass, flat glass, pressed and blown glass, and glass fibers, and in that order of decreasing value of production. Of these groups, the role of alumina in fibers has been discussed in Section 26. Alumina is a component of nearly all of the remaining types of glass, regardless of the source. About 90% of all glass produced contains at least 2% Al_2O_3. Silica-soda-lime glass, borosilicate glass, and lead glass contain up to 3% Al_2O_3; E-glass for continuous fiber production, about 15%; and hard aluminosilicate glasses for cooking ware and combustion tubes, about 20% (Simpson, 1959, anon Glass, 1964). Within this rather small range of concentration, alumina has a marked influence on the properties of the glass.

In accordance with Zachariesen's rules, alumina falls in the class of oxides having an anion to cation ratio of 1.5, hence is considered a glass former having substantially directional, covalent, anion-cation bonding. Each oxygen ion is linked to no more than two atoms of aluminum; the number of oxygens surrounding aluminum must not exceed 4; the oxygen polyhedra share only corners with each other; and at least three corners in each polyhedron must be shared. The arrangement of polyhedra is irregular and hence isotropic. The description of glass networks and some other theories of the glass structure have been presented by Rooksby et al. (crystallite theory); Porai-Koshitz, 1960; Weyl, 1960 (acid-base hypothesis); Stevels; and Appen, 1954 (skeletal-coordination theory). Salmang (1957) gave little attention to the crystallite theory. According to Weyl, the glasses are classified by their acidity or alkalinity, based on the degree of screening of the positive core of the molecules. Molecules with a well screened positive core have zero acidity; as the screening becomes less complete, the condition becomes acidic. Rindone, Day, and Caporali (1960) and Graham and Rindone (1964) have measured the acidities of glasses containing variable amounts of alumina, and find an increase in acidity for increasing substitution of Al_2O_3 for SiO_2 in the glass. The first additions increase acidity, supposedly by forming AlO_4 tetrahedra which use a part of the nonbridging oxygen ions to fulfill the coordination requirements. Dietzel and Scholze (1955) concluded from refractive index measurements that the aluminum ion in the mullite region of glass in the system $BeO\text{-}Al_2O_3\text{-}SiO_2$ is present in both 4-fold and 6-fold coordination, with the 4-fold coordination increasing with increasing silica content. Dubrova and Kefeli found that the coordination of aluminum in glasslike silicates is 4 and 6, based on absorption spectra.

Appen (1953) and Haray-Szabo correlated the properties of glasses with changes in the oxygen coordination number. In silicate glasses, alumina undergoes a change from 6 (corundum) to 4 (glass). Partial properties of alumina are calculated for crystalline silicates showing a coordination number of 4 in albite, nepheline anorthite, and celsian, and a coordination number of 6 in sillimanite and kyanite. Al^{3+} with a coordination number of 6 is unlikely in alkaline glass, but the contribution of Al_2O_3 to the refractive index of glass, melted from the compounds, $MgO \cdot Al_2O_3 \cdot 2SiO_2$ and $BeO \cdot Al_2O_3 \cdot 2SiO_2$, is that of corundum. In sillimanite, one-half of all aluminum atoms have 4-coordination, the other half 6-coordination.

Warburton and Williams (1963) found that the addition of alumina (whether coarse or fine-ground) to mixtures of lime and silica lowers the temperature required to yield a glass-forming liquid. Alumina improves glass with respect to resistance to water and acid. The simplest method of introducing it is to add it as feldspar; for hard glasses, high additions are usually introduced as mixtures of hydrated alumina and feldspar in the ratio of 10:1 in German practice (Arnot, 1937).

W. B. Silverman (1939) stated that alumina lowers the melting temperature of soda-lime-silica glasses when either wollastonite or tridymite is the primary phase. Silverman (1940) observed that a wide area of the sodium oxide-dolomite-lime-silica glass system cannot be devitrified if it contains up to 6.4% Al_2O_3. Parmelee and Harman (1937) found that alumina raised the surface tension of molten glasses from about 3.0 dynes/cm at $1200°C$ for 0% Al_2O_3 to 3.35 dynes/cm for 6% Al_2O_3. This was confirmed by Okhotin and Bazhbeuk-Melikova (1952). Barrett and Thomas (1959) correlated the surface tension and density with composition of glasses in the $CaO-Al_2O_3-SiO_2$ system.

Dietzel (1943) noted the phenomenally low expansion coefficients in silica glasses containing aluminum ions. Schweig observed the favorable effect on viscosity and liquidus temperature, and Hlavac (1950) noted that as little as 3% Al_2O_3 improved the viscosity, and liquidus temperature for glass produced by the Fourcault process. The chemical resistance was increased by alumina additions to the glass, but even the presence of aluminum ions in 0.5N NaOH solutions used in glass-solubility tests retarded the corrosion rate of 42 types of quartz glass, barium silicate glass, and antimony silicate glass (Molchanov and Prikhed'ko). The electrical conduction of sodium aluminosilicate glasses becomes a minimum for the atomic ratio Al/Na = 1, from which it was concluded that the size of the oxygen shells increases as AlO_4 tetrahedra are introduced into the network. Beyond this limit, aluminum is assumed to go into 6-fold coordination as a network modifier, allowing the coordination shell of the sodium ion to again decrease (Isard). Alumina increases the tensile strength of soda-lime-silicate glasses, even at low concentrations of about 1% (Watkins, 1950). The function of alumina in lead silicate glasses is to facilitate the reduction of the lead (Kitaigorodskii, Mendeleev et al., 1963).

Lacy concluded that aluminum is unlikely to occur as a truly interstitial cation in glasses and melts having a high proportion of basic oxides. Compositions with a molar excess of Al_2O_3 over the basic oxides do not contain AlO_6 groups; rather triclusters are formed in which an oxygen ion is shared between three tetrahedral groups. The properties of glasses support the Zachariasen-Warren theory (Lacy, 1963).

Lyle, Horak, and Sharp (1936) found that 1.5 to 2.5% Al_2O_3 markedly improved the durability of glasses in the $Na_2O-CaO-SiO_2$ system in terms of an expression: Log durability D = a + b log N, in which a and b are constants, and N = % $Na_2O + Al_2O_3$.

21-2 BOTTLE GLASS

Kotshmid (1959) stated that glass bottles produced on automatic machines initially contained 4% Al_2O_3, which was gradually increased at the expense of SiO_2. The high-alumina glasses have an extended working range and do not require higher melting or fining temperatures (bubble removal). Springer showed that alumina makes both melting and fining easier. W. Koenig (1943) found that green bottle glass melts easier with Al_2O_3 contents as

high as 17 to 20%, but a practical limit of 15% is dictated by the sensitivity to acids. Brown bottle glasses containing 3 to 4% pyrolusite ($MnO_2 \cdot nH_2O$) are readily melted with 15 to 16% Al_2O_3. The tough and lustrous surface of alumina bottle glass is affected less by mold imperfections.

References to glass containers and alumina in glass containers include Kuehl, Shaw, and Anon, Container Glass, 1940–1943.

21-3 DEVITRIFIED GLASSES CONTAINING ALUMINA

The devitrified glasses have been called "glass ceramics" (USA), "vitroceramics" (Europe), and "sitals" (USSR). Emrich (1964) referred to them as "devitrified ceramics." In general, they represent the more or less complete conversion of preformed glass articles by nucleation or crystallization. Devitrification has usually been undesirable in glass because of its adverse effect on transparency and other properties, but nucleation has been used for many years to produce ruby glass, opal glass, and lately, photo-sensitive glasses, which represent controlled applications of the process. The present interest in devitrified glasses stems from the ability to produce polycrystalline ceramics of almost ideal fineness of particles, randomly oriented, and homogeneously dispersed in a vitreous matrix, and without the defects of porosity and gross crystal growth inherent in conventional sintered polycrystalline forms.

Albrecht (1954) described alumina-rich silicate glasses that spontaneously crystallized at proper temperatures to form hard substances. Schneider (1926) discussed the use of aluminum hydroxide for the manufacture of opaque glass. Blair, Silverman, and Hicks detected the crystallites in opal glass as low-temperature quartz with traces of low-temperature cristobalite. W. B. Silverman (1939) discussed the devitrification of soda-lime-silica glasses, and Swift (1946) the effect of alumina on the rate of crystal growth. Zschacke claimed that Al_2O_3 and B_2O_3 actually reduce opacification. Alumina does not produce opacity unless fluorine is present. Okhotin and Ushanova (1959) found that additions of fluorine to a high-alumina, alkali-free, 3-component glass (23.3% CaO, 14.7% Al_2O_3, 62% SiO_2) and a 4-component glass (15.4% CaO, 4.2% MgO, 18.5% Al_2O_3, 61.9% SiO_2) reduced the viscosity. Callow (1952) found that the substitution of ZnO (up to 6%) for CaO was also effective with fluorine. Lungu and Popescu-Has (1958) developed homogeneous crystallization of special glass batches by addition of fluorides to the melt, and by careful reheating. Mechanical strengths higher than those of the original glasses were obtained, but the recrystallization temperature range was above the softening point of the glasses.

Nagaoka and Hara (1962) investigated the submicroscopic crystallization from $Li_2O-Al_2O_3-SiO_2$ glasses by additions of P_2O_5. With increasing Al_2O_3 (1 to 5%) the number of phosphate crystallites increased and the crystallized $Li_2O \cdot 2SiO_2$ apparently decreased. Cristobalite, quartz, spodumene-quartz solid solution, and β-spodumene were also crystallized products.

Stookey (1960, 1961) prepared crystalline or semi-crystalline ceramics of high mechanical strength and good dielectric properties by heat-treating glasses containing from 2 to 20% TiO_2 as a nucleator, and at least 50% crystallizable glass-making ingredients consisting of SiO_2, Al_2O_3, and one or more of the oxides of Li, Be, Mg, Ca, Zn, Sr, Cd, Ba, Pb, Mn, Fe, Co, Ge, B, P, and Ni. An original glass composition of 45.5% SiO_2, 30.5% Al_2O_3, 11.5% TiO_2 and 12.5% MgO had a coefficient of expansion of 37.5 $\times$ $10^{-7}/°C$ and a density of 2.75 g/ml. After reheating under controlled conditions at 870°C and finally at 1345°C (deformation temperature 1370°C), crystallization of phases other than TiO_2 occurred. The product had a coefficient of expansion of 14.1 $\times$ 10^{-7}, a density of 2.62 g/ml, and a flexural strength of 37,500 psi. The claimed strength was not exceptional in comparison with poly-crystalline alumina, but there was an advantage in lower density for air-borne ceramics. This is the basis of the trademarked Pyroceram 9606, announced by Corning Glass Works early in 1957. It can be formed by conventional methods in the glass industry: blowing, pressing, and drawing. It has been tested in radomes for guided missiles, and in electrical insulation. Kitaigorodskii and Bondarev (1962) developed similar materials from slags, various sital compositions for which a softening temperature as high as 1500°C has been claimed. These products have apparently been used mainly in cooking utensils.

The devitrified products from compositions containing MgO had large amounts of a-cordierite; those with CaO had anorthite; those with BeO yielded beryl; compositions with ZnO contained either zinc spinel or willemite (Hinz and coworkers).

King and Stookey (1960) developed a relatively weaker devitrified glass but of very low coefficient of expansion (10 $\times$ $10^{-7}/°C$ between 0 and 300°C) from a composition containing 71 wt % SiO_2, 2.5% Li_2O, 18% Al_2O_3, 4.5% TiO_2, 3% MgO, and 1% ZnO. A semicrystalline body having an average flexural strength of 21,000 psi was claimed. The devitrification temperature was 1150 to 1200°C. Compositions of this type are the basis of Pyroceram 9608. Okhotin and Ushanova investigated fluoride-crystallization from high-alumina glasses; Brown and Kistler examined the devitrification of high-SiO_2 glasses in the system Al_2O_3-SiO_2; and Mariya the devitrification of phosphate glasses. Devitrification to cristobalite was found to proceed more rapidly in the Al_2O_3-SiO_2 system than in the ultrapure silica glasses, but the fluidities (reciprocal viscosities) were lower for the same temperature. The devitrification growth rate, g, was of the zero order, and followed the temperature-dependent expression: ln (g/T) = (B/T) + ln A, in which A and B are constants and T is the absolute temperature in the range 1561 to 1730°K. Hayami, Ogura, and Tanaka (1960) crystallized hard bodies containing dark green chromia crystallites.

Stookey (1961) developed a new photothermally opacified glass containing 2 to 25 wt % Al_2O_3, and in which the ratio, Al_2O_3/Li_2O, is significant at 1.7:1. Olcott and Stookey (1961) developed a glass body having on its surface a thin semicrystalline layer (in compression) of the same oxide composition as the glass. The composition contains 65 to 72 parts SiO_2, 4 parts Li_2O, 22.5 to 30 parts Al_2O_3, and at least one crystallization catalyst (TiO_2, B_2O_3, Na_2O, or PbO). The formed body is heated from one to 40 hours at a temperature at which its viscosity is 10^7 to 10^{10} poises, until crystallites of beta-eucryptite form within the surface. The factor of crystal growth being initiated and controlled by surface crystallization and orientation (epitaxy) was investigated by Vogel (1961) and others. Flexural strengths obtained by the Chemcor process of Corning Glass Works, of the order of 100,000 psi, in comparison with 15,000 to 35,000 psi obtained by the older techniques of chill tempering, have opened new opportunities for this ceramic material. The first product, Centura Ware, has remarkable strength in comparison with conventional china. Presumably, the surface condition must be protected by a coating to take scratches and abrasion, otherwise the ware might be expected to suffer the same deterioration as is experienced with other types of tempered glass.

Corning Glass Works (June 1964) has developed a metal-ceramic body obtained by enveloping a silicate glass form in a molten bath of 20 to 35% silicon, the remainder being aluminum, at 900 to 950°C. By diffusion and thermitic reaction, the final form contains 72 to 82% Al_2O_3, 11 to 21% Al, 1 to 3% Si, and Al-Si alloy in the ratio around 27.3 Al-Si:72 Al_2O_3.

21-4 BORON GLASSES

Alumina appears in high concentration in at least one type of low-soda, borosilicate glass, Pyrex 1720 (Corning Glass Works). A composition containing 58 wt % SiO_2, 19% Al_2O_3, 12% MgO, and 5% B_2O_3 had the following properties: specific gravity 2.53 g/ml, linear coefficient of expansion (0 to 300°C) 0.0000042/°C, softening point 915°C, dielectric constant at 1 mc and 20°C 7.2, and log_{10} resistance at 250°C 11.4. Similar glasses were developed in Europe, as for example Kavalier glass (Czechoslovakia), described by M. B. Volt (1947). Their applications in chemical equipment, piping for corrosive liquids, and many other applications are well known. Difficulties in the melting of these glasses have been overcome by the use of fused block refractories, as for example, Corhart ZAC (Ishino). Ginsberg discussed Pyrex and Superpyrex types.

Chandappa and Simpson delineated the area of glass formation in the system Na_2O-Al_2O_3-B_2O_3 as the triangular field bounded by the compositions: (1) B_2O_3, (2) 60 mol. % B_2O_3-40 mol. % Na_2O, and (3) 60 mol. % B_2O_3-20 mol. % Na_2O-20 mol. % Al_2O_3. As Al_2O_3 replaced Na_2O, softening temperatures increased, but strain points, annealing points, thermal expansion, density, and refractive index all decreased.

Galant and Appen stated that the addition of alumina to boron-free glasses causes an increase in the index of refraction. It also causes an increase in the presence of up to 5% B_2O_3 in the glass. Beyond 6% Al_2O_3, however, the index decreases. The addition of alumina to glasses containing 11 to 17% B_2O_3 causes a sharp reduction in the index, the variation in index being in the range between 1.4817 and 1.5079. Tamura et al. found that alumina represses the volatility of B_2O_3 from alkaline earth-borate glasses. Danilov and Joffe prepared Uviol type borate glasses (high ultraviolet transmission of Hg wavelength 254 millimicrons) containing either 17.5 or 15% Al_2O_3.

Kitaigorodskii, Keshishyan et al. (1958) recommended glasses in the system $BaO-Al_2O_3-B_2O_3-SiO_2$ containing up to 30 mol. % Al_2O_3 for electrovacuum processes on the basis of their favorable coefficient of expansion, electrical resistance, and dielectric losses. Bills and Evett (1959) to the contrary, found borosilicate (and other glasses) to be undesirable construction materials in clean-surface experiments and measurements because the glass appears to decompose at temperatures above 350°C, liberating water vapor and decomposition products not readily pumped out of a vacuum system. These so-called Cabal glasses, show a definite 10% increase in the expansion coefficients. The temperature at which this occurs, the softening points, the dc conductivities, and the dielectric constant vary with the composition, suggesting that Abe's screening hypothesis is applicable and that the Cabal glasses segregate into borate and aluminate phases.

Cap (1950) found an anomalous coefficient of expansion for glasses in the system $Li_2O-Al_2O_3-B_2O_3$. Hirayama measured the electrical resistivities of glasses in the aluminoborate system, containing Group II metal oxides, as high as 10^{11} ohm-cm at 450°C, and activation energies ranging from 26 to 42 kcal/mole. The covalent nature of the group IIB aluminoborates was indicated.

21-5 LITHIUM GLASSES, PHOSPHATE GLASSES

Alumina sharply decreases the solubility of the lithium silicates in water (Williams and Simpson). Glasses in the system $Li_2O-Al_2O_3-B_2O_3$, containing 20 to 70 mol. % Al_2O_3 have expansion coefficients ranging from (63 to 101) $\times$ 10^{-7}. By addition of silica to the system, heat-resistant glasses stable against crystallization were obtained having a coefficient of 36.5×10^{-7} or less (Bezborodov and Ulazoskii). Tanigawa (1963) applied the low-expansion, thermal shock-resistant, and high-strength lithia-containing

glasses as a vitreous bond for natural petalite ($Li_2O \cdot Al_2O_3 \cdot 8SiO_2$). The flexural strength was about 8000 to 10,000 psi, and coefficient of linear expansion (5 to 10) $\times$ 10^{-7}/°C (40 to 900°C). The bodies withstood repeated cycles of water quenching from 1000°C.

Gelstharp; Takahashi et al.; and Drexler and Schütz prepared glasses in which silica was partially or wholly replaced by aluminum phosphates. The aluminum ortho-phosphate glasses are extremely corrosive toward tank block materials, and their stability in acid and alkali was found to be poor. Kumar prepared glasses in the system $PbO-B_2O_3-AlPO_4$ and in the system $Al_2O_3-B_2O_3-P_2O_5$, for which good electrical properties were obtained. Syritskaya and Yakubik prepared stable phosphate glasses in the system $ZnO-Al_2O_3-P_2O_5$.

21-6 OPTICAL GLASSES

Some examples of the application of alumina to widely different types of optical glass are shown in the following references. Sun and Callear (Eastman Kodak Co.) patented a glass consisting of the compatible fluorides, beryllium fluoride 10 to 25 mol. % and aluminum fluoride, the total of the two representing between 40 and 50 mol. %. Sun also claimed a nonborate, nonphosphate, nonsilicate optical glass of which at least 1.5 wt % is BeO, at least 15% is Al_2O_3, and the total of the two oxides is at least 30% of the glass. Loeffler prepared rare-earth glasses containing 50 wt % La_2O_3, 10% ThO_2, 3.1% BeO, 11.8% Al_2O_3, and 24.9% SiO_2 with a refractive index around 1.75 and a dispersion number (Abbe) about 50. Kreidl (1953) found that small amounts of Al_2O_3 are necessary to prevent devitrification of borosilicate optical glasses which are susceptible to discoloration on irradiation by X-rays, γ-rays, or electrons. Weissenberg and Ungemach added Al_2O_3 to flint glasses to hinder crystallization. Weissenberg and Meinert prepared a glass of about 93.5% TeO_2 and 6.5% Al_2O_3 (10 mol. % Al_2O_3). These glasses with small amounts of added fluorides (LiF, CaF_2, SrF_2, or BaF_2) had exceptionally high indices, 1.833 to 1.944. Elomanov et al. prepared clear glasses of TeO_2 containing up to 10% Al_2O_3, which were quite transparent to infrared transmission.

22 ALUMINA IN COATINGS

22-1 INTRODUCTION

Alumina finds application as coatings on metals, ceramics, and other materials as evaporated films, vitreous glazes and enamels, chemical coatings, flame-sprayed coatings, etc. There are many purposes for the coatings, among which are: decoration, prevention of corrosion, sealing against open porosity, and protection against erosion, abrasion, heat, and mechanical action. Coatings have become of prime interest in high-temperature uses in jet engines, and in rocket motors, as well as supersonic applications, in which severe limitations are placed on conventional materials. Some references to coatings include the following: Machu (1952); High-Temperature Materials Conference, Cleveland, Ohio, 1957, 3–15 (Pub. 1959); Klopp (1961); Krier (1961); Graziano (1962); Bergeron, Tennery, et al. (1962); Humenik (1963); Gangler (1963); Payne (1964); and Plunkett (1964).

Coatings are used to protect high-temperature metals against oxidation. The metals of most interest and their melting points are niobium (2420°C), molybdenum (2620°C), tantalum (2995°C), and tungsten (3410°C). Each of these melts above the temperature for aluminum oxide, yet aluminum compounds have received consideration in their protection. Among the many coating methods examined for high-temperature protection are electroplating, electrophoretic deposition, flame and plasma-arc spraying, vapor deposition, and cementation processes (the transfer of a coating and diffusion into the substrate with the formation of new compounds). An example of the last is hot-dipping in molten aluminum (Bartlett et al., 1961; Lawthers and Sama, 1961; Jefferys and Gadd, 1961).

The more successful coating systems for the refractory metals niobium, molybdenum, and tantalum, at present, appear to be complex intermetallics containing silicon, chromium, aluminum, boron, or titanium. Protection for several hours at 1650°C is realized. Tungsten silicides promise the best protection for tungsten and its alloys at 1650 to 2200°C (Gangler, 1963). The pack process (immersion of the substrate in heated, granulated, or powdered, cementation coating materials) is favored for producing reliable coatings. Metallic and ceramic coatings show less promise.

Some semicommercial coatings that involve alumina for refractory applications are prepared as follows:

General Electric GE 300. A flame-sprayed coating of alumina on chrome-plated molybdenum provides protection to about 1650°C for several hours.

General Electric GE 400. A flame-sprayed coating of alumina on niobium or tantalum is sealed by a vitreous frit, and then sintered at 1480°C. Protection to 1370°C for up to 100 hours is provided.

Mond Nickel. An alumina diffusion layer is applied over molybdenum. A coating of platinum over the alumina layer provides protection at 1180°C for several hundred hours.

Douglas investigated the emissivities of ceramic coatings on low alloy steels. Most dark refractories gave highly emissive coatings. The best white refractory materials for use in low emissive coatings at 650 and 815°C were found to be feldspar, zirconium spinel, and Treopax (National Lead Co.), with emissivities in the range 0.22 to 0.32. The comparable emissivities of aluminum oxide (in 5% water-glass as a binder) were 0.42 at 650°C and 0.35 and 815°C.

22-2 ANODIC COATINGS ON ALUMINUM

The naturally forming coating on aluminum, which effectively retards oxidation of the metal, and anodic coatings formed under applied voltages in specific solutions have been referred to previously (Section 3-6-3). It was found that the film is amorphous and porous. Certain electrolytic treatments have yielded either eta or alpha alumina, the latter having high resistance to chemical attack. Mott (1904) produced electrolytic coatings on aluminum in solutions of sodium hydrogen phosphate. Flick (1924) discovered that a dense adherent coating of oxide could be applied by making the aluminum an anode in a solution of NH_4OH or NH_2S at a potential of about 220 volts. The films had better corrosion resistance than for immersion in ammoniacal solutions containing traces of silver, nickel, or cobalt (Pacz). Bengough and Stuart developed an anodized treatment in chromic acid (3%) under gradually increasing voltage to 50 volts. This treatment reduced corrosion of aluminum-alloy airplane surfaces in marine service. Films in oxalic acid and sulfuric acids were subsequently developed. Campbell (1952) described a process in which an ac voltage was superposed on a dc voltage during electrolysis. "Hard" anodic coatings can be produced quickly, but it is difficult to obtain continuous coatings on some aluminum alloys, especially aluminum-copper (3%) and aluminum-silicon alloy (7.5%). Thick, hard, anodic coatings provide durable abrasion-resistant surfaces, used for wear resistance, and in architecture for aesthetic effects.

For best resistance to corrosion and discoloration, anodic coatings are sealed in boiling, deionized water. Raether (1959), Young (1959), and Holmes (1964) have offered explanations for the nature of anodic films and their conduction properties.

When aluminum is made the anode in certain electrolytes, such as the borates, it acquires a film which is unidirectional with respect to the passage of alternating current. The film can be subjected to potentials as high as 450 volts in some cases before its insulating ability becomes seriously impaired. Use is made of this property in electrolytic rectifiers. As the film has high dielectric strength and is extremely thin, another application is in electrolytic capacitors. The anodic film on aluminum sheets can adsorb a variety of organic dyes and mineral pigments.

22-3 GLAZES AND ENAMELS

Alumina is a component of glazes and enamels, satisfying about the same functions as in glass. The properties are modified to compensate for adjustments related to the thinness of the applied layer, as for example volatilization of constituents, reactions with the body, and factors related to the fit of the expansion characteristics to those of the body. King, Tripp, and Duckworth (1959) cite other factors: (1) the enamel at the interface must be saturated with an oxide of the metal, (2) this oxide must not be reduced by the metal when the oxide is in solution in the glass, and (3) the heavy scale that forms during the heat treatment of firing the enamel in air must be dissolved in the glass.

Lauchner and Bennett (1960) investigated the mechanically and thermally induced stresses in ceramic-coated metal composites. Considerations of the interfacial structure of the composite indicated that the glassy phase was saturated with oxides of the base metal, accompanied by residual strain gradients mainly in the interfacial zone. The maximum induced strain (thermal or mechanical) which coatings could withstand without fracture was found to be related to the residual stress.

Harrison, Moore, and Richmond (1947) developed ceramic (enamel) coatings for mild-steel aircraft exhaust systems, consisting of protective frits of high refractoriness and high thermal expansiveness mixed with refractory fillers. Alumina A-l (Alcoa) produced the most satisfactory filler for protection to about 680°C. Bennett (1950) developed ceramic coatings for ingot iron and 4140 alloy steel, good to 1090°C for 6 hours. Petzold (1958) claimed protection of ceramic coatings against lead bromide ($PbBr_2$) at 820°C, and against loss in strength to 1100°C.

Bennett (1955) predicted wide acceptance of enamels for high-temperature resistance of metals. Watts (1958) based the control of glazes, by the aid of published data, on the eutectics in the systems Na_2O-Al_2O_3-SiO_2 and PbO-Al_2O_3-SiO_2. Good glaze compositions, resistant to acid and alkali attack, contain about 12% Al_2O_3. The addition of Al_2O_3, CaO, and BaO in place of alkali is beneficial on resistance to attack (Lehmann and Bohme, 1958). Mehmel observed that glazes based on lithium aluminate have a high thermal expansion, while those based on β-eucryptite have a low thermal expansion. Budnikov and Shteinberg (1956) developed a glaze for porcelain with an Al_2O_3 content of about 11 to 29%, and small amounts of ZrO_2, CaO, MgO, Na_2O, and Fe_2O_3, the remainder being SiO_2.

Azarov and Chistova (1958) prepared vitreous compositions for enamels containing 10 to 20 mol. % each of SiO_2 and Al_2O_3, and 40 to 50% P_2O_5. Azarov and Serdyukova appraised the merits of using alumina gel to increase the stability of feldspar suspensions by making them thixotropic. Alumina introduced as feldspar opacifies enamels, but in excess, imparts a mat finish (Vielhaber, 1940). Karmaus developed enamels with melting points below 725°C containing 19 to 36% Al_2O_3. An example consists of $NaH_2PO_4 \cdot H_2O$ 66.3%, $NaNO_3$ 9.9%, H_3BO_3 21.2%, $Al_2O_3 \cdot 3H_2O$ 42.8%, TiO_2 7.8%, smelted at 1300°C. Every (1949) developed superopaque enamels of the ZrO_2 type containing 20 to 25% Al_2O_3. Alumina is necessary to opacify these enamels (Ref. Anon. Zircon needs alumina). Both require cryolite as a flux and as an opacifying agent (Maerker, 1957). Pavlushkin, Zhuralev, and Eitigon (1961) developed a satisfactory glaze for enameling decorative alumina ware, consisting of 50 wt % PbO, 25% SiO_2 10% B_2O_3, 10% K_2O, 2% Li_2O, and 8% Na_2O. Various inorganic pigments could be used to give color in melts at 950°C for 2.5 to 3 hours. Amberg and Harrison (Exolon) developed an enamel for difficultly wettable substances such as graphite, carbon, silicon carbide, fused alumina, and their mixtures.

Farnham and Burdick presented data on enamels for aluminum. Enamels for aluminum may contain substantial amounts of alumina. Donahey (1953) prepared an enamel containing 18 to 20 mol. % Al_2O_3, 8 to 13% Li_2O, 25 to 30% Na_2O, 24 to 28% P_2O_5, 2 to 12% B_2O_3, 4 to 6% of an easily reduced oxide from the group CuO, CoO, NiO, and CdO, and 2 to 15% $(NaF)_2$. Bezborodov, Mazo et al.; and Azarov and Christova (1963) investigated the phosphate enamels for aluminum. Hüttig (1961) prepared the base layer of aluminum and its alloys for improved adhesion of enamel layers by treating the aluminum surface with boiling water containing some alkali (about 0.01% KOH for an aluminum alloy containing 1% Mg, 0.04% Fe, 0.08% Si, and 0.01% Cu); but 0.005% LiOH is recommended for cast aluminum (99.8%).

Devrup (du Pont) developed a vitreous enamel suitable for use on aluminum and its alloys. Andrus (Croname, Inc.) produced an anodic film on aluminum electrochemically, coated the film with a fusible vitreous material, and heated to fuse the coating.

22-4 FLAME-SPRAYED COATINGS

Dickinson (1954) described a flame-spraying technique for applying refractory coatings of aluminum oxide onto aluminum, magnesium, iron, steel, or titanium surfaces. Fine, refractory powder is blown through a flame which melts it and causes it to stick to the article if the coefficient of thermal expansion of the coating matches the base material. Wheildon (Norton Co.) in 1955 demonstrated the process of spray-coating alumina (or zirconia) onto an iron base, comprising heating a rod of the stable oxide in a flame hot enough to melt it (oxyacetylene), which instantaneously atomized the molten oxide to form discrete particles. At the same time, the molten particles were projected with a blast of gas onto the heated base (about 430°C) before they could congeal. The coating, designated "Rokide A" in the case of alumina, could be applied to an iron backing to a thickness of about 0.1 inch, and although the expansion characteristics did not match closely, there was enough flexibility to partially compensate the strain. The coatings are porous, and do not give complete protection against oxidation, but even at a thickness of only 0.025 inch, a temperature differential of about 540°C is obtained when the face is subjected to a temperature of 1980°C. Ault (1959) specified that the aluminum oxide rod should have 8 to 40% open pores and a flexural strength above 2000 psi, and should contain at least 95 wt % Al_2O_3. Curtis (1959), Westerholm (1959), Meyer (1960), and Vasilos and Harris (1962) investigated the process of flame-spraying.

Bradstreet (Armour Research Foundation) in 1959 described a variation of the flame-spraying process in which a powder of crystalline particles of an oxide (refractory at 1200°C) is injected into a gaseous stream being fed into a continuous flame-generating device. A suitable powder consists of 200 parts by weight of 60-mesh alumina and 20 parts MgP_2O_7 with 5 parts $Al(OAc)_3$. Improved results were obtained when the surface of the base (steel) was pretreated with a solution of H_3PO_4. The coatings were claimed to be useful for skid-resistance for metal gratings and for refractory surfaces of brick.

Eisenberg (1964) injected into one of the gas streams entering the flame a solid powdered mix of the refractory oxide and a fluoride, and deposited the mixture on the object to be coated. Another variation, by Bradstreet and Griffith (1956), comprised spraying a solution, which is thermally decomposable to the refractory metal oxide, into the flame. B. W. King (1962) concluded that cermets of the graded type can be made by flame-spraying alternate layers of an oxide and a metal. Most metals, as well as graphite, glass, and various ceramics could be spray-coated. Aves and Hart (1962) obtained improved high-temperature protection by flame-spraying coatings of alternating metal and ceramic layers, obtained by impregnating the coatings with vaporization or sublimation materials such as tin, zinc, nickel, cobalt, halides, silicone, furan, phenolic, or epoxy resins, which upon heating pass from the solid to the gaseous state without significant formation of liquid. The metallic layers were selected from the group boron, niobium, iridium, molybdenum, tantalum, and tungsten, alternating with ceramic layers of metal oxides (Al_2O_3) or carbides.

The plasma arc is a powerful tool for obtaining temperatures as high as 11,000°C, which can be sustained for periods of an hour. Hedger and Hale (1961) described the use of the plasma torch for the spheroidization of alumina powders. Orbach (1962) cited various ceramic uses for the plasma jet. Stackhouse, Guidotti, and Yenni (1958) substituted the plasma torch for the oxyacetylene torch in flame spraying. Ten basic coatings that could be melted and deposited as very dense pure coatings were Ta, Pd, Pt, Mo, W, Al_2O_3, ZrO_2, and three combinations of W with ZrO_2, CrO_3, and Al_2O_3. Kramer and Levinstein (1961) reported that all coatings from the group Al_2O_3, Cr_2O_3, ZrB_2, and the carbides can be built up to thick coatings on a graphite substrate. The materials deposit in substantially the same form as the raw material. Westerholm, McGeary, Huff, and Wolff (1961) compared the principal flame and plasma spray methods. Sound coatings could be obtained with alumina by these methods. H. Meyer (1962) drew attention to the peculiarity that sprayed deposits of alpha alumina feed deposited as gamma alumina from plasma temperatures as high as 16,000°C. Huffadine and Hollands (Plessey Company, Ltd, 1962) were able to convert the gamma alumina (density 3.55 g/ml) into alpha alumina (density 3.95) by flame-spraying onto graphite at 1500°C. By transferring the assembly to a furnace at 1250°C, the graphite was oxidized, leaving the ceramic form. Meyer (1963) determined the extent of the conversion by adding chromia. Chromia dissolves up to 8 mol. % in corundum with a red color, but in gamma alumina it is green. Huffadine (1964) patented the chromia method of detecting the phase change in aluminum oxide.

Ault and Milligan (1959) flame-sprayed alumina radomes over a removable form. The forms were claimed to be highly refractory, dense, strong, and impervious. Owing to the slight shrinkage in the final firing and the strength of the original flame-sprayed shape, distortion and cracking were reduced to a minimum. Huffadine and Thomas commented on the advantages of flame-spraying as a means of fabricating dense bodies of alumina. By this method, porosities of only 8 to 12% at 0.125-inch thickness were attained, with an overall coating efficiency of 10 to 50%. Higher efficiencies and lower porosity levels were attained with metals than with ceramics. Spraying with alumina presents special problems because of the change from the low-temperature forms to stable alpha alumina. The normal spraying results in deposits of gamma alumina with about 10% porosity. At about 1000°C, these deposits shrink to alpha alumina. By keeping the temperature above 1000°C during spraying, cracking was markedly decreased. On parts of about 6 inches in diameter, porosity was reduced to 3.9% at a temperature above 1400°C. The average crystallites were apparently flattened particles 4.6 by 2.9 microns. At the higher cited temperatures, grain growth was considerable, 12.2 by 4.2 microns.

175

Dittrich (1965) flame-sprayed self-bonding nickel aluminide coatings, which could subsequently be used to provide good adhesion for oxide coatings. The powder consisted of an aluminum core, clad with a nickel sheath. An exothermic reaction of these in the heating zone of the flame-spray gun forms nickel aluminide.

Montgomery et al. (Ohio State Research Foundation) flame-sprayed a sintered mixture of a metal from the group aluminum, iron, nickel, and cobalt with a metal oxide from the group alumina, magnesia, and zirconia. During spraying, the particles were heated to about $1150°C$.

22-5 PAINTED, CAST, OR TROWELED COATINGS

The method of painting ceramic pastes on metals to form protective coatings is exemplified by the following procedures. Kleinert coated graphite or amorphous carbon electrodes with a slurry containing alumina, chromia, etc., stabilized zirconia, and silicone binders. The purpose was to prevent rapid burning of the electrodes. A protective coating for oxidizable metals may contain 20 to 50 wt % sodium metaborate, 20 to 40% plastic clay, and 5 to 40% alumina (Govan and Govan, 1957). Metallic castings and other shapes may be protected against excessive oxidation and loss of surface details in heat treatment by coating 1/64 to 1/32 inch-thick layers composed of 7 to 16 parts Al_2O_3, 29 to 60 parts SiO_2, and 3 to 7.5 parts sodium oxide. The coatings are applied by spraying, painting, or dipping from suspensions in water or glycerine (Cowles and Stedman, 1962). Satterfield used suspensions containing 60% alumina in resin to form cup-shaped retainers for holding base-metal parts such as combustion-engine exhaust valves, while cladding with hard metal facings.

Blocker, Hauck et al. (1960) claimed that a metal-reinforced ceramic coating withstands temperatures of $1930°C$ when the refractory medium is phosphate-bonded alumina aggregate against stainless steel. Eubanks and Moore (1959) applied coatings of various refractory oxides with aluminum phosphate in thicknesses of 0.002 to 0.20 inch by spraying, brushing, or dipping. The heavier coatings were reinforced with expanded SAE 1020 carbon steel, spot-welded to the base material. Many phosphate-bonded coatings are possible, as for example, a 50:50 mixture of alumina and zircon, bonded with phosphoric acid and colloidal silica (Noble, Bradstreet, and Rechter, 1961). The phosphate-bonded coatings are porous, and do not protect against oxidation at high temperatures.

Alexander (1963) specified conditions for bonding a refractory oxide coating to a cermet.

22-6 ELECTROLYTIC COATINGS

Grazen codeposited up to 50% by volume finely divided particles of aluminum oxide dispersed in nickel, copper, silver, cadmium and other metals by an electroplating process. The tough coatings greatly improved the life of gages, cutting tools, and other wear surfaces.

Ziegler and Ruthard (1958) produced coats of aluminum on copper, superficially converted to aluminum oxide, by electrolytic deposition from an organic solution containing an alkali metal halide, as for example, $NaF \cdot [Al(Et)_3]_2$.

22-7 EVAPORATED COATINGS

Deutscher (1961) deposited evaporated aluminum isopropylate $(235°C)$ in contact with nitrogen, ammonia and water vapor onto a ceramic substrate to obtain a coating of aluminum oxide. Oxley, Browning, Veigel, and Blocher (1962) vapor-deposited aluminum oxide on nuclear-fuel particles. Day and Stavrolakis (1962) evaporated aluminates of beryllium, strontium, yttrium, or lanthanum on molybdenum to form a coating to hold the cations on the surface of the article; they then oxidized the cations under controlled conditions to form a suboxide layer, and applied an oxide with heat to form a bond. Reichelt and Schmeken developed a heating element and holder for the vaporizing coating method consisting of 50 to 80% (wt) of a metal conductor (W, Ta, Mo, or Cr) with the remainder a metal oxide of Al, Zr, Be, Th, or Mg, or a spinel such as $MgO \cdot Al_2O_3$. These compositions had good wettability for the satisfactory evaporation of Fe, Co, Ni, Pt, Pd, and Al. In one example, 20% alumina and 80% molybdenum were mixed with an organic binder, were pressed into a shape having a recess to hold the material to be evaporated, and were sintered at a temperature below the melting point of at least one of the elements.

DaSilva and White (1962) found that aluminum oxide films prepared by the evaporation of aluminum at an oxygen pressure of 10^{-3} mm Hg had a dielectric constant of about 6 in vacuum or in nitrogen. On exposure to oxygen at 1 atm $(23°C)$, the dielectric constant increased to about 12, comparable to that of bulk Al_2O_3. It was inferred that the original film was a suboxide. Schmidt, Brodt, and Lampatzer found that coating metal parts, especially steel sheet and strip alloyed with silicon, with the oxides or hydroxides of alumina in an inert or protective atmosphere for several hours at 900 to $1050°C$ resulted in a surface coating of from 2 to 3.5 microns in thickness.

22-8 DIP COATINGS, CEMENTATION COATINGS

A very effective method for prevention of oxidation of steel and the refractory metals is by immersion in a bath of molten aluminum. The surface of the aluminum is protected from rapid oxidation by halide salt fluxes. Aluminum baths are operated in the range 650 to $1150°C$ and for various time periods to allow for some diffusion of the aluminum into the surface. Subsequent treatments in an inert atmosphere or in vacuum may be applied to increase the amount of diffusion. With aluminum these periods may extend from 0.5 to 2 hours at 930 to $1150°C$. Flame-sprayed coatings may be given a dip treatment in molten aluminum to seal the pores (Blanchard, 1954, 1955).

22-9 COATINGS ON ALUMINA AND OTHER CERAMIC BASES

The coating of ceramic alumina with metals is described in Section 25 as a metallizing prerequisite for forming seals. Different methods are presented here. Cline and Wulff

(1951) coated ceramic particles (alumina) with molybdenum deposited by hydrogen-reduction of molybdenum pentachloride. Nickel was deposited by decomposition of nickel carbonyl. The particles in either case must be kept in motion by gas flow or by mechanical agitation during the coating process. In the presence of titanium carbide, the decomposition of nickel carbonyl proceeds abnormally, leading to the formation of carbon. In another example, Luks (1961) applied a metallic coating to a refractory nonmetallic body by forming a molten mixture of a manganese component and a silica-containing glass capable of forming a flux with the manganese component at 700 to 1300°C. A ground frit prepared from this was mixed with nickel, cobalt, iron, or mixtures of them with molybdenum or tungsten. The mix was applied to the body and fired. Kirchner, Gruver, and Walker immersed high-purity sintered alumina articles in chromia-fluoride packing materials during high-temperature reheating to develop low-expansion, solid-solution surface layers which increased the strength. N. V. Philips (1965) coated high-purity, high-density alumina articles with a mixture of fine refractory metal (Pt, Mo, W, or Nb) and mixed oxides (CaO, Al_2O_3, or SiO_2) and heated at 1200 to 1600°C to form a refractory coating.

Methods for coating abrasive grain, so as to obtain firmer bonds in grinding wheels, are represented by the following: Amberg and Harrison (Exolon Co) produced ceramic coatings on difficultly wettable fused alumina by fusing on the surface of the grains a silicate melt, having a wetting agent present from the group: molybdenum, tungsten, uranium, vanadium, and their compounds. Walton and Bassett coated refractory grain with a glass of low melting point, and including molybdenum oxide to wet the surface of the grain. Masin and Tull (Monsanto Chemical Company) coated fused alumina grain with a substantially water-insoluble, unfused, unvitrified film, which bonds finely divided inert material to the grain surfaces. Aluminum, iron, and magnesium metaphosphates may be used as the film materials. The coated grains are calcined at a low temperature (550°C).

Robson and Tucker (1955) described a method for coating concrete articles with two vitreous coatings. The first coat, to fill the voids, is a glaze frit with 20 to 60 wt % of a refractory material (alumina) added. The frit contains 44.3% flint, 37.3% dehydrated borax, and 18.4% limestone. Alternatively it contains 36.1% flint, 30.4% borax, and 33.5% litharge. The firing range is from 540 to 820°C. A cover glaze coat containing pigments may then be applied.

Barnes and McCandless (1948) produced glossy surfaces on sapphire by applying a thin film of any oxide fulfilling one of the following requirements: having a melting point lower than sapphire; forming a eutectic with sapphire; forming a low-melting peritectic; and forming a solid solution. The film was then heat-bonded to the surface of the sapphire. Single-crystal silicon could be deposited epitaxially on sapphire by a process of reducing $SiCl_4$ with hydrogen at high temperature. The early stages of growth do not initiate the formation of monolayers but of very small nuclei. The silicon deposits are p-type and of high

resistivity. Novak (1961) observed that alumina could be gasplated on quartz or silica articles by heating them in an atmosphere containing the vapor of a trialkylaluminum compound or a trialcoholate above the decomposition temperature of the aluminum compound. The plating pressure is preferably 7 to 7.5 mm; the aluminum compound vapor may be entrained in 1 to 6 times as much carrier gas (air or nitrogen) at about 70 to 600 cc/minute. Thin adherent coatings were obtained from aluminum ethoxide vaporized at 200°C, and decomposed at 800°C; or from aluminum isopropylate vaporized at 177°C, and decomposed at 370°C with nitrogen and water vapor carrier. The process is useful for lining quartz tubes or crucibles to be used in handling semiconductor materials.

Attempts to evaporate quartz alone and deposit it upon a selective surface have usually been unsuccessful, owing to the difficulty in evaporating quartz. The difficulty is overcome by coating the entire quartz surface with aluminum (Ogle and Weinrich). The process of vaporizing aluminum onto glass and other surfaces for reflectors and astronomical mirrors has been described by K. Rose (1948).

Long and Furth (Solar Aircraft Co.) claimed the method of providing the surface of a metal, glass or ceramic article with a cermet coating having both a metal phase and a ceramic phase. The cermet coating consisted of 60 to 70 wt % of aluminum or aluminum base alloys and 30 to 40% of a smelted and fritted product consisting of $NaCaB_5O_9 \cdot 8H_2O$ (Gerstley borate) and Li_2O. The surface was heated at least to 670°C. Such coatings are useful on alloys such as Timken 1722-A(S) and General Electric Chromaloy, which are tempered at 677°C. The coating can also be used to seal woven glass fibers.

22-10 ALUMINA COATINGS FOR ELECTRICAL INSULATION

Alumina is used in compositions to insulate or fix in place electrical windings or heater elements. A recent example of this type of coating was disclosed by Larsh and Coppock (1962) for a slurry composition of density 35 to 60° Be' containing 50 to 75 parts of glass frit, fused at 760 to 870°C, 1 to 10 parts silica, and 1 to 10 parts tabular alumina, suspended in an organic liquid boiling at 10 to 38°C. The mixture is flowed over the electrical elements to be sealed, the vehicle is allowed to evaporate, and the part is then heated to the fusion point of the lower-fusing frit.

Junker (1962) used mixtures of alumina and a product which lowered the melting point, such as titania or silica, applied in the melted state by means of a sprayer, to insulate the electrical windings of copper-furnace bobbins. Marbaker (1945) described coatings for insulating wire-wound resistors. Alston and Strickland (1962) described the process of preparing insulating coatings of alumina on metals, in particular for coating the heaters for cathodes in electron tubes.

Vondracek (1963, 1964) examined the requirements for electrical insulation in space environments. Materials in which alumina serves to insulate include: paper coatings in motor slot liners, potting compounds, cements and adhesives, and anodized aluminum conductors.

22-11 ALUMINA COATINGS BY SPUTTERING

Sputtering is the nonevaporative, cold transfer of materials by radiofrequency, minute-particle bombardment of a surface in an evacuated assembly. Although sputtering had long been applied to the transfer of metal surfaces, it was not until recently that a method for applying it to such insulators as aluminum oxide was successful. The main advantage of sputtering is that it can be applied to thermally unstable substrates. Frieser found that two phases of deposited alumina could be prepared: either an amorphous HF-soluble form or a polycrystalline form of γ-Al_2O_3 which was considerably more HF-resistant. Dugdale and Ford observed that sputtering from thermal spikes contributes strongly to the sputtering mechanism. Single-phase alumina is etched under these conditions in a manner similar to thermal etching.

23 ALUMINA IN CERMETS AND POWDER METALLURGY

23-1 INTRODUCTION

The word "cermet" was coined by A. L. Berger, Wright Air Field to designate structures composed of an intimate mixture of a metal and a ceramic. The cermets were originally investigated in the hope that the mixture of metals and ceramics would develop the most desired properties of the two materials, that is, the refractoriness and chemical inertness of the ceramics coupled with the ductility and impact resistance of the metals. The prime objective was a suitable structure for airplane power plants. Although this has not been realized very completely, the cermets have developed certain uses. Alumina has figured conspicuously in the development of cermets as the ceramic phase in various combinations with refractory metals as well as with nonrefractory metals, including aluminum.

Initially, there was a lack of basic data for many of the chemical and physical properties involved in the general metal-ceramic field (Westbrook, 1952). Some attempt was made to explain the interfacial reactions between various metals with likely refractory oxides (Kingery, 1953, Economos and Kingery, 1953). Various types of cermets: oxide base, carbide, boride, and intermetallic types, were considered for high-temperature applications (J. T. Norton, 1956). Murray (1960) pointed out that oxide systems show little gain in strength, compared with that of the oxide itself, but do show an intermediate improvement in thermal shock resistance. The carbide cermets appeared to be adequate for many high-temperature applications in thermal-shock resistance, but poor resistance to mechanical shock was their main weakness. Tret'yachenko (1963) assessed the factors affecting the resistance of brittle cermets to thermal fatigue fracture. The sensitivity could be evaluated in terms of two criteria, one for nonsteady-state heating and the other for steady-state heating. J. T. Norton (Ref. Hehemann and Ault, 1959, Chapter 8 page 119) stated that cermets have found application in two important areas (1) tools as cemented carbides, wire-drawing dies, rock-drills, and cold heading tools, and (2) structural parts for high-temperature service in which resistance to oxidation is the important criterion. The future of cermets in the latter uses was questioned, however. Lavendel and Goetzel

(Ref. Hehemann and Ault, page 149) stated that cermets are quite brittle, have poor impact resistance unnotched (only 4 to 5 in.-lb) for a cobalt-alloy-infiltrated titanium carbide. Wambolt and Redmond (same reference, page 125) found no failure for a titanium carbide composition (K163B1) after 550 cycles of heat shock at 1200°C. A chromium-alumina cermet failed after 234 cycles at 1100°C.

Maziliauskas (1960) described various developments in the preparation of cermets. *Cermets*, eds. Tinklepaugh and Crandall (1960), is an attempt to present a cross-section of information published in the United States. Shevlin (1957) and Frechette (1956) discussed applications and uses of cermets.

23-2 CHROMIUM-ALUMINA CERMETS

In 1949, Blackburn, Shevlin, and Lowers investigated the mechanism of wetting and adherence of sintered alumina by the metals: nickel, cobalt, iron, chromium, and chromium-boron. The objective was to prepare well-bonded cermets as a possible structural material for jet engines. The tests consisted of firing the metals on sintered alumina tiles in reducing and oxidizing atmospheres in the range 1480 to 1930°C. It was concluded from the tests that there is no inherent tendency of the cited metals to wet alumina below 1650°C. Wetting was observed in the case of chromium in a slightly oxidizing atmosphere by the formation of chromic oxide, which readily forms solid solution with alumina. It was concluded that wetting and adherence may be developed through thermochemical reactions involving glass-free interfacial compounds in which the oxide of the metal combines with alumina. This precludes a strongly reducing atmosphere. Consolidation depends on the amount of oxide formed, its rate of formation, and its rate of reaction with alumina.

Blackburn and Shevlin (1951) found an optimum composition containing 30% electrolytic chromium, dispersed in tabular alumina, both components being finer than 10 microns in particle size. Test bars were pressed from mixtures of the two components that had been dry-pressed with organic lubricants ("Ceremul C," paraffin, "Sterotex," or "Carbowax") at pressures of about 55,000 psi. The forms, designed for various physical tests, were fired at 1705°C on a 30-hr schedule in a flowing atmosphere of hydrogen and water vapor, controlled by the dissociation of hydrated alumina. The following data characterized the product:

Fired shrinkage 13–14%
App. porosity < 0.5%
Sp.G. app. 4.60–4.65, true 4.68–4.72
Thermal expansion (25–800°C) 8.65×10^{-6}
                 (25–1315°C) 9.45×10^{-6}
Heat transfer–Slightly less than for sintered alumina
Thermal conductivity–0.023 (cgs)
Resistance to thermal shock–excellent, cycled from 1315°C
Resistance to oxidation–excellent at 1510°C
Hardness 1100–1200 (Vickers pyramid number)
Modulus of elasticity 5.23×10^7
Impact resistance at 30°C   1.05 in.-lb
                            18.95 in.-lb/in.^{2}
Compressive strength at 30°C   320,000 psi
Bending strength
          30°C   55,000 psi
          870   43,100
          1090   32,800
          1315   24,400
Tensile strength
          30°C   35,000
          870   21,500
          1090   18,400
          1315   14,000

It was concluded that the strengths may be adequate for jet-engine parts at above 980°C. The stress-rupture curves were higher than those for any known high-temperature alloy, and fell off less abruptly with rising temperature. The strength was not adversely affected by a 100-hr test of oxidation in air at 1510°C. The resistance to thermal shock was greater than that of sintered alumina in cycling from 1316°C in air. The cermet was found to be extremely hard, but brittle, and the resistance to mechanical shock was very low.

Shevlin (1952) reported on a cermet composition containing 72% chromium–28% alumina, which also appeared very promising for aircraft turbine applications in comparison with the then potentially useful carbide-base cermets. A thermal shock record of 540 cycles of simulated burner blow-out, and a 1000-hr stress-rupture life of 15,000 psi in air at 980°C were achieved.

Conant and Gillies (1955), Kiyoura and Sata (1958), Onitsch-Modl (1955), and A. E. S. White et al. (1956) confirm the general findings on the chromium-alumina cermet. N. C. Moore (Plessey Company, Ltd, 1957) patented a combination of fused alumina, 10 to 70% metallic chromium, and up to 5% Cr_2O_3, obtained by mixing alumina, chromia, and carbon, and heating so that most of the chromia was reduced to the metal. Matsumoto,

Yamauchi, and Nishiyama (1957) observed that the presence of chromium oxide in alumina aided in the diffusion and adherence between chromium and the alumina. The adhesion took place at a relatively low temperature. For optimum results, the sintering shrinkage of chromium should be as close as possible to that of alumina containing Cr_2O_3. This was accomplished by varying the amount of Cr_2O_3 added, changing the particle size distribution and compacting pressure, and by use of other additives.

Improvements were claimed by introducing molybdenum and chromium as the metal phase (Huffadine, Longland, and Moore), molybdenum, chromium, and titania, in addition to alumina (Frangos, 1958), and chromium and carbon (1 to 16 wt %) in alumina (Klingler and Dawihl, 1962). Conant (1957) claimed a cermet of improved thermal shock resistance and with very high strength at elevated temperatures by the presence of a small amount of a partially reducible oxide (molybdenum or tungsten). The ceramic component, alumina, also may contain small amounts of another oxide (TiO_2, Nb_2O_5, or Ta_2O_5). One such mixture consists of 58 wt % Cr, 19% Mo, 1% Ta_2O_5, and 21% Al_2O_3. Conant (1959) coated chromium-alumina thermocouple protection tubes against molten metal penetration by applying a continuous metal oxide coating on the surface, consisting of alumina and chromia. The coating should have an electrical resistance of at least 100 megohms for the purpose.

Sata and Kiyoura (1959) hot-pressed chromium-alumina cermets containing about 20 to 70 vol % Cr at 1400 to 1470°C under a pressure of 2100 psi for 20 to 30 minutes in argon atmosphere surrounding graphite molds. An increase in the chromium content lowered the pressing temperature. The bending strength ranged from 78,000 to 100,000 psi for specimens 4 × 4 × 15 mm. Compared with cermets prepared in hydrogen, these had smaller corundum crystals (1 to 3 microns) and a rougher boundary between the Al_2O_3 and Cr phases.

Schatt and Schulze found that the structure of the fracture of chromium-alumina cermets indicated intercrystalline fractures, determined by electron microscopy.

23-3 (IRON, NICKEL, COBALT)–ALUMINA CERMETS

Jellinghaus and Shuin (1958) prepared cermets of iron or iron alloys with alumina from partially oxidized Fe-Al alloys with the addition of iron carbonyl and other metal powders. Up to 22 vol % Al_2O_3, the density is 95% of theoretical; at higher contents (around 50 vol %) the density falls off, but with further addition of alumina, it again rises to 95%. The material can be rolled with alumina contents as high as 22 vol %. Recovery and recrystallization take place, but grain growth is inhibited by the alumina. The ultimate tensile strength passes through a maximum between 3 and 5 wt % Al_2O_3. The oxide-rich cermets have poor impact strength, but resist scaling at elevated temperatures. The bending strength for an alloy containing 50.7 wt % Al_2O_3 at 800°C was 14,200 psi for at least 100 hours, and at 1000°C it was 7100 psi; the bending strength for an alloy containing 68% Al_2O_3 was 17,000 psi at 800°C, and at 1000°C it was 11,400 psi. This may be compared with a chromium-bearing cermet (35.3 wt % Al_2O_3) which sustained a bending load of 26,900 psi for 100 hours at 800°C. Siede and Metcalfe investigated the fineness of distribution of iron cermets by metallography, thermal analysis, and mechanical testing. Gatti (1959) used powder metallurgy techniques to produce iron-alumina cermets. He found that alloys containing up to 16 wt % Al_2O_3 are ductile. These structures have a higher yield stress and an improved creep resistance over that of pure iron.

Eisenkolb and Richter found no practical applications for mixtures of iron and flint. A small addition of very fine iron powder to alumina, however, improved the bending strength after sintering above the melting point of the iron. Larger additions of iron caused consolidation of the iron particles into drops. Matsumoto, Yamauchi et al. (1958) found that replacement of part of the alumina by iron oxide, lime, magnesia, manganese dioxide, chromia, titania, or silica reduced the contact angle, hence improved adherence between the iron and the alumina.

R. F. Thomson (General Motors Corp., 1958) composed a sintered-powder, metal piston ring for an internal combustion engine, characterized by high wear resistance and formed of a sintered and worked mix of approximately 0.3 to 3% carbon, 1.5 to 6% crystalline alumina, and the remainder substantially all iron.

Baxter and Roberts concluded from comparison of the properties of the bonded carbides and of the metal-oxide cermets that a metal phase was needed to wet the nonmetal as well as to form a strong bond. The binary nickel (and cobalt) alloys were investigated but none was found to wet alumina effectively, although several formed high-strength bonds. Sintered compacts of the continuous alumina phase type were found to have increased thermal shock resistance but poorer strength compared with sintered alumina. Kiyoura and Sata (1956) found that a sintering temperature of 1670°C was necessary for various nickel concentrations in hydrogen. The oxide phase was continuous for metal concentrations to 70%, indicating that the properties of oxide cermets depend mainly on the properties of the oxide phase. This was believed to explain the superiority of cermets of the alumina series in mechanical strength and thermal shock resistance.

Cremens and Grant pointed out the necessity of obtaining very fine nickel powders (1 to 2 microns) and alumina (0.2 to 0.02 micron) in blended extrusions to produce strong cermets containing up to 45 vol % Al_2O_3. Creep-rupture tests at 700 to 980°C confirmed other investigations that still further strength improvements can be achieved if finer interparticle spacing is attained. Zwilsky and Grant attributed the increased stability of dispersion-strengthened composites to their resistance to recrystallization up to temperatures near the melting point of the base material. Dromsky et al. observed the growth of aluminum oxide crystals in a nickel matrix in the range 1170 to 1350°C. The instability of the dispersed alumina was independent of the crystal structure of the alumina. The activation energy for the growth of the dispersed alumina was 84.7 kcal. It was believed that the rate-controlling mechanism is the solution of aluminum and oxygen into the nickel lattice, rather than diffusion.

Mayfield (1960) prepared a cutting tool composition with a continuous matrix of 80 wt % alpha alumina and a discontinuous metallic phase consisting of 0.1 to 7% nickel.

Burney (1958) attempted to reinforce an 80 Ni-20 Cr alloy by the addition of reactive titania and refractory alumina. Armstrong, Chaklader, and Clarke (1962) measured the interfacial energy of Ni-Ti and Ni-Cr powder mixtures on sapphire surfaces by the sessile drop method at 1500°C under a pressure of 10^{-4} mm Hg. Two basic mechanisms were believed to operate: (1) a segregation of solute metal at the interface, and (2) an interfacial reaction. X-ray diffraction indicated the formation of a-Ti_2O_3.

Crandall and West (1956) used DTA, X-ray, and weight change in conjunction to derive the supposed mechanism by which oxidation proceeds in porous compacts of cobalt and alumina.

23-4 ALUMINUM-ALUMINA ALLOYS

The powder metallurgy of aluminum-alumina alloys was originated by Aluminum Industrie Actien Gesellschaft (AIAG) of Zurich, Switzerland, as a result of unusually high strengths observed in products wrought from milled aluminum powders. The product was called Sintered Aluminum Product (SAP). Alcoa Research Laboratories became interested in the product in 1948, and have designated the process Aluminum Powder Metallurgy (APM), and the alloys developed at present, XAP. While the properties of these materials are predominantly metallic, the materials are of interest ceramically as representing end members in the metal-ceramic series.

The high properties at elevated temperatures are due primarily to the dispersion-hardening effect of the aluminum oxide (Irmann, 1952; Lyle, 1952; Fisher, Hart, and Pry, 1953; Nock, 1953; Grant and Preston, 1957; Cremens, Bryan, and Grant, 1958). A Powder Metallurgy Symposium on the subject was held in London in 1954. During an International Conference held in New York in 1960 (Leszynski), progress in the development of the SAP process was discussed by Bloch and Hug. Bloch (1961) and Towner (1961) have provided comprehensive bibliographies on the subject.

The original products were prepared from flake aluminum powder (2-dimensional particles). Alcoa has developed four principal alloys of this type, three of which conform with SAP. A fifth alloy (XAP005) is a low-iron type for thin-wall tubing for atomic power applications. Types have also been developed from atomized aluminum alloy powders (3-dimensional particles). Alcoa types M486, M643, and M457 contain 15% $FeAl_3$, 29% $FeNiAl_9$, and 3% $FeNiAl_9$, respectively, as the dispersed phase.

Data for some of the XAP alloys are shown in Table 15.

Table 15

Properties of Aluminum-Alumina Alloys[1]

	XAP001 (M257)	XAP002 (SAP 930)	XAP003 (SAP 895)	XAP004 (SAP 865)	XAP005 —
Principal Components (%)					
Al_2O_3	6.0	8.0	11.0	14.0	6.0
Fe	0.20	0.28	0.27	0.29	0.05
C	0.40	0.20	–	0.40	0.40
Si	0.08	0.10	0.10	0.10	0.08
Cu	0.00	0.01	0.01	0.01	0.00
Sp.G.	2.74	2.74	2.75	2.77	2.74
Thermal Conductivity (cgs)					
24°C	0.43	0.40	0.36	0.32	0.43
200	0.44	0.43	0.39	0.36	0.44
400	0.45	0.44	0.42	0.39	0.45
540	0.45	0.44	0.42	0.40	0.45
Coefficient of Thermal Expansion ($°C \times 10^{-6}$)					
20-100	21.6	21.2	20.7	20.2	21.6
20-300	23.6	23.2	22.5	21.8	23.6
20-450	25.0	24.8	23.4	22.5	25.0
Specific Heat (cal/g)					
20°C	0.213	0.212	0.211	0.210	0.213
100	0.225	0.224	0.223	0.223	0.225
Typical Mechanical Properties (psi)[1] **Longitudinal Extrusions[2,3]**					
20°C					
Ultimate, psi	37,000	36,000		54,000	37,000
Yield, psi	27,000	21,000	–	38,000	27,000
Elongation, %	13	12		5	13
538°C					
Ultimate, psi	6,000	6,000		8,000	6,000
Yield, psi	5,000	5,000	–	7,000	5,000
Elongation, %	6	4		2	6

[1] Source-Alcoa Green Letter "Alcoa's Aluminum Powder Metallurgy (APM) Alloys" September 1962.
[2] These properties are independent of time at temperature. Properties of specimens heated as high as 538°C revert to room-temperature values on cooling to room temperature.
[3] Extrusions measured transverse to direction of extrusion are about 13 to 15% lower in strength at 20°C.

The strengths of the flake-powder alloys decrease with increasing temperature but revert to the original room temperature values following heating at temperatures as high as 538°C. The maximum permissible soaking times at temperature of the atomized-particle type, M486, that will not cause a loss in strength, is 6 hours at 441°C, 1 hour at 468°C, and 0.1 hour at 510°C. The strengths longitudinal to extrusion are higher than those transverse.

The products can be formed by compacting, extruding, forging, impacting, rolling, drawing, etc. The applications are primarily in the aircraft, missile, automotive, and atomic energy fields.

Langrod (1959) investigated the compatibility of the alloy APM 257 with uranium dioxide; Friske appraised the general applicability for atomic energy uses. Lyle (1962) described the properties of commercial alloys. Modl-Onitsch (1961) discussed the necessary stability of the phases present to provide strength at high temperatures. Sherby (1962) compared the creep resistance of these alloys with other polycrystalline solids, and concluded that the creep strength is probably only of secondary importance. Myers and Sherby (1962) showed no apparent discontinuity in compressive creep behavior in the range 540 to 735°C (at the melting point of aluminum). The microstructure of the alloys is stated to be a honeycomb of alumina filled with and interpenetrated by aluminum.

Goetzel stated that the percent by volume of the dispersed phase, the size of the dispersed particles, and their spacing are the most important variables. Where the dispersed phase is the oxide coating (eta alumina), its size is near 0.02 micron.

The addition of certain alloying elements to both types of APM alloys can develop desirable properties. Nachtman (1958, 1960) claimed a metallurgical composition of aluminum or its alloys containing strengthening metals (iron, molybdenum, vanadium, titanium, tantalum, niobium, chromium, manganese, nickel, cobalt, tungsten, boron, beryllium, thorium, iridium, or the rare-earth elements) with aluminum oxide or mixtures with various aluminates. The composition was sintered at a temperature between 370°C and the melting point of aluminum. Lenel discussed the dispersed phases in powder metallurgy as being primarily Al_2O_3 in aluminum. Wolf (1960) devised a method for assuring an aluminum oxide content of 6% in the powder, making it particularly suitable for the production of sintered bodies of high tensile strength and hardness at elevated temperatures.

23-5 MISCELLANEOUS CERMETS

Johnston, Stokes, and Li showed the effect of alumina particles on the nucleation and growth of cracks through a silver chloride matrix. Fibrous cracking could be promoted under notch-impact conditions at temperatures at which silver chloride alone shows brittle cleavage. Baxter and Roberts found that alumina particles coated with cuprous oxide and bonded with silver act as a "model system" in which the metal phase actually wets and bonds to the nonmetal. Specimens of the continuous metal-phase type gave high bond strengths and a linear stress/strain relation.

Jackson (1961) found that the porosity of hydrogen-sintered bodies composed of $MoSi_2$ and Al_2O_3 was affected by grain size and alumina content of the cermet. Oxidation-resistant bodies could be made with grain sizes not over 10 microns, but the bodies were not strong. Specimens sintered in dissociated ammonia rather than in hydrogen corroded more easily, probably because of the formation of Si_3N_4 during sintering.

The use of metallic aluminum powder in ceramic protective coatings for refractory linings has been mentioned (Schurecht, 1939). Powdered aluminum has also been recommended for use in ceramic binders for abrasives (The Carborundum Co., Ltd., 1954; Cantrell; Degg; and Nobes, 1957).

German patent 873,674 (1943) covers the production of metal-ceramics, which comprises heating formed silica articles for 5 hours at 900°C in molten aluminum, whereby a complete transformation of the silica articles into (Al,Si)-Al_2O_3 occurs, owing to the exothermic reaction with aluminum. Such articles were intended for measuring electrical resistance. Walton and Poulos (1959, 1960) prepared cermets by the thermitic reaction of powdered aluminum with the oxides: Co_3O_4, Cr_2O_3, Fe_2O_3, MnO_2, MoO_3, NiO, SiO_2, TiO_2, V_2O_5, and ZrO_2. The temperatures required to ignite the compositions ranged from 560°C for MoO_3 to 980°C for Fe_2O_3. A third component, clay, alumina, or magnesia was added to moderate the reaction. Metallic silicides and borides as the metallic phase of a cermet were produced from the appropriate silicates or borates, or metal oxides with silica or boric acid. $ZrSi_2$ was formed by the reduction of either $ZrSiO_4$ or ZrO_2 and SiO_2.

Meyer-Hartwig (1961) prepared a chromium-alumina cermet by the thermitic reaction of powdered aluminum and chromia. Fleming and Johnson (1963) heated Al-U_3O_8 compacts above the melting point of aluminum to produce a cermet containing about 65 wt % U_3O_8. At the ignition point, 816°C, the reaction was violent, the calculated maximum energy release being 1.2 megawatt-seconds/kg of reactant. The sintered compacts contained α-Al_2O_3, UO_2, UAl_2, and UAl_3, but the relative amounts of each compound differed greatly from sample to sample.

Cronin (1951) investigated cermets as likely refractory materials for thermionic cathodes in electron tubes. Unlike other cathodes, cermet bodies can be brought to elevated temperatures by impressing a voltage directly across the terminals. Resistivity can be altered by processing changes and control of the composition. Satisfactory cermet bodies were developed having low vapor pressure, good chemical stability, adequate thermionic and secondary electron emission, and high hot strength.

24 ALUMINA IN AIRBORNE CERAMICS

24-1 INTRODUCTION

Active interest in the United States in sintered alumina as an airborne ceramic material received an impetus from the reports of ceramic developments in Germany prior to 1945 (Watt et al., 1945; Keller and Herberger; de Witt; Kistler; and Schilling). These developments related mainly to ceramics in turbine blades and in jet propulsion accessories for hopefully prospective use at temperatures as high as 1200°C. High-temperature metal alloys were scarce in Germany during 1940 to 1945, and ceramic materials were being substituted wherever possible. Blackburn (1946); McRitchie; King; Graff; Hursh; and Slyh et al. (1946) reported on early progress in ceramic coatings for gas engine tubes, cermets of alumina, with nickel or chromium, and the testing of refractories for rocket motor liners and nozzles. Duckworth (1948) and Bressman (1948) described the potential lines of development as consisting of (1) solid ceramic shapes for high-temperature parts, and (2) ceramic coatings to protect the high-temperature metals from corrosion. Besides the refractory oxides, BeO, Al_2O_3, ZrO_2, ThO_2, MgO, and CaO, German developments using powder metallurgy of mixtures of metals and ceramics were mentioned, including Brown-Boveri disclosures of cermet compositions for turbine blades having a highly ceramic tip and increasing metal content toward the root.

24-2 GAS-TURBINE ACCESSORIES

Although the inherent mechanical shortcomings of oxide ceramics and cermets with respect to impact, thermal shock, and lack of ductility were realized early in the investigation of these materials for airborne applications, the gains to be realized by successfully circumventing these problems encouraged further investigation. The application of sintered alumina and alumina cermets as a complete substitution for refractory metals in the fabrication of gas turbine blades has been discussed in the following references: Geller (1948); Bole (1949); Bobrowsky (1949); Ault and Deutsch (1950); Duckworth and Campbell (1950); Rotherham et al. (1952); Carruthers and Roberts (1952); Thielke (1953); Richardson (1954); Waeser (1955); Puri (1959); Blakeley and Darling (1957); and Bragdon (1958). Flock and Halpern (1952) collected a bibliography of books and published reports on gas turbines, jet propulsion, and rocket power plants.

Bobrowsky stated that both ceramic and ceramic-metal compositions have operated as blades in gas turbines at temperatures above those in service use with alloy blades, although speeds were lower. The life of the ceramic materials was short, primarily because of mechanical design problems that could not be anticipated prior to research evaluation. Tacvorian and Levecque (1957) claimed that a cermet containing 56% Al_2O_3, 4% Cr_2O_3, 30% Cr, and 10% Mo, pressed at 28,000 psi, and sintered at 1700°C, was especially suitable for turbine blades or nozzles of jet power plants.

Lavendel and Goetzel (1957) found that failures occurring during engine testing of cermet buckets for high-temperature gas turbine use were generally located in the upper 25% of the airfoil section or in the neck of the root. In an attempt to correct difficulties with ductility in these regions, they investigated a "graded" cermet (variable in metal content). The cermet selected for this test contained no alumina, but was a titanium carbide skeleton infiltrated with Inconel-X. The test results indicated a substantial improvement in ballistic impact strength and in thermal shock resistance.

The use of ceramic materials as coatings for aircraft power plants and in high-temperature applications in general has been discussed by the following: Koenig et al. (1950), Schell and Neff (1951), Holland (1952), Garrett and Gyorgak (1953), Lynch et al. (1958), Graham and Zimmerman (1958), Miller and Wheeler (1958), Levy (1963), and Merry and Vondracek (1963).

Some of the coating applications for high-temperature metals appear more promising at present than the complete substitution of metals by ceramics in airplane power plants. Koenig found that Inconel and Haynes Stellite metals, coated with BaO-Al_2O_3-SiO_2 glass, were protected to 1000 to 1150°C. These coatings would not adhere to other high-temperature alloys and stainless steels. Lynch, Quirk, and Duckworth (1958) investigated 34 ceramic materials, including three alumina coatings on stainless steel, as nozzle liners in a laboratory rocket motor. The motor was operated with a 4:1 volume ratio of hydrogen to oxygen to give a combustion temperature of about 2480°C at 200 psig. The functional life of the motor with the various experimental nozzles ranged from less than 15 seconds to

107 seconds. The alumina coatings on stainless steel lasted 15 to 30 seconds, in comparison with over 60 seconds for complete nozzles of recrystallized silicon carbide, zirconium boride and beryllia. It was noted that the erosion in the case of the coated alumina probably resulted from melting of the stainless steel because of lack of thermal insulation from the coating. The alumina coatings provided an adherent refractory layer.

Cermets have found application as pump seals in rockets and in nuclear power plants in uses where their resistance to corrosive liquid metals makes them valuable (Graham and Zimmerman).

Truesdale, Swica, and Tinklepaugh investigated the use of molybdenum metal fibers as reinforcing for alumina ceramics. The main difficulty was that the metal fiber reinforcing did not shrink with the alumina during firing; the fibers opposed both the forming pressure and the sintering. Hot-pressing was the only likely solution, but this method affords only a few simple shapes. Cracks were not observed for alumina ceramics containing only 5 wt % molybdenum fibers, but fine cracks developed for 10 and 20 wt % loadings. There was some evidence of increased resistance to failure by thermal shock cycling, as determined by changes in flexural strength and modulus of elasticity. Atlas (1958) investigated metal reinforcing by the use of wire screens, sheets, and plates in sintered alumina for the prospective application as a ceramic radome. The structures were produced by hot-pressing at 1275 to 1450°C with sintering additives. Inconel alloys, stainless steels, titanium, molybdenum, and mild steel were investigated as the metallic components. Phosphate-bonded laminates were also formed from tabular alumina, bubble alumina and $P_2O_5-H_2O$, with different high-temperature metal components. Wetting of the alumina was accomplished by a light coating of titanium hydride in combination with dry hydrogen for brazing operations. Flexural loading caused failure at lesser loads than for an equal thickness of the homogeneous ceramic. The test bars, however, retained a significant integrity of form. The most significant advantage of metal reinforcing was shown by impact tests. The impact absorption of the reinforced specimens (Charpy pendulum) depended on the thickness and type of metal, but in some cases was 10 times or more that of the ceramic alone.

H. B. Porter (1955) appraised the applications of ceramics in rocket refractories meeting temperatures of 2200 to 2770°C (above the melting point of Al_2O_3), reached within a second after ignition. Although the general conditions are more severe with solid fuels, erosion and corrosion were stated to be more severe with liquid propellants. In addition to aluminum oxide, the hard metal aluminide, Mo_xAl_y was considered a possible rocket material. Its resistance to oxidation is good, but its strength and thermal shock resistance are poor, and it is hard to fabricate.

24-3 RADOMES AND ROCKET EQUIPMENT

When the speeds of aircraft approached Mach one, the plastic radomes for enclosing radar antenna equipment began to encounter difficulties with air friction and impact from dust and raindrops in the atmosphere (Wahl, 1957). It was claimed that velocities of 733 ft/sec (500 mph) cause serious damage to plastics. Raindrop impacts at velocities of about 6600 ft/sec do not significantly abrade sintered alumina surfaces, but can fracture sintered alumina surfaces. Engel (1960) estimated that plates of sapphire or hot-pressed alumina 0.125 inch thick can be expected to survive collision with a 0.2 cm water drop without damage at a velocity as high as 11,000 ft/sec (Mach 10).

The choice of materials for transparent windows for enclosing radar equipment in aircraft and guided missiles narrowed to a few sintered ceramic materials (mainly high-alumina) and special refractory glasses (Pyroceram). The size requirements and the necessity for close tolerances on the dielectric constant, wall thickness and shape of the radomes imposed new problems in their fabrication (Smyth, 1957). The problems involved forming, firing, and surface-machining techniques.

Smoke (1957) described the fabrication of dense ceramic radomes by casting. Dorsey (1958) and Ault and Milligan (1959) described flame-spraying methods. The former claimed that silica, alone, or with additions, was the only material that could be fabricated to a dense deposit. The latter prepared thin, dimensionally controlled alumina radomes by flame-spraying a coating of potassium chloride over an aluminum form, by flame-spraying a coating of alumina over the salt, by dissolving the salt to facilitate removal of the alumina shell, and by firing the shell to an impervious structure. Isostatic forming either over or within a nonyielding mandrel, appears to be a more favored method of forming at present.

Ward and Passmore (1960) described diamond-grinding techniques for finishing sintered alumina radomes. Thomas (1963) stated that a single machine has been developed to perform all the grinding operations, except finishing the outer diameter on alumina ogival nose cones. The surface, chamfer, inside diameter, contour and plunge grinding and core drilling operations were described, with details of feed rates and speeds. The claimed accuracy was 0.0005 inch for size and contour shape from piece to piece.

Smoke and Koenig (1962, 1963) developed high-alumina radome bodies containing 94 and 96% Al_2O_3, that had excellent engineering properties when matured in the range 1430 to 1540°C. The favorable maturing conditions were ascribed to a presintering process in which the fluxing agents were more homogeneously distributed than in usual one-fire processes. Magnesium carbonate and magnesium fluoride were the sources of magnesia. Substitution of aluminum oxide-oxychloride for the alumina resulted in a very high-density body (Smoke et al., 1962). Atlas (1958, 1960) prepared very high purity alumina from 99.999% aluminum in order to control the electrical properties of radomes made from it.

Metzger (1957) designed light-weight radomes, in one of which a half-wave wall construction, electrically homogeneous was proposed. A sandwich construction was used (dense skins, light cores). The radome wall thickness is related to the wavelength of the radar frequency by the expression:

$$d = \frac{n\lambda}{2(e-\sin^2 \theta)^{1/2}},$$

in which d = the approximate mechanical thickness, n = an integer, e = the dielectric constant, θ = the effective angle of incidence of the radiation, and λ = the wavelength of the radiation.

Poulos, Elkins et al. (1962) investigated fused silica as a radome structural material. Fused silica has the advantage of a very low coefficient of thermal expansion (less than one-tenth that of sintered alumina), but is inferior in strength, refractoriness, and hardness.

During the Seventh Symposium on Electromagnetic Windows, Ohio State University, 1964 (Ref. Slonin), Guarini stated that supersonic velocities to Mach 6 force designers to high-density, thin-walled, brittle ceramics. High-Q environments prohibit the use of most half-wave refractory walls, because of thermal shock. Two rain-erosion sled tests (NOTS and Sandia) showed that sintered alumina (Coors AD-99) survived the rain impact well beyond the peak velocity used in the tests. Competitive structural materials were not as satisfactory. Guarini concluded, however, that none of the structures, whether ceramic or reinforced plastic, could survive high-velocity flight (greater than Mach 2.0), unless the radome shape was sufficiently sharp.

Wheeler, Winslow, and Gates (Ref. Slonin) recommended sintered alumina for matching to niobium, but zircon and mullite for matching to molybdenum, in the design of slot closures for an antenna operating at 1370°C. The favored forming method was by spraying the alumina onto the mandrel from an alcohol vehicle containing an organic binder.

Loyet and Yoshitani (Ref. Slonin) described sandwich types of radomes in which the skins have either a higher (A-type) or a lower (B-type) dielectric constant than the core. Type A radomes have been developed at Interpace (International Pipe and Ceramics Corp.), that appear to be more suitable than any single-structure form that has been so far developed (Caldwell, 1962; LeClercq, 1962; Caldwell and Gdula, 1962).

Dense alumina is mandatory for the resistance to rainfall. The electrical and thermal properties are also better than those of the devitrified ceramics.

N. F. Dow (1961) showed that 25-micron filaments, 2 inches long, dispersed in a ceramic matrix of the same modulus of elasticity, require a bond strength 1/8000 the tensile strength of the fibers. B. D. Coleman (1958) demonstrated that the strengths of a group of parallel filaments is determined by the variation of individual strengths about the average. From statistical considerations, a bundle of filaments with individual variations 30% of the mean could support a load of only 60% of the average strength. If the filaments are not completely parallel, the reinforcement is still further reduced. Lockhart (1965) had difficulty in dispersing alumina whiskers in initial attempts at reinforcement of alumina ceramics matured at 1427°C to over 80% of maximum density by use of MnO_2 or $MgTiO_3$ used as sintering aids. The filament-loaded, slip-cast bars cracked severly during air-drying, and later failed in bending tests. High shear stresses between the matrix and filaments were obtained, however, at 50 vol % fiber loadings in the ceramic.

Aves (1959) claimed that the most satisfactory protection for backup plates of rockets for re-entry at 2760°C was alternate layers of flame-sprayed molybdenum and alumina. Trout (1960) compared these coatings with other prospective types. A. V. Levy (1959) recommended flame-sprayed or troweled alumina for insulating airborne power plants. Pellini and Harris (1960) gave detailed requirements and possible solutions to the problems involved in thermal protection of heat sink systems, ablation, sublimation, transpiration, and radiation systems in flight in the thermosphere. Grindle and Rosenbery (1964) simulated re-entry conditions at Mach 3 to 7 velocities in a plasma arc re-entry tunnel using nitrogen and oxygen as the operating gas. Miller and Wheeler (1958) described the potential applications of ceramic materials for airframes. Ramke and Latva (1963) presented the favorable properties of ceramics for nose cones, leading edges, and insulation. Nowak (1961) discussed the material property requirements and test methods for ceramic materials for leading edges of hypersonic aircraft.

M. Levy (1963) found that flame-sprayed alumina was successful in preventing erosion of titanium and aluminum rocket nozzle components for at least one firing. B. W. Lewis claimed that an alumina cermet was one of the least affected of 22 refractory materials investigated to withstand a Mach 2 jet at a stagnation temperature of 2095°C.

Bradstreet (1961) reviewed the problems encountered in attaching ceramic coatings for high-temperature service. The "tailoring" of protective coatings for refractory metals was suggested. Cavanaugh and Sterry (1961) suggested ablative protective coatings against thermal shock for radomes (on an advanced interceptor). Blome and Kummer (1964) described the use of alumina in the glide-re-entry vehicle ASSET (McDonnell Aircraft Corp.). In one antenna, a 99% alumina ceramic (Coors AD-99) was used, metallized and brazed to a Kovar retainer. Vincent (1960) described the testing and operation of Avco ceramic re-entry vehicles on the Martin Titan ICBM. W. H. Wheeler (1963) evaluated the properties of low-density ceramic foams, laminated composites, and high-density, high-strength materials as heat shields. High emissivity is required to keep wall temperatures within practical limits. Merry and Vondracek (1963) found three specific uses for flame-sprayed alumina coatings: electrical insulation on an aircraft generator, a mounting bar for an edge-wound resistor, and a tube furnace.

Wilkins, Lodwig, and Greene collected the free energy functions for 47 gases and 9 condensed species containing the elements, lithium, boron, and aluminum, free and in combination with hydrogen, oxygen, nitrogen, and fluorine. The theoretical flame temperatures and the primary metallic chemical species in flames containing these metals were determined because of their relation to rocket propulsion. Vasilos and Gannon concluded both from thermodynamic calculations and experimental measurements that SiO_2, MgO, BN, SiC, and graphite are potentially good ablators, whereas Al_2O_3, ZrO_2, NiO, and Cr_2O_3 have definite limitations. Those compounds which decomposed with little or no melting or whose melt viscosity was appreciable showed greater heat absorption capability than those having low vapor pressure at their melting points.

25 SEALS, METALLIZING, WELDING

Polycrystalline alumina and sapphire forms are subjected to joining operations of the types: alumina to metal, alumina to itself or other crystalline ceramic, and alumina to glass. Sintered alumina envelopes for high-wattage electron tubes, alumina cutting tools, cermets in gas turbines, and the use of sapphire windows in thermionic converters and other devices, are examples of equipment requiring rugged seals, capable of holding impermeable to leakage or rupture at high temperatures.

The German firms, Siemens and Halske, Telefunken, and AEG had begun investigations on ceramic sealing about 1934, and had developed disk seals by means of thin metal interfaces by 1943. A symposium on ceramics and ceramic-metal seals was held at Rutgers University in 1953 to discuss problems related generally to electron tubes. Metallizing and metal-to-ceramic seals have probably drawn the greatest attention. Palmour reviewed the high-temperature metal-ceramic seal literature prior to 1955; Van Houten surveyed the art in 1959.

Burnside classified bonding with an intermediate layer into nine divisions: (1) bonds produced by firing a metal oxide on the ceramic surface, and reducing it to the metal; (2) bonds produced by firing laminates bearing metal and ceramic powders in varying proportions; (3) bonds produced by hot-pressing, for example at $1000°C$, 3000 psi in nitrogen or vacuum; (4, 5) bonds produced with active metals, active alloys, or active metal hydrides; (6) bonds produced using glazes in an oxidizing atmosphere: (7) bonds produced using glazes and metal powders in a reducing atmosphere; (8, 9) bonds produced using refractory metal powders with or without manganese for continuity, copper or nickel for build-up, and solder for the final bond.

Bonding techniques in use circa 1959, according to Van Houten included the Mo-Mn, the hydride, and the active metals processes. The Mo-Mn process is generally ascribed to Pulfrich (1939), although Vatter (Siemens and Halske, 1935) also claimed the process. The Vatter patents describe vacuum sintering of the ceramic after coating with iron, chromium, nickel, or tungsten, with or without additions of manganese. After metallizing, the ceramics could be joined to gold, silver, copper, or their alloys. The Pulfrich disclosures describe painting a layer of molybdenum,

rhenium, or tungsten (in a suitable binder) onto a nonporous ceramic surface in a controlled atmosphere to obtain a platable or solderable surface. Nolte and Spurek (1950) substituted manganese powder for iron. The British Thomson-Houston Co. (1951) included nickel, and fired in a vacuum. The Mo-Mn process is claimed to be relatively insensitive to small changes in the process, hardly the case with the Ni-Fe-Mo process (Telefunken). This latter process has been modified by the substitution of tungsten for molybdenum, additions of iron or manganese, or the use of MoO_3, reduced directly on the ceramic. Coatings of nickel or copper are placed over the molybdenum to aid wetting by the silver. The method was performed in hydrogen (Bender, 1955).

The hydrides of titanium and zirconium were found particularly active in producing low-temperature bonds with ceramics (Bondley, 1951). A metallizing and bonding patent was issued to F. C. Kelley in 1951 for TiH_2-ZrH_2-Cu bonds that could be developed at $400°C$. The decomposition of the hydrides liberates active hydrogen which is effective in cleaning the surface to be bonded. A typical reference to use of a hydride involves the low-melting eutectic $(878°C)$ composed of 72 wt % copper–28% titanium hydride. Strong bonding is possible by the method, using silver, copper, lead, or other solders.

Active metal films of titanium or zirconium are easily welded to glass. Vapor deposition of metal films of chromium, titanium, and silicon had been deposited by spraying the volatile chlorides onto the work, and molybdenum, tungsten and tantalum films were deposited by reduction of the volatile chlorides (Van Houten).

Cronin (1956) discussed trends in the design of magnetrons and the transition from glass-bonded, through glass-bonded ceramic, to ceramic-to-metal seals. Hynes (1956) described the forming, firing, and grinding requirements of alumina for use in ceramic-to-metal seals. The sandwich-type seal was designed primarily for the construction of output windows for magnetrons. The hydride-glass seal and sinter-type processes also were used (Pryslak, 1955).

Knecht (1954) reviewed the Air Force projects on electrontube, metal-ceramic structures, and cermets for gas turbines at different laboratories. Gallup and Dingwell described some low-temperature lead borosilicate glasses

(70% PbO) for low-temperature electron-tube solders. Cole and Hynes concluded that higher seal strengths could be attained by closer control of firing temperature and flux content, and that the glass content is of prime importance in obtaining good seals. Denton and Rawson investigated the chemical reactions involved in metallizing high-alumina ceramics by two refractory-metal methods, the Mo-Mn and the MoO_3. The found that minor constituents in the ceramic are important, acidic oxides, such as SiO_2, being of great importance in the Mo-Mn technique, and basic oxides, such as MgO, being of more significances in the MoO_3 technique. Best results were obtained with fine crystal size.

Sterry investigated the brazing of alumina to titanium and molybdenum with silver and copper-silver eutectic. Brazed joints develop higher strengths than clamped or pinned joints. Suitable ceramics include, among others: sintered alumina, calcium aluminate and germanate glasses, sapphire, phosphate-bonded composites, and Pyroceram. In the nuclear and missile fields, most radio tube assemblies use high-alumina ceramics, and the brazing is best accomplished, by the Mo-Mn or the active alloy method (Lerman, 1961). In the former, the ceramic is cleaned, coated with the metal, sintered, cooled, and nickel plated before assembly with the metal component. In the active alloy method, titanium or zirconium is contacted with the two ceramic parts or one ceramic and one metal part in a vacuum or neutral atmosphere, and heated. The active metal reduces some of the ceramic oxide. Cole and Sommer (1961) claimed that the glassy phase in the high-alumina ceramic migrates into the interstices of the molybdenum-based coating when using Mo-Mn, Mo-Ti, and pure Mo metallizing agents on 94 and 99% Al_2O_3 compositions.

Pincus (1954) concluded that the experimental evidence on molybdenum-alumina interactions points to the need for a controlled amount of oxide on the metal, available at the metal-ceramic interface, as an essential step in the establishment of strong ceramic-to-metal bonds. When molybdenum metal is used as the starting material, adherence is not satisfactory until a sufficiently high temperature has been reached to cause dissociation of oxide of the ceramic. When molybdenum oxide is used as the initial coating, adherence and interpenetration occur at lower temperature, and more uniformly. Floyd investigated the mechanism of adherence of 80 Mo-20 Mn powdered metal to a high-alumina ceramic containing glass, and concluded that the weakest bonds were caused by formation of a modified $MnAl_2O_4$ spinel at the metal-ceramic interface. Greater bond strength was observed as the spinel was moved from the interface into the voids of the metal coating. The metal-alumina reaction theory and the glass-penetration theory were both believed to apply.

Some patents of interest include the following: KLG Sparking Plugs (1954) applied an oxide film to the bonding metal. Dijksterhuis and Hovingh (1957) added a thin layer of a metal such as Ag, Au, Cu, Ni, Fe, Co, or Mn, or alloys of these to the active-metal process, which dissolves part of the titanium or zirconium, when melted. Cohen, Herbst, Jenkins, and Wilkinson (1960) joined titanium or zirconium

to a suitably matching ceramic by brazing in an inert atmosphere with Cu, Ag, or Au, and other metals could then be brazed to the titanium or zirconium. Louden and Homonnay stacked in order an alumina backup ring, a thin titanium washer shim, an end cap, another shim, an alumina tube, and another assembly of shim, end cap, shim, and alumina backup ring. The end caps consisted of an Fe-Co-Ni alloy in approximately 48-26-26 proportions. The stacked parts were heated in a nonoxidizing atmosphere at 955 to 1060°C to obtain a sealed device. Ross bonded polycrystalline alumina shapes with Nb, Ti, Pt, Ta, Mo, W, Ni, Cr, or stainless steel by using a powdered glass consisting of equal parts of alumina and calcia. The assembly was heated above the melting point of the glass, but below the melting points of metal and the ceramic. Elms (1964) sprayed the surface of 99.5 to 99.9% Al_2O_3 bodies with a mixture containing from 5 to 40 wt % powdered TiN or TiC, the remainder being powdered molybdenum, suspended in a resin, gum, wax, or nitrocellulose lacquer base material. The assembly was fired at 1450 to 1900°C in hydrogen (dew point > 4°). The metallized layer was then nickel plated and brazed. Mackey (1963) sintered a paint coating on vacuum-tube elements to be joined, consisting of 1200 parts powdered molybdenum, 200 parts powdered manganese, ground in petroleum solvent containing manganese stearate 200 parts, and titania (100 g per gallon of ball mill capacity). Beggs (1958) inserted a metal shim (titanium, zirconium, hafnium, thorium, or an alloy of more than one metal of the group) between the ceramic surface and the other metal member to be joined, selected from the group: copper, nickel, iron, molybdenum, chromium, platinum, cobalt, and alloys of others in the group. The stack was heated in a nonreactive atmosphere to a temperature at least equal to the melting point of the eutectic alloy of the metal member and the metal shim member.

Mesick formed gas-tight seals between the ceramic and a metal surface by applying a thin coating (less than 15 mg/cm^2) of an oxide selected from the group, molybdenum, tungsten, and mixtures of these. The coated assembly was fired in a reducing atmosphere. Umblia (1958) applied a layer of powdered mixture, 60 to 95% from the group: molybdenum, tungsten or rhenium, 5 to 40% of an easily deoxidizable metal from the class: nickel or cobalt, and 0.05 to 0.5% of an oxide from one of these metals. The member and layer were heated to 1200 to 1350°C in an inert atmosphere. Luks and Powell (1958) coated the ceramic, refractory body with a metallic glass consisting of 5 to 30 wt % of a powdered metal (molybdenum or tungsten), 40 to 85% of a powdered heavy metal (nickel, cobalt, or iron), a powdered manganese component not exceeding 35% as elemental manganese, and 5 to 30% powdered glass. The glass was fused to the body at 815 to 1260°C, and soldered for subsequent bonding. Nolte (1959) metallized pure alumina bodies with a powder containing 20% Cr and 80% Mo, and fired at 1625°C for high heat-shock uses. Rhoads and Berg (1961) silk-screened a uniform layer of powdered molybdenum, tungsten, molybdenum-copper, or molybdenum trioxide, fired to sinter the material, silk-screened a second mixture of powdered nickel or copper, dried the second layer, applied another layer of a hard solder and pressed the metal member into contact, and finally heated to seal the members together. B. R. Smith metallized the ceramic by applying a coat of powdered tungsten, molybdenum, or manganese, and by firing in a reducing atmosphere at about 1700°C until the ceramic was bonded and all binders, vehicles and plasticizing agents were decomposed.

LaForge metallized vacuum tube envelopes by coating the ceramic with a metal molybdate or a metal tungstate, by firing in an oxidizing atmosphere at 1125 to 1040°C for 2 to 20 minutes to produce a glassy surface, and by firing in a reducing atmosphere at 1300 to 1225°C to provide an outer surface of metal integrally united with and overlying the glassy surface. Woerner (1962) coated the alumina ceramic with a suspension containing at least 1% finely divided borosilicate glass free from materials reducible in hydrogen, and manganese oxide molybdenum oxide, and optionally tungsten oxide. The bond is matured at 1600°C in a hydrogen atmosphere. Omley obtained a hermetic seal between sintered alumina and a solid pin of nickel-iron-cobalt alloy by use of an alloy having the approximate composition: 60% silver, 27% copper, and 13% indium.

McDonald and Whitemore metallized with a carbide (major portion tungsten carbide, minor portion titanium carbide). The film was fired in a hydrogen atmosphere. A metallized surface of nickel, cobalt, or iron was formed on the surface for brazing. Grattidge (1958) formed a metal surface on a high-alumina body by applying one or more layers of a mixture of alumina and an oxide of iron or chromium, the alumina constituting 30 to 90%, then applying a layer of either platinum or iridium and heating in an oxygen-containing atmosphere to 1200 to 1600°C for platinum and 1200 to 1750°C for iridium to obtain a bond between the body and the applied metal layer.

Some more recent investigations are as follows:

Cowan (1964) described the bonding of tungsten to sintered alumina containing from 0.05 to 2% yttria. Yttria was substituted for glassy fluxes in sintering alumina to be used for thermionic converters because of the attack by cesium vapor. At a metallizing temperature of 1600 to 1800°C, tungsten powder coatings in a mixed atmosphere, 96% hydrogen, 4% water vapor, developed bond strengths of 4,000 to 10,000 psi.

Bristow (1964) used active alloys such as TiN_3 to seal a thermionic converter at 700 to 900°C. Leaks were unsatisfactory above 600°C because of continued reaction of the sealing alloy. The alumina-nickel bond, sealed by TiN_3 deteriorates to a weaker bond above 900°C.

Fox and Slaughter observed that metallizing is unsatisfactory in nuclear environments. Most sealing operations require pre-metallizing–then brazing. Two experimental brazing alloys were developed, Ti 48 wt %-Zr 48%-Be 4%, and Ti 49%-Cu 49%-Be 2%. The latter provides good brazing–bonding to sintered alumina and uranium dioxide in a vacuum or inert atmosphere. The 5% beryllium in the alloy reduces the melting point of zirconium below 900°C. The fabrication of a synthetic sapphire Hall cell was demonstrated.

Mattox produced active-metal seals (chromium and titanium) on sintered alumina (Coors AD-99) by first applying a dc gas discharge at about 5000 ev to clean the surface and to ionize a portion of the metal atoms. The required time was about 20 minutes for cleaning and 20 minutes for deposition. Bond strengths of 3000 to 7000 psi were claimed.

Fulrath and Hollar sealed molybdenum to a 99% Al_2O_3 body by the use of a glass having the composition 50% MnO, 20% Al_2O_3, and 30% SiO_2, and a coefficient of thermal expansion of about 7.0×10^{-6} (20 to 600°C), between that of the molybdenum (5.9×10^{-6}) and that of alumina (7.2×10^{-6}). By proper adjustment of the concentration of molybdenum powder, tensile strengths of about 8000 psi were achieved. The addition of the glass to the metallizing solution allowed sealing to sapphire. Some devitrification to galaxite was observed.

Cowan (1965) metallized alumina ceramics with dense copper coatings that had been formed by reduction of a 0.001-inch thickness, approximately, of a powdered Cu_2O-Al_2O_3 eutectic during firing at 1250°C for 30 minutes. Kiwak (Bendix Corporation) sealed refractory metals to alumina ceramics with a composition including the eutectic of Y_2O_3-Al_2O_3 which melts at 1760°C, heated to at least partial fusion at 1800°C in a vacuum or inert atmosphere.

Williams and Nielsen, and Sutton and Feingold used the sessile-drop method to appraise the wetting obtainable by coatings and brazing solders during metallizing.

Tentarelli and White (1964) developed a low-temperature ceramic seal applicable to 99.5% Al_2O_3 bodies, using a metallizing paint of MoO_3-MnO_2 suspended in a butyl alcohol-butyl acetate-naphthenic acid mixture. A Cu-CuO braze was applied at 1100°C in a wet ammonium hydroxide atmosphere at 900 or 1150°C (not at 1000°C). Tensile strengths as high as 17,000 psi were attained.

Reed and McRae found that molybdenum can be evaporated onto 94% Al_2O_3 bodies and sapphire to provide vacuum-tight seals having about 20,000 psi tensile strength for 94% Al_2O_3 compositions and 10,000 psi for sapphire. The seal was resistive to potassium vapor at 550°C. L. Reed (1964) stated that the factors affecting the integrity of a ceramic-to-metal seal include sintering, surface energy, thermal expansion of the components involved, strength of the crystalline phase, and devitrification of the glassy phase. Crystallites precipitate from the glassy phase above 1100°C. Glassy phase disappears from the high-alumina compositions. The active metal-sealing types include Ti, Zr, V, Be, generally alloyed with Ni, for example, Ni-Ti. Brittle alloys are formed, as, for example, by the equation: $BeO + Ti = Be + TiO_x$. The refractory metal metallizing method uses glass bonding, for example, the system Al_2O_3-$Al_2O_3 \cdot SiO_2$-M_x, to provide tensile strength of about 5000 psi. Sapphire requires a paint (a glass melt for molybdenum). The metallizing paints consist of MnO, SiO_2, etc.

Anderson and Stepp (1962) described the fabrication of glass-to-sapphire end-window seals capable of holding 2×10^{-6} mm Hg vacuum with no noticeable effect of attack when exposed to sodium vapor at $400°C$ for 4 hours. Reinhart (1961) collected a review of American and European literature on ceramic-to-ceramic and ceramic-to-glass seals. Hokansen, Rogers, and Kern (1963) presented details on the welding of alumina to alumina, tungsten, molybdenum, niobium, and Kovar by the electron beam. Kessler described a novel method for sealing glass by pressing aluminum foil between the surfaces to be joined at high pressures and at a temperature lower than the melting point of the aluminum and the softening point of the glass. Diffusion of aluminum atoms appears to replace silicon atoms in the glass structure.

An example of an early spark plug porcelain seal between the ceramic and the electrode is a lead borosilicate glass around a nickel-iron-cobalt alloy (Bosch, 1942). More recent examples of seals for high-alumina spark plugs are the following. Schwartzwalder and Somers (1958) mixed from 50 to 90 parts by weight of calcium aluminate with 20 to 60 parts of phosphoric acid (25 to 75 wt %). The exothermic reaction that ensues should not be allowed to go above $21°C$ $(70°F)$. The mixture can be stored for several hours at about $4°C$, or for longer periods at $-12°C$. Somers and Schwartzwalder (1959) described a cement consisting of 3 parts calcium aluminate and 2 to 2.5 parts sodium silicate containing 50 to 65% water and having a ratio $Na_2O:SiO_2$ of 1:1.5 to 1:2.5.

26 FIBERS, WHISKERS, FILAMENTS

26-1 INTRODUCTION

Much interest has developed in applications of ceramic fibers, in particular synthetic continuous fibers and whiskers, because of their high tensile strengths. Fiber-reinforced structures are probably the solution to some problems in hypersonic aircraft and missiles. Since methods have been found to make crystalline aluminum oxide fibers, (alluded to in Section 3-9), alumina is a likely contender in applications of this type. It also is involved in high-alumina-silica glass fibers for the more general large-volume applications of special glasses in textiles, insulation, and reinforced plastics, an industry that totaled over 1.2 billion pounds in 1963 (Ref. "Glass," Chem. & Eng. News).

McCreight, Rauch, and Sutton (1965) have compiled a current report on the state of the art of ceramic fiber production. The following data, Table 16, taken from Figure 1 of that report (page 4) show the comparative tensile strengths for virgin glass fibers (S-glass and E-glass of Owens-Corning) and alpha alumina fibers (Brenner, 1962). Also included are data for Houze glass No. 29 (Ceramic Industry, April 11, 1964, p. 108).

Table 16

Tensile Strength of Ceramic Fibers (1000 psi)

Temp., °C	α-Al_2O_3	S-Glass	E-Glass	Houze No. 29-A
20	900	700	500	680
250	880	600	450	670
500	800	450	270	630
800	750	130	–	410
1000	570	–	–	
1500	300	–	–	
1900	150	–	–	

General references to the state of the art of fiber and whisker production and the progress in fiber metallurgy are discussed in the following: C. Z. Carroll-Porczynski, *Refractory Fibers, Fibrous Metals, and Composites*, 1962; Baskey

(1962); Anon. "Developments to Watch," 1964; Accountius (1963); Bradstreet (1964); Bartlett and Ogden (1965); Krock and Kelsey (1965); and McClure (1965).

26-2 ALUMINA FIBERS

Brenner (1962), from theoretical considerations, concluded that the upper limit of the strength of alumina (S_f) as a function of the modulus of elasticity is 0.17E, and has found about 0.16E experimentally. The lower limit, because of plastic deformation is about 0.03E, or an average S_f of about 0.1E. The conventional determination of tensile strength on rods is usually as low as 0.001E. His experimental room-temperature values for alumina fiber were about 0.099E. Room temperature values for different fiber types, assembled by McCreight et al., show only 0.018E to 0.055E for strong glasses, 0.081E for fused silica, 0.005E to 0.021E for metals of the continuous filament types, 0.002E to 0.066E for metals of the discontinuous (whisker) type, and 0.036E for graphite. Competitive ceramic fibers, including BeO at 0.47E, B_4C at 0.013E, and SiC at 0.024E, were also weaker than alumina. The best compositions for fabrication purposes have the lowest density/$E^{1/3}$ values and the highest use temperatures. Alumina falls in a favorable position in this respect, in comparison with other potential ceramic materials.

Watson et al. (1957); DeVries and Sears (1959); Cunningham; and Sears and DeVries studied the growth and properties of alumina fibers prepared by vapor deposition from aluminum and in some cases by rapid drying of colloidal suspensions that were subsequently heat-treated to develop strength. Du Pont de Nemours (British Patent 849,051) prepared corundum in ribbon-shaped fibers by heating aluminum in the presence of hydrogen and any inorganic silicon compound such as quartz, calcium silicate, etc., at 1100 to 1450°C.

Several methods have been found to form single-crystal alumina fibers from the vapor state. W. B. Campbell (Lexington Laboratories, 1963) evaporated, oxidized, and nucleated an aluminum halide by either batch or continuous process on a powder substrate. The partial pressures of gaseous H_2, CO_2, and $AlCl_3$ were controlled to obtain the desired degree of supersaturation of the growth species. Wainer and Cunningham (Horizons, Inc., 1963) produced fibers by passing hydrogen with a dew point of -30 to

193

-90°C over aluminum containing 25 mole percent Ti, Zr, Nb, Ta, or Si, and maintained at 1370 to 1510°C. Brisbin and Heffernan, 1963 deposited the fibers from aluminum chloride in a controlled mixture of CO_2 and H_2 at reduced pressure (25mm Hg) and at 1200°C. Hexagonal fibers averaging 1mm length and 0.05mm diameter were obtained. A typical procedure by this method and the growth characteristics of single-crystal alumina showing orientation along the C-axis are described in General Electric Company, Netherlands Application 6,411,302. Schmidt varied the size of the fibers by changing the degree of saturation and the temperature of the crystallizing zone.

Kerrigan was successful in transporting and depositing whiskers of alpha alumina in HCl. A black, oxygen-rich species was formed. Papkov and Berezhkova obtained whisker alumina by heating corundum crystals in a graphite oven to 2000°C in argon or nitrogen or in vacuum.

Lynch, Valdiek, and Robinson grew alumina whiskers in several distinct forms at 1700 to 1800°C in argon and helium atmospheres containing small amounts of hydrogen. It was claimed that the vapor deposition onto a heated resistance wire (W, Ni, Ti, or Mo), produces composite filaments containing a core surrounded by a polycrystalline (or amorphous) coating (Davies et al., 1965). Kelsey claimed the method of passing a controlled flow of dry hydrogen over a surface of molten aluminum containing titanium at 1370 to 1425°C. Greatly elongated crystals were claimed to be obtained by vapor-phase methods if the production space was filled substantially with a porous form or gauze to act as a base for whisker formation (Barratt). Webb, Wissler, and Forgeng (1961) reacted aluminum or an aluminide of Zr, Hf, V, Nb, Ta, Cr, Mo, or W at 1300 to 2050°C in hydrogen-water vapor atmosphere, and condensed fibrous single-crystals of alpha aluminum oxide onto an alumina base.

Polycrystalline fibers can be grown by extrusion of alumina pastes through spinnerets. Wainer and Mayer, and Beasley and Johns described methods for preparing inorganic fibers in which the structure depends on forming treatment rather than by vapor deposition. Wainer and Beasley, and Beasley and Johns spread colloidal suspensions of alumina in thin layers on a non-sticking base, and evaporated the moisture rapidly to form gels that cracked into fiber structures of rectangular cross-section. The tensile strength of such polycrystalline structures is quite low, particularly if the crystal size approaches the fiber diameter in magnitude. Bugosh (1963) used colloidal fibrous boehmite in producing fibers of gamma alumina at 300 to 850°C, without change in the particle size or shape, and to alpha alumina above 1000°C. Lockhart and Wainer (1966) claimed the production of alumina and monofilaments spun from the mixture of water, tartaric acid, aluminum formoacetate, oxalic acid, and lithium bromide, through spinnerets and fired at 565°C and finally at 1370°C.

Fibers produced by vapor deposition are almost devoid of flaws. The claimed tensile strengths are around 1,000,000 psi or higher (Mehan and Feingold, 1967). The tensile strength is found to be dependent on fiber diameter,

due to the greater probability of flaws with increasing size (Brenner, 1956, Regester, 1963). Metallic films, 0.5μ thick, do not impair the room-temperature tensile strength, and the hot bend strength follows the temperature dependency for uncoated whiskers (Mehan and Feingold, 1967). The room-temperature tensile value falls with increasing temperature. However, the fibers were found to be stronger at 20° below the melting point than bulk sapphire at room temperature. Lynch, Valdiek, and Robinson concluded that impurities appear to be necessary to obtain good whisker growth, since they were able to grow whiskers from less pure grades of alumina containing SiO_2 and Fe_2O_3 as major impurities under conditions which failed to produce fibers from pure grades of alumina. Additional references to apparatus for producing whiskers, and their growth on various substrates were furnished by Hoffman (1958), Hargreaves (1961), Edwards and Happel (1962), Mazer and Straughan (1963), Coleman (1964), Gatti (1964) and Levitt (1966). Brenner (1965) reviewed the factors influencing strength of whiskers. Bayer and Cooper (1967) found that chemical polishing of sapphire whiskers in hot orthophosphoric acid, for example, increased the strength of large whiskers as much as ten times, but not of small whiskers. Soltis (1964) found that the mechanical behavior of sapphire fibers was anisotropic, being weaker in the c-axis direction, than in a_1, $\langle 1120 \rangle$, or a_{11}, $\langle 1100 \rangle$. Regester, Gorsuch, and Girifalco stated that tensile strengths of fibers vary inversely with surface area; the strength-controlling defects are apparently located at the surface of the whiskers, although they are either submicroscopic or invisible. Unbroken whiskers invariably terminate in a small globule of aluminum and have the "drumstick" form. The most perfect whiskers are ribbons with principal surfaces parallel to (0001) planes; intense heating, as by the electron beam, causes etch pits. Many drumsticks are tubular (Barber). Sapphire whiskers grown on single-crystal alumina grow coherently with and in the direction of screw dislocations in the substrate (Edwards and Huang).

Aluminum borate whiskers were grown by the reaction of aluminum halide vapors with B_2O_3 vapor and moisture at 1000 to 1400°C, preferably from a melt of AlF_3 and B_2O_3 (Alley and Johnson, 1967). Fibrous ceramic structures were prepared by impregnation of felted ceramics with colloidal alumina and refiring (Harris and Poulos).

Lockhart (1964) reported on the use of whisker alumina in the fabrication of an electromagnetic window in an environment of 1370°C. At high fiber loading in the ceramic matrix (50 vol %), the shear stress between the matrix and filament reached high values near the broken filament values. In a previous report, it had been observed that the critical diameter was about 50 microns; below this size cracks were rare.

Brenner (1962) proposed the use of alumina fibers for producing whisker metal composites of high strength at elevated temperatures. Sutton (1962, 1963, 1964), Sutton and Chorne, and Sutton and Feingold investigated the bonding of high-strength, heat-resistant metals and alloys for space vehicle applications. Silver, reinforced with sapphire whiskers, is 20 times stronger at 760°C in tension than pure silver under similar conditions. The strengthening was retained to large extent to within about 20°C from the melting point of silver (960°C). Other alumina fiber-metal composites were investigated by Price and Wagner, Ellis (1964), Wagner (1964), Salkind, Morley, and Sicka et al. (1964). Wagner (1963) described techniques for reinforcing ceramic matrices with metal filaments as well as the reinforcement of a metal with ceramic whiskers. Hill discussed the theory of the mechanical properties with respect to the elastic and inelastic behavior of the composite.

Sutton, Rosen, and Flom found that reinforced epoxy resin (14 volume % alumina whiskers) developed tensile strength of about 140,000 psi, and Young's modulus of 6×10^6 psi. The original plastic had a tensile strength of about 12,000 psi and Young's modulus of 0.4×10^6 psi. Short-fiber-reinforced plastics usually develop about 50,000 psi tensile strength, but with continuous fibers, about 250,000 psi. Milewski also investigated whisker-reinforced plastics, and Schuster and Scala their interaction mechanically with a birefringent plastic, to develop the strain behavior.

26-3 GLASS FIBERS

The chemical compositions of some ceramic and glass fibers are shown in Table 18.

Ceramic fibers for refractory applications may contain about 50% Al_2O_3. Fiberfrax, a product of the Carborundum Company, is based on the investigations of J. C. McMullen. The optimum mix consists of 50 parts Bayer alumina, 50 parts flint, and 1.5 parts borax, melted and poured at 1870°C. Short and long staple fibers in diameters less than 3 to 40 microns (average 10 microns) have a maximum use temperature of 1260°C (2300°F), coarser fibers around 80 microns, about 870°C (1600°F). The tensile strength is about 180,000 psi (Straka, 1961). Kaowool, a product of Babcock and Wilcox, contains about 45% Al_2O_3, derived from clay minerals, and has a claimed use temperature of at least 1200°C. Both types have good insulating value in the form of loose wool or batts. The material is available also as short fibers, yarns, rovings, felts, papers, and coating cements. Blocks, boards, and castables can be prepared from the fibers. At a hot-face temperature of 1100°C, a 2-inch thickness of the Fiberfrax forms assures a cold-face temperature of about 220°C for the bulk wool, 400°C for the paper, 180°C for the blanket, and 190°C for the block. The bulk and blanket forms have been used in preheating and stress relief of large units during welding, in jet-engine acoustical dampers, and in furnace back-up insulation (Straka). The retention of resilience at high temperatures reduces the mechanical deterioration of the brick overlay in rotary kilns.

Walworth (1957) suggested several thermal, acoustical, and electrical insulating applications, and uses in filters, gaskets, and packings. Basic properties were listed, as well as those for the different forms. First, Graham et al. prepared a graded-fiber filter of 4, 8, and 20 microns fiber diameter, providing 99% efficiency at 760°C in filtering a specific aerosol. Filter performance data were used to solve other filtration problems.

Table 17

Types of Ceramic Fibers

Ref.	(1)	(2)	(3)	(4)	(5)
Types		29-A		"Fiberfrax" (charge)	"Kaowool"
			Analysis (%)		
Al_2O_3	14.0	35.0	52.8	49.3	45.1
SiO_2	54.0	46.8	0-12	49.3	51.9
Na_2O, K_2O	–	2.5	5.4	0.4	0.2
MgO	–	10.5	4.1		–
CaO	22.0		20.1		0.1
B_2O_3	10.0		0-5	1.0	0.08
ZnO					
ZrO_2		4.1	5.1		
TiO_2		0.09	3.3		1.7
BaO		–	9.2		
Fe_2O_3		0.88			1.3

Ref.: (1) Tiede   (4) Straka, 1961
     (2) Provance, 1962   (5) Babcock & Wilcox
     (3) Machlan, 1959   Circular 12-62

Lambertson, Aiken, and Girard (1961) performed melting and fining of refractory glass compositions, and the drawing of continuous-filament, ceramic fibers in a dry inert-atmosphere glovebox, which permitted the use of refractory metals, graphite, and boron nitride as crucible and heater materials, and prevented the adsorption of moisture by the filaments. Refractory glasses in the $BaO-Al_2O_3-SiO_2$, $CaO-Al_2O_3-SiO_2$, and $MgO-Al_2O_3-SiO_2$ systems were successfully drawn into continuous filaments. Tensile strengths as high as 138,000 psi at room temperature and 132,000 psi at 815°C were obtained.

Analytical data on a few glass fiber types are shown in Table 17.

The glass fibers also represent an application of alumina since some of the most successful types have fairly high Al_2O_3 contents, added as Bayer alumina. The major markets for glass fibers are in insulation, textiles, and reinforced plastics. Soda-lime-borosilicate glass is the leading insulation type, since it is fireproof, heat-resistant, inert, and a better insulator than mineral wool. Since Fiberglas (Owens-Corning) was introduced commercially in the early 1930's, research has aimed at stronger and finer fibers. In 1945 more than 4000 forms of fibers for hundreds of different applications had been developed (Ref. Anon. Research Responsible for Fiberglas).

De Dani (1955) described the manufacturing processes for fiber glass.

Alumina enters significantly into the compositions of the special fiber types: beta yarn, E, S, and M glasses. Beta fiber, aluminum borosilicate glass, is used for textiles, for curtains, and drapes. E-glass fibers (low-soda) develop high tensile strength (around 450,000 psi), but S-glass fibers, alkaline earth-aluminosilicate, develop tensile strengths as high as 750,000 psi. Hollow glass fibers can be formed with an outside diameter of 15 microns, and an inner diameter of 9 microns, but some strength has been sacrificed for lightness and flexibility.

The primary area of uses for high-strength fibers is in composite-reinforcement of metals and plastics. The continuous ceramic fibers are only 10 to 20% as strong as the best whiskers in tensile strength, their usable strengths are about equal to those of glass fibers, and, additionally, provide an increase in elasticity several times that contributed by fiber glass. The greatest gains in fiber-reinforcement of metals with continuous fibers has apparently been the combination, silica fiber-aluminum, as applied by Rolls Royce Ltd (Arridge, 1964). This, however, required the development of coatings for the silica fibers to reduce the thermitic reaction between them and molten aluminum. Cratchley and Baker (1964) obtained about twice the tensile strength for silica-reinforced aluminum as for a high-strength aluminum alloy RR-58, and even a greater superiority in comparison with 10% S.A.P. alloy. The strength of all three materials decreased beyond 300°C, but the reinforced aluminum retained its superiority in about the same relative ratio.

The glass fibers (and others) do not retain their pristine strength properties for several reasons. McCreight et al. (1965) classified the factors involved into four types: compositional, processing, structural, and testing. Compositional factors are corrosion, abrasion, atomic bonding, and the time-dependent properties, creep, stress-rupture, and stress-corrosion. The processing factors are those affecting surface perfection, degree of crystal size, if polycrystalline, etc. Structural factors relate to size and shape of the fiber, internal stresses, and defects. Testing methods vary and may themselves affect the results.

Based on theoretical estimates the tensile strength of glass should approach 3×10^6 psi. In bulk, the value may be about 10,000 psi. Anderegg (1939) presented data to show that strength increases with increasing fineness of the fibers. More recently, Otto (1955) and Thomas (1960) show no agreement with this over a wide range of fiber diameter (5 to 15 microns). The values were not affected very markedly by variation in the drawing temperature.

Flaws and chemical contamination of the fiber surfaces reduce the strength, but the damage can be largely reduced by chemical cleaning of the glass rod prior to drawing (Holloway). It seems imperative to prevent abrasive contact between the fibers. Low-alkali glasses do not necessarily have higher tensile strength than high-alkali glasses, but they resist abrasion relatively better. Coatings applied to the fibers as soon as formed aid in preventing abrasion, but at the expense of some loss in strength (Morrison et al., 1962). There appears to be little choice between chemical bonding and physical bonding, as by a resin, in protecting the surface of the fiber.

Rapid cooling was believed to explain the difference in strength between massive glass and fiber glass. Anderson (1958) deduced simplified equations for the cooling time of the fibers to show that the cooling time is about the same as the Maxwell relaxation time of the glass at about 900°C. Otto and Preston (1950) concluded that the glass fibers contracted equally in all directions, hence the so-called Griffith flaws cannot be oriented in any one direction.

Moisture was found to be a potent effect in causing a slow decrease in strength of the fibers. The rate of deterioration and the loss in strength increased with increasing temperature (Holland, 1964).

27 MISCELLANEOUS CERAMIC APPLICATIONS OF ALUMINA

Specific applications of alumina not given extensive consideration in the preceding sections, have been collected under the general classification, "Miscellaneous Ceramic Applications of Alumina." Some of these applications involve no very significant consumption of alumina, and are merely representative of the wide range of uses that are dependent on the properties of the different alumina phases.

Thread Guides

Ceramic forms in the shape of loops are used for changing the direction of textile threads during forming or weaving. The wear of these pieces is considerably accelerated by additives to synthetic organic fibers such as titania, added to opacify or to dull and roughen the surface. Sintered alumina has high resistance to wear and chemical reaction of acid and alkaline solutions of synthetic fiber-formers. Ryshkewitch (1960) stated that flame-polished synthetic sapphire is less satisfactory than smooth but not highly polished, polycrystalline alumina, because a too complete contact with the thread increases the friction. Acceptable surfaces should have small, rounded elevations. Alumina porcelain guides are satisfactory but inferior to pure sintered alumina. The service life is over 100 times that of the formerly used porcelain guides. Thurnauer (1945), and West (1964) described sintered oxide textile guides containing titania to increase the electrical conductivity in order to conduct off static electricity generated by thread friction. Ultraviolet irradiation might be used for a similar purpose.

Frictional Members

Herron (1955) and DuBois (1956) explored the field of frictional devices as applications of ceramics and cermets for service temperatures above 425°C. The desired characteristics in brake and clutch linings are: sufficient mechanical strength to withstand the operating stresses; a high and constant coefficient of friction with different speeds, loads, and temperatures; a low wear rate; low tendency to score the opposing surfaces; and smooth engaging characteristics. Thermal shock resistance is also a factor, since gradients as high as 800°C are possible. Thomson (1958) described a highly wear-resistant, sintered composition comprising up to 15% aluminum oxide particles dispersed in copper or copper-base alloy. Moore and Huffadine (1959) claimed a composition containing up to 40% aluminum oxide dispersed in a molybdenum-chromium alloy having substantially the same coefficient of thermal expansion as the alumina. Huntress (1963) dispersed the alumina in sintered powdered nickel or molybdenum aluminide, bonded with nickel.

Wear-Resistant Ceramic Articles

Erickson (1958) described applications of both hot- and cold-pressed ceramics for plug and ring gauges, thread plug gauges, and various wear applications. Meyerson et al. (1960) described alumina gauge blocks of superior stability. Wheildon (1963) presented data on the use of sintered alumina and flame-sprayed oxide coatings for mold linings, mold arbors, mill linings, hammer protection, extrusion dies, etc. Twells (1950) produced gauges containing 90% Al_2O_3, and fired at 1680°C, from slip-cast compositions.

Sandmeyer (1959) recommended a fused-cast mill liner consisting of alumina and zirconia in the approximate ratio 5:4 (wt), bonded with a glassy matrix containing 1 to 2.5% alkali oxide, 0.1 to 0.5% B_2O_3, and 15% SiO_2. Such compositions show improved resistance to corrosion and erosion.

Ryshkewitch (1960) stated that sintered alumina was not satisfactory in use as air blasting nozzles because of its low impact resistance. A high-alumina composition containing a vitreous phase and manganese oxide was found more resistant to erosion by impingement of refractory grains, however, than steel.

Besides jewel bearings, gauges and mortars and pestles are prepared from synthetic sapphire (Anon. Sapphire Mortar; Anon. Sapphire Plug Gauge). Mortars and pestles are also made from formed polycrystalline alumina. Sapphire ball bearings in small sizes have been used, although their poorer load-bearing properties relative to steel and anisotropic wear have been a disadvantage. Sulc described a method for producing corundum balls 0.02 to 0.5 mm in diameter for use as miniature ball bearings or for optical purposes, obtained by jetting corundum grains or solid solutions through the field of an electric arc or an O-H flame, saturated with vapors of glass-forming oxides to prevent crystallization. Sintered alumina has been used in fabricating gyros to precision in parts per million for airborne directional controls. Alumina ceramic valve seats for faucet assemblies have been recommended by Blodgett (1967) and others.

Welding Rods for Steel

A welding rod compound to improve the weld quality of high-strength low-alloy Cr-Mn-Si-Ni steel contains a fused flux of 21.7 to 24.9% Al_2O_3, 23.8 to 28.8% SiO_2, 14.4 to 18.4% CaO, 9 to 11% MgO, 20 to 22.5% CaF_2, and 1.8 to 2.2% MnO (Makara et al., 1959). This was intended specifically for the peripheral welding of thin-walled tubes.

Nuclear Fuel Elements

A nuclear fuel element composition of the refractory oxide type contains a 50:50 mixture (wt) of coarser than 325-mesh particles of uranium dioxide (92% U^{235}, 8% U^{238}) and calcined alumina. During pressing, surface coatings consisting of 99% fine-ground calcined alumina and 1% of a mixture of $CaCO_3$, MgO, and SiO_2 are applied on both surfaces. The composite is fired for 15 hours at $1750°C$ and for three hours at $1700°C$ before cooling (Norton Grinding Wheel Co., 1960). Sheath failures caused by expansion of uranium dioxide fuel elements on melting have been reported by Pashos.

Aluminum Borates

Scholze (1956); Gielisse, Rockett et al.; Kim and Hummel (1961, 1962); and Berry (1963) described the formation of the fibrous aluminum borates. Scholze found that $2Al_2O_3 \cdot B_2O_3$ forms at $1000°C$, while $9Al_2O_3 \cdot 2B_2O_3$ is formed at $1100°C$, the latter having marked similarity to mullite. Berry found the ratio of length to width of the flexible fibers to be at least 10:1, and the length at least 0.7 mm. Recommended applications were for thermal insulation and filter media, in the form of felted mats, and as reinforcing agents for plastics, ceramics, and metals. Knapp (1962) recommended lithium borate as a pore-sealing agent for sintered alumina articles, when melted on the surface at 800 to $930°C$.

Light Guide

The use and calibration of synthetic sapphire rods, for use in light-guide radiation pyrometry, have been described by Vollmer (1959).

Ozone Storage

Streng and Grosse (1961) described the storage and transportation of gaseous ozone under pressure in a container packed with hollow porous alumina spheres, to prevent explosions and to increase the stability against decomposition.

Oil-Well Fracturing

Tabular alumina sintered balls have been used for "fracturing" plugged oil wells.

Color Modifiers

Alumina is used as a diluent in reducing the color of ceramic stains. It serves as the host, directly or in the form of an aluminate of beryllium, magnesium, calcium, strontium, barium, zinc, or cadmium, as a phosphor with various activators: manganese, chromium, gallium, iron, silver, or cerium (Crosby et al., 1959).

Fillers

Hydrated alumina and calcined alumina are used as fillers in epoxy resins and other organic resins and cements. The dielectric properties depend greatly on the shape and chemical composition of the filler (Dobrer, 1962). Applications are in potting electrical equipment. Calcined alumina is also used in thermosetting resins in wear-resistant journals and bearings.

Aluminum Oxide Films

The preparation of thin alumina films for electron tube windows, infrared transmission, and electron microscope specimen supports has been described by Walkenhorst (1947), Pupko (1949), and others. Amorphous disks as large as 2 to 3 inches in diameter have been prepared from aluminum foil about 10 microns thick. Walkenhorst oxidized pure aluminum foil anodically in 3% ammonium citrate, and separated the film by dissolving the aluminum backing in 0.25% mercuric chloride solution. The film thickness is linear with voltage, being 70 A at 75 volts, and 275 A at 200 volts. No structure could be observed in the film heated as high as $600°C$ at magnifications of 40,000. Pupko anodized in an electrolyte consisting of 1 part HNO_3 (d 1.35 to 1.40) and 2 parts methyl alcohol, for 10 to 15 minutes at 0.3 to 0.35 amp/cm^2, 4.0 to 4.5 volts, and at $20°C$. The washed specimens were immersed in 4% aqueous solution (75% H_3PO_4:25% Cr_2O_3) and reanodized for 15 to 20 seconds before separating the metal backing. Nicholson described a technique for preparing unbacked alumina films having areas up to 100 cm^2, supported at the periphery by a residual portion of the original metal foil.

Harris and Piper (1962) measured the transmission and reflectance of aluminum oxide films in the far infrared.

Alumina in Greases

Marsden and Berstein (1960) thickened lubricating oil with a mixture of pyrogenic alumina and colloidal silica to form high-body greases. The pyrogenic alumina was prepared by the flame hydrolysis of an aluminum halide.

Prosthetic Applications

McLean and Hughes described dental ceramics composed of 60 to 95% Al_2O_3 or mixtures with other oxides, having over 15,000 psi tensile strength and modulus of rupture sufficient to replace gold and platinum. Artificial bone structures prepared from high-alumina porcelains of high compatibility with human body tissues have been developed.

REFERENCES

Abbey, A., "Articles Formed from Abrasive Grains Held in a Fibrous Matrix and Method of Making." Brit. 652,135, 2/21/51.

Abou-El-Azm, Abd-El-Moneim, "Study of the Reaction Rates Between Silica and Other Oxides at Various Temperatures: III, Reaction Rates in Binary and Ternary Mixtures Additional to Those Described in Parts I and II." J. Soc. Glass Technol. 37 (177) 168-81T (1953).

Abou-El-Azm, Abd-El-Moneim, "Study of the Reaction Rates Between (a) Silica, (b) Alumina, (c) Mullite, (d) Zirconia, and Other Oxides at Various Temperatures." J. Soc. Glass Technol. 37 (179) 269-301T (1953).

"Abrasive Grain Sizes. Simplified Practice Recommendation R118-36." National Bureau of Standards, Supt. of Documents, Govt. Printing Office, Washington, D.C.

Abrecht, H., Lukacs, J., and Plotz, E., "New Method for the Determination of the Toughness of Abrasive Grains." Ber. deut. keram. Ges. 37 (8) 355-61 (1960).

Accary, A., and Caillat, R., "Study of Mechanism of Reaction of Hot Pressing." J. Am. Ceram. Soc. 45 (7) 347-51 (1962).

Accountius, O.E., "Whisker Microcomposites." Summary of 7th Refractory Composites Working Group Meeting, pp. 728-62, Vol. III. ADS.TDR 63-4131 (Nov. 1963). AD 601266.

Achenbach, H., "Thermal Decomposition of Synthetic Hydrargillite (Gibbsite)." Chem. Erde 6, 307-56 (1931).

Ackermann, R.J., and Thorn, R.J., "Gaseous Oxides of Aluminum, Tungsten, & Tantalum." J. Am. Chem. Soc. 78, 4169 (1956).

Adam, C., Tableau Mineralogique, Paris, 73 (1869).

Adams, C.E., "The Condensation of Rubidium Vapor Onto Hot Oxide Surfaces." U.S. Dept. Com., Office Tech. Serv., PB Rept. 150,303, 29 pp. (1959).

Adams, J.M., "A Determination of the Emissive Properties of a Cloud of Molten Alumina Particles." J. Quant. Spectrosc. Radiat. Transfer 7, 273-7 (1967).

Adams, M., "Thermal Conductivity. III Prolate Spheroid Envelope Method;" J. Am. Ceram. Soc. 37, 74-9 (1954).

Akiyama, K., "Crystalline Modification of Alumina. I. Inversion of Crystalline Form of Aluminum Hydroxide by Heating." J. Soc. Chem. Ind. Japan 42, 394-5 (1939).

Akiyama, K., et al. "Preparation of Alumina and Barium Compounds from Barium Aluminate." Jour. Soc. Chem. Ind. Japan 41 (7) 218-19B (1938).

Alais, Froges Et Camargue, Cie de Produits, Chimiques et Electrometallurgiques. "Production of Pure Electrocast Spinel." French 898,463, 7/3/44.

Albrecht, F., "The Anisotropy of the Hardness of Synthetic Corundum." Z. Krist. 106, 183-90 (1954).

Albrecht, F., "New Hard Materials (Refractories) Made of Glass." Society for Advancement of Science, Munich (1955).

Aldred, F.H., "Ceramic Materials." U.S. 3,025,175, 3/13/62.

Aleixandre-Ferrandis, V., "Activated Alumina." Ion (Madrid); abstracted in Chem. Trade Jour. 111 (2892) 358 (1942).

Alexander, G.B., "Ceramic Material Bonded to Metal Having Refractory Oxide Dispersed Therein." U.S. 3,110,571, 11/12/63.

Alexander, P.P., "The Hydride Process, II Metals & Alloys" 9 (2) 45-8 (1938).

Alexanian, C., "Filiation of Alumina." Compt. rend. 240 (15) 1621-2 (1955).

Alford, W.J. and Bauer, W.H., "Radiofrequency Plasma Growth and Crystalline Perfection of Single-Crystal Alumina." J. Phys. Chem. Solids Suppl. 1, 71-4 (1967).

Alford, W.J. & Stephens, D.L., "Chemical Polishing & Etching Techniques for Al_2O_3 Single Crystals." J. Am. Ceram. Soc. 46, 193-4 (1963).

Alfred University "Study of Heat Transfer of Ceramic Materials." Final Report Jan. 1, 1955 to April 30, 1957, 93 pp.

Ali, D., "Hydrothermal Synthesis, Its Scientific and Technical Importance." Berg- u. huttenmann. Monatsh. montan. Hochschule Leoben 94 (4) 79-80 (1949).

Allen, R.D., "Spectral and Total Emissivities of Rokide C on Molybdenum Above 1800°F." J. Am. Ceram. Soc. 44, 374 (1961).

Allen, T.B., "Abrasive Electric Furnace Product." U.S. 1,187,225, 6/13/16.

Allen, W.C., "Dimensional Control of Ceramic Materials." Am. Ceram. Soc. Bull. 40 (6) 383-9 (1961).

Alley, J.K. and Johnson, R.C., (to U.S. Dept. Interior) "Aluminum Borate Whiskers." U.S. 3,350,166. 10/31/67.

Allison, E.B., Brock, P. & White, J., "Rheology of Aggregates Containing a Liquid Phase with Special Reference to the Mechanical Properties of Refractories of High Temperatures." Trans. Brit. Cer. Soc. 58 (9) 495-531 (1959).

Alper, A.M., Begley, E.R., Londeree, J.W., & McNally, R.N., "Alumina-Zirconia Cast Refractories." U.S. 3,132,953, 5/12/64.

Alper, A.M. & McNally, R.N., Fr. 1,356,151, 3/20/64.

Alper, A.M., & McNally, R.N., "Fused Cast Refractory." U.S. 3,140,955, 7/14/64.

Alper, A.M., McNally, R.N., Ribbe, P.G., & Doman, R.C., "The System $MgO-MgAl_2O_4$." J. Am. Ceram. Soc. 45 (6) 263-8 (1962).

Alston, M.W. & Strickland, F.G.W., "Electrically Insulating Coatings on Metals." Brit. 893,370, 4/11/62.

Altman, R.L., "Vaporization of Magnesium Oxide and Its Reaction with Alumina." J. Phys. Chem., 67 (2) 366-69 (1963).

Alumina Ceramic Manufacturers Association, 53 Park Place, New York 7, N.Y.

   (a) E2-63 "Specification for Impervious, High Strength Alumina Ceramics for Electrical and Electronic Application." (1963).

   (b) M1-63 "Specification for Impervious, High Strength Alumina Ceramics for Mechanical Applications" (1963).

   (c) Standards of the Alumina Ceramic Manufacturers Association, 2nd Ed. (1964).

Amberg, C.R. & Harrison, H.C., "Vitreous Coating Refractory Material." U.S. 2,422,215, 6/17/47.

Amberg, C.R. & Harrison, H.C., "Enameling Difficultly Wettable Substances such as Graphite, Carbon, SiC, Fused Al_2O_2, and Their Mixture." Fr. 947,701, 7/11/49.

Amelinckx, S., "Growth Spirals on Synthetic Rubies." Compt. rend. 234 (18) 1793-4 (1952).

American Society for Testing Materials, Philadelphia 3, 1960. "X-Ray Powder Data File: Sections 1 to 5, Revised." Am. Soc. Testing Materials, Spec. Tech. Pub. 48.

Amerikov, A.V. & Pirogov, Y.A., "The Production of Corundum Tubes" Ogneupory 25, 527-30 (1960).

Ampian, S.G., "X-Ray Diffraction and Optical Microscopic Data on Several Important Phases in the Binary Systems $CaO-Al_2O_3$, $CaO-SiO_2$, and $Na_2O-Al_2O_2$." U.S. Bur. Mines, Rept. Invest. 6428 (4) 53 pp. (1964).

Anaconda Aluminum Company. Chem. & Eng. News 41 (20) 29 (1963).

Anderegg, F.O., "Strength of Glass Fiber." Ind. & Eng. Chem. 31 (3) 290-8 (1939).

Anderson, D., "Manufacture of Articles, e.g., Brick, from Alumina Cement and Corundum." Ger. 657,770, 2/24/38.

Anderson, O.L., "Cooling Time of Strong Glass Fibers." J. Appl. Phys., 29 (1) 9-12 (1958).

Anderson, P.J. & Murray, P., "Zeta Potentials in Relation to Rheological Properties of Oxide Slips." J. Am. Ceram. Soc. 42 (2) 70-4 (1959).

Anderson, R.A. & Stepp, E.E., "Glass-to-Sapphire End Window Seals." Rev. Sci. Instr. 33 (1) 119-20 (1962).

Anderson, W.D., "Hydrostatic Pressing of Alumina Radomes." PB Rept. 131565. U.S. Govt. Res. Repts. 29 (3) 125 (1958).

Anderson, W.D., Brandt, W.O., LeClercq, L.J., & Fargo, J.J., "Method of Making Dense Refractory Objects." U.S. 3,016,598, 1/16/62.

Ando, F., "Powder Form of γ-Alumina." Japan 13,455, 9/10/62.

Andrus, J.M., "Enameled Aluminum & Process for Manufacture Thereof." U.S. 2,991,234, 7/4/61.

Angelides, P., "Relief Polishing of High-Alumina Ceramics for Metallographic Study." J. Am. Ceram. Soc. 44 (3) 145 (1961).

Anon., "Abrasive Grain in Bonded Abrasive Products: I." Grinding Finishing 9 (5) 36-39 (1963).

Anon., "Alumina from Andalusite." Anglo-Swedish Rev. (Jan. 1944).

Anon., "Alumina from Clays—A Current Summary of Products & Processes." Chem. Industries 54 (1) 65 (1944).

Anon., "Alumina Powders." Chem. Inds. 63 (2) 424 (1948).

Anon., "Ceramic Foam and Ceramic Honeycomb—Literature Survey." AD-282,465, U.S. Govt. Res. Repts. 37 (23) 84 (1962).

Anon., "Ceramic Wafers—Base Materials for Electronic Units." Cer. Age, 77 (2) 58-61 (1961).

Anon., "Container Glass of High Alumina Content." Glastech. Ber., (1940-1943). Glass Ind., 27 (6) 293 (1946).

Anon., "Cutting Hard Brittle Ceramics by Air Abrasion." Ceram. Age 78 (2) 48-9 (1962).

Anon., "Developments to Watch." Prod. Eng. 35 (4) 55-56 (1964).

Anon., "Fiber Glass Manufacturers." Glass Ind., 41 (6) 328-35, 366 (1960).

Anon., "History of Abrasive Grain." Grinding Finishing 9 (2) 22-27 (1963).

Anon., "How to Grind Glass and Ceramics on a Surface Grinder." Ceram. Ind. 81 (6) 67 (1963).

Anon., "Modern Refractory Practice." 3rd ed. Published by Harbison-Walker Refractories Co., Pittsburgh 22, Pa., 1950.

Anon., "Oxide Thermoelectric Materials." PB 160,988, U.S. Govt. Res. Repts. 37 (22) S-23-4 (1962).

Anon., "Research Responsible for Fiberglas Development in Last 15 Years." Amer. Glass Rev. 65 (4) 11, 18 (1945).

Anon., "Sapphire Mortar and Pestle." Ind. Diamond Rev. 7 (81) 235 (1947).

Anon., "Sapphire Plug Gauge." New Equipment Digest 13 (4) 26 (1948).

Anon., "Surface Preparation Abrasives for Industrial Maintenance Painting—NACE Technical Committee Report." Mater. Protect. 3 (7) 76-80 (1964).

Anon., "Trip to Varian Associates." Ceram. News, 10 (9) 18-19 (1961).

Anon., "Ultrafine Alumina." Chem. Inds. 63 (4) 626 (1948).

Anon., "Use of Abrasive Grain in Polishing." Grinding Finishing 9 (12) 26-28 (1963).

Anon., "Use of Abrasive Grain in Pressure Blasting." Grinding Finishing 9 (9) 41-44 (1963).

Antal, J.J. & Goland, A.N., "Study of Reactor-Irradiated α-Al_2O_2." Phys. Rev. 112 (1) 103-11 (1958).

Antonsen, R., "Process for the Production of Aluminum Oxide." U.S. 2,773,741, 12/11/56.

Appen, A.A., "Structure of Complex Silicate Glasses." Zhur. Priklad. Khim. 27 (2) 121-6 (1954).

Appen, A.A., "Relation Between the Properties & Constitution of Glass: V, State and Properties of Al_2O_3 & TiO_2 in Silicate Glasses." Zhur. Priklad. Khim. 26, 9-17 (1953).

Arakelyan, O.I. & Chistyakova, A.A., "Synthetic Boehmite. II. Phase Composition of Products Obtained in Firing Boehmite." Zh. Prikl. Khim. 35, 1653-6 (1962).

Aramaki, S. & Roy, R., "Mullite-Corundum Boundary in the Systems $MgO-Al_2O_3-SiO_2$ & $CaO-Al_2O_3-SiO_2$." J. Am. Ceram. Soc. 42 (12) 644-5 (1959)

Aramaki, S. & Roy, R., "Revised Phase Diagram for the System $Al_2O_3-SiO_2$." J. Am. Ceram. Soc., 45 (5) 229-42 (1962).

Arandarenko, T.T. & Poluboyarinov, D.N., "Use of Hydrated Alumina for Manufacturing High-Alumina Refractories." Ogneupory 23 (10) 467-76 (1958).

Archibald, R.C., "Preparation of Spheroidal Particles." U.S. 2,435,379, 2/3/48.

Arkel, A.F. van & Fritzius, C.P., "Infrared Absorption of Hydrates." Rec. Trav. chim. Pays-Bas. 50, 1035 (1931).

Arghiropoulos, B., Elston, J., Juillet, F. & Teichner, S., "A New Method for the Preparation of Colored Nonstoichiometric Alumina." Compt. rend. 249, 2549-51 (1959).

Arghiropoulos, B., Juillet, F., Prettre, M. & Teichner, S., "Structural Evolution and Electric Properties of Nonstoichiometric Colored Alumina." Compt. rend. 249, 1895-7 (1959).

Arizumi, T. & Tani, S., "On the Electrical Conductivity of Alumina." J. Phys. Soc. Japan 5, 442-7 (1950).

Armstrong, T.C., Jr. & Whitehurst, B.M., "Sulfoaluminate Cement." U.S. 3,147,129, 9/1/64.

Armstrong, W. M., Chaklader, A.C.D., & Clarke, J.F., "Interface Reactions Between Metals & Ceramics: I, Sapphire-Nickel Alloys." J. Am. Cer. Soc. 45 (3) 115-18 (1962).

Arndt, K. & Hornke, W., "Electrically Prepared Corundum." Tonind Ztg. 60, 212-4, 249-50, 273-4, 310-11 (1936).

Arnold, H., "Decrease of Internal Strains in Synthetic Spinels by Formation of Voids." Z. Krist., 114 (1-2) 23-27 (1960).

Arnot, E., "Fledspar in Glass." Sprechsaal 70 (29) 374 (1937).

Arridge, R.G., "Fiber Reinforced Metals, A New Technology." The Chem. Eng. CE 252-6 CE 256 (Oct. 1964).

Aschenbrenner, M., Z. physik. Chem. 127, 415 (1927).

Ashley, K.D. & Bruni, G.J., "Process of Preparing Alumina Sols from Aluminum Pellets." U.S. 2,859,183, 11/4/58.

Astbury, N.F., & Davis, W.R., "Internal Friction in Ceramics." Trans. Brit. Cer. Soc. 63 (1) 1-18 (1964).

ASTM Std., 1961, Part V, pp. 453-82. "Chemical Analysis of Refractory Materials." ASTM Designation C 18-60. Revised in 1960.

Atlas, L.M., "Metal-Reinforced Ceramic Radome." WADC Tech. Rept. 58-329 ASTIA No. AD 155872 (Oct. 1958).

Atlas, L.M., "Research & Development Services Leading to the Control of Electrical Properties of Materials for High Temperature Radomes" PB Rept. 161423, U.S. Govt. Research Repts. 33 (5) 522 (1960).

Atlas, L.M. & Firestone, R.F., "Application of Thermoluminescence & Reflectance Methods to Study of Lattice Defects in Alumina Ceramics." J. Am. Ceram. Soc. 43 (9) 476-84 (1960).

Atlas, L.M., Nagao, H., & Nakamura, H.H., "Control of Dielectric Constant & Loss in Alumina Ceramics." J. Am. Ceram. Soc. 45 (10) 464-71 (1962).

Atlas, L.M. & Sumida, K., J. Am. Ceram. Soc. 41 (5) 150-60 (1958).

Attinger, C., "Orientation & Hardness of Synthetic Corundum." Bull. Soc. Suisse Chronometrie (reprint), 4 pp. (June 1951).

Ault, G.M. & Deutsch, G.C., "NACA Studies Ceramals for Turbine Blading." S.A.E. Journal 58 (5) 40-2 (1950).

Ault, N.N., "Alumina Rods for Coating Articles." U.S. 2,882,174, 4/14/59.

Ault, N.N. & Milligan, L.H., "Alumina Radomes by Flame-Spray Process." Am. Cer. Soc. Bull. 38 (11) 661-4 (1959).

Ault, N.N. & Ueltz, H.F.G., "Sonic Analysis for Solid Bodies." J. Am. Cer. Soc. 36 (6) 199-203 (1953).

Auriol, A., Hauser, G. & Wurm, J.G., System CaO-Al₂O₃ Nov. 19, 1961. See Phase Diagrams for Ceramists (1964). Figure 232.

Aust, K.T., Hanneman, R.E., Niessen, P. & Westbrook, J.H., "Ceramic Microstructures," ed. Fulrath and Pask, Wiley, N.Y. Chap. 8 Analysis of Nature of Grain Boundaries in Ceramics.

Austin, C.R. & Rogers, E.J., "Ceramic Body & Batch for Making." U.S. 2,494,277, 1/10/50.

Austin, C. & Rogers, E.J., "Ceramic Batch." U.S. 2,494,276, 1/10/50.

Austin, C.R., Schofield, H.L., & Haldy, N.L., "Alumina in Whiteware." J. Am. Ceram. Soc. 29 (12) 341-54 (1946).

Austin, J.B., "The Thermal Expansion of Some Refractory Oxides." J. Am. Ceram. Soc. 14, 795-810 (1931).

Aves, W.L., "Coatings for Re-Entry." Metal Progr, 75 (3) 90-4, 189C, 190 (1959).

Aves, W.L., Jr. & Hart, R.A., "Metal-Ceramic Laminated Skin Surface." U.S. 3,031,331, 4/24/62.

Avgustinik, A.I., "Kinetics of Comminution of Some Refractory Materials." Poroshkovaya Met. Akad. Nauk Ukr. SSR, 3 (2) 3-7 (1963).

Avgustinik, A.I. & Kozlovskii, L.V., "Preparation of Thermo-couple Thimbles from Aluminum Oxide." Trudy Leningrad. Tekhnol. Inst. im. Lensoveta 1959, No. 57, 15-29.

Avgustinik, A.I. & Mchedlov-Petrosyan, O.P., "Kinetics of the Reaction of Barium Sulfate with Alumina in the Solid Phase." Zhur. Priklad. Khim. 20 (11) 1125-32 (1947).

Aylmore, D.W., Gregg, S.J. & Jepson, W.B., "The Oxidation of Aluminum in Dry Oxygen in the Temperature Range 400-650°." J. Inst. Metals 88, 205-8 (1960).

Azarov, K.P. & Chistova, E.M., "Phosphate Enamels." Zhur. Priklad. Khim. 31 (10) 1602-1604 (1958).

Azarov, K.P. & Chistova, E.M., "Phosphate Enamels for Aluminum." Tr. Rostovsk-Na-Donu Inzh.-Stroit. Inst. 1963 (25) 41-8.

Azarov, K.P. & Serdyukova, N.M., "Effect of Alumina Gel on the Properties of Enamel Slip: III." Kolloid. Zhur. 4, 699-704 (1938).

Baab, K.A. & Kraner, H.M., "Investigation of Abrasion Resistance of Various Refractories." J. Am. Ceram. Soc. 31 (11) 293-8 (1948).

Babcock, C.L. & McDavitt, M., "Refractory Glass Composition." U.S. 2,961,328, 11/22/60.

Babushkin, V.I. & Mchedlov-Petrosyan, O.P., "Thermodynamic Study of Solid-Phase Reactions in the System Calcium Oxide-Alumina." Zh. Prikl. Khim. 32 (1) 46-50 (1959).

Bachman, J.R. & Eusner, G.R., "Refractory Product & Method of Manufacture." U.S. 2,878,132, 3/17/59.

Backhaus, K., "Electrical Conductivity of Groundmasses, Binders, and Mixes for Preparing Insulators for Electric Heaters." Ber. deut. keram. Ges. 19 (11) 461-9 (1938).

Bacon, G.E., "X-ray Examination of the Spinel-Type Mixed Oxide, MgFeAlO₄." Acta Cryst. 7 (4) 361-3 (1954).

Bagley, R.D., "Effects of Impurities on the Sintering of Alumina." Univ. Microfilms (Ann Arbor, Mich) Order No. 64-8053, 223 pp.; Dissertation Abstr. 25 (2) 1064 (1964).

Bahn, R., "Electrophoretic Deposition of Ceramic Insulating Materials." Silikattech. 9 (7) 299-303 (1958).

Baimakov, Y.V. & Maizel, E.V., "Reduction of Silica and Alumina with Carbon." Elektromet Tsvetnykh. Metal. 1957 (188) 10-23.

Bakardiev, I., "Dependence of the Acidity & the Basicity of an Aluminum Oxide Surface on Its Curvature." Z. Physik. Chem. (Leipzig) 225 (5/6) 273-83 (1964).

Bakr, M.Y., "Extraction of Alumina from Egyptian Kaolins & Clays. I. Acid Process for the Recovery of Alumina." Sprechsaal 96 (24) 577 (1963).

Ballard, A.H., "Molding Refractory Oxides." U.S. 2,538,959, 1/23/51.

Ballman, A.A., "Method of Growing Corundum Crystals." U.S. 2,979,413, 4/11/61.

Banashek, E.I., Sokolov, V.A., Rubinchik, S.M., & Fomin, A.I., "Measurement of Enthalpy of Corundum from 1290 to 1673° K." Izv. Akad. Nauk SSSR, Neorgan. Materialy 1, 698-701 (1965).

Baque, H.W., "Glass-Refractories Symposium: II, Fused Cast Refractories & Their Application to Glass-Furnace Construction." Am. Ceram. Soc. Bull. 29 (1) 9-11 (1950).

Barany, R. & Kelley, K.K., "Heats & Free Energies of Formation of Gibbsite, Kaolinite, Halloysite, and Dickite." U.S. Dept. Interior, Bur. Mines, RI 5825 (1961).

Barber, D.J. (Aluminium Lab. Ltd.), "Electron Microscopy and Diffraction by Al₂O₃ Whiskers." Phil. Mag. 10, 75-94 (1964).

Barks, R.E. and Roy, D.M., "Single-Crystal Growth of R₂O₃ (Corundum Structure) Oxides by the Flux Method." J. Phys. Chem. Solids Suppl. 1, 497-504 (1967).

Barlett, H.B., "Occurrence & Properties of Crystalline Alumina in Silicate Melts." J. Am. Ceram. Soc. 15, 361-4 (1932).

Barnes, J.F., et al. "Making Ceramic-Bonded Articles." U.S. 2,293,099, 8/18/42.

Barnes, M.H., "Synthetic Gem Production." U.S. 2,634,554, 4/14/53.

Barnes, M.H. & McCandless, E.L., "Glossing Corundum & Spinel." U.S. 2,448,511, 9/7/48.

Barnitt, J.B., "Adsorbent Material & Method of Producing Same." U.S. 1,868,869, 7/26/32.

Baroody, E.M., Duckworth, W.H., Simons, E.M., & Schofield, H.Z., "Effect of Shape & Material on Thermal Rupture of Ceramics." AECD-3486, U.S. Atomic Energy Commission, May 22, 1951.

Baroody, E.M., Simons, E.M. & Duckworth, W.H., "Effect of Shape on Thermal Fracture." J. Am. Ceram. Soc. 38 (1) 38-43 (1955).

Barrer, R.M., "Migration in Crystal Lattices." Trans. Faraday Soc. 37 (11) 590-9 (1941).

Barrer, R.M., McKenzie, N. & Reay, J.S.S., "Capillary Condensation in Single Pores." J. Colloid Sci. 11, 479-95 (1956).

Barrett, E., Joyner, L.G. & Halenda, P.P., "The Determination of Pore Volume & Area Distributions in Porous Substances. I. Computations from Nitrogen Isotherms." J. Am. Chem. Soc. 73, 373-80 (1951).

Barrett, L.R. & Thomas, A.G., "Surface Tension and Density Measurements on Molten Glasses in the CaO-Al₂O₂-SiO₂ System." J. Soc. Glass Technol. 43 (211) 179-90T (1959).

Barrett, W.T. & Welling, C.E., "Method of Preparing Low Soda-Content Alumina Catalyst." U.S. 2,774,744, 12/18/56.

Barta, R. & Barta, C., "Study of the System Al₂O₂." Zhur. Priklad. Khim. 29 (3) 341-53 (1956).

Barta, R., Bartuska, M., Hlavac, J. & Prochazka, S., "Electrical Resistance of Sintered Alumina." Silikaty 1 (1) 77-83 (1957).

Bartlett, E.S., et al. "Coatings for Protecting Molybdenum from Oxidation at Elevated Temperatures." DMIC Rept 109 (Battelle) (1961).

Bartlett, E.S. & Ogden, H.R., "Summary of the 9th Meeting of Refractory Composite Working Group," Battelle DMIC Memo 200 (3/3/65).

Bartlett, R.W. and Hall, J.K., "Wetting of Several Solids by Al₂O₃ and BeO Liquids." Am. Ceram. Soc. Bull. 44, 444-8 (1965).

Bartos, L.H., "Control Tests for the Gel-Forming Properties of Bentonites." Ceram. Age 57 (5) 20-23 (1951).

Bartuska, M., "Sintering of Alumina." Silikaty 3, 139-53 (1959).

Baskey, R.H., "Fiber Reinforcement of Metallic & Nonmetallic Composites: I, State of Art & Bibliography of Fiber Metallurgy." AD 274379, U.S. Govt. Res. Repts. 37 (13) 40 (1962).

Basmadjian, D., Fulford, G.N., Parsons, B.I. & Montgomery, D.S., "The Control of the Pore Volume & Pore Size Distribution in Alumina & Silica Gels by the Addition of Water-Soluble Organic Polymers." J. Catalysis 1 (6) 547-63 (1962).

Bassi, A. & Camona, G., "Etchant for Sintered Alumina." Energia Nucl. (Milan) 10, No. 4, 215-6 (1963).

Bates, J.L. & Gibbs, P., "Some Optical Properties of Corundum." Univ. Utah Progress Rept. V ASTIA AD 140406 (1957).

Baudewyns, J., "Attack of Refractory Materials Used in Glass-making." Ing. Chim. 22 (131) 211-27 (1938).

Bauer, E., "Phenomenological Theory of the Crystallization on Surfaces: I." Z. Krist., 110 (5-6) 372-94 (1958).

Bauer, E. & Carlson, D.J., "Mie Scattering Calculations for Micron-Size Alumina & Magnesia Spheres." J. Quant. Spectry. Radiative Transfer 4 (3) 363-74 (1964).

Bauermeister, G. & Fulda, W., "The Bayer Process (for Purification of Bauxite)." Aluminum 25, 97-100 (1943).

Baumann, H.N., Jr., "Petrology of Fused Cast High Alumina Refractories." Am. Ceram. Soc. Bull. 35 (9) 358-60 (1956).

Baumann, H.N., Jr., "Petrology of Fused Alumina Abrasives." Am. Ceram. Soc. Bull. 35 (10) 387-90 (1956).

Baumann, H.N., Jr., "Crystal Habit of α-Alumina in Alumina Ceramics." Am. Ceram. Soc. Bull. 37 (4) 179-84 (1958).

Baumann, H.N., Jr. & Benner, R.C., "Aluminous Material." U.S. 2,360,841, 10/24/44.

Baumann, H.N., Jr. & Benner, R.C., "Fused Aluminum Oxide Abrasive Material." U.S. 2,424,645, 7/29/47.

Baumann, H.N., Jr. & Wooddell, C.E., "Aluminous Abrasive Material." U.S. 2,383,035, 8/21/45.

Baumann, H.N., Jr., "Microscopy of High-Temperature Phenomena." Am. Ceram. Soc. Bull. 27 (7) 267-71 (1948).

Bauple, R., Gilles, A., Romand, J. & Vodar, B., "Absorption Spectra of Quartz & Corundum in the Ultraviolet." J. Opt. Soc. Am. 40, 788-9 (1950).

Baxter, J.R. & Roberts, A.L., "Development of Metal-Ceramics from Metal-Oxide Systems." Iron Steel Inst. (London) Spec. Rept. 1956, No. 58, 315-24.

Bayer, Gerhard (to Owen-Illinois, Inc.), "Dense Alumina Refractories Containing an Oxide of Ta or Nb." U.S. 3,241,988, 3/22/66.

Bayer, P.D. and Cooper, R.E., "The Effects of Chemical Polishing on the Strength of Sapphire Whiskers." J. Mater. Sci. 2, 347-53 (1967).

Beals, R.J. & Cook, R.L., "Directional Dilatation of Crystal Lattices at Elevated Temperatures." J. Am. Ceram. Soc. 40 (8) 279-84 (1957).

Bearer, L.C., "Process for Manufacturing Mullite-Containing Refractories." U.S. 2,641,044, 6/9/53.

Beasley, R.M. & Johns, H.L., "Inorganic Fibers & Method of Preparation." U.S. 3,082,099, 3/63.

Beasley, R.M. & Johns, H.L., "Inorganic Fibers & Preparation Thereof." U.S. 3,110,545, 11/12/63.

Beauchamp, E.K., "Flash Etching." of Al$_2$O$_3$ Grain Boundaries." J. Am. Ceram. Soc. 43 (10) 552 (1960).

Beauchamp, E.K., "Impurity Dependence of Creep of Aluminum Oxide." Univ. Microfilms (Ann Arbor, Mich), Order No. 62-5696, Dissertation Abstr. 23, 3442 (1963).

Beauchamp, E.K., Baker, G.S. & Gibbs, P., "Impurity Dependence of Creep of Aluminum Oxide." AD-277,531, U.S. Govt. Res. Repts. 37 (20) 26 (1962).

Becart, M., "The Calculation of Molecular Constants by Means of Measurements Made on Band Heads." Colloq. Spectros. Intern. 9th Lyons 1961, 2, 164-9 (1962).

Becart, M. & Mahieu, J.M., "Band Spectra of Aluminum Oxide." Compt. Rend. 256 (26) 5533-4 (1963).

Becherescu, A. and Cristea, V., "Casting Al$_2$O$_3$ and ZrO$_2$ Used in Superrefractory Coatings by Plasma Melting." Bul. Stiint Tehnic Inst. Politehnic Timisoara, Vol. Spec. 9, 35-43 (1964).

Bechtel, H. & Ploss, G., "Bonding of Ceramic Raw Materials with Monoaluminum Phosphate Solution: I." Ber. deut. keram. Ges. 37 (8) 362-7 (1960).

Bechtel, H. & Ploss, G., "Bonding of Ceramic Raw Materials with Monoaluminum Phosphate Solution (Refractory Bond No. 32): II." Ber. deut. keram. Ges. 40 (7) 399-408 (1963).

Beck, W.R., "Crystallographic Inversions of the Aluminum Orthophosphate Polymorphs & Their Relation to Those of Silica." J. Am. Ceram. Soc. 32 (4) 147-51 (1949).

Beevers, C.A. & Brohult, S., "Formula of β-Alumina, Na$_2$O·11Al$_2$O$_3$." Z. Krist. 95. 472-5 (1936).

Beevers, C.A. & Ross, M.A.S., "Crystal Structure of "Beta Alumina," Na$_2$O·11Al$_2$O$_3$." Z. Krist. 97, 59-66 (1937).

Beggs, J.E., "Metallic Bond." U.S. 2,857,663, 10/28/58.

Beghi, G., Cazzaniga, & Piatti, G., "Thermal Transformations of Aluminas Extracted from Al-Al$_2$O$_3$ Composites." J. Nucl. Mater. 18, 237-46 (1966).

Beletskii, M.S. & Rapoport, M.B., "The Aluminum Compound Formed at High Temperatures." Doklady Akad. Nauk USSR 80, 751-4 (1951).

Bell, N., "Activated Bauxite & Catalyst Containing It." U.S. 3,011,980, 12/5/61.

Bell, W.C., Dillender, R.D., Lominac, H.R. & Manning, E.G., "Vibratory Compacting of Metal & Ceramic Powders." U.S. Air Force, Air Research & Dev. Command, WADC Tech. Rept. No. 53-193, 42 pp. (April 1953).

Bell, W.C., Dillender, R.D., Lominac, H.R. & Manning, E.G., "Vibratory Compacting of Metal & Ceramic Powders." J. Am. Ceram. Soc. 38 (11) 396-404 (1955).

Belova, E.N. & Ikornikova, N. Yu., "Laue Patterns of Synthetic Corundum." Doklady Akad. Nauk SSSR 81 (5) 829-32 (1951).

Beyankin, D.S., "Zeta-Al$_2$O$_3$ is Identical with Gamma-Al$_2$O$_3$." Zbl. Mineral. Geol. Palaont (A), 300 (1933).

Belyankin, D.S. & Lapin, V.V., "Zirconium Dioxide in Mullite Refractory." Doklady Akad. Nauk SSSR 73 (2) 367-69 (1950).

Bender, H., "High-Temperature Metal-Ceramic Seals." Ceram. Age 63 (4) 15-7, 20-1, 46-50 (1954).

Benedicks, C. & Wretblad, P.E., "Granulometric Determination of Fine Abrasives." Schleif- & Poliertech. 13 (1) 1-5 (1936).

Benedicks, C. & Wretblad, P.E., "Continuous Separation of Abrasives." Schleif- & Poliertech. 13 (3) 48-55 (1936).

Bengough, G.D. & Stuart, J.M. Brit. 223,994, 11/2/24.

Benner, R.C. & Baumann, H.N., Jr., "A Cast Refractory." U.S. 2,019,209, 10/29/35.

Benner, R.C. & Baumann, H.N., Jr., "Refractory & Method of Making." U.S. 2,154,318, 4/11/39.

Benner, R.C. & Baumann, H.N., Jr., "Abrasive." U.S. 2,318,360, 5/4/43.

Benner, R.C. & Easter, G.J., "Refractory & Method of Making." U.S. 2,203,770, 6/11/40.

Bennett, D.G., "High Temperature Resistant Ceramic Coatings." Proc. Porcelain Enamel Inst. Forum 17, 29-38 (1955) Ceram. Age, 66 (6) 24 (1955).

Bennett, D.G., et al. "Ceramic Coating Slip for Protection of Metal During Heat-Treating." U.S. Air Force, Air Material Command, AF Tech. Rept. No. 6079, 10 pp. (July 1950).

Bentley, F.J.L., & Feachem, C.G.P., "Alumina Catalysts for Organic Reactions." J.Soc. Chem. Ind. (London) 64, 148-9 (1945).

Berezhkova, G.V. & Rozhanskii, V.N., "Polysynthetic Twins in Threadlike Corundum Crystals." Fiz. Tverd. Tela 6 (9) 2745-9 (1964).

Berezhnoi, A.S., "Some Principles of Search for New Refractories." Ogneupory 28 (8) 341-7 (1963).

Bergeron, C.G., Tennery, V.J., et al. "Protective Coatings for Refractory Metals." PB Rept. 171193, 54 pp.; U.S. Govt. Research Repts. 35 (3) 329 (1961).

Bergmann, O. Thesis, Tech. Hochschule Graz No. 25000/428 (1958).

Bergmann, O.R. and Barrington, J., "Effect of Explosive Shock Waves on Ceramic Powders" J. Am. Ceram. Soc. 49, 502 (1966)

Bergmann, O. & Torkar, K., "Production of Hydrous Aluminas & Aluminas." U.S. 2,984,606, 5/16/61.

Berman, R., "Thermal Conductivities of Some Dielectric Solids at Low Temperatures." Proc. Roy. Soc. (London) A208,90-108 (1951).

Berman, R., "Thermal Conductivity of Some Polycrystalline Solids at Low Temperatures." Proc. Phys. Soc. (London) 65 (396A) 1029-40 (1952).

Berman, R., "Heat Conductivity of Nonmetallic Crystals at Low Temperatures." Cryogenics 5, 297-305 (1965).

Berman, R.M., Bleiberg, M.L. & Yeniscavich, W., "Fission Fragment Damage to Crystal Structures." J. Nuclear Materials 2 (2) 129-40 (1960).

Berman, R. & Foster, E.L., "Thermal Conduction in Artificial Sapphire Crystals at Low Temperatures: I, Nearly Perfect Crystals." Proc. Roy. Soc. (London) A231 (1184) 130-44 (1955).

Berman, R., Foster, E.L., Schneidmesser, B., & Tirmizi, S.M.A., "Effects of Irradiation on the Thermal Conductivity of Synthetic Sapphire." J. Appl. Phys. 31 (12) 2156-9 (1960).

Bernstein, R.B., "Hydrogen & Oxygen Isotopes Applied to the Study of Water-Metal Reactions. Exchange of D_2O^{18} with a-Alumina, "Monohydrate." U.S. Atomic Energy Comm. ANL-5889, 15 pp. (1958).

Berry, C.E. & Kamack, H.J., "Surface Activity in Fine Dry Grinding,", pp 196-202 in "Surface Activity-Solid Liquid Interface,", Editor J.H. Schulman, Academic Press, New York 1958, 352 pp.

Berry, K.L., "Fibrous Aluminum Borates." U.S. 3,080,242, 3/5/63.

Berthier, P., "Alumine Hydrates des Beaux." Ann. Mines (2) 6, 531 (1821).

Bertrand, A., "Application of the Ternary Eutectic SiO_2-Al_2O_3-CaO in Lead & Antimony Metallurgy." Chim. & Ind. (Paris) 77 (6) 1281-7 (1957).

Bevan, D.J.M., Shelton, J.P. & Anderson, J.S., "Properties of Some Simple Oxides & Spinels at High Temperatures." J. Chem. Soc. (London) 1948, 1729-41.

Bezborodov, M.A., Mazo, E.E., & Kaminskaya, V.S., "Enamels for Aluminum Based on the Lead-Phosphate-Silicate System." Steklo i Keram. 17 (1) 35-9 (1960).

Bezborodov, M.A. & Ulazovskii, V.A., "Investigation of the System Li_2O-Al_2O_3-B_2O_3-SiO_2 in the Vitreous State." Belorusskii Politekhnicheskii Institut, Nauchno-Issledovatel'skaya Laboratoriya Silikatov i Stekla, Minsk, 1957. 32 pp.

Bhimasenacher, J., "Elastic Constants of Corundum." Current. Sci. (India) 18, 372-3 (1949).

Bibbins, G.L., "Metal Bonded Abrasive Composition." U.S. 3,036,907, 5/29/62.

Bickford, F.A. & Smith, R.K., "Method of Making a Nonporous, Semicrystalline Ceramic Body." U.S. 2,887,394, 5/19/59.

Bielanski, A. & Burk, M., "Kinetics of Sorption of Water Vapor on Active Aluminum Oxide." Zeszyty Nauk. Akad. Gorniczo-Hutniczej, Ceram. No. 1, 97-106 (1956).

Bielanski, A. & Burk, M., "Influence of Temperature of Activitation on Porous Structure of Alumina." Roczniki Chem. 31, 969-81 (1957).

Bienstock, D., et al. "Removal of Sulfur Oxides from Flue Gas with Alkalized Alumina at Elevated Temperatures." J. Eng. Power 86 (3) 353-60 (1964).

Bierlein, T.K., Newkirk, H.W., Jr. & Mastel, B., "Etching of Refractories & Cermets by Ion Bombardment." J. Am. Ceram. Soc. 41 (6) 196-200 (1958).

Bills, D.G. & Evett, A.A., "Glass, A Disturbing Factor in Physical Electronics Measurements." J. Appl. Phys. 30 (4) 564-7 (1959).

Biltz, W. & Lemke, A., "Molecular and Atomic Volumes. XXII. Argillaceous Earth & Spinels." Z. anorg. allgem. Chem. 186, 373-86 (1930).

Binns, D.B., "The Testing of Alumina Ceramics for Engineering Applications." J. Brit. Ceram. Soc. 2, 294-308 (1965).

Binns, D.B., "Some Physical Properties of Two-Phase Crystal-Glass Solids." No. 23 "Science of Ceramics", Ed. G.H. Stewart Vol. I, Academic Press, New York (1962).

Birch, R.E., "Refractories of the Future." Ohio State Univ. Eng. Expt. Sta. News 17 (4) 3-7 (1945).

Birch, R.E., "The Future Refractories & Steel Making." J. Metals 16 (6) 512-5 (1964).

Birnbaum, M. & Stocker, T.L., "Multimode Oscillation of the Ruby Laser Near Threshold." Appl. Phys. Letters 3 (9) 164-6 (1963).

Bishay, A., "Gamma Irradiation Studies of Some Borate Glasses." J. Am. Ceram. Soc. 44 (6) 289-96 (1961).

Blackburn, A.R., "Ceramic Materials for Application in the Design of Jet-Propelled Devices—Progress Report." Ohio State Univ. Research Foundation Rept. 13 (July 1946). 5 pp. PB 60,664; abstracted in Bibliog. Sci. Ind. Repts. 4 (11) 964 (1947).

Blackburn, A.R. & Shevlin, T.S., "Fundamental Study & Equipment for Sintering & Testing of Cermet Bodies: V, Fabrication, Testing, and Properties of 30 Chromium-70 Alumina Cermets." J. Am. Ceram. Soc. 34 (11) 327-31 (1951).

Blackburn, A.R., Shevlin, T.S. & Lowers, H.R., "Sintering & Testing of Cermet Bodies." J. Am. Ceram. Soc. 32, 81-9 (1949).

Blackburn, A.R. & Steele, R.E., "Mold & Method for Molding Ceramic Ware." U.S. 2,584,109, 2/5/52.

Blaha, E., "Casting Slip." U.S. 2,527,390, 10/24/50.

Blakeley, T.H. & Darling, R.F., "Refractory Nozzle Blades for High-Temperature Gas Turbines." Engineer 203 (5273) 251-2 (1957).

Blanchard, J.R., "Oxidation-Resistant Coatings for Molybdenum." WADC TR 54-492, Dec. 1954, June 1965.

Blanchin, L. Thesis. Univ. Lyon, France (June 24, 1952).

Blanchin, L., Imelik, B. & Prettre, M., "Modifications in Texture & Structure of Hydrargillite During Dehydration." Compt. rend. 233 (18) 1029-31 (1951).

Blanpain, E., "Ceramic Tool Bits." Machine-Outil Franc. 21 (113) 67-75 (1956).

Blau, H.H., Silverman, A., & Hicks, V., "Opal Glass." IX Congr. intern. quim. pura aplicada, Madrid 3, 507-34 (1934).

Bloch, E.A., "Dispersion-Strenghtened Aluminum Alloys." Metallurgical Reviews 6, 193-239 (1961).

Bloch, E.A. & Hug, H., "The Latest Developments in the SAP Process." Symp. on Dispersion Strengthening, International Powder Metall. Conf. June 13-17, 1960 (New York).

Bloch, H.S. U.S. 2,758,011, 8/7/56. Cer.A., Nov. 1956, 248.

Bloch, H.S., "Production of Alumina." U.S. 2,867,505, 1/6/59.

Blocker, E.W., Hauck, C.A., et al. "Development & Evaluation of Insulating Type Ceramic Coatings." PB Rept. 149888, 74 pp. U.S. Govt. Research Repts., 34 (5) 598 (1960).

Blodgett, W.E., "High Strength Alumina Porcelains." Am. Ceram. Soc. Bull. 40 (2) 74-7 (1961).

Blodgett, Wm. E., "Alumina Ceramic Valve Seats and a Faucet Assembly for Their Use." Brit. 1,080,760, 8/23/67.

Blome, J.C. & Kummer, D.L., "Ceramics & Graphite for Glide Re-entry Vehicles." McDonnell Aircraft Corp. AF 33(616)-8106-ASSET & AF 33(657)-10996. Presented at 1964 Annual Fall Meeting Ceramic-Metal Systems Div. Am. Ceram. Soc.

Bobbitt, J.M., "Thin Layer Chromatography," Reinhold Book Div. N.Y. (1963).

Bobrowsky, A.R., "Applicability of Ceramics & Ceramals as Turbine-Blade Materials for the Newer Aircraft Power Plants." Trans. Am. Soc. Mech. Engrs. 71 (6) 621-9 (1949).

Boehm, J., "Aluminum Hydroxide and Iron Hydroxide." I.Z. anorg. allg. Chem. 149, 203-16 (1925).

Boehm, J. & Niclassen, H., "Amorphous Precipitates & Crystallized Sols." Z. anorg. allg. Chem 132, 1-9 (1924).

Boersma, S.L., "Theory of Differential Thermal Analysis & New Methods of Measurement & Interpretation." J. Am. Ceram. Soc. 38 (8) 281-4 (1955).

Bogoroditskii, N.P. & Fridberg, I.D., "Dielectric Loss in Ceramics at High Frequencies." Izvest. Akad. Nauk SSSR 1938, Ser. Phys. 289-97.

Bogoroditskii, N.P. & Polyakova, N.L., "Dielectric Losses of Alumina." Doklady Akad. Nauk SSSR, 95, 257-9 (1954).

Bogoyavlenskii, A.F. & Dobrotvorskii, G.N., "Anodizing Aluminum Alloys in Carbonate Solution. VIII. Incorporating Radioactive W^{185} in the Anodic Film of Al_2O_3 in the Process of its Formation." Zhur. Priklad. Khim. 33, 340-4 (1960).

Bole, G.A., "Cermets, Ceramic Metal Compounds." Ceram. Ind. 51 (3) 108 (1948).

Bolling, G.F., "Remarks on Sintering Kinetics." J. Am. Ceram. Soc. 48 (3) 168-9 (1965).

Bond, F.C., "Recent Advances in Grinding Theory & Practice." Brit. Chem. Eng. 8 (9) 631-34 (1963).

Bond, F.C. & Agthe, F.T., "Deleterious Coatings of the Media in Dry Ball-Mill Grinding." Amer. Inst. Mining Met. Engrs. Tech. Pub. No. 1160; Mining Tech. 4 (2) 10 pp. (1940).

Bondley, R.J., "Low-Melting-Temperature Solders in Metal-Ceramic Seals." Ceram. Age 58 (1) 15-18 (1951).

Bonem, Frank L., "Uses Laser Beam for Thin-Film Evaporation at $2^{\circ}K$" Res. & Develop., 15 (6) 50-52 (1964).

Bonnet, L., "Alumina Insulator." U.S. 2,571,526, 10/16/51.

Bonnet, L. & Marty, M., "Ceramic Mixtures." U.S. 2,436,708, 2/24/48.

Booth, A.E. & Hess, R.L., "Method of Making Porous Products from Volcanic Glass & Alumina." U.S. 2,956,891, 10/18/60.

Bor, L., "Note on Beta Alumina Occurrence in Glass Tank Refractories." J. Am. Ceram. Soc. 33 (12) 375-76 (1950).

Boreskov, G.K., Dzis'ko, V.A. & Borisova, M.S., "Effect of Ignition Temperature on Extent of Surface & on Water Content of Alumina & Magnesia." Zhur. Fiz. Khim. 27, 1176-80 (1953).

Boreskov, G.K., Dzis'ko, V.A., Borisova, M.S. & Krasnopol'-skaya, V.N. Zhur. Fiz. Khim 26, 492-9 (1952).

Boreskov, G.K., Shchekochikhin, Y.M., Makarov, A.D. & Filimonov, V.N., "Study by Infrared Absorption Spectra, of the Structure of the Surface Compounds Formed When Ethanol is Adsorbed on γ-Aluminum Oxide." Dokl. Akad. Nauk SSSR 156 (4), 901-4 (1964).

Bortaud, P. & Rocco, D., "Electrocast Refractories." J. Brit. Ceram. Soc. 1 (2) 237 (1964).

Bortz, S.A., Nelson, H.R., Weil, N.A., Daniel, I.M., Evans, P.R.V., Southgate, P.D., Petch, N.J., Orowan, E., Cutler, I.B., Charles, R.J., Stokes, R.J. & Murray, G.T., "Studies of the Brittle Behavior of Ceramic Materials." ASD Tech. Doc. Rept. 61-628 (1961).

Bosch, H., "Volume Stability of Corundum-Sillimanite Brick." Ber. deut. keram. Ges. 41 (5) 323-6 (1964).

Bosch, R., A.-G. "Compositions for Making Spark Plugs." Ger. 652,354, 10/29/37.

Bosch, R., A.-G. "Compostions for Making Spark Plugs." Ger. 655,082, 1/7/38.

Bosch, R., G.M.B.H. "Seal Between a Ceramic Insulating Body & an Electrode Passing Through It, e.g., in Spark Plugs." Ital. 397,612, 8/27/42.

Bosch, R., G.M.B.H. (Emil Klingler, inventor). "Firing of Ceramic Ware." Ger. 736,311, 7/28/43.

Bosch, R., G.M.B.H. (Emil Klingler, inventor). "Production of Ceramic Spark-Gap Insulators." Ger. 741,142, 11/5/43.

Bovensiepen, U., Wolf, F. & Schwarz, F., "Attack of Refractory Brick by Sodium Silicate Melts Containing Sulfate: I, II." Tonind.-Ztg. Keram. Rundschau 88 (2) 25-32; (4) 73-80 (1964).

Bowen, N.L. & Greig, J.W., "The System $Al_2O_3-SiO_2$." J. Am. Ceram. Soc. 7, 238-54 (1924).

Bowen, N.L. & Schairer, J.F., "The System $FeO-SiO_2$." Am. J. Sci (5th Series) 24, 177-213 (1932).

Bower, J.H., "Dehydrating Agents Used for Drying Gases." J. Res. N.B.S. 33, 199-200 (1944).

Bowie, D.M., "Microwave Dielectric Properties of Solids for Applications at Temperatures to 3000°F." IRE Natl. Convention Record 5, Pt. 1, 270-81 (1957).

Bradshaw, W.G. & Mathews, C.O., "Properties of Refractory Materials, Collected Data & References." LMSD-2466 (June 24, 1958).

Bradstreet, S.W., "Process of Coating & Hot Working of Metals." U.S. 2,869,227, 1/20/59.

Bradstreet, S.W., 'Refractory Compositions for Flame Spraying." U.S. 2,904,449, 9/15/59.

Bradstreet, S.W. "Ceramic Coatings for High Temperature Service." Corrosion 16, 309-11t (1960).

Bradstreet, S.W., "Principles Affecting High Strength to Density Composites with Fiber or Flakes." W-P AFB TDR 64-85, May 1964.

Bradstreet, S.W. & Griffith, J.A., "Spraying Process." U.S. 2,763,569, 9/18/56.

Bradt, R.C., "Todays Ceramics." Cutting Tool Engineering, May 6, 1964.

Bradt, R.C., "Cr_2O_3 Solid Solution Hardening of Al_2O_3." J. Amer. Ceram, Soc. 50, 54-5 (1967).

Bragdon, T.S., "Evaluation of Cermets for Jet Engine Turbine Blading." PB Rept. 134946, 64 pp.; U.S. Govt. Research Repts. 30 (6) 613 (1958).

Bragg, W.H. & Bragg, W.L., "X-rays & Crystal Structure." Bell & Sons, London 1915.

Bragg, W.L., Gottfried, C., & West, J., "The Structure of β-Alumina." Z. Krist. 77, 255-74 (1931).

Brand, F., et al. "Experiments in the Design of Ceramic Electron Tubes." Ceram. Age 63, (5) 18-23 (1954).

Brandes, E.A., "Ceramic Molding Process." U.S. 2,810,182, 10/22/57.

Brandt, D.J.O., "Use of Oxygen in the Ferrous & Nonferrous Metallurgical Industries." J. Soc. Glass Technol. 33 (151) 103-19T (1949).

Braniska, A., "Barium & Strontium Cements." Zement-Kalk-Gips. 10 (5) 176-84 (1957).

Braniski, A., "Refractory Concretes with Barium Aluminous Cement as Binder." Acad. rep. populare Romine, Studii cercetari met. 4 (3) 413-40 (1959).

Braniski, A., "Refractory Barium Aluminate Cements." Tonind.-Ztg. u. Keram. Rundschau 85 (6) 125-9 (1961)

Braniski, A., "Highly Refractory Concrete Containing Barium Aluminate Cement." Tonind.-Ztg. u. keram. Rundschau 85 (6) 129-35 (1961).

Braniski, A., "Refractory Barium-Aluminous Cement & Concrete." Natl. Bur. Std. (U.S.) Monograph 43 (2) 1075-91 (1962).

Braniski, A. & Ionescu, T., "Refractory Aluminous Cements Containing Barium Oxide." Zement-Kalk-Gips, 13 (3) 109-11 (1960).

Braun, O.L., "Furnace Door." U.S. 2,645,211, 7/14/53.

Bravinskii, V.G. & Reshetnikov, A.M., "Propagation of Microcracks in Ceramics." Izv. Akad. Nauk SSR, Ser. Fiz. 27 (9) 1219-23 (1963).

Bray, J.L., "Metal Production: High Alumina Refractory Field Merely Scratched." Steel 128 (1) 177, (1951).

Brcic, B.S., et al. "The System $CaO-Al_2O_3$ at Low Temperatures." Vestn. Solven. Kem. Drustva 9, 27-32 (1962).

Brenner, S.S., "Tensile Strength of Whiskers." J. App. Phys. 27 (12) 1484-91 (1956).

Brenner, S.S., "Mechanical Behavior of Sapphire Whiskers at Elevated Temperatures." J. App. Phys., 33 (1) 33-9 (1962).

Brenner, S.S., "Case for Whisker Reinforced Metals." J. Metals 14 (11) 809-11 (1962).

Brenner, S.S., "Factors Influencing the Strength of Whiskers." Fiber Compos. Mater. Pap. 1964, 11-36 (pub. 1965).

Bressman, J.R., "Ceramic Materials Show Promise for High Temperature Mechanical Parts." Materials & Methods 27 (1) 65-70 (1948).

Bretsznajder, S., "Production of Pure Al_2O_3 from Aluminum Sulfate." Przemysl Chem. 22 285-90 (1938).

Bretsznajder, S., Kawecke, W., Porowski, J., & Lis, J., "Hydrolysis of Alumina." Ger. 1,166,755, 4/2/64.

Bretsznajder, S. & Pysiak, J., "Thermal Decomposition of Basic Aluminum Ammonium Sulfate. III. Decomposition at 300-500°." Bull. Acad. Polon. Sci. Ser. Sci. Chim. 12 (5) 315-8 (1964).

Brewer, L. & Searcy, A.W., "Gaseous Species of the $Al-Al_2O_3$ System." J. Am. Chem. Soc. 73 (11) 5308-14 (1951).

Brewer, R.C., "Appraisal of Ceramic Cutting Tools." Engineers' Digest, 18 (9) 381-87 (1957).

Bridgman, P.W., "Effect of Hydrostatic Pressure on the Fracture of Brittle Substances." J. App Phys. 18, 246-58 (1947).

Bridoux, C., "Sintered Aluminum Powder as a Canning Material." Nuclear Eng. 6 (60) 189-92 (1961).

Brierley, E. & Smith, H.J., "Changes Occurring in Active Alumina on Storage." J. Chromatog. 14 (3), 499-502 (1964).

Brindley, G.W. & Choe, J.O., "Reaction Series Gibbsite to Chi-Alumina to Kappa Alumina to Corundum I." Am. Mineralogist, 46 (7-8) 771-85 (1961).

Brindley, G.W. & Nakahira, M., "X-ray Diffraction & Gravimetric Study of the Dehydration Reactions of Gibbsite." Z. Krist. 112, 136-49 (1959).

Brisbane, S.M. & Segnit, E.R., "Attack on Refractories in the Rotary Cement Kiln: I, Physical & Mineralogical Changes." Trans. Brit. Ceram. Soc. 56 (5) 237-52 (1957).

Brisbin, P.H. & Heffernan, W.J. "Alumina Fibers." U.S. 3,094,385, 6/18/63.

Bristow, R.H., "Active Alloy Ceramic-to-Metal Seals." Am. Ceram. Soc. Bull. 43 (8) 585 (1964).

British Thomson-Houston Co. Ltd. Brit. 656,724, 7/4/51.

Britsch, H., "Sintering of Corundum." Ber. deut. keram. Ges. 36 (5) 133-45 (1959).

Brockmann, H., "Chromatography of Colorless Substances & the Relation Between Constitution & Adsorption Affinity." Dis. Faraday Soc. 7, 58-64 (1949).

Brockmann, H. & Schodder, H., "Aluminum Oxide with Buffered Adsorptive Properties." Ber. 74B, 73-8 (1941).

Broge, R.W., " Alumina Abrasive Materials." U.S. 3,003,919, 10/10/61.

Bron, V.A., "Sintering of Alumina by Recrystallization." Ogneupory 16 (7) 312-23 (1951).

Bron, V.A., "Refractory with High Concentration of Alumina for Teeming Ladles." Tr. Vost. Inst. Ogneuporov No. 4, 98-105 (1963).

Bron, V.A., Savkevich, I.A., & Mil'shenko, I.A., "High Alumina Refractories from Slags (obtained) in the Production of Metallic Chromium." Ogneupory 22 (2) 49-55 (1957).

Brondyke, K.J., "Effect of Molten Aluminum on Alumina-Silica Refractories." J. Am. Ceram. Soc. 36 (5) 171-4 (1953).

Brown, K.W., Chirnside, R.C., Dauncey, L.A. & Rooksby, H.P., "Synthetic Sapphires." G.E.C. Jour. 13 (2) 53-9 (1944).

Brown, J.F., Clark, D. & Elliott, W.W., "Thermal Decomposition of the Alumina Trihydrate, Gibbsite." J. Chem. Soc. (London) 1953, 84-8.

Brown, L. & Coffin, L., "Use of Polyvinyl Alcohol in Ceramics." Ceram. Ind. 44 (5) 126-30 (1945).

Brown, R.W., "Alumina-Recovery Process." U.S. 2,280,998, 4/28/42.

Brown, R.W., "Extraction of Alumina from Ores." U.S. 2,375,342, 5/8/45.

Brown, R.W., "Recovery of Alumina." U.S. 2,375,343, 5/8/45.

Brown, R.W. & Landback, C.R., "Applications of Special Refractories in the Aluminum Industry." Am. Ceram. Soc. Bull., 38 (7) 352-5 (1959).

Brown, Roy W., "Role of Beta Alumina Fused Cast Refractories in Glass Tank Superstructures." J. Can. Ceram. Soc. 31 135-44 (1962).

Brown, S.D. & Kistler, S.S., "Devitrification of High-SiO$_2$ Glasses of the System Al$_2$O$_3$-SiO$_2$." J. Am. Ceram. Soc. 42 (6) 263-70 (1959).

Brown, W.R., Eiss, N.S., Jr., & McAdams, H.T., "Chemical Mechanisms Contributing to Wear of Single-Crystal Sapphire on Steel." J. Am. Ceram. Soc. 47 (4) 157-162 (1964).

Browning, Melvin F., Secrest, V.M. & Blocher, J.M., "Preparation and Evaluation of Alumina-Coated Fuel Particles." AEC Accession No. 15990, Rept. No. BMI-1708. Avail. CFSTI, 74 pp (1965).

Bruce, E.W., "Preparation of Fibrous Alumina Monohydrate & Aquasols Thereof." U.S. 3,031,417, 4/24/62.

Bruch, C.A., "Sintering Kinetics for the High Density Alumina Process." Am. Ceram. Soc. Bull. 41 (12) 799-806 (1962).

Bruch, C.A., "Treating Alumina with Magnesia for Spinel Formation." U.S. 3,155,534, 11/3/64.

Bruch, C.A., "Problems in Die-Pressing Submicron-Size Alumina Powder." Ceram. Age 83, 44-7, 50, 52-3 (1967).

Brun, E., "Electric Quadrupole Interactions of ^{27}Al & Distribution of Cations in Spinel MgAl$_2$O$_4$." Naturwissenschaften 47 (12) 277 (1960).

Brundin, N.H., & Palmqvist, S.R., "Microporous Powder of Aluminum Hydroxide or Oxide." Swed. 139,326, 3/8/47.

Brunetti, C. & Khouri, A.S., "Printed Electronic Circuits." Electronics 19 (4) 104-8 (1946).

Buckley, D.H., "Friction Characterisitics of Single-Crystal and Poly-crystalline Aluminum Oxide in Contact in Vacuum." NASA Accession No. N66-35207 Rept. No. NASA-TN-D-3599.

Buchner, S., "Manufacture of Abrasive Materials & Grinding Wheels." Keram. Z. 8 (8) 383-7 (1956).

Budnikov, P.P. and Karitonov, F.Ya., "Influence of High-Pressure Water Vapor on Some Properties of Corundum Ceramics." Silikattechnik 18, 243-8 (1967).

Budnikov, P.P. & Kravchenko, I.V., "Investigation of the Hydration Processes of Calcium Monoaluminate." Kolloid. Zhur. 21 (1) 9-17 (1959).

Budnikov, P.P., Marakueva, N.A. & Tresvyatskii, S.G., "Effect of the Composition of Binders on the Properties of Suspensions for Hot Castings of Ceramic Products." Zhur. Priklad. Khim. 34, 492-7 (1961).

Budnikov, P.P. & Shteinberg, Yu. G., "New Glaze for Porcelain." Zhur. Priklad. Khim., 29 (9) 1305-1309 (1956).

Budnikov, P.P. & Tresvyatskii, S.G., "Electrical Conductivity of Corundum Refractories at Elevated Temperatures." Ogneupory 20 (2) 70-1 (1955).

Budnikov, P.P. & Zlochevskaya, K.M., "Synthesis of Magnesium Alumina Spinel." Ogneupory 23 (3) 111-8 (1958).

Budnikov, P.P. & Zlochevskaya, K.M., "Synthesis & Properties of Mullite-Spinel Ceramics." Zh. Prikl. Khim. 37 (8), 1649-57 (1964).

Budnikov, P.P. & Zyvagil'skii, A.A., "Influence of BeO & Technical Al$_2$O$_3$ on the Basic Properties of Electrical Porcelain." Steklo i Keram. 16 (7) 3-7 (1959).

Budworth, D.W., "Measurement of Gas Permeation Through Disks of Hot-Pressed Alumina at Temperatures Up to 800°." Trans. Brit. Ceram. Soc. 62 (12), 975-87 (1963).

Budworth, D.W., Roberts, E.W. & Scott, W.D., "Joining of Alumina Components by Hot Pressing." Battelle Tech. Rev. 13 (4) 135a (April 1964).

Buehler, G. & Feulner, H., "Stable Aqueous Suspensions Based on Fused Alumina." Ger. 17,722, 10/21/59.

Buessem, W.R. & Bush, E.A., "Thermal Fracture of Ceramic Materials Under Quasi-Static Thermal Stresses (Ring Test)." J. Am. Ceram. Soc. 38 (1) 27-32 (1955).

Bussem, W. & Eitel, A., "Structure of Pentacalcium Trialuminate." Z. Krist. 95, 75-88 (1936).

Bugosh, J., "Process for Preparing Alumina Sols." U.S. 2,763,620, 9/18/56.

Bugosh, J., "Fibrous Alumina Monohydrate & Its Production." U.S. 2,915,475, 12/1/59.

Bugosh, J., "Alumina Bodies Produced from Colloidal Fibrous Boehmite." U.S. 3,108,888, 10/29/63.

Bugosh, J., "Chemically Modified Alumina Monohydrate, Dispersions Thereof, & Processes for Their Preparation." U.S. 3,031,418, 4/24/62.

Bugosh, J., "Strong, Dense Alumina Bodies from Colloidal Fibrous Boehmite with a Grain-Growth Inhibitor." U.S. 3,141,786, 7/21/64.

Bugosh, J., Brown, R.L., McWhorter, J.R., Sears, G.W. & Sippel, R.J., "Novel Fine Alumina Powder, Fibrillar Boehmite." Ind. Eng. Chem. Prod. Res. Develop 1 (3) 157-61 (1962).

Bulashevich, E.A. & Tolkachev, S.S., "Preparation of Aluminum Oxide Dihydrate." Vestn. Leningr. Univ. 19 (10), Ser. Fiz. i Khim. No. 2, 123-4 (1964).

Bulavin, I.A., "The Kinetics of Sintering Corundum Ceramics in the Presence of a Liquid Phase." Tr. Mosk. Khim.-Tekhnol. Inst., No. 37, 123-34 (1962).

Bulavin, I.A. & Zakharov, I.A., "Sintering of Alumina with Additions of Talc & Titanium Dioxide & the Properties of the Sintered Body." Trudy Moskov. Khim.-Tekhnol. Inst. im. D.I. Mendeleeva 1956, No. 21, pp. 86-8.

Bunting, E.N., "Phase Equilibrium in the System Cr$_2$O$_3$-Al$_2$O$_3$." Bur. Stds. J. Res. 6 (6) 947-9 (1931). RP-317.

Bunting, EN., "Phase Equilibria in the System: SiO$_2$-ZnO-Al$_2$O$_3$." Bur. Std. J. Res. 8 (2) 279-87 (1932). RP-413.

Burdese, A., "Systems of Vanadic Anhydride and Sesquioxides of Chromium, Iron & Aluminum." Ann. chim. (Rome) 47, (7,8) 804 (1957).

Burdick, J.N. & Glenn, J.W., Jr., "Synthetic Star Rubies & Star Sapphires & Process for Producing Same." U.S. 2,488,507, 11/15/49.

Burdick, J.N. & Jones, R.A., "Synthetic Corundum Crystals & Process for Making." U.S. 2,690,062, 9/28/54.

Burk, M., "Thermal Expansion of Ceramic Materials at -200° to 0°C." J. Am. Ceram. Soc. 45 (6) 305-6 (1962).

Burkart, W., "Grinding & Polishing Agents." Metalloberflache 6 (1) Al-4 (1952).

Burke, J.E., "Role of Grain Boundaries in Sintering." J. Am. Cer. Soc. **40** (3) 80-5 (1957).

Burney, J.D., " Study of the Possibility of Reinforcing High-Temperature Alloys by Addition of Refractory Powders." PB Rept. 131768, 49 pp.; U.S. Govt. Research Repts. 30 (1) 23 (1958).

Burnham, J. & Robinson, P., "Properties of Insulating Alumina Films." National Research Council, Div. Eng. and Ind. Res. Ann. Rept. Conf. on Elect. Insul. 1948, 68-71 (1949).

Burns, R.P., Jason, A.J. & Inghram, M.G., "Discontinuity in the Rate of Evaporation of Aluminum Oxide." J. Chem. Phys. 40 (9), 2739-40 (1964).

Burnside, D.G., "Ceramic-Metal Seals of the Tungsten-Iron Type." RCA Rev. **15** (1) 46-61 (1954).

Burrows, H.O., "Effect of Molten Aluminum on Various Refractory Brick." J. Am. Ceram. Soc. 23 (5) 125-33 (1940).

Busby, T.S. & Eccles, J., "Reactions Between Refractory Materials." Glass Technol., 2 (5) 201-208 (1961).

Busby, T.S. & Eccles, J., "Corrosion of Corundum Crystals by Commercial Glasses." Nature 196 (4850) 165-66 (1962).

Busby, T.S. & Partridge, J.H., "Improved Materials for Tank Blocks." J. Soc. Glass Technol. 36 (170) 131-6T (1952).

Busing, W.R. & Levy, H.A., "Single-Crystal Neutron-Diffraction Study of Diaspore." Acta Cryst. 11, 798-803 (1958).

Butcher, M.M., and White, E.A.D., "Vapor-Phase Growth of Thin Corundum." J. Am. Ceram. Soc. 48, 492-3 (1965).

Buttler, F.G., et al. "Studies on 4CaO·Al_2O_3·13H_2O & the Related Natural Mineral Hydrocalumite." J. Am. Ceram. Soc. 42 (3) 121-6 (1959).

Byalkovskii, V.I. & Kudinov, I.A., "Effect of the Medium on the Efficiency of Grinding in Ball Mills." Keram. Sbornik 1940, No. 10, pp.8-13; Khim. Referat. Zhur. 4 (5) 114 (1941).

Bye, G.C. & Robinson, J.G., "Ageing of Alumina Hydrates." Chem. & Ind. 1961, 433.

Bye, G.C. & Robinson, J.G., "Preparation of Bayerite & a Modified Form of Pseudoboehmite." Chem. & Ind. 1961, 1363.

Bye, G.C. & Robinson, J.G., "Crystallization Processes in Aluminum Hydroxide Gels." Kolloid Z. 198 (1-2) 53-9 (1964).

Cabannes-Ott, C., "Structure of Natural Hydroxides of the Type XO·OH-Diaspore." Compt. rend. 244, 2491-5 (1957).

Caddes, D.E. and Wilkinson, C.D.W., "Large Photoelastic Anisotropy of Sapphire." J. Quantum Electron. 2, 330-1 (1966).

Cadwell, D.E. & Duwell, E.J., "Evaluating Resistance of Abrasive Grits to Comminution." Am. Ceram. Soc. Bull. 39 (11) 663-7 (1960).

Caglioti, V. & D'Agostino, O., "Researches on Aerogels. I. The Structure of Metallic Oxides." Gazz. chim. ital. 66, 543-8 (1936).

Cahoon, H.P. & Christensen, C.J., "Sintering & Grain Growth of Alpha-Alumina." J. Am. Ceram. Soc. 39 (10) 337-44 (1956).

Caillat, R. & Pointud, R., "Sintering of Alumina & Beryllia Under Pressure." Rev. met., 54, 277-82 (1957).

Caldwell, O.G., "Ceramic Porous Core Laminates for Broadband Radar Radomes." Proc ASD-OSU Symposium on Electromagnetic Windows, I. July 1962.

Caldwell, O.G. & Gdula, R.A., "Alumina Laminates for Broadband Radome Applications." International Pipe & Ceramics Corporation, Glendale, Calif., Oct. 1962.

Callis, C.C., "Heat-Insulation Block." U.S. 2,396,246, 3/12/46.

Callow, R.J., "Influence of Al_2O_3, ZnO, & K_2O on the Opacity of Fluoride-Opacified Glasses." J. Am. Ceram. Soc. 35 (5) 120-2 (1952).

Calvet, E., "Thermokinetic Study of Adsorption. Soluble Bodies & Macromolecular Substances." J. Polymer Sci. 8, 163-71 (1954).

Calvet, E., Boivinet, P., Noel, M., Thibon, H., Maillard, A. & Tertian, R., "Alumina Gels." Bull. Soc. chim. France, 99-108 (1953).

Calvet, E. & Thibon, H., "Thermokinetic Study of the Rehydration of the Products of the Controlled Thermal Decomposition of Several Trihydrates of Alumina I." Bull. soc. chim. France 1954, 1343-6.

Calvet, E., Thibon, H. & Dozoul, J., "Barium Aluminates." Bull. soc. chim. France 1964 (8) 1915-6.

Calvet, E., Thibon, H. & Gambino, M., "Microcalorimetric Study of Various Aluminum Hydroxide Samples." Bull. soc. chim. France 1964 (9) 2132-6 (Fr).

Campbell, Wm., "Anodized Aluminum Surfaces for Wear Resistance." J. Electrodepositors Tech. Soc. 28, 273-91 (1952).

Campbell, W.B., "Feasibility of Forming Refractory Fibers by a Continuous Process." 4th Quarterly Rept. AMRA CR 63-03/4 (Lexington Laboratories, Inc.) (June 1963).

Campbell, W.B., "Continuous Whisker Formation." Chem. Eng. Prog. 62, 68-73 (1966).

Campbell, W.J. & Grain, C., "Thermal Expansion of a-Alumina." U.S. Bur. Mines. Rept. Invest. No. 5757, 16 pp. (1961).

Cantrell, J., Degg, E.P., & Nobes, F.L., "Metal-Ceramic Bonded Granular Material." U.S. 2,782, 110, 2/19/57.

Cap, M., "Glasses of the System B_2O_3-Al_2O_3-Li_2O & Their Network." Zpravy Ceskoslou. Keram. a Sklarske Spolecnosti 26 (3-4) 159-93 (1950).

Caprio, M.J., "Private Communication." Jan. 13, 1965.

Carborundum Co., Ltd. "Metal-Ceramic Bonded Granular Material." Brit. 715,528, 9/15/54.

Carborundum Co. "Alumina-Zirconia Abrasives." Brit. 993,891. 6/2/65.

Carlisle, S.S., "Sira (Scientific Instrument Research Association)." Instr. Pract. 18 (5) 461-7 (1964).

Carlson, E.T., "Some Observations on Hydrated Monocalcium Aluminate & Monostrontium Aluminate." J. Res. Natl. Bur. Stds. 59 (2) 107-11 (1957).

Carlson, E.T., "System Lime-Alumina-Water at 1°." J. Res. NBS 61 (1) 1-11 (1958). RP-2877.

Carlson, E.T., Chaconas, T.J. & Wells, L.S., "Study of the System Barium Oxide-Aluminum Oxide-Water at 30°C." J. Res. Natl. Bur. Stds. 45 (5) 381-98 (1950).

Carlson, E.T. & Wells, L.S., "Barium Aluminate Hydrates." J. Res. Natl. Bur. Stds. 41 (2) 103-9 (1948); RP-1908.

Carnahan, R.D., "Some Observations on the Wetting of Al_2O_3 by Aluminum." J. Am. Cer. Soc. 41 (9) 343-7 (1958).

Carnahan, R.D. and Knapp, W.J., "Oxide Ceramics." Ceram. Advan. Technol. 1965, 76-106 (1965).

Carniglia, S.C., "The Grain Boundary: Bridge Between Single-Crystal & Polycrystalline Behavior." Bull. Am. Ceram. Soc. 42 (9) 510 (1963).

Carniglia, S.C., "Petch Relation in Single-Phase Oxide Ceramics." J. Am. Ceram. Soc. 48, 580-3 (1965).

Carroll-Porczynski, C.Z., "Advanced Materials—Refractory Fibers, Fibrous Metals, & Composites." Chemical Publishing Co. New York 10, 286 pp., (1962).

Carruthers, T.G. & Gill, R.M., "Properties of Calcined Alumina: I, Behavior of Alumina Hydrates During Calcination." Trans. Brit. Ceram. Soc. 54 (2) 59-68 (1955).

Carruthers, T.G. & Gill, R.M., "Properties of Calcined Alumina: II, Behavior of Calcined Alumina During Fine Grinding." Trans. Brit. Ceram. Soc. 54 (2) 69-82 (1955).

Carruthers, T.G. & Roberts, A.L., "High Temperature Steels & Alloys for Gas Turbines—Symposium: Ceramics as Gas-Turbine Blade Materials—Survey of the Possibilities." Iron Steel Inst. (London) Spec. Rept. 1952, No. 43, pp. 268-73.

Carter, P.T. & MacFarlane, T.G., "Thermodynamics of Slag Systems: I, Thermodynamic Properties of CaO-Al_2O_3 Slags." J. Iron Steel Inst. (London) 185 (1) 54-66 (1957).

Carter, R.E., "Mechanism of Solid-State Reaction Between Magnesium Oxide & Aluminum Oxide & Between Magnesium Oxide & Ferric Oxide." J. Am. Ceram. Soc. 44 (3) 116-20 (1961).

Cass, W.J., "Forms of Alumina." Chem. Age, 54 (1393) 259 (1946).

Caton, M.W., "Sagger & Batch for Making." U.S. 2,531,397, 11/28/50.

Caton, M.W., "Refractory Material." U.S. 2,559,343, 7/3/51.

Cavanaugh, J.F. & Sterry, J.P., "Heat Protective Ablative Coatings for Radomes." PB Rept. 171416, 85 pp.; U.S. Govt. Res. Repts. 35 (4) 422 (1961).

Ceramic Tool Symposium. Am. Soc. Tool Engineers, Detroit 38, Mich. (1957).

Chaklader, A.C.D., "Effect of Trace Al₂O₃ on Transformation of Quartz to Cristobalite." J. Am. Ceram. Soc. 44 (4) 175-80 (1961).

Challande, R., Hantel, L. & Montgomery, D., "The Origin of Acquired Electrical Charges by Crystals, Induced by Friction with Metals." Compt. rend. 254, 3534-5 (1962).

Champion, J.A. (Natl. Phys. Lab., Teddington, Engl.), "The Electrical Conductivity of Single-Crystal Alumina." Brit. J. Appl. Phys. 15 (6), 633-8 (1964).

Chandappa, N. & Simpson, H.E., "Study of the Physical Properties of Some Sodium-Boroaluminate Glasses." Glass Ind. 32 (10) 505-7 (1951).

Chang, R., "High-Temperature Creep & Anelastic Phenomena in Polycrystalline Refractory Oxides." J. Nuclear Materials 1, 174-81 (1959).

Chang, R., "Creep of Al₂O₃ Single Crystals." J. Appl. Phys. 31, 484-7 (1960).

Chang, R., "Creep & Anelastic Studies on Alumina." U.S. At. Energy Comm. NAA-SR-2770, 24 pp. (1958).

Chang, R., "Dislocation Relaxation Phenomena in Oxide Crystals." J. Appl. Phys. 32 (6) 1127-32 (1961).

Chang, R., "Electrical Resistivity Changes of Al₂O₃ Crystals During Creep." J. Appl. Phys. 34 (5) 1564-65 (1963).

Chang, R. & Graham, L.J., "Transient Creep & Associated Grain-Boundary Phenomena in Polycrystalline Alumina & Beryllia." NAA-SR-6483, Atomics International, Sept. 30, 1961.

Chang, R. and Graves, P.W., "The Effect of Creep Deformation on the D.C. Conductivity of Undoped and Cr-Doped Alumina Crystals." Brit. J. Appl. Phys. 16, 715-19 (1965).

Charles, R.J. & Shaw, R.R., "Delayed Failure of Polycrystalline & Single-Crystal Alumina." NASA (Natl. Aeron. Space Admin.) Doc. N63-16396, 44 pp. (1962).

Charles, R.J., "Static Fatigue: Delayed Fracture, Task 9 in Studies of the Brittle Behavior of Ceramic Materials." ASD Rept. 61-628, April 1962.

Charrier, J. & Papee, D. "Decomposition Products of Aluminum Hydrates." Compt. rend. 237, 897-900 (1953).

Charvat, F.R. & Kingery, W.D., "Thermal Conductivity: XIII, Effect of Microstructure on Conductivity of Single-Phase Ceramics." J. Am. Ceram. Soc. 40 (9) 306-15 (1957).

Chase, A.B., "Habit Modification of Corundum Crystals Grown from Molten PbF-Bi₂O₃." J. Am. Ceram. Soc. 49, 233-6 (1966).

Chatelain, P., "Dehydration of Hydrargillite." Compt. rend. 241, 46-8 (1955).

Chatterjee, S.K. & Chakravarty, S.N., "Slip Casting of Refractory Oxides in Aqueous or Acid Media." Indian Cer. 2 (1) 33-57 (1955).

Chatterjee, N.B. & Panti, B.N., "Chromium-Alumina Refractories." Central Glass Cer. Res. Inst., Bull. (India) 10 (4) 97-101 (1963).

Chekhovskoi, V. Ya., "Thermodynamic Properties of Corundum: A Reference Material in Calorimetry." Teplofiz. Vysokikh Temperature, Akad. Nauk SSSR 2 (2) 296-302 (1964).

Chiochetti, V.E.J. & Henry, E.C., "Electrical Conductivity of Some Commercial Refractories in the Temperature Range 600° to 1500°C." J. Am. Ceram. Soc. 36 (6) 180-4 (1953).

Chirnside, R.C. & Dauncey, L.A., "Free-Flowing Powdered Alumina." U.S. 2,431,370, 11/25/47.

Choi, D.M., Palmour, Hayne, III, & Kriegel, W.W., "Slip Processes in Fine Grained Polycrystalline Spinel (MgAl₂O₃). Mechanical Properties & Microstructure of Hot-Pressed Magnesium Aluminate." NASA (Natl. Aeron., Space Admin.) Doc. N6216638, 86 pp. (1962).

Chretien, A. & Papee, D., "The porosity of Some Adsorbent Products." Compt. rend. 234, 214-16 (1952).

Chung, D.H., "Elastic & Anelastic Properties of Fine-Grained Polycrystalline Alumina at Elevated Temperatures." Bull. Res. Dept. Monthly Rept. No. 297, State Univ., New York, Alfred University, March 1961.

Chung, D.H., Terwilliger, G.R., Crandall, W.B. & Lawrence, W.G., "Elastic Properties of Magnesia & Magnesium Aluminate Spinel." AD 432276, Avail. OTS (Eng.).

Cibis, P., "Eye Surgeon." Washington University Magazine 35 (2) 2-7 (1965).

Ciccarello, I.S. & Dransfield, K., "Ultrasonic Absorption of Microwave Frequencies & at Low Temperatures in MgO & Al₂O₃." Phys. Rev. 134 (6A), A1517-A1520 (1964).

Cirilli, V., Hydrated Calcium Aluminates." Ricerca Sci. 10 (6) 559-62 (1939).

Cirilli, V., "Solid Solutions of γ-Fe₂O₃ and γ-Al₂O₃." Gazz. chim. ital. 80, 347-51 (1950).

Cirilli, V., & Brisi, C., "β-Fe₂O₃ & the Solid Solutions it Forms with β-Al₂O₃." Gazz. chim. ital. 81, 50-4 (1951).

Clark, A. & Holm, V.C.F., "The Nature of Silica-Alumina Surfaces. III. Statistical Interpretation of the Adsorption of Ammonia on Alumina." J. Catalyst 1 (3), 21-32 (1962).

Clark, G.L. & Hawley, G.G. (Editors), "Encyclopedia of Chemistry." Reinhold Publishing Corp., New York, 1957, 1037 pp.

Clark, H.N., "Furnace Lining." U.S. 2,407,135, 9/3/46.

Clark, P.W., & White, J. "Some Aspects of Sintering." Trans. Brit. Ceram. Soc. 49, 305 (1950).

Clark, P.W., et al. "Further Investigations on the Sintering of Oxides." Trans. Brit. Ceram. Soc. 52 (1) 1-49 (1953).

Clay, W., "High Density Grinding Media." Finish 9 (11) 33-4 (1952).

Cline, J.E. & Wulff, J., "Vapor Deposition of Metals on Ceramic Particles." J. Electrochem. Soc. 98 (10) 385-7 (1951).

Coated Abrasives Manufacturer's Institute, "Coated Abrasives—Modern Tool of Industry." McGraw-Hill Book Co., New York, 426 pp. (1958).

Cobaugh, G.D., "Use of Castable Refractories in Ceramics & Glass." Ceram. Ind. 73 (6) 60-63 (1959).

Coble, R.L., "Initial Sintering of Alumina & Hematite." J. Am. Ceram. Soc. 41 (2) 55-62 (1958).

Coble, R.L., "Sintering Crystalline Solids: I, Intermediate & Final State Diffusion Models." J. Appl. Phys., 32 (5) 787-92 (1961).

Coble, R.L., "Sintering Alumina—Effect of Atmospheres." J. Am. Ceram. Soc. 45 (3) 123-27 (1962).

Coble, R.L., "Transparent Alumina & Method of Preparation." U.S. 3,026,210, 3/20/62.

Coble, R.L. & Ellis, J.S., "Hot-Pressing Alumina-Mechanisms of Material Transport." J. Am. Ceram. Soc. 46 (9) 438-41 (1963).

Coble, R.L. & Guerard, Y.H., "Creep of Polycrystalline Aluminum Oxide." J. Am. Ceram. Soc. 46, 353-4 (1963).

Coble, R.L. & Kingery, W.D., "Effect of Porosity on Thermal Stress Fracture." J. Am. Ceram. Soc. 38 (1) 33-7 (1955).

Coble, R.L. & Kingery, W.D., "Effect of Porosity on Physical Properties of Sintered Alumina." J. Am. Ceram. Soc. 39 (11) 377-85 (1956).

Coblentz, W.W., J. ber. Rad. 3, 397 (1906).

Cockayne, B., Chesswas, M., and Gasson, D.B., "Single-Crystal Growth of Sapphire" J. Mater. Sci. 2, 7-11 (1967).

Cochrane, H. and Rudham, R., "Heats of Immersion of Alpha and Gamma Alumina." Trans. Faraday Soc. 61, (514), 2246-54 (1965).

Coffin, L.F., Jr., "Fundamental Study of Synthetic Sapphire as a Bearing Material." Lubrication Eng. Trans. 1, 108-14 (1958).

Cohen, E.R., Crowe, K.M., & Dumond, J.W.M., "The Fundamental Constants of Physics." Interscience Publishers, New York (1957). JANAF Interim Thermochemical Tables, Vol. I.

Cohen, E., Herbst, H., Jenkins, D.E.P. & Wilkinson, K.H., "Sealing Metals to Ceramic Bodies." Brit. 832,251, 4/6/60.

Cohen, J., "Electrical Conductivity of Alumina." Am. Ceram. Soc. Bull. 38 (9) 441-6 (1959).

Coheur, F.P. & Coheur, P., "Spectral Method of Determining Temperatures Beginning from the Rotation Lines of Molecular Bands." Rev. universelle mines 86 (4) 86-9 (1943).

Cole, S.S., Jr. & Hynes, F.J. Jr., "Some Parameters Affecting Ceramic-to-Metal Seal Strength of a High-Alumina Body." Am. Ceram. Soc. Bull. 37 (3) 135-8 (1958).

Cole, S.S., Jr. & Sommer, G., "Glass-Migration Mechanism of Ceramic-to-Metal Seal Adherence." J. Am. Ceram. Soc. 44 (6) 265-71 (1961).

Coleman, B.D., "On the Strength of Classical Fibres & Fibre Bundles." J. Mech. & Phys. Solids 7, 60 (1958).

Coleman, R.V., "The Growth & Properties of Whiskers." Met. Rev. 9, 261-304 (1964).

Colligan, G.A., Phipps, C., and Parille, D., "Liquid-Metal-Refractory Reaction Studies." Mod. Castings 47, 134-43 (1965).

Comeforo, J.E. & Hursh, R.K., "Wetting of Al_2O_3-SiO_2 Refractories by Molten Glass: II, Effect of Wetting on Penetration of Glass into Refractory." J. Am. Ceram. Soc. 35 (6) 142-8 (1952).

Compton, W.D. & Arnold, G.W., Jr., "Radiation Effects in Fused Silica and a-Al_2O_3." Discussions Faraday Soc., 1961, No. 31, pp. 130-9.

Comstock G.E., III., "Molded Alumina." U.S. 2,618,567, 11/18/52.

Conant, L.A., "Metal Ceramic Products." U.S. 2,783,530, 3/5/57.

Conant, L.A., "Laminated Metal Ceramic." U.S. 2,843,646, 7/15/58.

Conant, L.A., "Oxidized Chromium-Alumina Metal Ceramic Protective Tube." U.S. 2,872,724, 2/10/59.

Conant, L.A. & Gillies, D.M., "Chromium-Alumina Metal Ceramics." U.S. 2,698,990, 1/11/55.

Congleton, J. and Petch, N.J., "Surface Energy of a Running Crack in Alumina. . . ." Intern. J. Fracture Mech. 1, 14-19 (1965).

Conrad, H., "Mechanical Behavior of Sapphire." J. Am. Ceram. Soc. 48 (4) 195-201 (1965).

Conrad, H., Janowski, K., & Stofel, E., "Additional Observations on Twinning in Sapphire (a-Al_2O_3 Crystals) During Compression." Aerospace Corp., El Segundo, Calif., Materials Sci. Lab. Contract No.: ATN-64 (9236)-16, 9 pp. (1964), NASA Accession No. N64-19242.

Conrad, H., Stone, G. & Janowski, K. "Yielding & Flow of Sapphire (a-Al_2O_2 Crystals) in Tension & Compression." Aerospace Corp., El Segundo, Calif., Materials Sci. Lab. Contract No. AT (11-1-GEN-8, ATN-64 (9236-15)), 37 pp. (1964), NASA Accession No. N64-19481.

Conrad, R. & Lenne, H.U., "Peptizable Aluminum Hydroxide from Bayerite." Ger. 1,068,233, 11/5/59.

Continental Oil Co. "E-Alumina." Belg. 616,224, 4/30/62.

Cook, M.D., Cook, C.P., & King, D.F., "Pneumatic Placement of Refractory Castables: II." Bull. Am. Ceram. Soc. 42 (11) 694 (1963).

Cook, M.D., Cook, C.P. & King, D.F., "Pneumatic Placement of Refractory Castables: III." Am. Ceram. Soc. Bull, 43 (5) 380-82 (1964).

Cooke, P.W. & Harcsnape, J.N., "The Effect of Steam on Some Alumina Transitions." Trans. Faraday Soc. 43, 395-8 (1947).

Coombs, W.E., Jr., "Ceramic Process Water Conditioning by Ion Exchange." Am. Ceram. Soc. Bull. 35 (11) 430-2 (1956).

Coop, W.H. & Hammond, J.A., "Phosphorescence of Fused Quartz & Sapphire." J. Opt. Soc. Am. 52 (7) 835 (1962).

Cooper, A.R., Jr. & Goodnow, W.H., "Density Distributions in Dry-Pressed Compacts of Ceramic Powders Examined by Radiography of Lead Grids." Am. Ceram. Soc. Bull. 41 (11) 760-61 (1962).

Coppola, P.P., et al. "Thermionic Cathode & Method of Making." U.S. 2,769,708, 11/6/56.

Corhart Refractories Co. "Heat-Cast Refractory Material & Process for Making." Brit. 665,209, 11/21/51.

Corhart Refractories Co. "Fused & Cast Alumina Refractories." Brit. 966,269, 8/6/64.

Corhart Refractories Co. "Fused & Cast Alumina Refractories Resistant to Thermal Shock." Brit. 972,266, 10/14/64.

Corning Glass Works. "Manufacture of Refractory Blocks." Fr. 862,662, 12/16/40.

Corning Glass Works. "Refractory Blocks & Their Manufacture." Fr. 862,743, 12/16/40.

Corning Glass Works. "Metallic-Ceramic Body Production." Brit. 947,355, 1/22/64.

Cossa, A. Nuovo Cimento 3, 230 (1870).

Cotton, F.A. Ed., "Progress in Inorganic Chemistry: Vol. I." Interscience Publishers, Inc., New York, 1959, 566 pp.

Counts, W.E., et al. "Semiconductor Composition." U.S. 2,864,773, 12/16/58.

Counts, E.E., et al. "Resistor & Spark Plug Embodying It." U.S. 2,864,884, 12/16/58.

Courtial, R. & Trambouze, Y., "Composition of Solids Obtained by Dehydration of a Fine Hydrargillite." Compt. rend. 244, 1764-66 (1957).

Courtial, R., Trambouze, Y., & Prettre, M., "Influence of Conditions of Heat-Treatment on Boehmite Content of Products of Partial Dehydration of Hydrargillite." Compt. rend. 242, 1607-10 (1956).

Courtial, R., Trambouze, Y., & Prettre, M., "Composition of Solids Obtained by the Dehydration of Hydrargillite." Compt. rend. 242 (16) 1976-9 (1956).

Cowan, R.E., "Mechanism of Adherence of Tungsten to Bodies Containing Yttria." Am. Ceram. Soc. Bull. 43 (8) 584 (1964).

Cowan, R.E., Herrick, C.C. & Stoddard, S.D., "Mechanism of Adherence of Tungsten to Bodies Containing Yttria." Electronics Div. Am. Cer. Soc. Meeting Sept. 16-18, 1964. Bull. Am. Ceram. Soc. 43 (8) 584 (1964).

Cowan, Robert E. (to U.S. Atomic Energy Comm.), "Metallizing Alumina Ceramics with Copper." U.S. 3,180,756, 4/27/65.

Cowles, R.J. & Stedman, H.F., "Ceramic Coatings to Decrease Oxidation of Metallic Articles During Heat Treatment." U.S. 3,037,878, 6/5/62.

Cowley, J.M., "Stacking Faults in γ-Alumina." Acta Cryst., 6 (1) 53-4 (1953).

Cowley, J.M., "Structure Analysis of Single Crystals by Electron Diffraction III. Modifications of Alumina." Acta Cryst. 6, 846-53 (1953).

Cowling, K.W., Elliott, A. & Hale, W.T., "Note on the Relationship Between Bulk Density & Thermal Conductivity in Refractory Insulating Brick." Trans. Brit. Ceram. Soc. 53 (8) 461-73 (1954).

Cox, J.H. & Pidgeon, L.M., "Reduction of Calcined Dolomite by Al_4C_3." Can. J. Chem. 41, 671 (1963).

Coxey, J.R., "Refractories." Pennsylvania State College, State College, Pa. 1950, 162 pp.

Coykendall, W.E., "Annular Ceramic Tube." Ceram. Age 63 (3) 33-6 (1954).

Cramer, R.H. & Jenkins, E.E., "Preparation of Alumina Gels." U.S. 2,898,306, 8/4/59.

Crandall, W.B., & Bryant, C.A., "Sonic Method of Measuring Young's Modulus of Elasticity at High Temperature." Office of Naval Research Tech. Rept. NR-032-022. Alfred Univ. Dec. 1953.

Crandall, W.B., Chung, D.H., & Gray, T.J., "Mechanical Properties of Ultrafine Hot-Pressed Alumina." pp 349-79 in "Mechanical Properties of Engineering Ceramics," Ed. W.W. Kriegel & Hayne Palmour III, Interscience Publishers, Inc. New York, 1961, 646 pp.

Crandall, W.B. & Ging, J., "Thermal Shock Analysis of Spherical Shapes." J. Am. Ceram. Soc. 38 (1) 44-54 (1955).

Crandall, W.B. & West, R.R., "Oxidation Study of Cobalt-Alumina Mixtures." Am. Ceram. Soc. Bull. 35, (2) 66-70 (1956).

Cranston, R.W. & Inkley, F.A., "The Determination of Pure Structures from Nitrogen Adsorption Isotherms." Advances in Catalysis 9, 143-54 (1957).

Cratchley, D. & Baker, A.A., "The Tensile Strength of a Silica Fibre Reinforced Aluminum Alloy." Metallurgia 69, 153-9 (1964).

Cremens, W.S., Bryan, E.A. & Grant, N.J., "Temperature & Time Stability of M-257 & SAP Aluminum-Aluminum Oxide Alloys." ASTM Proc. 58, 753 (1958).

Cremens, W.S. & Grant, N.J., "Preparation & High-Temperature Properties of Nickel-Al_2O_3 Alloys." Am. Soc. Testing Materials Proc. 58, 714-30 (1958).

Crepaz, E. & Raccanelli, A., "The Initial Products of Hydration of Aluminate Cements." Ind. Ital. Cimento 33, 519-26 (1963).

Cronin, J.D., "Adaptation of Ceramic Tube Types." PB Rept. 151922, 166 pp.; U.S. Govt. Res. Repts. 33, (5) 494 (1960).

Cronin, L.J., "Refractory Cermets." Am. Ceram. Soc. Bull. 30 (7) 234-8 (1951).

Cronin, L.J., "Trends in Design of Ceramic-to-Metal Seals for Magnetrons." Am. Ceram. Soc. Bull. 35 (3) 113-6 (1956).

Crosby, G.E., Smith, A.L., & Whitmer, L.E., "Activated Alumina Dominated Phosphors." U.S. 2,919,363, 12/29/59.

Croskey, C.D., "Production & Control Method for Casting Slip." Am. Ceram. Soc. Bull. 28 (7) 265-6 (1949).

Cross, C.F. J. Chem. Soc. 35, 796 (1879).

Crowley, M.S., "Improved High-Alumina Concrete." U.S. 3,150,992, 9/29/64.

Csordas, I., et al. "Sintered Alumina Articles & A Process for the Production Thereof." U.S. 2,947,056, 8/2/60.

Csordas, I., et al. "Process for Producing Alumina." U.S. 2,982,614, 5/2/61.

Cuer, J.P., "The Production & Properties of Finely Divided Solids, Prepared in a Flame Reactor." Comm. energie at. (France), Rappt. CEA 1669 7℃ ℘ (1960).

Cuer, J.P., Elston, J. & Teichner, S.J., "Methods of Production & Properties of Finely Divided Solids Prepared in A Flame Reactor. II. Formation of Alumina." Bull. soc. Chim. France 1961, 81-8.

Cuer, J.P., Elston, J. & Teichner, S.J., "Methods of Production & Properties of Finely Divided Solids Prepared in a Flame Reactor. III. Properties of Alumina." Bull. soc. chim. France 1961, 89-93.

Cuer, J.P., Elston, J. & Teichner, S.J., "Methods of Production & Properties of Finely Divided Solids Prepared in a Flame Reactor. IV. Use of the Reactor with a Reactive Solid." Bull. soc. Chim. France 1961, 94-101.

Cumming, P.A. and Harrop, P.J., "Bibliographic Review of Self-Diffusion in Metal Oxides." U.K. At. Energy Auth., Res. Group, At. Energy Res. Estab., Bibliog. AERE-BIB 143, 22 pp (1965).

Cunningham, A.L., "Mechanism of Growth & Physical Properties of Refractory Oxide Fibers." PB Rept. 171520, 57 pp.; U.S. Govt. Res. Repts. 35 (5) 589 (1961).

Cunningham, J.G., "Silicosis in the Ceramic Industry." Jour. Can. Ceram. Soc. 9, 4-8 (1940).

Currier, A.E., "Glass-Refractories Symposium: VI, Refractories for Use in Glass-Furnace Regenerators." Am. Ceram. Soc. Bull. 29, (3) 90-5 (1950).

Curtis, F.L., "Abrasive Rods." Ger. 1,066,905, 10/8/59.

Curtis, T.S., "Methods of Heat Treatment of Alumina & Other Materials." U.S. 1,662,739, 5/13/28.

Curtis, T.S., "Converter." U.S. 2,065,566, 12/29/36.

Cutler, I.B., "Strength Properties of Sintered Alumina in Relation to Porosity & Grain Size." J. Am. Ceram. Soc. 40 (1) 20-3 (1957).

Cutler, I.B., Bradshaw, C., Christensen, C.J. & Hyatt, E.P., "Sintering of Alumina at Temperatures of 1400°C & Below." J. Am. Ceram. Soc. 40 (4) 134-9 (1957).

Czaplinski, A. & Zielinski, E., "Adsorption of Helium, Neon, & Hydrogen on Silica Gel & Alumina at the Temperature of Liquefied Nitrogen & High Pressure." Przemysl Chem. 38, 87-8 (1959).

Dalmai, G., et al. "Catalytic Properties of Alumina Obtained from Hydrargillite when Irradiated with γ-Rays or Neutrons." Compt. rend. 256, 3468-70 (1963).

Damerell, V.R., Hovorka, F., & White, W.E. "Surface Chemistry of Hydrates. II. Decomposition Without Lattice Rearrangements." J. Phys. Chem. 36, 1255-67 (1932).

Dana, J.D., "System of Mineralogy I." 7th Ed. Revised & Edited by Palache, C., Berman, H., & Frondel, C., New York, Wiley (1944).

Danilov, V.P. & Joffe, Z.A., "Uviol Borate Glass for Bactericidal Lamps." Compt. rend. Acad. Sci. U.R.S.S., 39 (6) 234-6 (1943).

DaSilva, E.M. & White, P. "Electrical Properties of Evaporated Aluminum Oxide Films." J. Electrochem. Soc. 109 (1) 12-5 (1962).

Datta, R.K. & Roy, R., "Phase Transitions in $LiAl_5O_8$." J. Am. Ceram. Soc. 46 (8) 388-90 (1963).

Dau, G.J. and Davis, M.V., "Gamma-Induced Electrical Conductivity in Al_2O_3." Nucl. Sci. Eng. 25, 223-6 (1966).

Daubenmeyer, H.W., "Elastic Mold & Method of Molding Material." U.S. 1,983,602.

Davies, L.G., et al. "A Study of High Modulus, High Strength, Filament Materials by Deposition Techniques." General Tech. Corp. Bu-Weps NOw-64.0176-C, Final Rept. (Jan. 1965).

Davies, L.M., "The Effect of Heat Treatment on the Tensile Strength of Sapphire." Proc. Brit. Ceram, Soc. 6, 29-35 (1966).

Davies, M.O., "Transport Phenomena in Aluminum Oxide." NASA Accession No. N65-20999, Rept. No. NASA-TN-D-2765.

Davis, T.A. and Vedam, K., "Photoelastic Properties of Sapphire." J. Appl. Phys. 38, 4555-6 (1967).

Dawihl, W. & Doerre, E., "Strength & Deformation Properties of Sintered Alumina Bodies as a Function of Their Composition & Structure." Ber. deut. keram. Ges. 41 (2) 85-96 (1964).

Dawihl, W. and Doerre, E., "The Shear Strength of Sintered Alumina as a Basis for the Interpretation of Friction Action." Ber. deut. keram. Ges. 43, 280-3 (1966).

Dawihl, W. & Klingler, E., "Performance & Economy of Oxide Ceramic Cutting Tools." V.D.I. Zeitschrift 100 (13) 559-63 (1958).

Dawihl, W. and Klinger, E., "Relation between Temperature and Vibration Resistance of Sintered Alumina Bodies." Ber. deut. keram. Ges. 42, 311-13 (1965).

Dawihl, W. and Klingler, E., "Corrosion Resistance of Al_2O_3 Single Crystals and of Al_2O_3-Base Material Against Inorganic Acid." Ber. deut. keram. Ges. 44, 1-4 (1967).

Dawihl, W. & Kuhn, K.D., "Production of Pure Corundum Powder." Ber. deut. keram. Ges. 41 (6), 365-8 (1964).

Dawson, R., "A High-Temperature Calorimeter; The Enthalpies of a-Aluminum Oxide & Sodium Chloride." J. Phys. Chem. 67 (8) 1669-71 (1963).

Day, A.L., Shepherd, E.S. & Wright, F.E. Am. J. Sci. 22, 265-302 (1906).

Day, F., Jr. & Ambrosone, J.P., "Corrosion Resistance of Pure Alumina as a Glass Refractory." J. Am. Ceram. Soc. 34 (5) 163-4 (1951).

Day, M.J. & Stavrolakis, J.A., "Self-Healing Coatings for Refractory Metals & Method for Applying Them." U.S. 3,038,817, 6/12/62.

Day, M.K.B. & Hill, V.J., "Thermal Transformations." Nature 170, 539 (Sept. 1952).

Day, M.K.B. & Hill, V.J., "Thermal Transformations of the Aluminas & Their Hydrates." J. Phys. Chem. 57, 946-50 (1953).

Day, O.L., "Refractories—Recent Types, Uses, & Unit Costs." Metal Progress 30 (4) 59-63, 72, 82 (1936).

de Boer, J.H., "Investigations on Microporous Salt- and Oxide-Systems. Angew. Chem. 70, 383-9 (1958).

de Boer, J.H., Fortuin, J.M.H., & Steggerda, J.J., "Dehydration of Alumina Hydrates: I." Koninkl. Ned. Akad. Wetenschap. Proc. 57B, 170-80 (1954).

de Boer, J.H., Fortuin, J.M.H., Lippens, B.C. & Meijs, W.H., "The Nature of Surfaces with Polar Molecules II. The Adsorption of Water on Aluminas." J. Catalysis 2 (1), 1-7 (1963).

de Boer, J.H., Fortuin, J.M.H., & Steggerda, J.J., "Dehydration of Alumina Hydrates: II." Koninkl. Ned. Akad. Wetenschap, Proc. 57B, 434-43 (1954).

de Boer, J.H., Heuvel, A. van der, & Linsen, B.G., "Pore Systems in Catalysts IV The Two Causes of Reversible Hysteresis." J. Catalysis 3 (3) 268-73 (1954).

de Boer, J.H., & Houben, G.M.M. Proc. Intern. Sympos. Reactivity of Solids (Gothenberg I, 237 (1952)).

de Boer, J.H., & Lippens, B.C., "Pore Systems in Catalyst. II. Shapes of Pores in Aluminum Oxide Systems." J. Catalysis 3 (1) 38-43 (1964).

de Boer, J.H., Steggerda, J.J. & Zwietering, P., "Dehydration of Alumina Hydrates: III, Formation of Pore System During Dehydration of Gibbsite." Koninkl. Ned. Akad. Wetenschap. Proc, 59B, 435-44 (1956).

Decker, R.F., Rowe, J.P. & Freeman, J.W., "Relations of High-Temperature Properties of Titanium + Aluminum Hardened Nickel-Base Alloy to Contamination by Crucibles." Trans. AIME 212, 686-94 (1958).

Decker, A.R. & Royal, H.F., "Use of Torsional Stresses for Observing High-Temperature Characteristic of Ceramic Materials." J. Am. Ceram. Soc. 31 (12) 332-7 (1948).

De Dani, A., "Glass Fibers—Their Manufacture, Properties, & Uses." Chemistry & Industry 1955 No. 18, pp. 482-90.

Deflandre, M., "The Crystal Structure of Diaspore." Bull. Soc. Franc. Min. 55, 140-65 (1932).

Degtyareva, E.V., "Growth of Corundum Crystals During Sintering." Dokl. Akad. Nauk SSSR. 165, 372-5 (1965).

Degtyareva, E.V. & Kainarskii, I.S., "Sintering Kinetics of Corundum." Dokl. Akad. Nauk SSR 156 (4) 937-40 (1964).

Degtyareva, E.V. and Kainarskii, I.S., "Kinetics of Sintering of Corundum under Pressure." Izv. Akad. Nauk SSSR Neorgan. Materialy, 2, 239-44 (1966).

Degtyareva, E.V., Kainarskii, I.S., Karyakin, L.I., & Alekseenko, L.S., "Dielectric Properties of Corundum Ceramics and Their Microstructure." Izv. Akad. Nauk SSSR, Neorgan. Materialy 1, 816-22 (1965).

Degtyareva, E.V., Kainarskii, I.S. & Totsenko, S.B., "Sintering of Corundum with Additives." Ogneupory 29 (9) 400-11 (1963).

Deissler, and Boegli, "Effective Thermal Conductivity of Powders in Various Gases" Trans. A.S.M.E. 80, 1417 (1958).

de Josselin de Jong, G., "Verification of Use of Peak Area for Quantitative Differential Thermal Analysis." J. Am. Ceram. Soc. 40 (2) 42-9 (1957).

de Keyser, W.L., "Mullite" Ber. deut. keram. Ges. 40, (5) 304-15 (1963).

De Keyser, W.L., "Reactions in the Interfaces between SiO_2 and Al_2O_3." Sci. Ceram. 2, 243-57 (1965).

De Keyser, W.L. & Wollast, R., "Action of CaO on Aluminosilicate Refractories." Bull. Soc. Franc. Cer. 1961, No. 50, pp. 25-33.

Dekker, A.J. & Geel, W.C. van, "The Amorphous & Crystalline Oxide Layers of Aluminum." Phillips Research Reports 2, 313-20 (1947).

De Maria, G., Drowart, J., & Inghram, M.G., "Mass Spectrometric Study of Alumina." J. Chem. Phys. 30, 318-9 (1959).

De Ment, Jack., "Fluorochemistry." Chemical Publishing Co., Inc., Brooklyn, 1945, 796 pp.

Denton, E.P. & Rawson, H., "Metalizing of High-Al_2O_3 Ceramics." Trans. Brit. Ceram. Soc. 59 (2) 25-37 (1960).

DePablo-Galan, L. & Foster W.R., "Investigation of Role of Beta Alumina in the System Na_2O-Al_2O_3-SiO_2." J. Am. Ceram. Soc. 42 (10) 491-8 (1959).

Deportes, C., "High Temperature Electrochemistry II" Rev. Gen. Elec. 76, 50-7 (1967).

Derfler, F., "Ball Making Machine." U.S. 2,480,716, 8/30/49.

Detwiler, D.P., & Tallan, N.M., "Dielectric Losses Due to Dislocations in Sapphire." PB Rept. 152544, 177 pp; U.S. Govt. Res. Repts. 35 (3) 345 (1961).

Deutscher, F.O., "Gas Plating of Alumina." U.S. 2,972,555, 2/21/61.

Devlin, G.E., McKenna, J., May, A.D. & Schawlow, A.L., "Composite Rod Optical Masers." Appl. Optics, 1 (1) 11-15 (1962).

DeVries, R.C., "System Al_2SiO_5 at High Temperatures & Pressures." J. Am. Ceram. Soc. 47 (5) 230-7 (1964).

DeVries, R.C. & Sears, G.W., "Growth of Aluminum Oxide Whiskers by Vapor Deposition." J. Chem. Phys. 31, 1256-7 (1959).

Devrup, A.J., "Vitreous Coatings for Light Metals." U.S. 2,467,114, 4/12/49.

Dew, R.J. Jr., "Damping Capacity Measurements on Refractory Oxides Under Varying Stress & Temperature Conditions." S.D. Thesis MIT 1950 (unpublished).

Dewey, C., "Gibbsite." Am. J. Sci. (1) 2, 249 (1820).

Dewey, J.L., "Refractory Lining for Alumina Reduction Cells." U.S. 3,093,570, 6/11/63.

deWitt, G.H., "Development of Ceramic Materials for Use in Gas-Turbine Engines." Combined Intelligence Objectives Sub-Committee Rept., File No. XXV-9. H.M. Stationery Office, London, 1945.

Diamond, J.J. and Dragoo, A.L., "Molten Alumina in the Arc-Image Furnace." Rev. Int. Hautes Temp. Refract. 3, 273-9 (1966).

Diamond, J.J. & Schneider, S.J., "Apparent Temperatures Measured at Melting Points of Some Metal Oxides in a Solar Furnace." J. Am. Ceram. Soc. 43 (1) 1-3 (1960).

Dickinson, T.A., "Atomic Energy & the Ceramic Industries." Ceram. Age 57 (4) 21-22, 58 (1951).

Dickinson, T.A., "Processing Ceramics with Ultrasound." Ceram. Age 58 (2) 21-3, 37 (1951).

Dickinson, T.A., "Ceramic Coatings for Aluminum Products." Ceram. Age 64 (5) 9-12 (1954).

Dickinson, T.A., "Flame-Spraying Ceramics." Ceramics 5 (59) 512-5 (1954).

Dickinson, T.A., "Ultrasonic Dispersion of Slips & Glazes." Ceram. Age 65 (3) 13-4, 27 (1955).

Diepschlag, E. & Wulfestieg, F., "Electrical Conductivity of Magnesite & Some Other Refractory Materials." J. Iron & Steel Inst. 120 (2) 297-321 (1929).

Diesperova, M.I. & Bron, V.A., "Influence of Additions on the Sintering Processes & on the Spinel Formation in the Systems of MgO-Al_2O_3 & MgO-Cr_2O_3." Tr. Vost. Inst. Ogneuporov No. 4, 164-83 (1963).

Dietzel, A., "Explanation of the Remarkable Expansion Phenomena in Silicate Glasses & Special Glasses." Naturwissenschaften 31 (1/2) 221-3 (1943).

Dietzel, A. & Mostetzky, H., "Mechanism of the Dewatering of a Ceramic Slip by the Plaster Mold: I, Experimental Investigation of the Diffusion Theory of the Slip-Casting Process." Ber. deut. keram. Ges. 33 (1) 7-18 (1956).

Dijksterhuis, P.R. & Hovingh, A.R., "Bonding Metal to Ceramics." Ger. 1,013,216, 8/1/57.

Dillingham, R.P., "Apparent Microscopic Structure of High Alumina Materials." Cer. Ind. 66 (4) 152-5, 176 (1956).

Dils, R.R., "Cation Interdiffusion in the Chromia-Alumina System." Univ. Microfilms Order No. 65-12, 767.

Dils, R.R., Martin, G.W. & Huggins, R.A., "Chromium Distribution in Synthetic Ruby Crystals." App. Phys. Letters 1 (4) 75-6 (1962).

Dinescu, R., "Behavior of Some Refractory Oxides Wetted by a Melted Mass." Acad. Rept. Populare Romine, Studii Cercetari Met. 7, 337-49 (1962).

Dingman, R.W., et al. "Process of Making Aluminum Oxide from Hydrated Alumina." U.S. 2,801,901, 8/6/57.

Dingman, R.W., et al. "Process & Apparatus for the Production of Aluminum Oxide." U.S. 2,828,186, 3/25/58.

Dittrich, F.J., "New Flame Spray Technique for Forming Nickel Aluminide Ceramic Systems." Am. Ceram. Soc. Bull. 44 (6) 492-6 (1965).

Dixon, R.W. & Bloembergen, N., "Linear Electric Shifts in the Nuclear Quadruple Interaction in Al_2O_3." Phys. Rev. 135 (6A) 1669-75 (1964).

Dobias, B., Spurny J., & Freudlova, E., "Mineral Solution-Phase Boundary Potentials in Flotation Research: III. Behavior of Anionic & Cationic Collector in Interphase Between Alpha-Corundum & Collector Solution." Coll. Czechoslov. Chem. Commun., 24, 3668-77 (1959).

Dobrer, E.K., "The Effect of Filler on the Dielectric Properties of Epoxide Insulation Resins." Plasticheski Massy 1962, No. 3, 32-7.

Doelph, L.C., Jr., "Preparing Alumina Beta Trihydrate." U.S. 3,092,454, 6/4/63.

Doelter, C., "Electrical Conduction in Crystals at High Temperatures." Z. anorg. Chem. 67, 387 (1910).

Dolkart, F.Z. & Guil'ko, N.V., "Changes in the Mineralogical Composition of Ti-Al_2O_3 Slag by Calcination." Doklady Akad. Nauk S.S.S.R. 98, 137-9 (1954).

Dolph, J.L., "Aluminum Melting Furnace Refractory." U.S. 3,078,173, 2/19/63.

Donahey, J.W., "Low-Temperature Vitreous Enamels." U.S. 2,608,490, 8/26/52.

Dontsova, E.I., "Exchange of Oxide-Mineral Oxygen with Carbon Dioxide." Doklady Akad. Nauk SSSR 105, 305-8 (1955).

Dorn, J. editor., "Mechanical Behavior of Materials at Elevated Temperatures." McGraw-Hill Book Co., New York (1961).

Dorsey, J.L., "Development & Evaluation Services on Ceramic Materials & Wall Composites for High-Temperature Radome Shapes." PB Rept. 131987, 21 pp U.S. Govt. Res. Repts. 30 (4) 260-1 (1958).

Douglas, E.A., "Ceramic Coatings for the Control of Radiant Heat." Am. Ceram. Soc. Bull. 38 (1) 20-3 (1959).

Dow, N.F., "Development of Composite Structural Materials for High-Temperature Applications." Contract NOw 60-0465d. Third Quarterly Rept., Jan. 1961.

Dragsdorf, R.D. & Webb, W.W., "Detection of Screw Dislocations in a-Alumina Whiskers." J. Appl. Phys. 29, 817-9 (1958).

Drake, L.C. & Ritter, H.L., "Macropore Size Distributions in Some Typical Porous Substances." Ind. Eng. Chem. Anal. Ed. 17, 787-91 (1945).

Dressler, M., "Ceramic High Frequency Insulating Materials & Their Importance in Electron Tube Engineering." STEMAG Nachr., 1962, No. 35, pp. 955-60.

Drexler, F. & Schutz, W., "Aluminum Orthophosphate Glasses." Glastech. Ber., 24 (7) 172-6 (1951).

Dreyling, L.J. & Dreyling, A.P., "Lightweight Refractories Resistant to High Temperatures." U.S. 3,141,781, 7/21/64.

Dromsky, J.A., et al. "Growth of Aluminum Oxide Particles in a Nickel Matrix." Trans. AIME 224 (2) 236-9 (1962).

Droscha, H., "Advantages of Oxide-Ceramic Cutting Tools." Ber. deut. keram. Ges. 41 (4), 253-5 (1964).

Drucker, D.C. & Gilman, J.J. eds., "Fracture of Solids." Proceedings of an International Conference Sponsored by Institute of Metals Div. American Institute of Mining, Metallurgical & Petroleum Engineers, Aug. 1962. Interscience Publishers, N.Y. (1963) 708 pp.

Druzhinina, N.K., "Preparation of Artifical Diaspore." Doklady Akad. Nauk SSSR 88 (1) 133-4 (1953).

DuBois, W.H., "Metal-Ceramic Friction Materials." Machine Design 28 (24) 139-41 (1956).

Dubrovo, S.K. & Kefeli, A.A., "Coordination of Gallium & Aluminum in Glasslike Silicates." Zhur. Priklad. Khim., 35 (2) 441-3 (1962).

Duckworth, W.H., "Ceramic Developments for Aircraft Power Plants." Am. Ceram. Soc. Bull. 27 (3) 93-6 (1948).

Duckworth, W.H., "Precise Tensile Properties of Ceramic Bodies." J. Am. Ceram. Soc. 34 (1) 1-9 (1951).

Duckworth, W.H., Johnston, J.K., Jackson, L.R., & Schofield, H.Z., "Mechanical Properties of Ceramic Bodies." Rand Rept. No. R-209, Battelle (1950).

Duckworth, W.H. & I.E. Campbell, "Outlook for Ceramics in Gas Turbines." Mech. Eng. 72 (2) 128-30, 144 (1950).

Duckworth, W.H. & Rudnick A., "Strength of Ceramic Materials." Battelle Tech. Rev. 10 (4) 3-8 (1961).

Duckworth, W.H., Schwope, A.D., Salamssy, O.K., Carson, R.L., & Schofield, H.Z., "Mechanical Property Tests on Ceramic Bodies." WADC Tech. Report No. 52-67.

Dugdale, R.A. and Ford, S.D., "Etching of Alumina and Fused Silica by Sputtering." Trans. Brit. Ceram. Soc. 65, 165-80 (1966).

Dugdale, R.A., McVickers, R.C. & Ford, S.D., "Alumina, Polishing & Etching Techniques for Dense." J. Am. Ceram. Soc. 45 (4) 199 (1962).

Dumas, L.P., "How More Accurate Sizing Improves Corundum Quality." Grinding & Finishing 6 (11) 59-62 (1960).

Dunay, S., et al. "Production of Alumina Hydrate of High Purity in Bayer Process Plants." Femipari Kutato Intezet Kozlemenyei (Proc. Research Inst. Non-Ferrous Metals), 1956 Nov. 1, pp. 77-95.

duPont de Nemours, E.I., "Special Forms of Corundum." Brit. 849,051, 9/21/60.

Durum, W.H., "Available Quality & Quantity of Water for the Ceramic Industries." Am. Ceram. Soc. Bull. 35 (11) 417-21 (1956).

Duwell, E.J. & McDonald, W.J., "Some Factors that Affect the Resistance of Abrasive Grits to Wear." Wear 4 (5) 372-83 (1961).

Dykstra, L.J., "X-ray Study of the Ternary System U-Al-O." U.S. At. Energy Comm. GA-1479 21 pp (1960).

Elbert, G., "Dielectric Resonance & Dispersion Phenomena in The System γ-Aluminum Oxide-Water." Kolloid-Z 184, 148-54 (1962).

Ebner, M., "Stability of Refractories in Hydrogen-Fluorine Flames." J. Am. Ceram. Soc. 44 (1) 7-12 (1961).

Economos, G., "Behavior of Refractory Oxides in Contact with Metals at High Temperatures." Ind. Eng. Chem. 45 (2) 458-9 (1953).

Economos, G. & Kingery, W.D., "Metal-Ceramic Interactions: II, Metal-Oxide Interfacial Reactions at Elevated Temperatures." J. Am. Ceram. Soc. 36 (12) 403-9 (1953).

Edwards, J.D., "Combination Process for Alumina." Amer. Inst. Mining Met. Engrs. Tech. Pub. No. 1833; Metals Tech. 12 (3) 5 pp. (1945).

Edwards, J.D., Frary, F.C. & Jeffries, Z., "The Aluminum Industry—Aluminum & Its Production." McGraw-Hill Book Co., Inc. New York (1930).

Edwards, P.L. & Happel, R.J., "Alumina Whisker Growth on a Single-Crystal Alumina Substrate." J. App. Phys. 33, 826-7 (1962).

Edwards, P.L. and Huang, S., "Comparison of Whisker Growth Sites and Dislocation Etch Pits on Single-Crystal Sapphire." J. Am. Ceram. Soc. 49, 122-5 (1966).

Ehman, P.J., "Activated Alumina Free of Fines." U.S. 2,720,492, 10/11/55.

Eigeles, M.A., "Stabilization of Corundum Suspensions in Classification by Cones." Mineral. Syr'e 11 (4) 29-38 (1936).

Eisenberg, M., "Flame Spraying of Oxidation-Resistant Adherent Coatings." U.S. 3,121,643, 2/18/64.

Eisenkolb, F. & Richter, W., "Investigations of Sintered Materials of Metallic & Nonmetallic Constituents." Silikattech. 8 (4) 140-7 (1957).

Eitel, W., "Ceramic Spark Plug Insulators Rich in Alumina." Ger. 655,082.

Eitel, W., "Comparative Microscopic Investigations on the Corrosion of Different Refractories by Fluoride-Silicate Melts." Radex Rundschau 1955, No. 3/4, pp. 440-59.

Eliasson, B.A.E., "Production of Fine Crystalline Aluminum Oxide." Brit. 661,935, 10/3/51.

Elleman, T.S., et al. "Fission-Fragment Induced Expansion in Single-Crystal Alumina." U.S. At. Energy Comm. B.M.I. 1646, 10 pp. (1963).

Ellingham, H.J.T., "Reducibility of Oxides & Sulfides in Metallurgical Processes." Jour. Soc. Chem. Ind. (London) 63 (5) 125-33 (1944).

Elliott, N.W., "Use of Alumina for Bedding." Pottery & Glass Record 19 (8) 198-200 (1937).

Ellis, C., "Printing Inks, Their Chemistry & Technology." Reinhold Publishing Corp., New York (1940).

Ellis, R.B., "Dispersion Strengthening of Metals." The American Scientist 52, 476-87 (1964).

Elms, W.A., "Bonding High-Purity Alumina to Metal." Brit. 967, 452, 8/19/64.

Elomanov, V.P., et al. "Glass Formation in the System TeO_2-Al_2O_3." Zhur. Neorg. Khim. 5 (7) 1632-3 (1960).

Elyard, A.C., "Discussion of Robinson & Gardner Note on Preparation of Highly Dense Al_2O_3 for Microscopic Examination." J. Am. Ceram. Soc. 45 (1) 47 (1962).

Emrich, B.R., "Technology of New Devitrified Ceramics—A Literature Review." ML-TDR-64-203, AF Materials Laboratory, W-P AFB, Ohio (Sept. 1964).

Emschwiller, G. et al., "The Infrared Study of Hydrated Calcium Aluminates." Compt. rend. 259 (6) 1329-32 (1964).

Enamels. "Zircon Needs Alumina to Opacify." Ceram. Ind. 51 (2) 60 (1948).

Endell, K., Fehling, R. & Kley, R., "Influence of Fluidity, Hydrodynamic Characteristics & Solvent Action of Slags on the Destruction of Refractories of High Temperature." J. Am. Ceram. Soc. 22, 105-16 (1939).

Engberg, C.J. & Zehms, E.H., "Thermal Expansion of Al_2O_3, BeO, MgO, B_4C, SiC, and TiC above 1000°." U.S. At. Energy Comm. NAA-SR-3086, 25 pp. (1958).

Engberg, C.J. & Zehms, E.H., "Thermal Expansion of Al_2O_3, BeO, MgO, B_4C, SiC, and TiC above 1000°C." J. Am. Ceram. Soc. 42 (6) 300-5 (1959).

Engel, O.G., "Resistance of White Sapphire & Hot-Pressed Alumina to Collision with Liquid Drops." J. Res. Natl. Bur. Stds., 64A (6) 499-512 (1960).

Engel, W.F. & Krijger, P., "Process for Decolorizing a Hydrocarbon Oil Using an Alumina Adsorbent." U.S. 2, 926, 135, 2/23/60.

Entress, K. & Skatulla, W., "Sintering Oxide-Ceramic High-Temperature Materials." Silikattechnik 15 (6) 188-92 (1964).

Entress, Kurt and Steiner, W., "Binders for Compression-Molded Alumina and Zirconia" Ger. (East) 35, 848 1/1/65.

Eremenko, V.N. & Beinish, A.M., "Electrical Conductivity of Binary Systems Composed of Refractory Oxides." Zhur. Neorgan. Khim. 1 (9) 2118-30 (1956).

Erickson, C.G., "Why Not Ceramic as a Material for Gauges." Machine & Tool Blue Book 53 (9) 129-31 (1958).

Ermolaeva, E.V., "Relationship of Surface Tension, Viscosity, & Density of Certain Three-Component Melts of Refractory Oxides." Ogneupory 20 (5) 221-8 (1955).

Ermolenko, N.F. & Efros, M.D., "Influence of Thermal Treatment on the Structure & Specific Surface of Aluminum Oxide Prepared from the Oxychloride." Ionoobmen i Sorbtsiya iz Rastvorov, Akad. Nauk Belorussk. SSR, Inst. Obshch. i Neorgan. Khim. 1963, 144-8.

Ervin, G., Jr., "Structural Interpretation of the Diaspore-Corundum & Boehmite-γAl_2O_3 Transitions." Acta Cryst. 5 (1) 103-8 (1952).

Ervin, G. & Osborn, E.F., "The System Aluminum-Water." J. Geol. 59, 381-94 (1951).

Eubanks, A.G. & Hunkelen, R.E., "Low Temperature Ceramic Solves High Temperature Problems." Ceram. Ind. 83 (2) 52-6 (1964).

Eubanks, A.G. & Moore, D.G., "Investigation of Aluminum Phosphate Coatings for Thermal Insulation of Airframes." NASA (Natl. Aeronaut. Space Admin.) Tech. Note D-106, 28 pp. (1959).

Evans, E.B., "Methods of Producing Metal Oxide Gels." U.S. 2,822,337, 2/4/58.

Evans, P.E., Hardiman, B.P., Mathur, B.C. and Rimmer, W.S., "Alumina-Based Cutting Tools." Trans. Brit. Ceram. Soc. 66, 523-40 (1967).

Evers, J., "Synthetic Chromium-Colored Spinel." Thesis, Konigsberg, 1936, 40 pp.; Neues Jahrb. Mineral. Geol. Referat. I, 1937, p. 199.

Every, C.E., "Vitreous Enamels & Compositions." Brit. 633,178, 12/21/49.

Ewing, C.T. & Baker, B.E., "Thermal & Related Physical Properties of Molten Materials." Part I. Thermal Conductivity & Heat Capacity of Molybdenum Disilicide, Naval Research Lab. WADC Tech. Rept. No. 54-185, 27 pp. (1954).

Ewing, F.J., "Crystal Structure of Diaspore." J. Chem. Phys. 3, 203-7 (1935).

Eyraud, C. & Goton, R., "Kinetics of the Thermal Dissociation of the Hydrates of Alumina." J. chim. phys. 51, 430-3 (1954).

Eyraud, C., Goton, R. & Prettre, M., "Kinetic Characteristics of the Thermal Dehydration of Hydrargillite." Compt. rend. 238, 1028-31 (1954).

Eyraud, C., Goton, R. & Prettre, M., "A Study of the Dehydration of Gibbsite by the Simultaneous Use of Thermogravimetry Under Reduced Pressure & Differential Thermal Analysis." Compt. rend. 240, 1082-4 (1955).

Eyraud, C., Goton, R., Trambouze, Y., Tran-Huu, T. & Prettre, M., "The Decomposition of Alumina Hydrates by Differential Enthalpy Analysis." Compt. rend. 240, 862-4 (1955).

Fabianic, W.L., "Glass-Refractories Symposium: VII, Refractories for the Glass-Furnace Superstructure." Am. Ceram. Soc. Bull. 29 (3) 96-8 (1950).

Faktor, M.M., Fiddyment, D.G., and Newns, G.R., "Preliminary Study of the Chemical Polishing of Alpha Corundum Surfaces with Vanadium Pentoxide." J. Electrochem. Soc. 114, 356-9 (1967).

Fallon, F.J., "Dielectric Materials for High Temperature Use: I." Ceram. Age 75 (3) 24-8 (1960).

Fang, J.H., "Comment on 'Comment on Vegard's Law ....'." J. Am. Ceram. Soc. 46 (5) 247 (1963).

Farnham, R. & Burdick, R.B., "Preliminary Study of Enamels for Aluminum." Presented at meeting of Am. Cer. Soc., Atlantic City, N.J., April 1947; abstracted in Am. Ceram. Soc. Bull. 26 (3) 68 (1947).

Feagin, R.C., "Molds for Casting Metals." Brit. 650,532, 1/17/51.

Feagin, R.C., et al. "Calcium Aluminate Binder." U.S. 2,911,311, 11/3/59.

Feichter, H.R., "Making Fired Vitreous Product." U.S. 2,482,580, 9/20/49.

Feitknecht, W., "Ordering Processes in Colloidally Dispersed Hydroxides & Hydroxy Salts." Koll. Z. 136, 52-6 (1954).

Felten, E.J., "Hot-Pressing of Alumina Powders at Low Temperatures." J. Am. Ceram. Soc. 44 (8) 381-5 (1961).

Fendo, T., Ono, S. & Onaka, K., "Zirconia-Alumina Cast Refractories." Japan 3896, 6/8/62.

Fenerty, M.J., "Process of Producing Alumina of Low Soda Content." U.S. 2,961,297, 11/22/60.

Fenerty, M.J., "Purification of Alumina." U.S. 2,887,561, 5/19/59.

Fenstermacher, J.E. & Hummel, F.A., "High Temperature Mechanical Properties of Ceramic Materials: IV, Sintered Mullite Bodies." J. Am. Ceram. Soc. 44 (6) 284-9 (1961).

Fessler, A.H., "Alkali-Free Ceramic Materials & Method of Making Same." U.S. 2,069,060, 1/26/37.

Fessler, A.H. & Navratiel, H., "Ceramic Body." U.S. 2,154,069, 4/11/39.

Fessler, A.H. & Schwartzwalder, K., "Ceramic Body." U.S. 2,272,618, 2/10/42.

Fessler, A.H. & Schwartzwalder, K., "Ceramic Body Especially Adapted for Use as a Spark-Plug Insulator." U.S. 2,272,338, 2/10/42.

Fichter, R., "Composition & Infrared Spectrum of the Aluminum Oxide Layer Produced Electrolytically in H_2SO_4." Helv. Chim. Acta 30, 2010-13 (1947).

Fiegel, L.J., Mohanty, G.P., & Healy, J.H., "Equilibrium Studies of Refractory Metal Oxides." TID 15,382, 23 pp.; U.S. Govt. Res. Repts. 37 (19) S-34 (1962).

Field, T.E., "Cast Refractory Product." U.S. 2,271,366, 1/27/42.

Field, T.E., "Cast Refractory Product." U.S. 2,408,305, 9/24/46.

Field, T.E., "Basic Cast Refractory." U.S. 2,409,844, 10/22/46.

Field, T.E., "Alumina Low Silica Refractory." U.S. 2,424,082, 7/15/47.

Field, T.E., "Cast Refractory Products." U.S. 2,467,122, 4/12/49.

Field, T.E. & Smyth, H.T., "Producing Cast Refractory." U.S. 2,381,945, 8/14/45.

Field, T.E. & Smith, H.T., "Producing Fusion-Cast Refractories." U.S. 2,603,914, 7/22/52.

Fieldhouse, I.B., Hedge, J.C. & Lang, J.L., "Measurements of Thermal Properties." PB Rept. 151,583, 108 pp. (1958).

Filby, J.D., "Chemical Polishing of Single-Crystal Alpha Alumina Using Silicon." J. Electrochem. Soc. 113, 1085-7 (1966).

Filonenko, N.E. & Borovkova, L.A., "Investigation of Electrocorundum in Reflected Light." Oneupory 17 (3) 124-33 (1952).

Filonenko, N.E. & Kuznetsova, O.S., "Anomalous Expansion of Electrocorundum." Ogneupory 17 (10) 470-4 (1952).

Filonenko, N.E. & Lavrov, I.V., "Calcium Hexaluminate in the System $CaO-Al_2O_3-SiO_2$." Doklady Akad. Nauk SSSR 66 (4) 673-6 (1949).

Filonenko, N.E. & Lavrov, I.V., "Investigation of the Equilibrium Conditions in the Alumina Corner of $CaO-Al_2O_3-SiO_2$." J. Applied Chem. (USSR) 23 (10) 1040-6 (1950).

Filonenko, N.E., Lavrov, I.V., Andreeva, O.V. & Pevzner, R.L., "Aluminous Spinel $AlO \cdot Al_2O_3$." Dokl. Akad. Nauk SSSR 115 (3) 583-5 (1957).

Finnigan, G., "Machining Ceramics is no Problem when the Industrial Diamond is used." Ceramics 14 (178) 12-5 (1963).

First, M.W., Graham, J.B., et al. "High Temperature Dust Filtration." Ind. Eng. Chem. 48 (4) 696-702 (1956).

Fischer, W.A. & Hoffmann, A., "Phase Diagram Ferrous Oxide-Aluminum Oxide." Arch. Eisenhuttenw. 27 (5) 344 (1956).

Fischer, W.H., Stone, E.E. & McGowan, E.J., "Development of Ceramic Insulated Magnet Wire: I, Wire Fabrication & Insulation." Electrochem. Technol. I (1-2 28-34 (1963).

Fisher, E. & Twells, R., "High-Temperature Glazing of Alumina Bodies." J. Am. Ceram. Soc. 40 (11) 385-8 (1957).

Fisher, J.C., Hart, E.W. & Pry, R.H., "The Hardening of Metal Crystals by Precipitate Particles." Acta Met. 1, 336 (1953).

Fitzgerald, A.E., "High-Alumina Refractory." U.S. 2,325,181, 7/25/43.

Fitzgerald, T.M., Chick, B.B., & Truell, R., "Ultrasonic Attenuation in Al_2O_3 at Ultrahigh Frequencies & Low Temperatures." J. Appl. Phys. 35 (9) 2647-8 (1964).

214

Fitzsimmons, E.S., "Thermal Diffusivity of Refractory Oxides." J. Am. Ceram. Soc. 33 (11) 327-32 (1950).

Fleck, R. & Gappish, M., "Structure & Edge Life of Ceramic Cutting Materials." Ind. Anzeig 80 (47) 689-92 (1958).

Fleming, J.B., Getty, R.J., & Townsend, F.M., "Drying with Fixed-Bed Desiccants." Chem. Eng., Aug. 31, 1964.

Fleming, J.D. & Johnson, J.W., "Aluminum-U_3O_8 Exothermic Reactions." Nucleonics 21 (5) 84-7 (1963).

Fleshman, W.S., "Rain Water Leaching of Buried Contaminated Alumina." U.S. At. Energy Comm. GAT-P-20, 11 pp. (1960).

Flick, F.B. U.S. 1,526,127, 2/10/25.

Flock, E.F. & Halpern, C., "Bibliography of Books & Published Reports on Gas Turbines, Jet Propulsion, & Rocket Power Plants." Natl. Bur. Stds. (U.S.) Circ. C 509, 1951.

Florio, J.V., "Dielectric Properties of Alumina at High Temperatures." J. Am. Ceram. Soc. 43 (5) 262-7 (1960).

Floyd, J.R., "Effect of Composition & Crystal Size of Alumina Ceramics on Metal-to-Ceramic Bond Strength." Am. Ceram. Soc. Bull. 42 (2) 65-70 (1963).

Floyd, J.R., "Apparent Effects of Alumina Crystal Size on Various Properties of Dense High Al_2O_3 Bodies." American Lava Corp. Presented at Fall Meeting, Electronics Div., Gatlinburg, Tenn., Oct. 4, 1963.

Floyd, J.R., "Effects of Firing on the Properties of High-Alumina Bodies." Trans. Brit. Ceram. Soc. 64, 251-65 (1965).

Foerland, Katrine S., "Sintering of Especially Pure Magnesia and Alumina." Tidsskr. Kjemi Bergv. Met. 27, 62-3 (1967).

Folk, H.F. and Bohling, W.C., "High-temperature Strength of High-Alumina Refractories." Am. Ceram. Soc. Bull. 47 580-3 (1968).

Folweiler, R.B., "Creep Behavior of Pore-Free Polycrystalline Aluminum Oxide." J. Appl. Phys. 32, 773-7 (1961).

Folweiler, R.C., "Ceramic Articles." Belg. 619,512, 12/28/62.

Foot, D.G., " 'Bubble' Wheels for Soft Nonmetallics." Grinding & Finishing 6 (1) 30-1 (1960).

Forchheimer, O.L., "Grit Hardness & Toughness—Their Effect on Grinding Operations." Grinding & Finishing 6 (11) 34-9 (1960).

Ford, W.F., "Phase Equilibrium Principles in the Corrosion of Acid Refractories." Glass Technol., 1 (1) 17-24 (1960).

Ford, W.F. & White, J., "Electrical Properties of Oxide Mixtures as an Index of Structural & Phase Changes at High Temperatures." Trans. Brit. Ceram. Soc. 51 (1) 1-79 (1952).

Forrester, A.T., Parkins, W.E. & Gerjuoy, E., "On the Possibility of Observing Beat Frequencies Between Lines in the Visible Spectrum." Phys. Rev. 72 728 (1947).

Forrester, J.J., Jr., "Abrasive Grains for Vitrified Grinding Wheels." Cer. Age 68 (6) 16-9 (1956).

Fortuin, J.M.H. Thesis, Delft (1955).

Foster, A.G., "Pore Size & Pore Distribution." Disc. Farady Soc. No. 3, 41-51 (1948).

Foster, L.M., Long, G. & Hunter, M.S., "Reactions Between Aluminum Oxide & Carbon—The Al_2O_3-Al_4C_3 Phase Diagram." J. Am. Ceram. Soc. 39 (1) 1-11 (1956).

Foster, L.M. & Stumpf, H.C., "Analogies in the Gallia & Alumina Systems—The Preparation & Properties of Some Low-Alkali Gallates." J. Am. Chem. Soc. 73 (4) 1590-5 (1951).

Foster, P.A., Jr., "Nature of Alumina in Quenched Cryolite-Alumina Melts." J. Electrochem. Soc. 106 (11) 971-5 (1959).

Foster, W.R., "Contribution to the Interpretation of Phase Diagrams by Ceramists." J. Am. Ceram. Soc. 34 (5) 151-60 (1951).

Foster, W.R., Leipold, M.H. & Shevlin, T.S., "Simple Phase Equilibrium Approach to the Problem of Oil-Ash Corrosion." Corrosion 12 (11) 539-48T (1956).

Fox, C.W. & Slaughter, G.M., "Ceramic Brazing Development at Oak Ridge National Laboratory." Am. Ceram. Soc. Bull. 43 (8) 586 (1964).

Francis, R.K., McNamara, E.P. & Tinklepaugh, J.R., "The Bonding of Ceramic-Metal Laminates." Prog. Rept. 5 AD154-872, 15 p. (1958).

Francis, R.K., Swica, J.J., et al. "Hot-Pressing of Fine Grained Ceramic Oxides for Use in Thermal Conductivity Studies." PB Rept. 161851 332 pp.; U.S. Govt. Research Repts., 34 (4) 447 (1960).

Francl, J. & Kingery, W.D., "Thermal Conductivity: IV, Apparatus for Determining Thermal Conductivity by a Comparative Method—Data for Pb, Al_2O_3, BeO, and MgO." J. Am. Ceram. Soc. 37 (2, part II) 80-4 (1954).

Francl, J. & Kingery, W.D., "Thermal Conductivity: IX, Experimental Investigation of Effect of Porosity on Thermal Conductivity." J. Am. Ceram. Soc. 37 (2, part II) 99-107 (1954).

Francombe, M.H. & Rooksby, H.P., "Structure Transformations Effected by the Dehydration of Diaspore, Goethite, & Delta Ferric Oxide." Clay Minerals Bull. 4 (21) 1-14 (1959).

Frangos, T.F., "New Alumina-Type Cermets." Materials in Design Eng. 47 (12) 112-5 (1958).

Frary, F.C., "Electrolytic Production of Aluminum." U.S. 1,534,031, 4/21/25.

Frary, F.C., "Adventures with Alumina." Ind. Eng. Chem. 38 (2) 129-31 (1946).

Frechette, V.D., "Cermet Development, U.S.A." Cer. Ind. 63 (3) 79-80 (1956).

Frederickson, L.D., Jr., "Characterization of Hydrated Aluminas by Infrared Spectroscopy-Application to Study of Bauxite Ores." Anal. Chem. 26 (12) 1883-5 (1954).

Free, G., "Manufacture of Iron-Free Activated Alumina for Catalytic Purposes." Ger. 850,450, 9/25/52.

Fremy, E. & Feil, E. Compt. rend. 85, 1029 (1877).

Freund, F., "Spontaneous Diaspore Formation in Partially De-hydrated Single Crystals of Gibbsite." Ber. deut. keram. Ges. 44, 241-4 (1967).

Freymann, M. & Freymann, R., "Microwave Absorption & Hydroxyl Bond; Water of Crystallization." Compt. rend. 232, 401-3 (1951).

Fricke, R., "The Crystalline Alumina Hydrate of von Bonsdorff." Z. Anorg. allgem. Chem. 175, 249-56 (1928).

Fricke, R., "The Transformation of Oxide Hydrates." Koll. Z. 49, 229-43 (1929).

Fricke, R., "Properties of Alumina made by the v. Bondsdorff Process." Z. anorg. allg. Chem. 179, 287-92 (1929).

Fricke, R. & Eberspacher, O., "Small- and Wide-Angle X-ray Scattering by γ-Al_2O_3." Z. anorg. u. allgem. Chem. 265, 21-40 (1951).

Fricke, R. & Jockers, K., "Simple Preparation of Boehmite." Z. Naturforsch. 2b, 244 (1947).

Fricke, R. & Jockers, K., "Porosity & Surface of Various Aluminum Hydroxides & Oxides." Z. anorg. allg. Chem. 265, 41-8 (1951).

Fricke, R. & Jucaitis, P, "Study of Equilibria in the Systems; Al_2O_3-Na_2O-H_2O & Al_2O_3-K_2O-H_2O." Z. anorg. allgem. Chem. 191, 129-49 (1930).

Fricke, R. & Keefer, H., "Position of Isoelectric Point of Gamma Alumina & of Various Modifications of Aluminum Hydroxide." Z. Naturforsch. 4a, 76-7 (1949).

Fricke, R., Neugebauer, W. & Schafer, H., "Inorganic Chromatography: I, Reaction of Aqueous Copper Chloride Solution with γ-Al_2O_3." Z. anorg. u. allgem. Chem. 273 (3-5) 215-26 (1953).

Fricke, R. & Schmah, H., "Water Solubility & Alkali & Zeolitic Properties of $Al(OH)_3$." Z. anorg. Chem. 255, 253-68 (1948).

Fricke, R. & Severin, H. Z. anorg. allg. Chem. 205, 287-308 (1932).

Fricke, R. & Wever, F., "X-Ray Spectrographic Investigation of the Effect of Aging on Precipitated Metal Hydroxides." Z. anorg. allg. Chem. 136, 321-4 (1924).

Fricke, R. & Wullhorst, B., "Energy of Different Modifications of Crystalline Aluminum Hydroxides." Z. anorg. allg. Chem. 205, 127-44 (1932).

Fridman, S.M., "Activated Aluminum Oxide for Continuous Regeneration of Transformer & Turbine Oils." Naladochnye i Eksptl. Raboty Gosudarst. Tresta Organizatsii i Ratsional Elektrostantsii 1954, No. 9, 37-42; Referat. Zhur. Khim. 1956, Abstr. No. 48123.

Friederich, E., "The Hardness of Inorganic Compounds & of Elements." Fortschr. Chem. Physik 18, 5-44 (1926).

Frieser, R.G., "Phase Changes in Thin Reactivity Sputtered Alumina Films." J. Electrochem. Soc. 113, 357-60 (1966).

Frisch, B., "The Hydration of Alpha Alumina." Ber. deut. keram. Ges. 42, 149-60 (1965).

Frischbutter, E. and Schroeder, W., "Beta Corundum and Its Use as an Electrocast Refractory Material." Silikat Tech. Tech-Wiss. Z. Keram., Glas, Email, Kalk Zem. 17, 317-22 (1966).

Friske, W.H., "Interim Report on the Aluminum Powder Metallurgy Product Development Program." AEC Res. & Develop. Rept. NAA-SR-4233 (Jan. 15, 1960).

Froelich, H.C. & Margolis, J.M., "Aluminum Phosphate Phosphor." U.S. 2,455,414, 12/7/48.

Frost, L.J., "Electric Furnace Product." U.S. 2,849,305, 8/26/58.

Fryer, G.M., Budworth, D.W. & Roberts, J.P., "Influence of Microstructure on the Permeability of Sintered Alumina Materials to Gases at High Temperatures." Trans. Brit. Ceram. Soc. 62 (6) 525-36 (1963).

Fuchs, E., "Determination of Lattice Distortions in Corundum Powder by X-ray Diffraction." Kohasz. Lapok 96 (4) 150-3 (1963).

Fujii, K. et al., "Toshiko" or Manganese Pink: I." Bull. Govt. Res. Inst. Cer. (Kyoto) 3 (1) 21-6 (1949).

Fulcher, G.S., "Refractory Zirconia-Alumina Casting." U.S. 2,271,369, 1/27/42.

Fulcher, G.S. & Field, T.E., "Refractory Zirconia Casting." U.S. 2,271,367, 1/27/42.

Fulda, N. & Ginsberg, H., "Tonerde und Aluminium." Walter de Gruyter & Co., Berlin W 35, (1951).

Fulrath, R.M. & Hollar, E.L., "Manganese Glass for Metallizing Ceramics." Am. Ceram. Soc. Bull. 43 (8) 584 (1964).

Funabashi, W., "Study on Abrasives: VII." Nagoya Kogyo Gijutsu Shikensho Hokoku 1 (2) 31-5 (1952).

Funabashi, W., "Study on Abrasives: XI, Finishing Abrasives for Synthetic Resins & Celluloid." Nagoya Kogyo Gijutsu Shikensho Hokoku 2 (11) 5-8 (1953).

Funabashi, W. & Terada, S., "Melted Media for Barrel Finishing." Nagoya Kogyo Gijutsu Shikensho Hokoku 11 (6) 313-7 (1962).

Funaki, K., "The Sulfuric Acid Process for Obtaining Pure Alumina from Its Ores." Bull. Tokyo Inst. Tech. B-1 (1950).

Funaki, K. & Shimizu, Y.J. Chem. Soc. Japan, Ind. Chem. Sect. 55, 194-6 (1952), 56, 53-6 (1953).

Funaki, K. & Shimizu, Y., "Thermal Transformations & Rehydrations of Amorphous Aluminas Produced by Thermal Decomposition of Aluminum Salts." Kogyo Kagaku Zasshi 62, 788-93 (1959).

Furnas, C.C., "Grinding Aggregates." Ind. Eng. Chem. 23 (9) 1052-8 (1931).

Furukawa, G.T., Douglas, T.B., McCoskey, R.E., & Ginnings, D.C., Thermal Properties of Aluminum Oxide from 0° to 1200°K." J. Res. Natl. Bur. Stds. 57 (2) 67-82 (1956).

Gabrysh, A.F., Eyring, H., LeFebre, V. & Evans, M.D., "Thermoluminescence & the Influence of γ-Ray Induced Defects in Single-Crystal Alpha Alumina." J. App. Phys. 33 (12) 3389-91 (1962).

Gabrysh, A.F., Lo, Mei-Kuo, Lin-Sen, Pan, & Eyring, H., "Some Mechanical and Optical Properties of Gamma-Irradiated Alpha Alumina." J. Franklin Inst. 282, 135-46 (1966).

Gad, G.M. & Barrett, L.R., "Action of Heat on Alunite & Alunitic Clays." Trans. Brit. Ceram. Soc. 48 (9) 352-74 (1949).

Galakhov, F.Y., "Alumina Region of Ternary Aluminosilicate Systems. I. MnO-Al_2O_3-SiO_2." Izvest. Akad. Nauk SSSR., Otdel Khim. Nauk, 5, 525-31 (1957).

Galakhov, F. Ya. "Study of the Alumina Region in Ternary Aluminosilicate Systems: II, Beo)Al_2O_3-SiO_2 System." Izv. Akad Nauk SSSR, Otd. Khim, Nauk, 1957 No. 9, pp. 1032-6.

Galant, E.I., & Appen, A.A., "Aluminate-Borate Anomaly in the Optical Properties of Silicate Glasses." Zhur. Priklad. Khim. 31 (11) 1741-44 (1958).

Galdina, N.M. & Deri, A., "Production & Service of 'Korvishit' (Fused Corundum) Refractory in Hungary." Steklo i Keram. 19 (6) 41-4 (1962).

Gallet, G., "Ceramics in Electron Tubes." Vide 11, No. 65, 420-3 (1956).

Gallet, G., "Ceramics in Electron Tubes." Vide 12 (65) 420-3 (1957).

Gallup, J.J., "The Transformation of Aluminum Oxide from the β- to the a-Form." J. Am. Ceram. Soc. 18, 144-8 (1935).

Gallup, J. & Dingwell, A.G.F., "Properties of Low-Temperature Solder Glasses." Am. Ceram. Soc. Bull. 36 (2) 47-51 (1957).

"Gamal—New Polishing Material for the Metallographer." Laboratory 13 (5) 120-2 (1942).

Gangler, J.J., "Resistance of Refractories to Molten Lead-Bismuth Alloy." J. Am. Ceram. Soc. 37 (7) 312-6 (1954).

Gangler, J.J., "A Review of High-Temperature Coatings for Refractory Metals." "High-Temperature Materials II." p. 719, Ed. G.M. Ault, et al., Interscience Publ. Co. (1963).

Gaodu, A.N. & Kainarskii, I.S., "Kinetics of Expanding Alumina Suspensions for the Preparation of Lightweight Corundum." Ogneupory 29 (6) 270-5 (1964).

Garn, P.D. & Flaschen, S.S., "Detection of Polymorphic Phase Transformations by Continuous Measurement of Electrical Resistance." Anal. Chem. 29 (2) 268-71 (1957).

Gatti, A., "Iron-Alumina Materials." Trans. Met. Soc. AIME 215 (5) 753-5 (1959).

Gatti, A., "Apparatus for Growing Whiskers." U.S. 3,147,085, 9/64.

Gaudin M. Compt. rend. 4, 999 (1837).

Gaunt, J., "The Infrared Transmission of Some HF Resistant Materials." U.S. AEC Publ. AERE-C/M-120, 4 pp (England) 1951.

Gayer, K.H., Thompson, L.C. & Zajicek, O.T., "Solubility of Aluminum Hydroxide in Acidic & Basic Media at 25°C." Can. J. Chem. 36, 1268-71 (1958).

Gedeon, T.G., "Bayerite in Hungarian Bauxite." Acta Geol. Acad. Sci. Hung. 4, 95-105 (1956).

Geiger, C.F., Turner, A.A., & Stach, O.R., "Applications of Superrefractories Made from Furnace Products." Chem. Eng. Prog. 44 (12) 933-6 (1948).

Geiling, S. & Glocker, R., "The Atom Arrangement in Aluminum Hydroxide Gel." Z. Elektrochem. 49, 269-73 (1943).

Geller, R.F., "Ceramics May Serve in Turbines & Jets - for Turbine Blades." S.A.E. Journal 56 (6) 46, 48-9 (1948).

Geller, R.F. & Yavorsky, P.J., "Melting Point of Alpha-Alumina." Jour. Res. Nat. Bur. Stds. 34 (4) 395-401 (1945). RP-1649.

Geller, R.F., Yavorsky, P.J., Steierman, B.L. & Creamer, A.S., "Binary & Ternary Combinations of MgO, CaO, BaO, BeO, Al_2O_3, ThO_2, & ZrO_2." J. Res. NBS 36 (3) 289 (1946).

Geller, S., "Crystal Structure of β-Ga_2O_3." J. Chem. Phys. 33, 676-84 (1960).

Gelstharp, F., "Phosphate Glass Batch." U.S. 2,294,844, 9/1/42.

General Electric Co. "Single Crystals of Aluminum Oxide." Neth. Appl. 6,411,302. 5/10/65.

Gentile, A.L. & Foster, W.R., "Calcium Hexaluminate & Its Stability Relations in the System CaO-Al_2O_3-SiO_2." J. Am. Ceram. Soc. 46 (2) 74-6 (1963).

Gerard, G. & Stroup, P.T., "Extractive Metallurgy of Aluminum, Vol. I. Alumina." Interscience Publishers (1963). Based on Internat. Sympos., New York, Feb. 18-22, 1962, sponsored by Extractive Metallurgy Div., Metallurgical Society, A.I.M.M. & PE.

Gerasimov, E.A. and Kovachev, I., "Effect of Cr_2O_3 and of Certain Combined Additions on the Sintering of Corundum." Compt. Rend. Acad. Bulgare Sci. 18, 931-4 (1965).

Gerasimov, E. and Kovachev, Iv., "Effect of Some Additives Made During Sintering of Corundum." Stroit. Materiali Silikat Prom. 6, 25-9 (1965).

Gerdian H., "Aluminum Oxide as a Highly Refractory Material." Z. Electrochem 39, 13-20 (1933).

Getty, R.J. & Armstrong, W.P., "Drying Air with Activated Alumina Under Adiabatic Conditions." Ind. Eng. Chem. Process Design Develop. 3 (1), 60-5 (1964).

Getty, R.J., Lamb, C.E., & Montgomery, W.C. Pet. Eng. 25 B-7-B-10 (1953).

Gibbs, N.E., "Ultrasonic Cutting of Quartz Wafers." J. Acoust. Soc. Amer. 27 (5) 1017 (1955).

Gibbs, P., "Kinetics of High-Temperature Processes." John Wiley & Sons, New York, 21-37 (1959).

Gibbs, P., et al., "Ceramic Studies Summary Technical Report." PB Rept. 161146, 20 pp. U.S. Govt. Res. Repts. 33 (5) 522-3 (1960).

Gibbs, P., Baker, G.S. et al., "Surface & Environmental Effects on Ceramic Materials." AD 267247, 34 pp. U.S. Govt. Res. Repts. 37 (8) S-15-16 (1962).

Gielisse, P.J., Rockett, T.J. et al., "Research on Phase Equilibria Between Boron Oxides & Refractory Oxides, including Silicon and Aluminum Oxides." PB 162,292-3, 24 pp.; U.S. Govt. Res. Repts. 37 (24) S-5-6 (1962).

Giesen, K., "Investigation of the Tenacity of Abrasives." Radex Rundschau 1959 No. 5, pp. 640-59.

Giess, E.A., "Solubility & Crystal Growth of $ZnAl_2O_4$ & Al_2O_3 in Molten PbF_2 Solutions." J. Am. Ceram. Soc. 47 (8) 388-89 (1964).

Gift, E.H., "HRT (Homogeneous Reactor Test) Iodine Removal Bed." U.S. At. Energy Comm. CF-57-9-50, 39 pp. (1957).

Gill, R.M., "Ceramic Oxide Cutting Tools." Machinery (London) 91 (2351) 1341-6 (1957).

Gill, R.M., "Hard Ceramic Materials." Ger. 1,058,917, 6/4/59.

Gill, R.M., "Hard Alumina Ceramics." Brit. 829,799, 3/9/60.

Gill, R.M. & Spence, G., "Ceramics for Machine Tools: I, II." Ceramics 9 (114) 30, 32-4; (115) 27-31 (1958).

Gilles, A., "Absorption at High Temperatures of Several Optical Materials in the Schuman Ultraviolet." J. phys. radium 13, 247 (1952).

Ginnings, D.C. & Furukawa, G.T., "Heat Capacity Standards for the Range 14° to 1200°K." J. Am. Chem. Soc. 75 (3) 522-7 (1953).

Ginsberg, A.C., "Concerning Wolfranic Pyrex & Superpyrex." Keram. i Steklo 8, 17-8 (1932).

Ginsberg, H., Huttig, W. & Stiehl, "The System H_2O-Al_2O_3, II. The Formation of Crystalline $Al(OH)_3$ and the Conversion of Bayerite into Hydrargillite." Z. anorg. Allgem. Chem. 318, 238-56 (1962).

Ginsberg, H., Huttig, W. & Strunk-Lichtenberg, G., "The Influence of the Starting Material on the Crystalline Forms Arising in the Thermal Decomposition & Conversion of Aluminum Hydroxides." Z. anorg. allgem. Chem. 293 33-46, 204-13 (1957).

Ginsberg, H. & Koster, M., "Note on the Aluminum Oxide Monohydrate." Z. anorg. u. allgem. Chem. 271 (1-2) 41-8 (1953).

Gion, L. & Perrin, L., "French Developments in Sintered Ceramic Cutting Tools." Machinery (London) 91 1420-30 (1957).

Gion, M., "Sintering, Especially of Oxides, in Ceramic Cutting Tools." Metallurgia ital 52, No. 3, 120-7 (1960).

Gitlesen, G. and Motzfeldt, K., "Melting Point of Alumina and Related Observations." Rev. Int. Hautes Temp. Refract. 3, 343-9 (1966).

Gitzen, W.H., "Treatment of Aluminum Hydrate." U.S. 1,950,883, 3/13/34.

Gitzen, W.H., "Production of Low-Soda Alumina." U.S. 3,092,452 and 3,092,453, 6/4/63.

Gitzen, W.H. & Hart, L.D., "Explosive Spalling of Refractory Castables Bonded with Calcium Aluminate Cement." Am. Ceram. Soc. Bull. 40 (8) 503-10 (1961).

Gitzen, W.H., Hart, L.D. & MacZura, G., "Properties of Some Calcium Aluminate Cement Compositions." J. Am. Ceram. Soc. 40 (5) 158-67 (1957).

Gitzen, W.H., Hart, L.D., & MacZura, G., "Phosphate-Bonded Alumina Castables—Some Properties & Applications." Am. Ceram. Soc. Bull. 35 (6) 217-23 (1956).

Gitzen, W.H., Heilich, R., & Rohr, F.J., "Carbon Monoxide Disintegration of Calcium Aluminate Cements in Refractory Castables." Am. Ceram. Soc. Bull. 43 (7) 518-22 (1964).

Gitzen, W.H. and MacZura, G., "A Mixture of Aluminas, and Molded Ceramic Material Obtained with This Mixture." Fr. 1,451,018. 8/26/66.

Giulini Bros. G.m.b.H., "Aluminum Oxide with Low Zinc Oxide and Sodium Oxide Content" Fr. 1,385,327. 1/8/65.

"Glass." Chem. & Eng. News 42 (46) 80-96 (Nov. 16, 1964).

Glemser, O., "Dielectric Constants of Some Oxides, Hydroxides, & Hydrated Oxides." Z. Elektrochem. 45, 865-70 (1939).

Glemser, O. & Rieck, G., "Aluminum Hydroxides & Their Dehydration Products." Z. angew. Chem. 67 652 (1955).

Glenny, E. & Royston, M.G., "Transient Thermal Stresses Promoted by the Rapid Heating & Cooling of Brittle Circular Cylinders." Trans. Brit. Ceram. Soc. 57 (10) 645-77 (1958).

Glenny, E. & Taylor, T.A., "High-Temperature Properties of Ceramics & Cermets." Powder Met. 1958, No. 1-2, pp. 189-226.

Glezin, B., "Preparation of Corundum." Novosti Tekhniki 1937, No. 12, p. 31.

Gmelin, "Aluminum Vol. II. The Compounds of Aluminum." Ed. Nachod, G., Blinnoff-Achapkin, G. et al. (In German). 613 pp. Verlag Chemie GmbH, Berlin (1934).

Godbee, H.W., "Thermal Conductivity of Magnesia, Alumina, and Zirconia Powders in Air at Atmospheric Pressure from 200°F to 1500°F." AEC Accession No. 23652 Rept. No. ORNL-3510.

Godbee, H.W. and Ziegler, W.T., "Thermal Conductivities of MgO, Al_2O_3,. . . to 850°" I Experimental. J. Appl. Phys. 37 40-55 (1966). II Theoretical. J. Appl. Phys. 37, 6-65 (1966).

Goetzel, C.G., "Dispersion Strengthened Alloys: Properties & Applications of Light Metals." J. Metals (4) 276 (1959).

Goldschmidt, Z., Low, W., & Foguel, M., "Fluorescence Spectrum of Trivalent Vanadium in Corundum." Phys. Letters 19, 17-18 (1965).

Goldsmith, A., Waterman, T.E., & Hirschhorn, H.J., "Handbook of Thermophysical Properties of Solid Materials." Macmillan Co. New York 11, 1961, 5 Vol. 4268 pp.

Goliber, E.W., "Sintered Hard Compositions." U.S. 2,872,726, 2/10/59.

Goliber, E.W., "Hard Refractory Composition." U.S. 2,873,198, 2/10/59.

Goncharov, V.V. & Shmitt-Fogelevich, S.P., "Mullite Formation in Clay Sintered with Technical Aluminum Oxide." Tr. Vses. Gos. Inst. Nauchno-Issled. i Proektn. Rabot Ogneuporn. Prom. 1963 (35) 73-104.

Gorbunova, O.E. & Vaganova, L.I., "X-Ray Investigation of the Transformation of γ-Aluminum Oxide into a-Aluminum Oxide." Trudy Tsentral. Nauch-Issledovatel. Lab. Kamnei Sanotsvetov Tr. "Russkie Samotsvety," 1938, No. 4, pp. 66-7; Khim. Referat. Zhur. 2 (5) 31-2 (1939).

Gorodishcher, Z. Ya. & Mashneva, N.I., "Contact Coagulation Method for Removing P^{32} from Drinking Water." Tr. Konf. po Radiats. Higiene, Leningrad 1959, 51-5 (Pub. 1960).

Gorter, E.W., "Saturation Magnetization & Crystal Chemistry of Ferrimagnetic Oxides." Philips Research Repts. 9 (4) 295-320; (5) 321-65; (6) 403-43 (1954).

Gorum, A.E. et al., "Effect of Surface Conditions on Room-Temperature Ductility of Ionic Crystals." J. Am. Ceram. Soc. 41 (5) 161-4 (1958).

Goton, R. Thesis No. 146. University of Lyon (July 9, 1955)

Gould, R.F., "Alumina from Low-Grade Bauxite." Ind. Eng. Chem. 37 (9) 796-802 (1945).

Gourge, G. & Hanle, W., "New Results on the Exoelectron Emission of Nonmetals." Acta Phys. Austriaca 10, 427-47 (1957).

Govan, J.F. & R.H., "Method of Protecting Metal Surfaces, Composition Therefor, & Article Resulting therefrom." U.S. 2,785,091, 3/12/57.

Gow, K.V., "Reaction of Vaporized Sodium Sulfate with Aluminous Refractories." J. Am. Ceram. Soc. 34 (11) 343-7 (1951).

Graff, W.A., "Ceramic Bodies for Airplane Power Systems. Univ. Illinois Dept. Ceram. Eng. Rept. No. 17, Sect. II.; PB 39,672, 1 p. n.d.; abstracted in Bibliog. Sci. Ind. Repts. 3 (7) 498 (1946).

Graff, W.A., "Ceramic Coatings & Bodies for Aircraft Power Systems." Univ. Illinois Dept. Ceram. Eng. Rept. 19 (Oct. 1946), 130 pp. PB 60405; abstracted in Bibliog. Sci. Ind. Repts. 4 (11) 965 (1947).

Graham, J.W. & Kennicott, W.L., "Sintered Carbides—New Tool of Ceramics." Ceram. Ind. 55 (6) 93-4, 96 (1950).

Graham, J.W. & Zimmerman, W.F., "What's Happened to Cermets?" Metal Progr. 73 (3) 89-91 (1958).

Graham, P.W.L. & Rindone, G.E., "Properties of Soda Aluminosilicate Glasses: IV, Relative Acidies & Some Thermodynamic Properties." J. Am. Ceram. Soc. 47 (1) 19-24 (1964).

Graham, R.P. & Horning, A.E., "Interaction of Hydrous Alumina with Salt Solutions." J. Am. Chem. Soc. 69, 1214-5 (1947).

Graham, R.P. & Thomas, A.W., "Reactivity of Hydrous Alumina Toward Acids." Jour. Amer. Chem. Soc. 69 (4) 816-21 (1947).

Grant, N.J. & Preston, O., "Dispersed Hard Particle Strengthening of Metals." Trans. AIME 209, (1957).

Grasshof, H., "The Adsorption of Inorganic Ions on Alkali-Free Aluminum Oxide." Angew, Chem. 63, 96-7 (1951).

Grattidge, W., "Method of Metalizing Ceramic Bodies." U.S. 2,820,727, 1/21/58.

Gray, T.J., Detwiler, D.P., Rase, D.E., Lawrence, W.G., West, R.E., Jennings, T.J., "Defect Solid State." Interscience Publishers, Inc., N.Y., 1951, 511 pp.

Grazen, A.E., "Codepositing Oxides or Carbides." Iron Age 183, No. 5, 94-6 (1959).

Graziano, E.E., "Oxidation-Resistant Refractory Coatings for Metals Tested at 3000° to 6000°F—an Annotated Bibliography." AD 271940, 26 pp.; U.S. Govt. Res. Repts. 37 (9) 55 (1962).

Greaves, J.C. & Linnett, J.W., "Recombination of Oxygen Atoms at Surfaces." Trans. Faraday Soc. 54, 1323-30 (1958).

Grebenyuk, A.A. & Zhuravleva, Z.I., "Production of High-Alumina Large-Volume Crucibles & Their Field-Testing." Sbornik Nauch. Trudov Vsesoyuz. Nauch.-Issledovatel. Inst. Ogneuporov 1958, No. 2, 159-76.

Green, H., "Industrial Rheology & Rheological Structures." John Wiley & Sons, Inc., New York (1949) 311 pp.

Greenblatt, J.H., "The Structure of Oxide Films Formed on Aluminum After Exposure to High-Temperature Pure Water." J. Electrochem, Soc. 109, 1139-42 (1962).

Greenwald, S., Pickart, S.J. & Grannis, F.H., "Cation Distribution and g Factors of Certain Spinels Containing Ni, Mn, Co, Al, Ga, & Iron." J. Chem. Phys. 22, 1597-1600 (1954).

Greger, H.H., "New Bonds for Refractories." Brick & Clay Record 117 (2) 63, 68 (1950).

Gregg, S.J., "Preparation of Highly Dispersed Solids." Kolloid-Z. 169 (1-2) 5-11 (1960).

Gregg, S.J. & Sing, K.S.W. J. Phys. & Coll. Chem. 55, 592-7, 597-604 (1951).

Gregg, S.J. & Sing, K.S.W., "Effect of Heat-Treatment on the Surface Properties of Gibbsite: III, Chemical Nature of the Product Formed." J. Phys. Chem. 56, 388-91 (1952).

Gregory, J.N. & Moorbath, S., "The Diffusion of Thoron in Solids." Trans. Faraday Soc. 47, 844-59 (1951).

Greville, C., "Corundum or Corindon." Royal Soc. London Phil. Trans. (1798).

Griffin & Tatlock, Ltd., "Microid Polishing Alumina." Ind. Diamond Rev., 4 (47) 218 (1944).

Griffith, A.A., "The Phenomena of Rupture & Flow in Solids." Phil. Trans. Royal Soc. 221A, 163 (1920).

Griffith, J.S., Olsen, R.S. & Rechter, H.L., "Compositions & Processes for Making Foamed Alumina Refractory Products & Articles So Produced." U.S. 3,041,190, 6/26/62.

Grime, G. & Eaton, J.E., "Determination of Young's Modulus by Flexural Vibration." Phil. Mag. 23 (152) 96-8 (1937).

Grindle, S.L. & Rosenbery, J.W., "Plasma-Arc Re-Entry Evaluation Techniques." Presented at Ceramic-Metal Systems Division, Sept. 20, 1964. Am. Ceram. Soc. Bull. 43 (8) 595 (1964).

Grodzinski, P., "Hardness Scale." Gemmologist 12 (138) 23; (139) 31 (1943).

Gross, S. & Heller, L., "A Natural Occurence of Bayerite." Mineral. Mag. 33, 723-4 (1963).

Grossman, L.N., "Niobium-Al₂O₃ Reactions Yielding Condensed and Volatile Products." J. Chem. Phys. 44, 4127-31 (1966).

Grossman, L.N., "High-Temperature Thermal Analysis of Ceramic Systems." AEC Accession No. 39437, Rept. No. AED-CONF-66-103-9

Groszek, A.J., "Heat of Preferential Adsorption of Surfactants on Porous Solids & Its Relation to Wear of Sliding Steel Surfaces." ASLE Trans. (Am. Soc. Lubication Engrs.) 5, 105-14 (1962).

Grum-Grzhimaile, S.V. & Lyamina, A.N., "Optical Testing of Ceramics Colors." J. Applied Chem. (USSR) 21 (12) 1228-41 (1948).

Grunberg, L. & Wright, K.H.R., "The Structure of Abraded Metal Surfaces." Proc. Roy. Soc. (London) A232, 403-23 (1955).

Gruner, E., "Theory of Plasticity." Ber. deut. keram. Ges. 31 (5) 135-42 (1954).

Guard, R.W. & Romo, P.C., "X-Ray Microbeam Studies of Fracture Surfaces in Alumina." J. Am. Ceram. Soc. 48 (1) 7-11 (1965).

Guareschi, P., "Electrolytic Production of Alumina." U.S. 2,833,707, 5/6/58.

Gulbransen, E.A. & Wysong, W.S., "Thin Oxide Films on Aluminum." J. Phys. & Colloid Chem. 51, 1087-1103 (1947).

Gumilevskii, A.A., "Microstructure of the Surface of Boules of Artifical Corundum." Zapiski Vsesoyuz. Mineral. Obshchestva 86 (6) 731-5 (1957).

Gurney, C., "Delayed Fracture in Glass." Proc. Phys. Soc. (London) 59, 169-85 (1947).

Gurr, W., "Laboratory Porcelains for Temperatures up to 2000°." Osterr. Chem.-Ztg. 41, 411-8 (1938); Brit. Chem. & Phys. Abs-B 58 (3) 264) (1939).

Guzman, I. Ya & Poluboyarinov, D.N., "Light-Weight Refractories from Alumina." Ogneupory 24 (2) 71-9 (1959).

Gvelisiani, G.G. & Pazukhin, V.A., "Reduction of Oxides of Strontium & Barium by Aluminum." Sbornik Nauch. Trudov Moskov. Inst. Tsvetnykh Metal. i Zolota 1954, No. 24, 184-201; Referat Zhur., Met. 1956, No. 1018.

Haase, T., "Attempt to Derive a General Theory of the Plasticity of Ceramic Bodies." Silikattech. 3 (6) 265-7 (1952).

Haase, T., "Workability of Plastic Ceramic Bodies." Ber. deut. keram. Ges. 34 (2) 27-35 (1957).

Haase, T., "Mechanism of Dry Pressing." Ber. deut. keram. Ges. 37 (3) 97-101 (1960).

Haber, F., "The Hydroxides of Aluminum & Trivalent Iron." Naturwiss. 13, 1007-12 (1925).

Hagg, G. & Soderholm, G., "Crystal Structure of Mg-Al Spinels with an Al₂O₃ Surplus & a-Al₂O₃." Z. Physik. Chem. B29, 88-94 (1935); abstracted in Chem. Zentr. 1935, ii, 1133.

Hafner, S. & Laves, F., "Comment on 'Phase Transitions in LiAl₅O₈'." J. Am. Ceram. Soc. 47 (7) 362 (1964).

Haglund, T.R., "Purifying Bauxite." U.S. 1,569,483, 1/12/26.

Hague, J.R., Lynch, J. F., Rudnick, A., Holden, F.C. & Duckworth, W.H. eds., "Refractory Ceramic for Aerospace - A Materials Selection Handbook." Published by the American Ceramic Society, Columbus, Ohio (1964) 511 pp.

Hahn, F.L. & Thieler, E., "Aluminium Amalgam, Hydroxide, & Oxide." Ber. deut. chem. Ges. 57, 671-9 (1924).

Hahn, H., Frank, G., Klinger, W., Storger, A.D. & Storger, G., "Ternary Chalcogenides: VI." Z. anorg. u. allgem. Chem. 279 (5-6) 241-70 (1955).

Hahn, H. & de Lorent, C., "Ternary Chalcogenides: VIII, Preparation of Ternary Oxides of Aluminum, Gallium, & Indium with Monovalent Copper & Silver." Z. anorg. u. allgem Chem. 279 (5-6) 281-8 (1955).

Hahn, H.T. & Vander Wall, E.M., "Salt-Phase Chlorination of Reactor Fuels. I. Solution of Zirconium Alloys in Lead Chloride." U.S. At. Energy Comm. IDO-14478, 23 pp. (1959).

Haidt, H., "Sintered Carbide & Oxide Ceramic Cutting Materials." Industrieblatt 60 (4) 247-53 (1960). Ind. Diamond Abstr. 17, A175 (1960).

Halden, F.A. and Sedlacek, R., "Ceramic Microstructures," ed. Fulrath and Pask, Wiley, N.Y., p 439.

Haldy, N.L., Schofield, H. Z. & Sullivan, J.D., "High-Alumina Porcelains." Am. Ceram. Soc. Bull. 35 (9) 351-5 (1956).

Hall, J.L., "Secondary Expansion of High-Alumina Refractories." J. Am. Ceram. Soc. 24 (11) 349-56 (1941).

Hall, W.K., Leftin, H.P., Cheselske, F.J., & O'Reilly, D.E., "Hydrogen Held by Solids. IV. Deuterium Exchange & NMR (nuclear magnetic resonance) Investigations of Silica, Alumina, & Silica-Alumina Catalysts." J. Catalysis 2 (6) 506-17 (1963).

Halm, L., "Particular Form of the Crystallization of Mullite." Trans. Intern. Ceram. Congr. 1948, pp. 82-97 (in French).

Hallse, R.L., "High-Energy Forming of Glasses & Ceramics." Bull. Amer. Ceram. Soc. 42 (11) 711 (1963).

Haltmeier, A., "Comminution by Successive Explosions." U.S. 2,826,369, 3/11/58.

Halversen, R.A., "Process for Making Aluminum Oxide." U.S. 2,643,935, 6/30/53.

Hamano, Y., Kinoshita, M. & Oishi, Y., "Studies on Sintering of Alumina: I, Enhanced Grain Growth & Orientation in Hot-Pressed Alumina." Osaka Kogyo Gijutsu Shikensho Kiho 14 (1) 108 (1963).

Hamano, Yoshio and Kinoshita, Makoto, "Sintering of Alumina III. Effect of Pores on the Grain Growth of Sintered Alumina." Osaka Kogyo Gijutsu Shikensho Kiho 15, 246-52 (1964).

Hamilton, R., "How to Control Variables in Slip Casting." Ceram. Ind. 73 (1) 64-7 (1959).

Hanemann, H. & Bernhardt, E.O., "A Microhardness Test Method." Z. Metalk. 32 (2) 35-8 (1940).

Hanford, C.M., "Coated Abrasive Article & Method of Making." U.S. 2,873,181, 2/10/59.

Hannahs, W.H. & Stein, N., "Printed Unit Assemblies for TV." Tele-Tech. 11 (6) 38-40, 112-20 (1952).

Hannay, N.B. ed., "Semiconductors." Reinhold Publishing Corp., New York (1959) 767 pp.

Hannon, J.W.G., "Method of Making Alumina Powder." U.S. 2,837,451, 6/3/58.

Hannon, J.W.G., "Aluminum Powder." U.S. 2,861,880, 11/25/58.

Hansen, K.W. & King, D.F., "Unshaped High Temperature Refractory." U.S. 2,852,401, 9/16/58.

Hensen, W.C. & Livovich, A.F., "Thermal Conductivity of Refractory Insulating Concrete." Am. Ceram. Soc. Bull. 37 (7) 322-8 (1958).

Harder, E.C., "Bauxite, Industrial Minerals & Rocks." 2nd Ed. American Institute of Mining & Metallurgical Engineers (1949).

Harders, F. & Kienow, S., "Refractories." (Feuerfestkunde). Springer-Verlag, Berlin (1960) xvi + 981 pp.

Hargreaves, C.M., "Growth of Sapphire Microcrystals." J. Appl. Phys. 32, 936-8 (1961).

Harris, J.N. and Poulos, N.E., "Fibrous Ceramic Structures." Amer. Ceram. Soc. Bull. 45, 1075-7 (1966).

Harris, L., "Preparation & Infrared Properties of Aluminum Oxide Films." J. Opt. Soc. Am. 45 (1) 27-9 (1955).

Harris, L. & Piper, J., "Transmittance & Reflectance of Aluminum Oxide Films in the Far Infrared." J. Opt. Soc. Am. 52 (2) 223-24 (1962).

Harris, M.R. & Sing, K.S.W., "Surface Properties of Precipitated Alumina: I, Preparation of Active Samples & Determination of Nitrogen Adsorption Isotherms." J. Appl. Chem. (London) 5 223-7 (1955).

Harris, M.R. & Sing, K.S.W., "Activated Alumina." Chem. & Ind. (London) 1957, No. 48, p. 1573.

Harris, M.R. & Sing, K.S., "Surface Properties of Precipitated Alumina: II, Aging at Room Temperature." J. Appl. Chem. (London) 7 (7) 397-401 (1957).

Harris, M.R. & Sing, K.S.W., "The Surface Properties of Precipitated Alumina. III Samples Prepared from Aluminum Isopropoxides." J. App. Chem. 8, 386-9 (1958).

Harris, R.C., "Spark Gap Semiconductor." U.S. 2,861,961, 11/25/58.

Harrison, A.W.C., "The Manufacture of Lakes & Precipitated Pigments." Leonard Hill, Ltd. (London WC2) (1930).

Harrison, W.N., Moore, D.G., & Richmond, J.C., "Ceramic Coatings for High-Temperature Protection of Steel." Presented at meeting of Amer. Ceram. Society, Buffalo, N.Y., April-May, 1946; abstracted in Amer. Ceram. Soc. Bull. 25 (4) 127 (1946).

Harrop, P.J., "Intrinsic Electrical Conductivity in Alumina" Brit. J. Appl. Phys. 16, 729-30 (1965).

Harrop, P.J. & Creamer, R.H., "High-Temp. Electrical Cond. of Single-Crystal Alumina." Brit. J. Appl. Phys. 14 (6) 335-9 (1963).

Hart, T.F., "Thermal Shaping of Corundum & Spinel Crystals." U.S. 2,517,661, 8/8/50.

Hart, L.D. & Hudson, L.K., "Grinding Low-Soda Alumina." Am. Ceram. Soc. Bull. 43 (1) 13-17 (1964).

Hart, H.V., & Drickhamer, H.G., "Effect of High Pressure on the Lattice Parameters of Al_2O_3." J. Chem. Phys. 43, 2265-6 (1965).

Hartmann, F., "Melting & Softening of Refractories as a Viscosity Problem." Ber. deut. keram. Ges 19, 367-82 (1938).

Hartmann, F. & Schulz, E.H., "The Viscosity of Slags & Its Importance for the Production of Steel." Z. Elektrochem. 43, 518-24 (1937).

Hartmann, W., "Electrical Investigation of Oxide Semiconductors." Z. Physik. 102 (11/12) 709-33 (1936).

Hartwig, J., "Experiences with Corhart ZAC Blocks for Borosilicate Glass-Melting." Sprechsaal 89 (1) 1-3; (2) 26-8 (1956).

Harvey, F.A. & Birch, R.E., "Silica Refractory." U.S. 2,351,204, 6/13/44.

Hasapis, A.A., Panish, M.B. & Rosen, C., "Vaporization & Physical Properties of Certain Refractories: I, Techniques & Preliminary Studies." PB Rept. 171413, 73 pp.; U.S. Govt. Res. Repts., 35 (4) 445 (1961).

Hass, G., "Growth & Structure of Thin Oxide Layers in Aluminum." Optik 1, 134-43 (1946).

Hasselman, D.P.H., "Thermal Shock by Radiation Heating." J. Am. Ceram. Soc. 46 (5) 229-234 (1963).

Hasselman, D.P.H., "Relation Between Effects of Porosity on Strength and on Young's Modulus of Elasticity of Polycrystalline Materials." J. Am. Ceram. Soc. 46 (11) 564-5 (1963).

Hasselman, D.P.H. & Shaffer, P.T.B., "Factors Affecting Thermal Shock Resistance of Polyphase Ceramic Bodies." AD-277,605, 155 pp., illus.; U.S. Govt. Res. Repts. 37 (20) 26-7 (1962).

Hatch, R.A., "Phase Equilibrium in the System $Li_2O-Al_2O_3-SiO_2$." Am. Mineral. 28, 471-96 (1943).

Haupin, W.E., "Hot Wire Method for Rapid Determination of Thermal Conductivity." Am. Ceram. Soc. Bull. 39 (3) 139-41 (1960).

Hauser, E.A., "Modern Colloidchemical Concepts of the Phenomenon of Coagulation." J. Phys. & Colloid Chem. 55, 605-11 (1951).

Hauser, E.A. & LeBeau, D.S., "Surface Structure & Properties of Colloidal Silica & Alumina." J. Phys. Chem. 56, 136-9 (1952).

Hauth, W.E., Jr., "Slip Casting of Aluminum Oxide." J. Am. Ceram. Soc. 32 (12) 394-8 (1949).

Hauth, W.E., "Behavior of the Alumina-Water System." J. Phys. & Colloid Chem. 54, 142-56 (1950).

Hauschild, U., "Nordstrandite, γ-Al(OH)₃." Z. Anorg. allg. Chem. 324 (1-2) 15-30 (1963).

Hauschild, U., "Alkali metal-free Hydrargillite, Al(OH)₃." Naturwissenschaften 51 (10) 238-9 (1964).

Hauschild, U., "Manufacture of Pure Aluminum Hydroxide Modifications." Ger. 1,162,337, 2/6/64.

Hauschild, U. & Nicolaus, H., "Pure Bayerite (n-Al₂O₃·3H₂O)" Ger. 1,138,749, 10/31/62.

Hauschild, U., "Nordstrandite Production." Brit. 950,165. 2/19/64.

Hauttmann, H., "Sintering of Ceramic Materials." Austrian 180,231, 11/25/54.

Hauy, R.J., "Traite de Mineralogie." Paris 4, 358 (1801).

Havestadt, L. & Fricke, R., "Dielectric Behavior of Oxide Hydrates." Z. anorg. allg. Chem. 188, 357-95 (1930).

Hawkes, W.H., "Symposium on High-Alumina Refractories: Production of Synthetic Mullite." Trans. Brit. Ceram. Soc. 61 (11) 689-703 (1962).

Haxel, O., Houtermans, F.G. & Seeger, K., "Electron Emission from Metallic Surfaces as an After Effect of Mechanical Working or Glow Discharge." Z. Physik 130, 109-23 (1951).

Hay, R., McIntosh, R., Rait, A.B. & White, J., "Slag Systems." J. West Scot. Iron Steel Inst. 44, 85-92 (1936-7).

Hay, R., White, J., & McIntosh, A.B., "Slag Systems." J. West Scot. Iron Steel Inst. 42, 99 (1934-35).

Hayami, R., Ogura, T., & Tanaka, H. Osaka Kogyo Gijutsu Shikenso Kiho 11 (4) 235-40 (1960).

Hayes, D., Budworth, D.W. & Roberts, J.P., "Selective Permeation of Gases Through Dense Sintered Alumina." Trans. Brit. Ceram. Soc. 60 (7) 494-504 (1961).

Hayes, D., Budworth, D.W. & Roberts, J.P., "Permeability of Sintered Al_2O_3 Materials to Gases at High Temperatures." Trans. Brit. Ceram. Soc. 62 (6) 507-23 (1963).

Hayes, J.C., "Calcium & Magnesium Exempt Alumina." U.S. 3,104,944, 9/24/63.

Hayes, J.C., "Alumina Free from Alkaline Earth Metal Impurities." U.S. 3,105,739, 10/1/63.

Heath, D.L., "Valve Resistors." U.S. 3,040,282, 6/19/62.

Hebert, G.R. & Tyte, D.C., "Intensity Measurement on the $A^2\Sigma$-$X^2\Sigma$ System of Aluminum Oxide." Proc. Phys. Soc. (London) 83 (534) 629-34 (1964).

Heckel, R.W. and Youngblood, J.L., "X-Ray Line Broadening Study of Explosively Shocked MgO and Alpha Alumina." J. Am. Ceram. Soc. 51 398-401 (1968).

Hedger, H.J. & Hale, A.R., "Preliminary Observations on the Use of the Induction-Coupled Plasma Torch for the Preparation of Spherical Powder." Powder Met. 1961, No. 8, pp. 65-72.

Hedley, T. & Co. Ltd., " Alumina Abrasive Materials for Cleaning Teeth." Brit. 833,057, 4/21/60.

Hedvall, J.A., "Sintering & Reactivity of Solids." Ceram. Age 65 (2) 13-7 (1955).

Hedvall, J.A., "Effects of Dissolved or Adsorbed Inactive Gases on the Reactivity of Oxides." Trans. Brit. Ceram. Soc. 55 (1) 1-12 (1956).

Hedvall, J.A., Alfredsson, S., Runehagen, O. & Akerstrom, P., "The Effect of Normally Inactive Gases on the Chemical Activity of Solids." Iva. 1942, 48-64; Chem Zentr. 1942,II, 1101-2.

Hedvall, J.A. & Blomkvist, M., "Arkiv. Kemi. Mineral. Geol." 19A (22)1-14 (1945).

Hedvall, J.A. & Loeffler, L., "Influence of Transition Points Upon Speed of Formation of the Cobalt Spinel from Oxides in the Solid State." Z. anorg. & allgem. Chem. 234 (3) 235-6 (1937).

Hedvall, J.A. & Lundberg, A., "Effects of the Atmosphere in which Powders are Prepared on Their Chemical Activity & Surface Properties." Kolloid-Z. 104, 198-203 (1943).

Heeley, R. & Moore, H., "Experimental Investigation of Alumino-silicate Refractories of High Purity for Use in Glassmelting: II, Effect of Soda (Na_2O) on Materials of Sillimanite & Mullite Composition when Fired to Different Temperatures." J. Soc. Glass Technol. 36 (171) 242-65T (1952).

Hegedus, A.J. & Fukker, K., "Thermogravimetric Studies of the Decomposition & Reduction of Sulfates: I." Z. anorg. u. allgem. Chem. 284 (1-3) 20-30 (1956).

Hegedus, A.J. & Kurthy, J., "Aluminum Oxide of the Following Types." J. pract. Chem. 14 (14) 113-8 (1961).

Hehemann, R.F. & Ault, G.M. eds., "High Temperature Materials." John Wiley & Sons, Inc., N.Y. (1959).

Heilich, R.P., "Refractory-Grade Bauxite." Unpublished Report, Alcoa Research Laboratories (1962).

Heilman, R.H., "Emissivities of Refractory Materials." Mech. Eng. 58, 291-2 (1936).

Heinemann H. et al., "Some Physical Properties of Activated Bauxite." Ind. Eng. Chem. 38 (8) 839-42 (1946).

Held, H.E., "Capacitor." U.S. 3,086,150, 4/16/63.

Heldt, K. & Haase, G., "Electrical Resistance of Pure, Vacuum-Sintered Aluminum Oxide." Z. angew. Phys. 6 (4) 157-60 (1954).

Henrie, J.O., "Properties of Nuclear Shielding Concrete." J. Am. Concrete Inst. 31 (1) 37-46 (1959).

Henry, E.C., "Ceramics in Electronics: I, How Big is the Electronic Ceramic Field?" Ceram. Ind., 70 (4) 160-5 (1958).

Hensler, J.R. & Henry, E.C., "Electrical Resistance of Some Refractory Oxides & Their Mixtures in the Temperature Range 600° to 1500°C." J. Am. Ceram. Soc. 36 (3) 76-83 (1953).

Hermann, E.R. & Cutler, I.B., "Kinetics of Slip Casting." Trans. Brit. Ceram. Soc. 6 (4) 207-11 (1962).

Herold, P.G. & Hoffman, J.L., "Pure Alumina Refractories Bonded with Lumnite Cement." Bull. Amer. Ceram. Soc. 20 (10) 336-8 (1941).

Herold, P.G. & Dodd, C.M., "Thermal Dissociation of Diaspore Clay." J. Am. Ceram. Soc. 22 (11) 388-91 (1939).

Herring, C., "Effect of Change of Scale on Sintering Phenomena." J. App. Phys. 21, 301-3 (1950).

Herron, R.H., "Friction Materials—A New Field for Ceramics & Cermets." Am. Ceram. Soc. Bull. 34 (12) 395-8 (1955).

Hervert, G.L. & Bloch, H.S., "Production of Alumina." U.S. 2,855,275, 10/7/58.

Hervert, G.L. & Bloch, H.S., "Production of Alumina." U.S. 2,871,095 & 2,871,096, 1/27/59.

Hervert, G.L. & Bloch, H.S., "Production of Alumina." U.S. 2,958,581, 2,958,582, 2,958,583, 11/1/60.

Hessel, F.A. & Rust, J.B., "Abrasive Bodies & Methods of Making." U.S. 2,559,122, 7/3/51.

Heuer, A.H. and Roberts, J.P., "The Influence of Annealing on the Strength of Corundum Crystals." Proc. Brit. Ceram. Soc. 6, 17-27 (1966).

Heuer, A.H. and Roberts, J.P. "Thermal Etching of Single-Crystal Corundum" Trans. Brit. Ceram. Soc. 65, 219-32 (1966).

Hewson, C.W. and Kingery, W.D., "Effect of MgO and $MgTiO_3$ Doping on Diffusion-Controlled Creep of Polycrystalline Aluminum Oxide." J. Am. Ceram. Soc. 50, 218-19 (1967).

Hickman, B.S. and Walker, D.G., "The Effect of Neutron Irradiation on Aluminum Oxide." J. Nucl. Mater. 18, 197-205 (1966).

Hicks, J.C., "Basic Refractories in the Glass Industry." Glass Ind. 36 (6) 313-5, 336-7 (1955).

Hiester, N.K. et al., "Tools for Tomorrow—High Temperatures." Chem. Eng. 63 (12) 171-8 (1956).

Higgins, J.K., "Reaction of Alumina with Cesium Vapor." Trans. Brit. Ceram. Soc. 65, 643-59 (1966).

High-Temperature Materials Conference. Cleveland, Ohio, 1957 3-15 (Pub. 1959).

Hijikata, K. & Miyake, K., "Sintering of Alumina: I, Effect of Particle Size & Atmosphere on Sintering." Funtai Oyobi Funmatsuyakin 7 (1) 9-14 (1960).

Hill, R., "Theory of Mechanical Properties of Fiber-Strengthened Materials." Part I. Elastic Behavior. Part II. Inelastic Behavior." J. Mechanics & Physics of Solids 12, 199-219 (1964).

Hinz, W., "Vitrokeram." Silikattech. 10 (3) 119-22 (1959).

Hinz, W. & Baiburt, L., "Glass-Ceramics Products from $MgO-Al_2O_3-SiO_2$ System with TiO_2 Additions." Silikat Technik. 11 (10) 455-9 (1960).

Hinz, W. & Wihsmann, F., "Cordierite Products from Glass." Silikat Technik. 10 (8) 408 (1959).

Hippel, A.R. von., "Table of Dielectric Materials." Vol. IV. Tech. Rept. 57 (Jan. 1953), Vol. VI Tech. Rept. 126 (June 1959) MIT Laboratory for Insulation Research.

Hirayama, C., "Properties of Aluminoborate Glasses of Group II Metal Oxides: II, Electrical Properties." J. Am. Ceram. Soc. 45 (6) 288-93 (1962).

Hlavac, J., "Influence of Small Quantities of Al_2O_3 on Some Properties of Glass Produced by the Fourcault Process." Sklarske Rozhledy 26 (8-9) 125-8 (1950).

Hoch, M. & Johnston, H.L., "Formation Stability & Crystal Structure of the Solid Aluminum Suboxides Al_2O & AlO." J. Am. Chem. Soc. 76, 2560-1 (1954).

Hoch, M. & Johnston, H.L., "The Heat Capacity of Aluminum Oxide from 1000 to 2000° & of Thorium Dioxide from 1000 to 2500°." J. Phys. Chem. 65, 1184-5 (1961).

Hoch, M. and Silberstein, A., "Thermal Conductivity of Aluminum Oxide." U.S. Air Force Syst. Command, Res. Technol. Div., Tech. Rept. AFML 67-12 (1966).

Hoepli, Max H. and Klasse, F., "Sintered High-Grade Refractories Based on Aluminum Oxide." Belg. 648,166 9/5/64.

Hogberg, E. & Heden, S., "Destruction of Refractory Brick by Carbon Monoxide Atmosphere." Jernkontorets Ann. 138 (10) 655-64 (1954).

Hoekstra, J., "Manufacture of Shaped Alumina Particles." U.S. 2,680,099, 6/1/54.

Hoffman, G.A., "The Exploitation of the Strength of Whiskers." The Rand Corporation, Rept. P-1294 (March 1958).

Hofmann, U., Scharrer, E., Czerch, W., Fruhauf, K., & Burck, W., "Fundamentals of Dry-Pressing & the Causes of Compacting of the Dry Body." Ber. deut. keram. Ges. 39 (2) 125-30 (1962).

Hokanson, H.A., Rogers, S.L. & Kern, W.J., "Electron Beam Welding of Alumina." Ceram. Ind. 81 (2) 44-7 (1963).

Holder, G., Helmboldt, O., & Vogt, P., "Process for Preparing Sodium-Free Alumina." Ger. 1,092,457, 10/10/60.

Holder, G. & Thome, R., "Process of Purifying Sodium Aluminate Solutions. Ger. 1,172,246, 6/18/64.

Holder, G., Vogt, P., & Helmboldt, O., "Procedure for Preparation of Low-Soda Alumina for Forming Pressed Bodies of Constant Weight." Ger. 1,159,417, 12/19/63.

Holland, F.G.C., "Ceramics & Glass: I, Use of Ceramic Coatings in Gas Turbine Combustion Chambers." Selected Govt. Res. Repts. (Gt. Brit.) 10, 1-7 (1952).

Holland, I.J., "Porous Refractory Bodies." Brit. 923,862, 4/18/63.

Holland, L., "The Properties of Glass Surfaces." John Wiley & Sons, Inc., New York, 546 pp. (1964).

Holland, M.G., "Thermal Conductivity of Several Optical Maser Materials." J. Appl. Phys. 33 (9) 2910-1 (1962).

Holley, C.E., Jr. & Huber, E.J., Jr., "Heats of Combustion of Magnesium & Aluminum." J. Am. Chem. Soc. 73 (12) 5577-9 (1951).

Holloway, D.G., "The Strength of Glass Fibres." Phil. Mag. 4 1101-6 (1959).

Holm, V.C.F. & Blue, R.W. Ind. Eng. Chem. 43, 501-5 (1951).

Holmes, P.J., "Thin Films of Aluminium Oxide (RP 9-31) ACSIL No. 2058. British Admiralty (June 24, 1964).

Hopkins, R. W., "Unusual Sources of Alumina in Glass Manufacture." Glass Ind. 38 (5) 266-9 (1957).

Hoppe, W., "Crystal Structure of a-AlOOH (Diaspore)." Z. Krist. 103 73 (1941).

Hoppe, W., "Crystal Structure of a-AlOOH (Diaspore): II, Fourier Analysis." Z. Krist, 104 11-7 (1942).

Horibe, T. & Kuwabara, S., "Oxidation of Ti_3^+ Ion Dissolved in a-Alumina of Brown Electrofused Aluminous Abrasive Grains During the Firing of Grinding Wheels." Nagoya Kogyo Gijutsu Shikensho Kokoku 10 (4) 262-7 (1961).

Horibe, T. & Kuwabara, S., "Preliminary Investigation of Phase Equilibriums in Al_2O_3-Ti_2O_3 Systems." Kogyo Kagaku Zasshi 67 (2), 276-81 (1964).

Hornstra, J., "Dislocations, Stacking Faults, & Twins in the Spinel Structure." J. Phys. Chem. Solids 15, 311-23 (1960).

Hoskins, R.H., "Two-Level Maser Materials." J. Appl. Phys. 30, 797 (1959).

Houben, G.M.M., Thesis, Delft (1951).

Houben, G.M.M., & deBoer, J.H., "Constitution of γ-Alumina." Trans. Intern. Ceram. Congr., 3rd Congr., Paris, 1952, pp. 77-9 (in English).

Houchins, H.R., "Abrasive Articles & Method of Making." U.S. 2,730,439, 1/10/56.

Howard, P. & Roberts, A.L., "Study of Alumina Suspensions: I, Effect of Electrolytes on the Stability." Trans. Brit. Ceram. Soc. 50 (88) 339-47 (1951).

Howard, P. & Roberts, A.L., "Study of Alumina Suspensions: II. Effect of Aging on the pH." Trans. Brit. Ceram. Soc. 52 (7) 386-403 (1953).

Howatt, G.N., "Method of Injection Molding Ceramic Bodies Using Thermoplastic Binder." U.S. 2,434,271, 1/13/48.

Howatt, G.N., Breckenridge, R.G. & Brownlow, J.M., "Fabrication of Thin Ceramic Sheets for Capacitors." J. Am. Ceram. Soc. 30 (8) 237 (1947).

Howatt, G.H. et al., "Fabrication of Thin Ceramic Sheets for Capacitors." J. Am. Ceram. Soc. 30 (8) 237-41 (1947).

Howie, J., "Radiation Damage in Ceramics." Nuclear Eng. 6 (62) 299-304 (1961).

Hsu, P.H. & Bates, T.F., "Formation of X-Ray Amorphous & Crystalline Alumina Hydroxides." Mineral. Mag. 33 (264) 749-68 (1964).

Hsu, S.E., Kobes, W., and Fine, M.E., "Strengthening of Sapphire by Precipitates Containing Titanium." J. Am. Ceram. Soc. 50, 149-51 (1967).

Huber, R.J., "Internal Friction in Aluminum Oxide Single Crystals." Univ. Microfilms (Ann Arbor, Mich) L.C. Card No. Mic 61-1132, 85 pp.; Dissertation Abstr. 22, 299 (1961).

Huegel, F.J. & Li, C.H., "Dispersion Hardened Materials." U.S. 3,137,927, 6/23/64.

Hueter, T.F. & Bolt, R.H., "Sonics." John Wiley & Sons, Inc. New York, 456 pp. (1955).

Huttig, G.F., "Intermediate States in the Transformation of (Allotropic) Modifications & the Catalytic Effects of Gases." Angew. Chem. 53, 35-9 (1940).

Huttig, G.F. & Kolbl, F., "Active Oxides, LXVII. Aluminum Oxides & Their Addition Products with Water." Z. anorg. allgem. Chem. 214, 289-306 (1933).

Huttig, G.F. & Markus, G., "Active Oxides & Reactions of Solid Substances. CXVI. The Transition of γ-Aluminum Oxide into a-Aluminum Oxide, and the Effect of the Presence of Foreign Gases on This Transition." Kolloid-Z 88, 274-88 (1939).

Huttig, G.F. & Peter, A., "Oxide Hydrates. XXXV. The Crystallized Oxides & Oxide Hydrates of Aluminum As Adsorbents for Organic Dyestuffs." Koll Z. 54, 140-7 (1931).

Huttig, G.F. & von Wittgestein, E., "Information on the System, Al_2O_3-H_2O." Z. anorg. allg. Chem. 171, 323-43 (1928).

Huttig, W., "Base Layer for Improved Adhesion of Enamel Layers to Aluminum & Its Alloys." Ger. 1,098,318, 1/26/61.

Huttig, W., & Ginsberg, H., "Crystal Morphology of Dehydration Products of Aluminum Oxide Trihydrate." Z. anorg. u. allgem. Chem. 278 (1-2) 93-107 (1955).

Huffadine, J.B. & Hollands, J., "Flame Spraying of Ceramic Materials." Brit. 903,709, 8/15/62.

Huffadine, J.B., "Detecting by Color the Crystal Form of Flame-Sprayed Alumina." Brit. 954,298, 4/2/64.

Huffadine, J.B. & Hollands, E.J., " Flame Spraying Alumina to Form Stable Dense a Forms." Brit. 967,952, 8/26/64.

Huffadine, J.B. Longland, L. & Moore, N.C., "Fabrication & Properties of Chromium-Alumina & Molybdenum-Chromium-Alumina Cermets." Powder Met. 1958, No. 1-2, pp. 231-52.

Huffadine, J.B. & Sanders, R.W., "Resistors & Resistor Materials." U.S. 2,901,442, 8/25/59.

Huffadine, J.B. & Thomas, A.G., "Flame Spraying as a Method of Fabricating Dense Bodies of Al_2O_3." Powder Met. 7 (14), 290-9 (1964).

Huffman, H.C., "Catalytic Treatment of Hydrocarbons." U.S. 2,437, 531, 3/9/48.

Huminik, J. Jr. ed., "High Temperature Inorganic Coatings." Reinhold Publishing Corp., New York 22, 310 pp, 8 chapters. (1963).

Hummel, F.A., "Ceramics for Thermal Shock Resistance: II." Ceram. Ind. 65 (6) 84-6 (1955).

Humphreys, C.J., "Sapphire Spectrum Tube for Microwave Excitation." U.S. 3,042,829, 7/3/62.

Huneck, S., "Production of Fibrous Aluminum Oxide." Ger. (East) 34,081. 12/5/64.

Hunt, J.M., et al., "Infrared Absorption Spectra of Minerals & Other Inorganic Compounds." Anal. Chem. 22 (12) 1478-97 (1950).

Hunter, M.S. & Fowle, P., "Natural & Thermally Formed Oxide Films on Aluminum." J. Electrochem. Soc. 103, 482-5 (1956).

Huntress, H.B., "Powder Metallurgy." U.S. 3,074,152, 1/22/63.

Hursh, R.K., "Development of a Porcelain Vacuum Tube." J. Am. Ceram. Soc. 32 (3) 75-80 (1949).

Hursh, R.K., "Rammed, Solid Cast, & Pressure Cast Ceramic Combustion Tubes." Univ. Illinois Dept. Ceram. Eng. Rept. 16 Sect. IV, n.d. 7 pp. PB 60,670; abstracted in Bibliog. Sci. Ind. Repts. 4 (11) 965 (1947).

Hutchins, O., "Refractory Material & Method of Making." U.S. 2,362,825, 11/14/44.

Hutchison, T.S., "Ultrasonic Absorption in Solids." Science 9 (1960).

Hynes, F.J., "III, Alumina Bodies." Ceram. Ind. 66 (2) 87-91, 102 (1956).

Hyslop, J.F., "Chemical Notes on Chrome Magnesite." Trans. Brit. Ceram. Soc. 52 (10) 554-64 (1953).

Hyslop, J.F., "Examination of Some Refractory-Slag Reactions by Contraction Tests." Trans. Brit. Ceram. Soc. 58 (6) 329-40 (1959).

Hyslop, J.F., et al., "Corrosion & The Fluxing of Refractory-Glass Mixtures." Trans. Brit. Ceram. Soc. 46 (9) 377-86 (1947).

Hyslop, J.F. & McLeod, M., "Acid Oxides & Some Ceramic Reactions." Trans. Brit. Ceram. Soc. 50 (6) 265-8 (1951).

Ikornikova, N. Yu. & Popova, A.A., "Preparation of Uniaxial Crystals of Synthetic Corundum." Doklady Akad. Nauk SSSR 106 (3) 460-1 (1956).

Iler, R.K., "Fibrillar Colloidal Boehmite–Progressive Conversion to Gamma, Theta, & Alpha Aluminas." J. Am. Ceram. Soc. 44 (12) 618-24 (1961).

Iler, R.K., "Adsorption of Colloidal Silica on Alumina & of Colloidal Alumina on Silica." J. Am. Ceram. Soc. 47 (4), 194-8 (1964).

Iler, R.K., "Effect of Silica on Transformations of Fibrillar Colloidal Boehmite & Gamma Alumina." J. Am. Ceram. Soc. 47 (7) 339-41 (1964).

Imelik, B., Mathieu, M.V., Prettre, M. & Teichner, S., "Preparation & Properties of Amorphous Alumina." J. chim. phys. 51, 651-2 (1954).

Imelik, B., Mathieu, M.V. & Prettre, M., "Study of An Alumina Gel with Large Specific Surface." Compt. rend. 242, 1885-8 (1956).

Imelik, B., Petitjean, M. & Prettre, M., "The Dehydration of Bayerite." Compt. rend. 238, 900-2 (1954).

Ingalls, A.G., "Optical Glass Polishing." Sci. Amer. 176, 191 (1947).

Innes, K.K., "Characterization of the Aluminum Oxides & of Diatomic Aluminum." PB Rept. 126396, 3 pp.; U.S. Govt. Res. Repts. 29 (6) 321 (1954).

Innes, W.B., "Use of a Parallel Plate Model in Calculation of Pore-Size Distribution." Anal. Chem. 29, 1069-73 (1957).

Insley, R.H. & Barczak, V.J., "Thermal Conditioning of Poly-crystalline Alumina Ceramics." J. Am. Ceram. Soc. 47, (1) 1-4 (1964).

Ioffe, V.A., et al., "Nonlinear Properties of Cerium Aluminate." Fiz. Tverd. Tela 6 (8) 2405-10 (1964).

Irmann, R., "Sintered Aluminum with High Strength at Elevated Temperatures." Metallurgia (9) 125 (1952).

Isaev, A.I., et al., "Use of Ceramic Materials in Cutting Metals." Stanki i Instr. 23 (4) 12-4 (1952).

Isard, J.O., "Electrical Conduction in the Aluminosilicate Glasses." J. Soc. Glass Technol. 43 (211) 113-23T (1959).

Ishibashi, W., "Method and Apparatus for Manufacturing Microfine Metallic Powder." U.S. 3,355,279. 11/28/67.

Ishino, Kozo, "Melting of Borosilicate Glass in a Small Electric Tank Furnace." J. Ceram. Assoc. Japan 63 (704) 11-3 (1955).

Ishizaka, N., "The Relation Between Colloidal Precipitation & Adsorption." Z. physik. Chem. 83, 97 (1913).

Ivanov, B.V. & Polinkovskaya, A.I., "Service of High-Alumina Refractory in the Checkers of a Glass Furnace Regenerator." Ogneupory 23 (7) 307-12 (1958).

Jaccodine, R.J., "Nodule Growth on Al_2O_3 Coatings." J. Electro-chem. Soc. 107, 62-3 (1960).

Jackman, E.A. & Roberts, J.P., "Strength of Polycrystalline & Single Crystal Corundum." Trans. Brit. Ceram. Soc. 54 (7) 389-98 (1955).

Jackman, E.A. & Roberts, J.P., "Strength of Single-Crystal & Poly-crystalline Corundum." Phil. Mag. 46, 809-11 (1955).

Jackson, J.S., "Hot Pressing High-Temperature Compounds." Powder Met. 1961, No. 8, pp. 73-100.

Jackson, J.S., "Studies on the Sintering of $MoSi_2$-Al_2O_3 Cermets." Powder Met. 1961, No. 8, pp. 101-12.

Jacobs, H., "Dissociation Energies of Surface Films of Various Oxides, as Determined by Emission Measurements of Oxide-Coated Cathodes." J. App. Phys. 17 (7) 596-603 (1946).

Jacobs, L.J., "Refractory Materials." U.S. 2,949,704, 8/23/60.

Jacquet, P.A., "Structure of Polished Metallic Surfaces." Tech. Moderne 31 (12) 427-37 (1939).

Jaeger, G., et al., "Determination of the Thermal Conductivity of Oxide Ceramics." Ber. deut. Keram. Ges. u. Ver. deut. Email-fachleute 27 (5/6) 202-5 (1950).

Jaeger, G. & Krasemann, R., "Resistance of Sintered Alumina to Corrosion." Werkstoffe u. Korrosion 3, 401-15 (1952).

Jaffe, H.W., "Application of the Rule of Gladstone & Dale to Minerals." Am. Mineralogist 41 (9/10) 757-7 (1956).

Jaffee, R.I. & Maykuth, D.J., "Refractory Materials." PB Rept. 161194, 35 pp.; U.S. Govt. Res. Repts. 33 (6) 653 (1960).

Jagodzinski, H. & Saalfeld, H., "Distribution of Cations & Structural Relations in Mg-Al Spinels." Z. Krist. 110 (3) 197-218 (1958).

Jamieson, J.C. & Lawson, A.W., "High-Temperature Heat Con-ductivity of Some Metal Oxides." J. Appl. Phys. 29 (9) 1313-4 (1958).

Janowski, J.R. & Conrad, H., "Dislocations in Ruby Laser Crystals." Trans. AIME 230 (4) 717-25 (1964).

Janssen, W.F., "Precision Ceramics." Bell Labs. Record 33 (10) 369-71 (1955).

Janssen, W.F., "Ceramic Solves Problems in Casting Cores." Ceram. Ind. 69 (2) 87-8 (1957).

Jeffery, B.A., "Method of and Apparatus for Shaping Articles." U.S. 1,863,854, 6/21/32.

Jeffery, B.A., "Method of & Apparatus for Molding Materials." U.S. 2,152,748, 4/4/39.

Jeffery, B.A., "Glazing Ceramic Ware." U.S. 2,209,624, 7/30/40.

Jeffery, B.A., "Spark-Plug Insulator." U.S. 2,274,067, 2/24/42.

Jefferys, R.A. & Gadd, J.D., "Development & Evaluation of High Temperature Protective Coatings for Columbium Alloys." ADS TR 61-66, Part I. (May 1961).

Jellinek, M.H. & Fankuchen, I., "X-Ray Diffraction Examination of Gamma Alumina." Ind. Eng. Chem. 37 (2) 158-63 (1945).

Jellinek, M.H., et al., "Measurement & Analysis of Small-Angle X-ray Scattering." Ind. Eng. Chem. Anal. Ed. 18 (3) 172-5 (1946).

Jellinek, M.H. & Fankuchen, I., "X-Ray Examination of Pure Alumina Gel." Ind. Eng. Chem. 41 (10) 2259-65 (1949).

Jellinghaus, W. & Shuin, T., "Cermets of Alumina & Iron or Iron Alloys." Stahl u. Eisen 78 (7) 419-29 (1958). abstracted in Met. Abstr. J. Inst. Metals 26 (1) 71 (1958).

Jephcott, C.M., Johnston, J.H. & Finlay, G.R., "Fume Exposure in the Manufacture of Alumina Abrasives from Bauxite." J. Ind. Hyg. Toxicol. 30 (3) 145-59 (1948).

Johannsen, Klaus, "Molten Ceramic Oxides as Possible Fuel Matrix for Homogeneous Thermal Nuclear Reactors." Keram. Z. 17, 218-20 (1965).

Johansen, P.G. & Buchanan, A.S., "Application of Microelectro-phoresis Method to Study of Surface Properties of Insoluble Oxides." Austral. J. Chem. 10, (4) 398-403 (1957).

Johnson, A.F., "(Aluminum Nitride) Refractory." U.S. 2,480,473, U.S. 2,480,475, 8/30/49.

Johnson, D.L. & Cutler, I.B., "Diffusion Sintering I. Initial Stage Sintering Models & Their Application to Shrinkage Powder Compacts." J. Am. Ceram. Soc. 46 (11) 541-5 (1963).

Johnson, J.R., "Radiation Effects on Ceramics–Basic Science Division Research Committee Report, 1956-57." Am. Ceram. Soc. Bull. 36 (9) 372-4 (1957).

Johnson, O.W. & Gibbs, P., "Fracture of Ge & Al_2O_3." J. Appl. Phys. 34 (9) 2852-62 (1963).

Johnson, P.D., "Behavior of Refractory Oxides & Metals, Alone & In Combination, in Vacuo at High Temperatures." J. Am. Ceram. Soc. 33 (5) 168-71 (1950).

Johnson, P.W., Peters, F.A. & Kirby, R.C., "Methods for Producing Alumina from Clay–An Evaluation of a Nitric Acid Process." U.S. Bur. Mines Rept.Invest. 1964, No. 6431, 25 pp.

Johnston, T.L., Stokes, T.R., & Li, C.H. (appendix by K.H. Olsen)., "Fracture Behavior of Silver Chloride-Alumina Composites." Trans. AIME, 221 (4) 792-802 (1961).

Jones, C., "Process for Manufacturing Synthetic Inorganic Silicates, etc." U.S. 2,678,282, 5/11/54.

Jones, J.T., Maitra, P.B., & Cutler, I.B., "Role of Structural Defects in the Sintering of Alumina & Magnesia." J. Am. Ceram. Soc. 41 (9) 353-7 (1958).

Jones, M.C.K. & Hardy, R.L., "Petroleum Ash Components & Their Effect on Refractories." Ind. Eng. Chem. 44 (11) 2615-9 (1952).

Jones, S., "Alumina-Lithia-Iron Oxide Phosphor." J. Electrochem. Soc. 95 (6) 295-8 (1949).

Jorgensen, P.J. & Westbrook, J.H., "Role of Solute Segregation at Grain Boundaries During Final-Stage Sintering of Alumina." J. Am. Ceram. Soc. 47 (7) 332-8 (1964).

Jourdain, A., "Studies of the Constituents of Refractory Clays by Means of Thermal Analysis." Ceram. 40 (593) 135-41 (1937).

Juillet, F., "Porous or Very Finely Divided Aluminas." Comm. Energie At. (France) Rappt. No. 2257, 54 pp. (1963).

Juillet, F., Prettre, M. & Teichner, S., "Textural Change of Non-stoichiometric Colored Alumina." Compt. rend. 249, 1356-8 (1959).

Junker, O., "Electric Insulation of a Copper Furnace Bobbin." Ger. 1,128,580, 4/26/62.

Kaempfe, "Testing Corundum & Silicon Carbide." Ber. deut. keram. Ges 18 (7) 321 (1937).

Kainarskii, I.S. & Gaodu, A.N., "Light-Weight Corundum Refractories." Ogneupory 28 (5) 218-23 (1963).

Kainarskii, I.S. et al., "Interrelation Between the Electrical & Mechanical Strengths of Corundum Ceramics." Dokla. Akad. Nauk SSSR 157 (1), 168-70 (1964).

Kainarskii, I.S., Degtyareva, E.V. & Alekseenko, L.S., "Effect of Modifying Additives on the Dielectric Properties of Corundum Ceramics" Izv. Akad. Nauk SSSR Neorgan. Materialy 1, 810-15 (1965).

Kainarskii, I.S., Orlova, I.G., and Degtyareva, E.V., "Relation between Shrinkage and Deformation in Sintering Corundum." Dokl. Akad. Nauk SSSR 164, 1283-5 (1965).

Kainarskii, I.S., Orlova, I.G., & Degtyareva, "Deformation and Shrinkage of Corundum during Sintering." Poroshkovaya Met., Akad. Ukr. SSR 5, 82-6 (1965).

Kalyanram, M.R. and Bell, H.B., "Activities in the System CaO-MgO-Al_2O_3" Trans. Brit. Ceram. Soc. 60 (2) 135-46 (1961).

Kandykin, Yu. M., "The Mechanism of Formation & Crystallization of Aluminum Hydroxide." Kolloidn. Zh. 26 (3) 318-23 (1964).

Kanolt, C.W., "The Melting Point of Some Refractory Oxides." J. Wash. Acad. Sci. 3, 315-8 (1913).

Kao, C.-H., "Corrosion of High-Al_2O_3 Bricks in the Electro-furnace Roofs." Kuei Suan Yen Hsueh Pao 2 (2) 66-73 (1963).

Kappmeyer, K.K., Lamont, J.A. & Manning, R.H., "High-Alumina Plastics & Ramming Mixes." Am. Ceram. Soc. Bull. 43 (6) 452-6 (1964).

Kappmeyer, K.K. & Manning, R.H., "Evaluating High-Alumina Brick." Am. Ceram. Soc. Bull. 42 (7) 398-403 (1963).

Karklit, A.K. & Gruzdeva, N.V., "High-Duty Refractories from Technical Alumina." Ogneupory 15 (11) 504-10 (1950).

Karklit, A.K. & Timofeev, N.N., "High Alumina Ladle Brick." Ogneupory 24 (11) 490-5 (1959).

Karmaus, H.J., "New Enameling Processes." Glas-Email-Keramo-Tech. 4 (11) 409-11 (1953).

Karpacheva, S.M. & Rozen, A.M., "Oxygen Exchange Between Oxide Catalysts & Water Vapor." Doklady Akad Nauk SSSR 75, 55-8 (1950).

Karyakin, L.I. & Margulis, O.M., "Formation of Mullite in an Unusual Form." Doklady Akad. Nauk SSSR 109 (4) 821-3 (1956).

Kassel, R.E., "Manufacture of Large Alumina Cylinders." Am. Ceram. Soc. Bull. 43 (4) 317 (1964).

Kato, Katsuo, "Corrosion of Fusion-Cast Beta Alumina Refractory." Asahi Garasu Kenkyu Hokoku 14, 95-105 (1964).

Kato, S. & Yamauchi, T., "Studies on β-Alumina: VII, Synthesis in the CaO-Al_2O_3 System." J. Japan Ceram. Assoc. 52 (614) 47-50 (1944).

Kato, S. & Yamauchi, T., "Studies on β-Alumina: III, Synthesis in the System K_2O-Al_2O_3." J. Japan Ceram. Assoc. 51 (610) 586-93 (1943).

Kato, S. & Yamauchi, T., "Studies on β-Alumina: IV, Synthesis in the Fused System K_2O-Al_2O_3." J. Japan Ceram. Assoc. 51 (611) 640-6 (1943).

Kato, S. & Yamauchi, T., "Studies on β-Alumina: V, Synthesis in MgO-Al_2O_3 System." J. Japan Ceram. Assoc. 52 (613) 11-3 (1944).

Kato, S. & Okuda, H., "Alumina Porcelain as a High Frequency Insulator: I, Effects of Aluminum Oxide as a Raw Material on Physical Properties." Nagoya Kogyo Gijutsu Shikensho Hokoku 9 (8) 46-52 (1960).

Kato, S. & Okuda, H., "Alumina Porcelain as a High-Frequency Insulator. I. Effects of Kinds of Alumina on Physical Properties." Nagoya Kogyo Gijutsu Shikensho Hokoku 9, 402-8 (1960).

Kato, S., Okuda, H., Iga, T. & Okawara, S., "Alumina Porcelain as a High Frequency Insulator: II, Effects of Metal Oxides." Nagoya Kogyo Gijutsu Shikensho Hokoku 11 (8) 490-8 (1962).

Kawamura, H. & Azuma, K., "Dielectric Breakdown of Thin Alumina Film." J. Phys. Soc. Japan 8, 797-8 (1953).

Kawashima, C., et al., "Porous Sintered Refractories." Japan 14,343, 9/18/62.

Kearby, K.K. & Giblert, G.R., "Production of Alumina. U.S. 2,893,837, 7/7/59.

Kebler, R.W., "Optical Properties of Synthetic Sapphire." Linde Air Products Company Brochure July 7, 1955.

Keith, C.D., "Process for Preparing Alumina Trihydrate." U.S.2,874,130, 2/17/59.

Keith, C.D., "Alumina Preparation." U.S. 2,894,915, 7/14/59.

Keith, C.D., "Mixtures of Hydrated Alumina." Ger. 1,162,338, 2/6/64.

Keith, C.D. & Hauel, A.P., "Process for Preparing Alumina Catalyst Compositions." U.S. 2,867,588, 1/6/59.

Keith, C.D. & Seligman, B., "Calcination of Macrosize Alumina Hydrate." U.S. 2,916,356, 12/8/59.

Keith, W.P. & Whittemore, O.J., Jr., "High Temperature Laboratory Kiln." Am. Ceram. Soc. Bull. 28 (5) 192-4 (1949).

Keller, A. & Herberger, C., "Turbine Ceramic Bucket & Wheel Development Project at M.A.N. Plant, Augsburg, Germany." U.S. Naval Tech. Mission in Europe Tech. Rept. 166-45; PB 22,801, 7 pp. (Sept. 1945); abstracted in Bibliog. Sci. Ind. Repts. 2 (1) 56 (1946).

Kelley, F.C., "Metallizing & Bonding Nonmetallic Bodies." U.S. 2,570,248, 10/9/51.

Kelly, G.D., "Effects of Hydrostatic Forming." Am. Ceram. Soc. Bull. 40 (6) 378-82 (1961).

Kelly, P.J. and Laubitz, M.J., "Thermoluminescence of Alumina." J. Am. Ceram. Soc. 50, 540-2 (1967).

Kelman, L.R., Wilkinson, W.D. & Yaggee, F.L., "Resistance of Materials to Attack by Liquid Metals." Argonne National Laboratory Report ANL-4417 (July 1950).

Kelsey, R.H. (to Horizons, Inc.), "Alumina Wool." U.S. 3,341,285. 9/12/69.

Keltz, F.O., "(Silica Brick with Added Alumina)." U.S. 2,573,264, 10/30/51.

Kennedy, G.C., "Phase Relations in the System Al_2O_3-H_2O at High Temperatures & Pressures." Am. J. Sci. 257 (8) 563-73 (1959).

Kerper, M.J. & Scuderi, T.G., "Modulus of Rupture of Glass in Relation to Fracture Pattern." Am. Ceram. Soc. Bull. 43 (9) 622-5 (1964).

Kerr, E.C., Johnston, H.L. & Hallet, N.C., "Low-Temperature Heat Capacities of Inorganic Solids: III Heat Capacity of Aluminum Oxide (Synthetic Sapphire) from 19° to 300°K." J. Am. Chem. Soc. 72 (10) 4740-2 (1950).

Kerrigan, J.V., "Studies on the Transport & Deposition of Alpha Aluminum Oxide." J. Appl. Phys. 34 (11) 3408-10 (1963).

Keski, J.R., "Effects of Manganese Oxide on the Sintering of Alumina." Univ. Microfilms Order No. 66- 10,484.

Keski, J.R. and Cutler, I.V., "Effect of Manganese Oxide on Sintering of Alumina." J. Am. Ceram. Soc. 48, 653-4 (1965).

Keski, J.R. and Cutler, I.B., "Initial Sintering of Mn_xO·Al_2O_3." J. Am. Ceram. Soc. 51, 440-444 (1968).

Kessler, S.W., Jr., "Glass Sealing." U.S. 2,876,596, 3/10/59.

Kharitonov, F.Ya., "Kinetics of the Recrystallization of Corundum." Ogneupory 31, 53-8 (1966).

Khemelvskii, V.I. & Minakov, A.C., "Service of Refractories for Melting Technical Glass in Continuous Glass Tanks." Ogneupory 22 (6) 275-82 (1957).

Khvostenkov, N.I. & Tararin, A.A., "Chromite-Alumina Linings at the Bryanski Cement Works." Tsement 6 (7) 1-3 (1939).

Kick, F., "The Law of Proportional Resistance & Its Application. Arthur Felix, Leipzig (1885).

Kiley, L.D., "Technical Inspection of Robert Bosch Plant re Ceramics." Army Air Forces Tech. Intelligence Rept. P-21; Rept. PB 2012, 2 pp. (1945); abstracted in Bibliog. Sci. Ind. Repts. 1 (5) 173 (1946).

Kilham, J.K., "Third Symposium on Combustion Flame & Explosion Phenomena," 733-40. The Williams & Wilkins Company, Baltimore, Md. (1949).

Kim, K.H. & Hummel, F.A., "System Al_2O_3-B_2O_3, Tentative." Dec. 20, 1961. Private communication "Phase Diagrams for Ceramists (1964)," Fig. 308.

Kim, K.H. & Hummel, F.A., "Studies in Lithium Oxide Systems: XII, Li_2O-B_2O_3-Al_2O_3." J. Am. Ceram. Soc. 45 (10) 487-9 (1962).

Kimberlin, C.N., Jr., "Preparation of Alumina by Burning." U.S. 2,754,176, 7/10/56.

Kimberlin, C.N., Jr. & Gladrow, E.M., "Hydroforming Catalysts & Method." U.S. 2,773,842, 12/11/56.

Kimberlin, C.N. Jr. & Gladrow, E.M., "Process for Making n-Alumina from Aluminum Alcoholates." U.S. 2,796,326, 6/18/57.

Kimpel, R.F., "Tensile Tests of Ceramic Bodies - Spinning in Vacuo." Univ. Ill. Dept. Ceram. Eng. Rept. 16, Sect. III; PB 27,589, 3 pp n.d.; abstracted in Bibliog. Sci. Ind. Repts. 2 (5) 362 (1946).

Kimpel, R.F., "Cement Composition." U.S. 3,069,278, 12/18/62.

King, A.G., "Influence of Microstructure on the Mechanical Properties of Dense Polycrystalline Alumina." Mech. Properties Eng. Ceramic Proc. Conf. Raleigh, N.C. 1960, 333-47 (Pub. 1961).

King, A.G., "Chemical Polish and Strength of Alumina." Mater. Sci. Res. 3, 529-38 (1966).

King, B.W., "Flame-Sprayed Oxides." Ind. Heating 29 (10) 2013-4, 2023 (1962).

King, B.W., Tripp, H.P. & Duckworth, W.H., "Nature of Adherence of Porcelain Enamels to Metals." J. Am. Ceram. Soc. 42 (11) 504-25 (1959).

King, D.F. & Renkey, A.L., "Explosion Resistant Refractory Castable." U.S. 2,845,360, 7/29/58.

King, E.G. & Weller, W.W., "Low-Temperature Heat Capacities & Entropies at 298.15°K of Diaspore, Kaolinite, Dickite, & Halloysite." U.S. Bur. Mines Rept. Invest., 1961, No. 5810, 6 pp.

King, C. & Stookey, S.D., "Method of Making a Semicrystalline Ceramic Body." U.S. 2,960,801, U.S. 2,960,802, 11/22/60.

King, R.M., "Ceramic Materials for Application in the Design of Jet Propelled Devices. Progress Report." Ohio State Univ. Research Foundation Rept. 15 (July 1946). 3 pp. PB 60,668; abstracted in Bibliog. Sci. Ind. Repts. 4 (11) 965 (1947).

Kingery, W.D., "Fundamental Study of Phosphate Bonding in Refractories: I, Literature Review." J. Am. Ceram. Soc. 33 (8) 239-41 (1950).

Kingery, W.D., "Fundamental Study of Phosphate Bonding in Refractories: IV, Mortars Bonded with Monoaluminum & Monomagnesium Phosphate." J. Am. Ceram. Soc. 35 (3) 61-3 (1952).

Kingery, W.D., "Metal-Ceramic Interactions: I, Factors Affecting Fabrication & Properties of Cermet Bodies." J. Am. Ceram. Soc. 36 (11) 362-5 (1953).

Kingery, W.D., "Thermal Conductivity, VI, Determination of Conductivity of Al_2O_3 by Spherical Envelope & Cylinder Methods." J. Am. Ceram. Soc. 37 (2, part II), 88-90 (1954).

Kingery, W.D., "Symposium on Thermal Fracture: Recommended Letter Symbols for Thermal Stress Analysis." J. Am. Ceram. Soc. 38 (1) 1-2 (1955).

Kingery, W.D., "Factors Affecting Thermal Stress Resistance of Ceramic Materials." J. Am. Ceram. Soc. 38 (1) 3-15 (1955).

Kingery, W.D., "Note on Thermal Expansion & Microstresses in Two-Phase Compositions." J. Am. Ceram. Soc. 40 (10) 351-2 (1957).

Kingery, W.D., Ed., "Ceramic Fabrication Processes." Technical Press of Massachusetts Institute of Technology & John Wiley & Sons, Inc., New York, xi + 235 pp. (1958).

Kingery, W.D., "Surface Tension of Some Liquid Oxides & Their Temperature Coefficients." J. Am. Ceram. Soc. 42 (1) 6-10 (1959).

Kingery, W.D., "Densification During Sintering in the Presence of a Liquid Phase: I, Theory." J. Appl. Phys. 30 (3) 301-6 (1959).

Kingery, W.D. ed., "Kinetics of High-Temperature Processes—Report of Endicott House Conference on Kinetics of High-Temperature Processes, June 1958." Technology Press of Massachusetts Institute of Technology, Cambridge, & John Wiley & Sons, Inc., New York (1959). xvi + 326 pp.

Kingery, W.D., "Property Measurements at High Temperatures." John Wiley & Sons, Inc., New York (1959). xii + 416 pp.

Kingery, W.D., "Introduction to Ceramics." John Wiley & Sons, New York, xv + 781 pp. (1960).

Kingery, W.D. & Berg, M., "Study of the Initial Stages of Sintering Solids by Viscous Flow, Evaporation-Condensation, and Self-Diffusion." J. Appl. Phys. 26 (10) 1205-12 (1955).

Kingery, W.D., Francl, J., Coble, R.L., & Vasilos, T., "Thermal Conductivity: X, Data for Several Pure Oxide Materials Corrected to Zero Porosity." J. Am. Ceram. Soc. 37 (2, part II) 107-10 (1954).

Kingery, W.D., Klein, J.D. & McQuarrie, M.C., "Development of Ceramic Insulating Materials for High-Temperature Use." Trans. ASME 80, 705-10 (1958).

Kingery, W.D. & Norton, F.H., "The Measurement of Thermal Conductivity of Refractory Materials." NYO-6451 Quarterly Progress Rept. AD 80699, 16 pp. (1955).

Kingery, W.D. & Norton, F.H., "The Measurement of Thermal Conductivity of Refractory Materials." NYO-6477 Quarterly Progress. Rept. AD-55595 (1955).

Kingery, W.D. & Pappis, J., "Note on Failure of Ceramic Materials at Elevated Temperatures Under Impact Loading." J. Am. Ceram. Soc. 39 (2) 64-6 (1956).

Kingery, W.D., Sidhwa, A.P., & Waugh, A., "Structure & Properties of Vitrified Bonded Abrasives." Am. Ceram. Soc. Bull. 42 (5) 297-303 (1963).

Kipling, J.J. & Peakall, D.B., "Reversible & Irreversible Adsorption of Vapors by Solid Oxides & Hydrated Oxides." J. Chem. Soc. 834-42 (1957).

Kirby, D., "Pure Oxide Refractories." Refractories J. 27 (1) 11-4 (1951).

Kircher, J.F. & Bowman, R.E. eds., "Effects of Radiation on Materials & Components." Reinhold Publishing Corp., New York, xi + 690 pp. (1964).

Kirchner, H.P. and Gruver, R.M., "Chemical Strengthening of Polycrystalline Ceramics." J. Am. Ceram. Soc. 49, 330-3 (1966).

Kirchner, H.P., Gruver, R.M., and Walker, R.E., "Chemical Strengthening of Polycrystalline Alumina." J. Am. Ceram. Soc. 51, 251-55 (1968).

Kirchner, H.P., Gruver, R.M., and Walker, R.E., "Strengthening Alumina by Glazing and Quenching." Am. Ceram. Soc. Bull. 47, 798-802 (1968).

Kirk, R.E. & Othmer, D.F. eds., "Encyclopedia of Chemical Technology—First Supplement." Interscience Encyclopedia, Inc., New York, 974 pp. (1957).

Kirillin, V.A., Sheindlin, A.E. & Chekhovskoi, V.Ya., "Experimental Determination of the Enthalpy of Corundum (Al_2O_3) at 500° to 2000°." Dokl. Akad. Nauk SSSR 135 (1) 125-8 (1960).

Kirillova, G.K., "Electrical Properties of Mullite." Zhur. Tekh. Fiz. 28 (10) 2186 (1958).

Kirshenbaum, A.D. & Cahill, J.A., "Density of Liquid Aluminum Oxide." J. Inorg. & Nuclear Chem. 14 (3-4) 283-7 (1960).

Kirshenbaum, I. & Hinlicky, J.A., "Preparation of Eta-Alumina by Acid Hydrolysis." U.S. 2,903,418, 9/8/59.

Kisliuk, P. & Krupke, W.F., "Biquadratic Exchange Energy in Ruby-0.5% Cr_2O_3)." Appl Phys. Letters 3 (12) 215-6 (1963).

Kistler, S., "Heat-Exchange Pebble." U.S. 2,624,556, 1/6/53.

Kistler, S.S., "Refractories in Turbine Blades." Combined Intelligence Objectives Sub-Committee Paper, File No. XXXI-22. H.M. Stationery Office, London, 1945.

Kistler, S.S. & Barnes, C.E., "Making Abrasive Articles." U.S. 2,218,795, 10/22/40.

Kitaigorodskii, I.I., "Corundum Microlit & Its Structure." Doklady Akad. Nauk. SSSR 90 (2) 225-6 (1953).

Kitaigorodskii, I.I. & Bondarev, K.T., "New Glass Crystal Materials Made from Slag." Priroda (Moscow) (9) (1962).

Kitaigorodskii, I.I. & Gurevich, Ts.N., "The Influence of Roasting Temperature on the Strength of Corundum Material." Tr. Mosk. Khim.- Tekhnol. Inst. 1959, No. 27, 73-7.

Kitaigorodoskii, I.I. & Gurevich, Ts.N., "Intensifying the Grinding of Alumina in the Glass & Ceramic Industries." Steklo i Keram. 16 (5) 5-9 (1959).

Kitaigorodskii, I.I. & Gurevich, Ts.N., "New Developments in the Synthesis of Corundum Mikrolit." Steklo i Keram. 17 (2) 10-2 (1960).

Kitaigorodskii, I.I. & Keshishyan, T.N., "Calcined Mullite Refractory." Compt. rend. Acad. Sci. U.R.S.S. 23, 152-4 (1939). Brit. Chem. & Phys. Abs. B, 58 (10) 1038 (1939).

Kitaigorodskii, I.I., Keshishyan, T.N., & Fainberg, E.A., "Study of Glasses in the System SiO_2-Al_2O_3-B_2O_3-BaO." Steklo i Keram. 15 (3) 1-5 (1958).

Kitaigorodskii, I.I., Mendeleev, D.J., Fainberg, E.A. & Grechanik, L.A., "Lead Silicate Glasses, The Influence of Certain Oxides on Reduction of, in Hydrogen." Glass & Ceramics, Moscow 19 (11-12) 645 (1963).

Kitaigorodskii, I.I. & Pavlushkin, N.M., "Mikrolit—An Artificial Superstrong Stone." Steklo i Keram. 10 (11) 4-7 (1953).

Kitaigorodskii, I.I. & Pavlushkin, N.M., "Characteristics of Corundum Mikrolit." Steklo i Keram. 12 (11) 16-21 (1955).

Kiwak, R.S., "Bonding Refractory Metals to Ceramics with Molybdenum, Yttria, and Alumina." U.S. 3,340,026. 9/5/67.

Kiyoura, R. & Sata, T., "Preliminary Study on Oxide Cermets of Alumina, Zirconia & Magnesia Series Containing Nickel & Iron." J. Ceram. Assoc. Japan 64 (728) 183-92 (1956).

Kiyoura, R. & Sata, T., "Effects of the Addition of Various Oxides on the Strength of Sintered Alumina Bodies at High Temperature." Yogyo Kyokai Shi 66 (746) 23-7 (1958).

Kiyoura, R. & Sata, T., "Selection of Metals for Cermet Bodies of Alumina Series." Yogyo Kyokai Shi 66 (747) 49-59 (1958).

Klassen-Neklyudova, M.V., "Mechanical Properties of Corundum Crystals." J. Tech. Phys. (USSR) 12, 519-51 (1942).

Klein, A.A., "Properties of Fused Alumina Grain." Metal Ind. (N.Y.) 35 (8) 401-3 (1937).

Klein, A.A., "Abrasive Grain." U.S. 2,301,123, 11/3/42; U.S. 2,303,284, 11/24/42.

Klein, A. & Ridout, G.T., "New Alpha-Alumina Abrasive." Machinery (London) 74 (1893) 145-7 (1949).

Klein, D.J., "Measurement of the Crystallographic Thermal Expansion of α-Alumina & Beryllium Oxide to Elevated Temperatures Emphasizing Anisotropic Effects." U.S. At. Energy Comm. NAA-SR-2542, 24 pp. (1958).

Kleinert, A., Kleinert, U. & Kleinert, A., "Protection of Graphite or Amorphous Carbon Electrodes from Fire Loss." Ger. 1,105,656, 4/4/57.

Kleppa, O.J. & Yokokawa, T., "Enthalpy of Transformation, Delta Al_2O_3 to Alpha Al_2O_3." J. Am. Chem. Soc. 86 (13) 2749 (1964).

Klerk, J.D., "Behaviour of Coherent Microwave Phonons at Low Temperature in Al_2O_3, Using Vapor-Deposited Thin-Film Piezoelectric Transducers." Phys. Rev. 139, 1635-9 (1965).

Klevens, H.B. & Platt, J.R., "Ultraviolet-Transmission Limits of Some Liquids & Solids." J. Am. Chem. Soc. 69, 3055-62 (1947).

K.L.G. Sparking Plugs, Ltd., "Ceramic-to-Metal Seals." Brit. 706,183, 1/27/54.

Klingler, E. & Dawihl, W., "Sintered Product Consisting of An Aluminum Oxide Lattice & A Metallic Component Filling the Interstices of the Lattice." U.S. 3,032,427, 5/1/62.

Klopp, W.D., "Oxidation Behavior & Protective Coatings for Niobium." U.S. Dept. Com. Office Tech. Serv. PB Rept 151080, 97 pp. (1960).

Klopp, W.D., "Review of Recent Developments on Oxidation-Resistant Coatings for Refractory Metals." AD 255278, 3 pp., U.S. Govt. Res. Repts. 36 (1) 80 (1961).

Klopp, W.D., "Review of Recent Developments in Oxidation-Resistant Coatings for Refractory Metals." AD 266469, 4 pp.; U.S. Govt. Res. Repts. 37 (3) 91 (1962).

Knapp, W.J., "Method of Treating an Alumina Ceramic Article with Lithium Borate." U.S. 3,049,447, 8/14/62.

Knauft, R.W., "Zircon Refractories for Aluminum Melting Furnaces." Metals & Alloys 18 (6) 1326-30 (Dec. 1943); abstracted in Bull. Brit. Non-Ferrous Metals Res. Assn. No. 180, p. 154 (June 1944).

Knauft, R.W., Smith, K.W., Thomas, E.A. & Pittman, W.C., "Bonded Mullite & Zircon Refractories for the Glass Industry." Am. Ceram. Soc. Bull. 36 (11) 412-5 (1957).

Knecht, W., "Application of Pressed Powder Technique for Production of Metal-to-Ceramic Seals." Ceram. Age 63 (2) 12-3 (1954).

Knizek, J.O. & Fetter, H., "Refractory Properties of Alunite." Trans. Brit. Ceram. Soc. 49 (5) 202-23 (1950).

Knoop, F., Peters, C.G. & Emerson, W.B. J. Res. N.B.S. 23, 39-61 (1939).

Knudsen, F.P., "Effect of Porosity on Young's Modulus of Alumina." J. Am. Ceram. Soc. 45 (2) 94-5 (1962).

Knuth, K., "High-Quality (Corhart) Tank Blocks." Glastech. Ber. 13 (4) 116-25 (1935).

Koberstein, E., "Laminated Structure of Sintered Corundum." Z. anorg. u. allgem. Chem. 279 (3-4) 194-204 (1955).

Kocher, D.W., "Refractory Concrete." Can. 463,920, 3/21/50.

Koehler, E.K. & Leonov, A., "Influence of a Gaseous Medium on the Shrinkage, Sintering, & Recrystallization of Oxides During Firing." Bull. Soc. Franc. Ceram. 1961, No. 50, pp. 7-16.

Koening, H., "The Lattice Constants of γ-Alumina." Naturwiss. 35, 92-3 (1948).

Konig, W., "Acid Stability of High Alumina Bottle Glasses." Glastech. Ber. 21 (12) 255-7 (1943).

Konig, W., "Melting of High Alumina Containing Brown Bottle Glass." Glastech. Ber. 21 (12) 260 (1943).

Kohl, H., "Sinterkorund, A New Ceramic Material of Pure Alumina." Ber. deut. keram. Ges. 13, (2) 70-85 (1932).

Kohl, H. & Rice, P., "Electron Tubes for Critical Environments." Stanford Research Institute (March 1958) WADC Technical Rept. 57-434, ASTIA Document No. AD 151158.

Kohlmeyer, E.J. & Lundquist, S., "Thermal Reduction of Argillaceous Earth." Z. anorg. Chem. 260, 208-30 (1949).

Kohlschutter, V., "Disperse Aluminum Hydroxide." Z. anorg. Chem. 105, 1-25 (1919).

Kohlschutter, V., & Beutler, W., "IV Chemistry & Morphology of Aluminum Hydroxides." Helv. Chim. Acta 14, 305-54 (1931).

Kohlschutter, V., Beutler, W., Sprenger, L. & Berlin, M., "Principles of the Genetic Formation of Materials." Helv. Chim. Acta 14, 3-49 (1931).

Kohn, J.A., Katz, G. & Broder, J.D., "β-Ga_2O_3 & Its Alumina Isomorph, θ-Al_2O_3. Am. Mineral. 42, 398-408 (1957).

Koifman, M.I., "New Definition of Hardness." Compt. rend. Acad. Sci. URSS 30, 830-1 (1941); abstracted in Chem. Zentr. 1943, I (6) 604.

Kolbl, F., "Can Ceramic Tools Replace Cemented Carbide?" Planseeber, Pulvermet. 6 (2) 48-66 (1958); Met. Abstr. J. Inst. Metals 26 (4) 326 (1958).

Koldaev, B.G. et al., "Technology of Producing High Alumina Blocks for the Structure of Glass Tanks." Ogneupory 22 (8) 340-5 (1957).

Kolosova, N.I., Kharitonov, F.Ya., et al., "Testing the Stability of Corundum Ceramics in a Liquid Alloy of K and Na." Teplofiz. Vysokikh Temperatur, Akad. Nauk SSSR 4, 115-119 (1966).

Klingsova, V.A., "Infrared Absorption Spectra of the Silicates Containing Aluminum & of Certain Crystalline Aluminates." Optics & Spectroscopy (USSR) (English Translation), 6 (1) 20-4 (1959).

Kolesova, V.A. & Ryskin, Ya.I., "Infrared Absorption Spectrum of Hydrargillite, $Al(OH)_3$." Optics & Spectroscopy (USSR) (English Translation), 7 (2) 165-7 (1959).

Kolesova, V.A. & Ryskin, Ya.I., "Infrared Spectra of Diaspore (a-AlOOH), Boehmite (γ-AlOOH), & GaOOH." Zh. Strukt. Khim. 3, 680-4 (1962).

Komarek, K.L., Coucoulas, A., & Klinger, N., "Reactions Between Refractory Oxides & Graphite." J. Electrochem. Soc. 110, 783 (1963).

Konopicky, K., "Contribution to the Study of the Alumina-Silica Equilibrium Diagram." Bull. soc. franc. ceram. 1956, No. 33, pp. 3-6.

Konopicky, K., "Generalities on the Structure of Fire-Clay Brick." Ber. deut. keram. Ges. 36 (11) 367-71 (1959).

Konoval'chikov, O.D., Galich, P.N., Musienko, V.P., Skarchenko, V.K., & Petro, A.A., "Effect of the Pore Structure of an Aluminochrome Catalyst on the Conversion of n-hexane." Kinetika i Kataliz 5 (2), 350-4 (1964).

Konrad, H.E. & Stafford, W.L., "Light-Weight Castable Refractories." U.S. 3,079,267, 2/26/63.

Konta, J., "Proposed Classification & Terminology of Rocks in the Series Bauxite-Clay-Iron Oxide Ore." J. Sediment. Petrol., 28 (1) 83-6 (1958).

Kordes, E., "Crystal & Chemical Studies of Aluminum Compounds with a Spinel Lattice & of Gamma-Fe_2O_3: Notes on Mineralizing Effects of Fluorides." Z. Krist. 91, 193-228 (1935).

Korelova, A.I., "Polishing Glass with Fine Abrasive Powders." Steklo i Keram. 13 (12) 5-9 (1956).

Kose, S. and Hamano, Y., "Dielectric Properties of Hot-Pressed Alumina." Osaka Kogyu Gijutsu Shikenjo Kiho 18, 78-86 (1967).

Kostkowski, H.J., "The Accuracy & Precision of Measuring Temperatures Above 1000°K." "High Temperature Technology." McGraw-Hill Book Co., Inc., New York (1960).

Kotshmid, F., "Manufacture of Bottles from High-Alumina Glass in Czechoslovakia." Steklo i Keram. 16 (9) 41-3 (1959).

Kourimsky, J., "Morphological Crystallography of Corundum." Sb. Narod. Musea Praze 20B (2), 109-18 (1964).

Kovalev, "Measurement of the Moduli of Elasticity of Refractory Materials at High Temperatures under Vacuum." Zavodsk. Lab. 28 (6) 729-31 (1962).

Kovaschev, I., "Sintering of Corundum in the Presence of Manganese Dioxide-Titanium Dioxide Mixture." C.R. Acad. Bulg. Sci. 20, 1279-81 (1967).

Kovatschev, I. and Serbezova, R., "Effect of Nb_2O_5, MnO_2, ZrO_2, and Na_3AlF_6 on the Sintering of Corundum." C.R. Acad. Bulg. Sci. 20, 685-8 (1967).

Koz'mina, Z.P. & Dobrynina, V.A., "Investigation of Zeta-Potential of Bayerite & of the Products of its Thermal Treatment." Kolloidn. Zh. 26 (5) 592-4 (1964).

Kramer, B.E. & Levinstein, M.A., "Effect of Arc Plasma Deposition on the Stability of Nonmetallic Materials." AD 255945, 9 pp. U.S. Govt. Res. Repts. 36 (1) 68 (1961).

Kramer, J., "Point Counter & Counter Tube in Metallographic Investigations of Surfaces." Z. Physik 125, 739-56 (1949).

Kramer, J., "Applications of Exoelectrons." Acta Phys. Austriaca 10, 392-8 (1957).

Krannich, R., "Ethyl Silicate 40 As Hydraulic & Ceramic Binder." Silikattech. 12 (2) 78-80 (1961).

Kraut, H., Flake, E., Schmidt, W., & Volmer, H., "Aluminum Ortho-hydroxide & Its Transformation Into Bayerite." Ber. 75B, 1357-73 (1942).

Kraut, H. & Humme, H. Ber. 64, 1697 (1931).

Krebs, J.J., "Effect of Applied Electric Fields on the Electron Spin Resonance (ESR) of Fe^{3+} & Mn^{2+} in a-Al_2O_3." Phys. Rev. 135 (2A), 396-401 (1964).

Kreidl, N.J., "Borosilicate Optical Glass." U.S. 2,657,146, 10/27/53.

Kreidl, N.J., "Irradiation Damage to Glass." Contract AT(30-1) 1312, Rept. Nos. NYO 3777, 3778, 3779, 3780, & 3781). (1955).

Kreidl, N.J. & Hensler, J.R., "Formation of Color Centers in Glasses Exposed to Gamma Radiation." J. Am. Ceram. Soc. 38 (12) 423-32 (1955).

Kretzschmar, H., "Process of Making Alumina." U.S. 2,951,743, 9/6/60.

Kriegel, W.W. & Palmour, H. III. eds., "The Mechanical Properties of Engineering Ceramics." Interscience Publishers, New York (1961).

Kriegel, W.W., Palmour, H. III, & Choi, D.M., "The Preparation & Mechanical Properties of Spinel." Special Ceramics Symposium, Stoke-on-Trent, England, June 1964.

Krieger, F.J. (Rand Corp., Santa Monica, Cal.), "Thermodynamics of the Alumina/Aluminum-Oxygen Vapor System." AD636808.

Krieger, K.A. & Heinemann, H., "Regeneration of Alumina Absorbents by Oxidation." U.S. 2,457,566, 12/28/48.

Krier, C.A., "Coatings for the Protection of Refractory Metals from Oxidation." DMIC Rept. 162 (Battelle) 226 pp (1961).

Krischner H. & Torkar, K., "Microcrystalline Active Corundum." Sci. Ceram. 1, 63-76 (1962).

Krishnan, R.S., "Progress in Crystal Physics: Vol. I. Thermal, Elastic, & Optical Properties." Interscience Publishers, New York (1960).

Krishnan, R.S., "Raman Spectrum of Alumina & The Luminescence & Adsorption Spectra of Ruby." Nature 160, 26 (1947).

Krleza, F., "The Mutual Flocculation of Aluminum Hydroxide, Iron Hydroxides, & Silicic Acid Sols." Kolloid Z. 197 (1-2), 154-9 (1964).

Krock, R.H. & Kelsey, R.H., "Whiskers—Promise & Problems." Ind. Res. p. 47-57 (Feb. 1965).

Kroger, F.A., "Some Aspects of the Luminscence of Solids." Elsevir Publishing Co., New York (1948).

Kroger, F.A. & Vink, H.J., "Relations Between Concentration of Imperfections in Crystalline Solids." p. 307. "Solid State Physics—Advances in Research & Applications. Vol. 3." eds. F. Seitz & D. Turnbull. Academic Press Inc., New York, 588 pp. (1956).

Krokhina, A.I., Spivak, G.V., Reshetnikov, A.M. & Zhelninskaya, R.I., "Electron Microscopic Study of the Structure of Ceramic Materials Exposed by Ionic Etching." Izv. Akad. Nauk. SSSR, Ser. Fiz. 27 (9) 1224-7 (1963).

Kroll, W.J. & Schlechten, A.W., "Reactions of Carbon & Metal Oxides in Vacuum." Trans. Electrochem. Soc. 93, 247-58 (1948).

Kronberg, M.L., "Plastic Deformation of Single Crystals of Sapphire: Basal Slip & Twinning." Acta Met. 5, 507-24 (1957).

Kronberg, M.L., "Dynamical Flow Properties of Single Crystals of Sapphire: I." J. Am. Ceram. Soc. 45 (6) 274-9 (1962).

Kronberg, M.L. et al., "Deformation of Refractory Crystals." U.S. Dept. Com., Office Tech. Serv., AD 269,220, 4 pp. (1961).

Kroon, D.J. & Stolpe, C. van der., "Positions of Protons in Aluminum Hydroxides Derived from Proton Magnetic Resonance." Nature 183, 944 (1959).

Kryukova, V.G., "Comparative Study of Synthetic Boehmite & Gibbsite, Formed During Hydrothermal Synthesis of Corundum, & the Analogous Natural Minerals." Tr. Vses. Nauchn.-Issled. Inst. P'ezooptich. Mineral'n. Syr'ya 1961, No. 5, 91-9.

Kryukova, V.G. & Nozdrina, V.G., "Results of Mineralogic-Petrographic & X-ray Studies of Crystalline Phases Detected During the Hydrothermal Synthesis of Corundum." Tr. Vses. Nauch.-Issled. Inst. P'ezooptich. Mineral'n. Syr'ya 4, No. 2, 101-9 (1960).

Kubli, H., "Separation of Anions by Adsorption on Alumina." Helv. Chim. Acta 30, 453-63 (1947).

Kuczynski, G.C., "Self-Diffusion in Sintering Metal Powders." Trans. AIME 185, 169 (1949).

Kuczynski, G.C., "Effect of Oxygen on Sintering of Oxides." Plansee Proc. 4th Seminar, Reutte/Tyrol, 1961, pp. 166-80 (1962).

Kuczynski, G.C., "Sintering Phenomena in Ceramic Systems." 14 pp. (1964), AD 603267. Avail. OTS (Eng.).

Kuczynski, G.C., Abernethy, L. & Allan, J., "The Chemical Aspects of the Sintering of Aluminum Oxide." Plansee Proc., 3rd Seminar, Reutte/Tyrol 1958, 1-12 (Pub. 1959).

Kuczinski, G.C., Abernethy, L. & Allan, J., "Sintering Mechanisms of Aluminum Oxide," pp. 163-72. "Kinetics of High Temperature Process." Technology Press of MIT & John Wiley & Sons, Inc., N.Y. (1959).

Kuehl, H., "Alumina in Glass Batches." Glashutte 64, 479-81 (1934).

Kuhn, G. & Weinhold, H., "The Light Coagulator." Jena Review, Special Issue 1965, pp. 47-51.

Kulpmann, F., "Grinding Wheels of 'Noble Corundum' as Universal Grinding Wheels." Oberflachentech. 17 (4) 23-4 (1940).

Kukolev, G.V., & Karaulov, A.G., "The Properties of Water Suspensions of Technical Alumina & Rational Conditions for Slip Casting." Ogneupory 28, 168-74 (1963).

Kukolev, G.V. & Karaulov, A.G., "The Properties of Aqueous Suspensions of Technical-Grade Alumina." Refractories (USSR) 3-4, 184 (1963).

Kukolev, G.V. & Leve, E.N., "Sintering of Alumina in Various Systems." Zhur. Priklad. Khim. 28 (8) 807-16 (1955).

Kukolev, G.V. & Mikhailova, K.A., "The Effect of Some Additives on the Sintering of Aluminous Materials." Ogneupory 24 (1) 39-44 (1959).

Kumar, S., "Glasses in Systems Containing Lead Oxide, Boric Oxide, Alumina, & Phosphorus Pentoxide." Central Glass & Ceram. Res. Inst. Bull. (India) 6 (1) 13-22 (1959).

Kumar, S., "Properties of Glasses in the System Al_2O_3-B_2O_3-P_2O_5." Central Glass & Ceram. Res. Inst. Bull. (India), 7 (3) 117-22 (1960).

Kuntzsch, E. & Rabe, G., "Fire Concrete Building Elements, a New Manufacturing Branch of the Refractory Industry of the DDR (East Germany)." Silikattech. 10 (9), 448-9 (1959).

Kupffer, E., "Experience with Ceramic Skid Rails in a Pusher Furnace." Stahl u. Eisen. 82 (13) 825-36 (1962).

Kushida, T. & Silver, A.H., "Electrically Induced Nuclear Resonance in Al_2O_3 (Ruby)." Phys. Rev. 130, 1692-1702 (1963).

Kuznetsov, S.I., et al., "The Processes of Solution & Growth of Al Hydroxide Crystals in Alkaline Aluminate Solutions." Kristallizatsiya i Fazovye Perekhody, Otd. Fiz. Tverd. Tela i Poluprov., Akad. Nauk Belorussk. SSR 1962, 321-6.

Kuznetsov, V.A., "Kinetics of Hydrothermal Crystallization of Corundum." Krystallografiva 10. 663-7 (1965).

Kvapil, J., "Low-Temperature Tempering of Corundum Single Crystals." Vitezlav-Kement and Vaclav-Dolejs. Czech 113,420. 1/15/65.

Labusca, E. & Labusca, N., "Experimental Studies on the Specific Propertis of Romanian Hard Mineraloceramic Materials Intended for the Machining of Metals." Acad. rep. populare Romine, Studii cercetari metalurgie 2 (1-2) 121-36 (1957).

Lacy, E.D., "Aluminum in Glasses & in Melts." Phys. Chem. Glasses 4 (6) 234-8 (1963).

Lafarge, J. & A., "(First commercial calcium aluminate cement). Fr. 320,290, 1908.

La Forge, L.H., Jr., "Method of Making Metalized Ceramic Structures for Vacuum Tube Envelopes." U.S. 2,780,561, 2/5/57.

Lafuma, H., "Expansive Cements." Proc. Intern. Symposium Chemistry of Cement, 3rd Symposium, London, 1952, pp. 581-97 (1954).

Lake, W.O., "Refractory Cements in Industry." Sands, Clays & Minerals 3 (1) 51-2 (1936).

Lambertson, W.A., "Ceramics & Atomic Energy." Am. Ceram. Soc. Bull. 30 (1) 18 (1951).

Lambertson, W.A., "Reactions of Sodium Sulfate with Alumina-Silica Refractories." J. Am. Ceram. Soc. 35 (7) 161-5 (1952).

Lambertson, W.A., Aiken, D.B., & Girard, E.H., "Continuous Filament Ceramic Fibers." PB Rept. 171061, 84 pp.; U.S. Govt. Res. Repts. 35 (1) 69 (1961).

Lambertson, W.A., & Gunzel, F.H. Jr., "Refractory Oxide Melting Points." U.S. AEC Publ. AECD-3465, 4 p.

Lamy, A. "Corhart Standard & Cohart ZAC." Glasteck. Tid. 7 (3) 66-72 (1952).

Lang, J.I. "Specific Heat of Materials." Thermodynamic Transport Properties Gases, Liquids, Solids, Papers Symposium, Lafayette, Ind. 1959, 405-14.

Lang, S.M., "Properties of High-Temperature Ceramics & Cermets—Elasticity & Density at Room Temperature." N.B.S. Monograph No. 6, 45 pp. (1960).

Lang, S.M., Fillmore, C.L., & Maxwell, L.H., "Phase Relations & General Physical Properties of 3-Component Porcelains." J. Res. N.B.S. 48 (4) 298-312 (1952).

Langrod, K., "Compatibility of UO_2 & APM 257." AEC Res. & Dev. Rept. NAA-TDR-4270 (August 1959).

Langrod, K., "Making Porous Refractory Alumina Material." U.S. 2,463,979, 3/8/49.

Lanyi, B., "The Reaction of Alumina with Hydrogen." Magyar Kem. Folyoirat 56, 51-4 (1950).

Lapparent, J. de., "The Hydrated Alumina of Bauxites." Compt. rend. 184, 1661-2 (1927).

Lappe, H. & Ahrens, F., "Abrasive Bodies." Ger. 812,294, 8/27/51.

LaRocque, A.P., "Coated Cathode Assemblies for Electron Tubes." U.S. 2,895,854, 7/21/59.

Larsh, E.P. & Coppock, F.E., "Insulation Coating & Method of Application Thereof." U.S. 3,025,188, 3/13/62.

Lasch, H., "Investigation of the Properties of Corundum Brick with Special Reference to Their Thermal Conductivity." Tonind.-Ztg. 66, 89 (1942); abstracted in Trans. Brit. Ceram. Soc. 42 (1) 6A (1943).

Laubengayer, A.W. & Weisz, R.S., "Hydrothermal Study of Equilibria in the System Alumina-Water." J. Amer. Chem. Soc. 65 (2) 247-50 (1943).

Lauchner, J.H. & Bennett, D.G., "Fatigue & Internal Stress Analysis of Ceramic Coated Metal Composites." PB Rept. 138624, 37 pp.; U.S. Govt. Res. Repts. 33, (1) 67 (1960).

Lauchner, J.H., Bennett, D.G. & Morgan, G.L., "Nondestructive Tests for Ceramic, Cermet, & Graphite Materials." PB Rept. 161815, 61 pp.; U.S. Govt. Res. Repts. 34 (3) 309-10 (1960).

Laudise, R.A. & Ballman, A.A., "Hydrothermal Synthesis of Sapphire." J. Am. Chem. Soc. 80 (11) 2655-7 (1958).

Lavendel, H.W. & Goetzel, C.G., "A Study of Graded Cermet Components for High-Temperature Turbine Applications." WADC Tech. Rept. 57-135 ASTIA No. AD-131031 (Aug. 1957).

Lawson, L.R., Jr. & Keilen, J.J., "Pine-Wood Lignin as a Ceramic Deflocculant." Am. Ceram. Soc. 30 (4) 143-7 (1951).

Lawthers, D.D. & Sama, L., "High Temperature Oxidation-Resistant Coatings for Tantalum Base Alloys." ASD-TR 61-233 (1961).

Lay, K.W., "Grain Growth in UO_2-Al_2O_3 in the Presence of a Liquid Phase." J. Am. Ceram. Soc. 51, 373-6 (1968).

Layden, G.K., "System Al_2O_3-Nb_2O_3." J. Am. Ceram. Soc. 46 (10) 506 (1963).

Layng, E.T., "Silica Gel-Alumina Supported Catalyst." U.S. 2,487,564, 11/8/49.

Lea, F.M. & Desch, C.H., "The Chemistry of Cement & Concrete. 2nd Ed. p. 52. Edward Arnold & Co., London (1956).

Le Chatelier, H., "The Action of Heat on the Clays." Bull. Soc. Min. 10, 204 (1887); Compt. rend. 104, 1443-6, 1517-20 (1887).

Le Clercq, L.J., "Fabrication of Ceramic Radomes to Close Tolerances." ASD-OSU Symposium on Electromagnetic Windows. I. July 1962.

Lecompte, J., "The Infrared Spectrum of Water in Solids. I. General Introduction." J. chim. phys. 50, C53-64 (1953).

Lederer, E. & Lederer, M., "Chromatography." Elsevier Publ. Co., New York (1955).

Leduc, R., "Manufacture of Bauxite Refractory Brick." Ceramica (Sao Paulo) 2 (6) 56 (1956).

Lefrancois, P.A., "Improved Method of Preparing Alumina." U.S. 2,787,522, 4/2/57.

Lefrancois, P.A., "Alumina-Containing Catalyst." U.S. 2,840,529, 6/24/58.

Lehman, "Procedure for Freeing Aluminum Oxide of Impurities." Ger. Application T59,861, 12/10/43. PB 9349, 3 pp.; Abstracted in Bibliog. Sci. Ind. Repts. 2 (11) 855 (1946).

Lehmann, H. & Bohme, G., "Dependence of the Alkali & Acid Resistance of Ceramic Glazes on Their Composition." Tonind-Ztg. u. Keram. Rundschau 82 (10) 190-9 (1958).

Lehmann, H. & Mitusch, H., "Refractory Concrete from Fused Alumina Cement." (Feuerfester Beton aus Tonerde-Schmelzzement). Hermann Hubener Verlag, Goslar, Germany, 71 pp. (1959).

Lemonnier, J.C., Stephan, G. and Robin, S., "Optical Properties of Corundum and Ruby Single Crystals in the Far Ultraviolet." Compt. Rend. Ser. A,B 262B, 355-8 (1966).

Lenel, F.V., "Observation of SAP & Present Theories for Its Remarkable High-Strength." High Temp. Materials, Conf. Cleveland, Ohio, 1957, 321-31 (Pub. 1959).

Lenel, F.V., "Dispersed Phases in the Powder Metallurgy of Nuclear Materials." Plansee Proc. 4th Seminar, Reutte/Tyrol, 196, pp. 529-42 (1962).

Leonov, A.I. & Keler, E.K., "Reactions Between Ce_2O_3 & Al_2O_3 at High Temperatures & the Properties of the Formed Ce Aluminates." Izv. Akad. Nauk SSSR. Otd. Khim. Nauk. 1962, 1905-10.

Le Peintre, M., "Method of Forming Boehmite by Electrolysis." Compt. rend. 226 (17) 1370-1 (1948).

Lepp, J.M. & Slyh, J.A., "(Refractory of Alumina & Aluminum)." U.S. 2,568,157, 9/18/51.

Lepp, J.M. & Slyh, J.A., "(Refractory of Aluminum & Alumina)." U.S. 2,599,185, 6/3/52.

Lerman, L., "Brazing Ceramics to Metals." Metal Progr. 79 (3) 126-8 + 4 pp. (1961).

Lesar, A.R. & Charles, G.W., "Lightweight Insulating Firebrick & Method of Manufacture." U.S. 2,865,772, 12/23/58.

Lesar, A.R. & McGee, T.D., "Abrasion of Fire-Clay Refractories." Refractories Inst. Tech. Bull., No. 95, 6 pp. (April 1956).

Leszynski, W. ed., "Powder Metallurgy." Proceedings of An International Conference, New York, 1960. Interscience Publishers, N.Y. 847 pp. (1961).

Leum, L.N., Connor, J.F., Rothrock, J. J. & Shipley, C.S., "Purifying Alumina." U.S. 3,073,675, 1/15/63.

Levin, E.M., Robbins, C.R. & McMurdie, H.F., "Phase Diagrams for Ceramists." The American Ceramic Soc., Inc., Columbus, Ohio, 601 pp. (1964).

Levin, E.M. & Roth, R.S., "System Bi_2O_3-R_2O_3 near Bi_2O_3 Component." J. Res. N.B.S. 68A (2) 200 (1964).

Levitt, A.P., "Recent Advances in Alumina Technology." Mater. Res. Std. 6, 64-71 (1966).

Levy, A.V., "Thermal Insulation Ceramic Coatings." S.A.E. (Soc. Automotive Engrs.) J. Preprint 4T, 5 pp (1959).

Levy, A.V., "Ceramic Coatings for Insulation." Metal Progr. 75 (3) 86-9 (1959).

Levy, M., "Radiation Coloration in Sodium Aluminum Silicate Glasses." J. Soc. Glass Technol. 40 (196) 462-9T (1956).

Levy, M., "Evaluation of Flame-Sprayed Coatings for Army Weapons Applications." Am. Ceram. Soc. Bull. 42 (9) 498-500 (1963).

Levy, P.W., "Color Centers & Radiation Induced Defects in Alumina." Phys. Rev. 123, 1226-33 (1961).

Lewis, B.W., "Investigations of the Deterioration of 22 Refractory Materials in a Mach 2 Jet at a Stagnation Temperature of 3800°F." AD 258762 17 pp., Illus.; U.S. Govt. Res. Repts. 36 (5) 47 (1961).

Lewis, H.R., "Lasers, An Introduction." RCA Engr. 8 (5) 6-7 (1963).

Lewis, H.D. and Goldman, A., "Theorems for Calculation of Weight Ratios to Produce Maximum Packing Density of Powder Mixtures" J. Am. Ceram. Soc. 49, 323-27 (1966).

Lewis, D. and Lindley, M.W., "Enhanced Activity and the Characterization of Ball-Milled Alumina." J. Am. Ceram. Soc. 49, 49-50 (1966).

Libbey, F.W., "Alumina for Northwest Aluminum Plants." Mining Congr. Jour. 31 (3) 34-9 (1945).

Lichtenberger, E., "Mechanism of Anodic Oxidation of Aluminum & the Structure of the Oxide Film." Zhur. Priklad. Khim. 34, 1286-91 (1961).

Lifshits, M.A., "Use of High-Alumina Brick in Regenerators." Steklo i Keram 6 (7) 14-7 (1949).

Lima-de-Faria, J., "Dehydration of Goethite & Diaspore." Z. Krist. 119 (3/4) 176-203 (1963).

Linde Air Products Co., "Synthetic Sapphire Production Reaches Commercial Scales in U.S." Product Eng. 14, 668 (1943).

Linde Air Products Co., "Deformation of Unicrystalline Bodies of Corundum & Spinel & Articles such as Rods, Filaments, & Filament Guides made Thereby." Brit. 596,326, 1/14/48.

Lindsay, J.G., "The Effect of Fluorides on Crystal Growth in Alpha Alumina During Calcination of Gibbsite." Amer. Ceramic Society Meeting, Dallas, Tex., May 8, 1957.

Lindsay, J.G. & Gailey, H.J., "Process for Reducing the Soda Content of Alumina." U.S. 3,175,883, 3/30/65.

Lippens, B.C., "Structure & Texture of Aluminas." Doctorate Dissertation Technical School, Delft, April 26, 1961.

Lippens, B.C., & de Boer, J.H., "Pore Systems in Catalyst. III. Pore-Size Distribution Curves in Aluminum Oxide Systems." J. Catalysis 3 (1) 44-9 (1964).

Lippens, B.C. & de Boer, J.H., "Study of Phase Transformations During Calcination of Aluminum Hydroxides by Selected Area Electron Diffraction." Acta Cryst. 17 (10) 1312-21 (1964).

Litvakovskii, A.A., "Zircon-Mullite Refractories." Steklo i Keram. 7 (9) 6-9 (1950).

Livey, D.T., "Ceramics & The Development of Nuclear Power." Trans. Brit. Ceram. Soc. 56 (9) 482-98 (1957).

Livey, D.T. & Murray, P., "Surface Energies of Solid Oxides & Carbides." J. Am. Ceram. Soc. 39 (11) 363-72 (1956).

Livey, D.T. & Murray, P., "Stability of Refractory Materials." "Chapter IV in Physico-Chemical Measurements at High Temperatures". J.O.M. Bockris, J.L. White, & J.D. Mackenzie. Butterworths Scientific Publications (London) 394 pp. (1959).

Livey, D.T., Murray, P., Scott, R., & Williams, T., "Processes for the Production of Ceramic Bodies." U.S. 3,141,782, 7/21/64.

Livovich, A.F., "Portland vs. Calcium Aluminate Cements in Cyclic Heating Tests." Am. Ceram. Soc. Bull. 40 (9) 559-62 (1961).

Lockhart, R.J., "Experimental Research on Filamentized Ceramic Radome Materials." Interim Eng. Progress Report April 1, 1964 to June 30, 1964.

Lockhart, R.J. and Wainer, E., "Preparation of Alumina Monofilaments." U.S. 3,271,173. 9/6/66.

Locsei, B., "Laminar-Crystalline, Alkali-Free Corundum." Veszpremi Vegyip. Egyet. Kozlemeny. 6, 29-35 (1962).

Locsei, B., "Lamilarly Crystallizing a-Al_2O_3." Hung. 150,839, 9/63.

Locsei, B., "Thermogravimetry of Mullite Formation in the System Kaolinite-Aluminum Fluoride." Acta Chim. Acad. Sci. Hung. 40 (1), 79-97 (1964).

Loeffler, J., "Glasses with High Content of Rare-Earth Oxides." Ger. 840,297, 5/29/52.

Loewenstein, E.V., "Optical Properties of Sapphire in the Far Infrared." J. Opt. Soc. Am. 51 (1) 108-12 (1961).

Loh, E., "Ultraviolet Reflectance of Al_2O_3, SiO_2, & BeO." Solid State Commun. 2 (9), 269-72 (1964).

Lohre, W. & Urban, H., "Morphology of Mullite." Ber. deut. keram. Ges. 37 (6) 249-51 (1960).

Lomer, P.D., "Electric Strength of Alumina Films." Proc. Phys. Soc. (London) B-63, 818-9 (1950).

Lommel, J.M. & Kronberg, M.L., "X-Ray Diffraction Micrography of Aluminum Oxide Single Crystal," pp. 543-59. "Direct Observation of Imperfections in Crystals." eds. J.B. Newkirk & J.H. Wernick, Interscience Publishers, Inc., New York, 617 pp. (1962).

Long, G. & Foster, L.M., "Crystal Phases in the System Al_2O_3-AlN." J. Am. Ceram. Soc. 44 (6) 255-8 (1961).

Long, J.V. & Furth, J.V., "Protective Cermet Coating Method & Materials." U.S. 2,898,236, 8/4/59.

Longuet, P. & Tournadre, M. de., "Hydration of Calcium Aluminates in Presence of Water Vapor." Compt. rend. 256, 2830-3 (1963).

Lorenz, C., A-G., "Method of Depositing Solderable Metallic Layers on Ceramics." Ger. Application L116,059, 10/30/44. PB 12, 537, 6 pp.; abstracted in Bibliog. Sci. Ind. Repts. 2 (8) 617 (1946).

Louden, W.C. & Homonnay, E., "Method of Making a Ceramic-to-Metal Seal." U.S. 3,088,201, 5/7/63.

Luck, G.C., "Some Factors Affecting Applicability of Optical-Band Radio (Coherent Light) to Communication." RCA Review 22, 359-409 (1961).

Lucks C.F. & Deem, H.W., "Thermal Conductivities, Heat Capacities, & Linear Thermal Expansion of Five Materials." Battelle Memorial Institute, Columbus, Ohio, WADC Tech. Rept. No. 55-496, 65 p. (1956).

Ludvigsen, C.F. & Andsager, H.L., "Influence of Grain Size, Concentration, Time, & pH Value in (Nonagitated) Suspensions of Plastic & Nonplastic Materials on the Unmixing Tendency of the Particles." Trans. Intern. Ceram. Congr., 3rd Congr., Paris, 1952, pp. 243-50.

Ludwik, P., "The Determination of Hardness by the Brinell Ball Indentation & an Analogous Impression Method." Z. Ost. Ing. Architekt. Ver. 59 (11) 191-6 (1907).

Lukesh, J.S., "An X-Ray Study of Several Irradiated Glasses." Oak Ridge Nat. Lab. Rept. No. 1945 (August 30, 1955).

Lukesh, J.S., "Neutron Damage to the Structure of Vitreous Silica." Phys. Rev. 97, 345-6 (1955).

Luks, D.W., "Vitreous High-Alumina Porcelain." U.S. 2,290,107, 7/14/42.

Luks, D.W., "How They Make Al_2O_3 Ceramics." Ceramic Industry, June, July, August 1956.

Luks, D.W., "Method for Preparing Coated Bodies." U.S. 2,985,547, 5/23/61.

Luks, D.W. & Powell, J., "Coated Nonmetallic Refractory Bodies, Composition for Coating Such Bodies, & Method for Bonding Such Bodies by Means of the Composition." U.S. 2,857,664, 10/28/58.

Lungu, E. & Nagy, V., "New Superaluminous Ceramic Compositions with Superior Electrical Properties at Elevated Temperatures." Ind. Usoara (Bucharest) 10 (4) 143-8 (1963).

Lungu, S.N. & Popescu-Has, D., "Fine Ceramic Masses Obtained by the Crystallization of Glass." Ind. usoara (Bucharest) 5, 63-5 (1958).

Lyapunov, A.N., Khodokova, A.G. & Galkina, Z.G., "Solubility of Hydrargillite in NaOH Solutions Containing Soda & Common Salt at 60° & 95°." Tsvetn. Metal. 37 (3) 48-51 (1964).

Lyle, A.K., Horak, W. & Sharp, D.E., "Effect of Alumina on the Durability of Sand-Soda-Lime Glasses." J. Am. Ceram. Soc. 19, 142-7 (1936).

Lyle, J.P., Jr., "Excellent Products of Aluminum Powder Metallurgy." Metal Progress (12) 109 (1952).

Lyle, J.P., Jr., "Aluminum-Aluminum Oxide Wrought Products." ASTM Meeting, New York (June 28, 1962).

Lynam, T.R., et al., "Low-Alumina Silica Brick—Symposium: II, Manufacture." Trans. Brit. Ceram. Soc. 51 (2) 113-35 (1952).

Lynch, C.T., Vahldiek, F.W. & Robinson, L.B., "Growth & Analysis of Alumina Whiskers." AD-278,806, 34 pp., illus.; U.S. Govt. Res. Repts. 37 (2) 26-7 (1962).

Lynch, J.F., Quirk, J.F., & Duckworth, W.H., "Investigation of Ceramic Materials in a Laboratory Rocket Motor." Am. Ceram. Soc. Bull. 37 (10) 443-5 (1958).

Lyon, R.N. (Ed. in Chief)., "Liquid Metals Handbook." 2nd Ed. Supt. Doc. U.S. Government Printing Office (June 25, 1952).

McAleer, H.J., "Polishing, Buffing, & Burring Compounds in the War Program." Metal Finishing 41 (3) 144-5 (March, 1943); abstracted in Bull. Brit. Non-Ferrous Metals Research Assn. No. 168, p. 190 (June, 1943).

McAuliffe, J.F., "Ceramics Cut High-Speed Steel & Show Their Weaknesses." Am. Machinist, 101, (Sept. 9) 130-3 (1957); abstracted in Wear 1 (4) 356 (1958).

McCallum, N. & Barrett, L.R., "Some Aspects of the Corrosion of Refractories." Trans. Brit. Ceram. Soc. 51 (11) 523-48 (1952).

McCandless, E.L. & Yenni, D.M., "Stretching Corundum Crystals." U.S. 2,485,978, 10/25/49.

McCandless, E.L. & Yenni, D.M., "Bending Single Crystals of Corundum & Spinel." U.S. 2,485,979, 10/25/49.

MacChesney, J.B. & Johnson, G.E., "Room-Temperature Dielectric Properties of Fast-Neutron-Irradiated Fused Silica & a-Alumina." J. Appl. Phys. 35 (9) 2784-5 (1964).

McClelland, J.D., "Kinetics of Hot Pressing." Rept. No. NAA-SR-5591 Atomics International, Div. North American Aviation, Inc. Jan. 1, 1961.

McClelland, J.D., "A Plastic Flow Model of Hot Pressing." J. Am. Ceram. Soc. 44 (10) 526 (1961).

McClelland, J.D. & Zehms, E.H., "End-Point Density of Hot-Pressed Alumina." J. Am. Ceram. Soc. 46 (2) 77-80 (1963).

McCullough, J.A. & Williams, P.D., "Ceramic Type Electron Tube." U.S. 2,910,607, 10/27/59.

Machlan, G.R., "Glass Composition." U.S. 2,876,120, 3/3/59.

McClure, L., "Whiskers, A Selected Bibliography." Martin Co. Tech. Info. Center. RB No. 104 (Mar. 1965), AD-458-434

McCreight, D.O. & Birch, R.E., "Synthetic Spinel Refractory Products." U.S. 2,805,167, 9/3/57.

McCreight, L.R., "Processing Studies of Pure Oxide Bodies." J. Am. Ceram. Soc. 37 (8) 378-85 (1954).

McCreight, L.R. & Sowman, H.G., " Ceramics for Nuclear Power Applications." Am. Ceram. Soc. Bull. 35 (5) 176-9 (1956).

McCreight, L.R., Rauch, H.W., Sr. & Sutton, W.H., "A Survey of the State of the Art of Ceramic & Graphite Fibers." Tech. Rept. AFML-TR-65-105 (May 1965).

McCune, S.E., Greaney, T.P., Allen, W.C. & Snow, R.B., "Reaction Between K_2O & Al_2O_3-SiO_2 Refractories as Related to Blast-Furnace Linings." J. Am. Ceram. Soc. 40 (6) 187-95 (1957).

McDonald, H.A. & Dore, J.E., "Refractory Brick & Preparation Thereof." U.S. 2,997,402, 8/22/61.

McDonald, H.A., Dore, J.E. & Peterson, W.S., "How Molten Aluminum Affects Plastic Refractories." J. Metals, 10 (1) 35-7 (1958).

MacDonald, K.A. & Whitmore, E.J., "Method of Making Ceramic-to-Metal Seals." U.S. 2,836,885, 6/3/58.

McDougal, T.G., Fessler, A., & Schwartzwalder, ·K., "Spark-Plug Insulator." U.S. 2,120,338, 6/14/38.

McDougal, T.G., et al., "Spark-Plug Insulator." U.S. 2,152,655, 4/4/39.

McDougal, T.G., et al., "Insulator for Spark Plugs." U.S. 2,152,656, 4/4/39.

McDougal, T.G., et al., "Ceramic Body for Spark-Plug Insulators." U.S. 2,177,943, 10/31/39.

McDougal, T.G., Fessler, A.H., & Barlett, H.B., "Ceramic Body for Spark-Plug Insulators." U.S. 2,214,931, 9/17/40.

McDougal, T.G., et al., "Ceramic Body for Spark-Plug Insulators." U.S. 2,232,860, 2/25/41.

McDowall, I.C. & Vose, W., "Determination of the Pyroplastic Deformation in the Firing of Ceramic Bodies." Trans. Brit. Ceram. Soc. 50 (11) 506-16 (1951).

McDowell, J.S., "Refractories in Nonferrous Metallurgical Furnaces." Mining Congr. Jour. 25 (5) 17-20 (1939).

McDowell, J.S., Scott, R.K. & Clark, C.B., "Mineral Composition of Refractory Materials." A chapter in "Modern Refractory Practice." Published by Harbison-Walker Refractories Co., 1961; revised & updated in Refractories J., 38 (12) 438 + 15 pp. (1962).

McGee, T. & Dodd, C.M., "Mechanism of Secondary Expansion of High-Alumina Refractories Containing Calcined Bauxite." J. Am. Ceram. Soc. 44 (6) 277-83 (1961).

McGill, L.A., & McDowell, J.S., "Reaction Temperatures Between Refractories." Am. Ceram. Soc. Bull. 30 (12) 425-31 (1951).

McGill, L.A. & McDowell, J.S., "Behavior of Refractories in Mixed Linings." Radex-Rundschau 1952, No. 4, pp. 187-8.

McGrue, W.M., "Refractory Concretes for Coke Ovens & Furnaces." Blast Furnace & Steel Plant 25 (6) 624-7 (1937).

Machu, Willi., "Nonmetallic Inorganic Coatings." (Nichtmetallische Anorganische Ueberzuege). Springer Verlag, Vienna, 1952, 404 pp., 153 illus.

McHugh, C.O., Whalen, T.J., & Humenik, M. Jr., "Dispersion-Strengthened Aluminum Oxide." J. Am. Ceram. Soc. 49, 486-9 (1966).

McIntosh, A.B., Rait, J.R. & Hay, R., "The Binary System FeO-Al_2O_3." J. Roy. Tech. Coll. (Glasgow) 4 (Part 1) 76-2 (1937).

MacIver, D.S., et al., "Catalytic Aluminas. I. Surface Chemistry of n- & γ-Alumina." J. Catalysis 2 (6), 485-97 (1963).

McIntire, H.O., "Refractories & Mold Materials for Vacuum Melting & Casting." Foundry Trade J. 103 (Nov 7) 543-8 (1957); abstracted in J. Iron Steel Inst. (London), 189 (2) 174 (1958).

McKee, R.L., "Standard Marking System: Grain Types." Grinding Finishing 7 (2) 30-2 (1961).

McKee, R.L., "Bonded or Coated Abrasives - A Comparative Tool Analysis." Grinding Finishing 9 (10) 28-30 (1963).

McKee, W.D. and Aleshin, E., "Aluminum Oxide-Titanium Oxide Solid Solutions" J. Am. Ceram. Soc. 46, 54-58 (1963).

Mackenzie, J., "Low-Alumina Silica Brick - Symposium: III, Properties & Performance." Trans. Brit. Ceram. Soc. 51 (2) 136-71 (1952).

Mackenzie, J.K. & Shuttleworth, R., "Phenomenological Theory of Sintering." Proc. Phys. Soc. (London) 62, 833-52 (1949).

Mackey, R.J., "Metallizing Ceramic Surfaces." U.S. 3,093,490, 1/11/63.

McLaren, J.R. & Richardson, H.M., "Action of Vanadium Pentoxide on Alumino-silicate Refractories." Trans. Brit. Ceram. Soc. 58 (3) 188-97 (1959).

McLean, J.W. and Hughes, T.H. (Nat. Res. Develop. Corp.), "Strong, Tough, Sintered Ceramic Artificial Teeth." Brit. 1,051,735. 12/21/66.

McMullen, J.C., "Refractory Glass Wool." U.S. 2,557,834, 6/19/51.

McMullen, J.C., "Mineral Fiber Compositions. " U.S. 2,710,261, 6/7/55.

McMullen, J.C., "Method & Apparatus for Fiberizing Refractory Materials." U.S. 2,714,622, 8/2/55.

McMullen, J.C., "Refractory Fibrous Material." U.S. 2,873,197, 2/10/59.

McNally, R.N., Yeh, H.C., & Balasubramanian, N., "Surface Tension Measurements of Refractory Liquids Using the Modified Drop Weight Method." J. Mater. Sci. 3, 136-8 (1968).

McNamara, E.P. & Comeforo, J.E., "Classification of Natural Organic Binders." J. Amer. Ceram. Soc. 28 (1) 25-31 (1945).

McQuarrie, M., "Thermal Conductivity: VII, Analysis of Variation of Conductivity with Temperature for Al_2O_3, BeO, & MgO." J. Am. Ceram. Soc. 37 (2, part II) 91-5 (1954).

McQuarrie, M., "Thermal Conductivity, V, High-Temperature Method & Results for Alumina, Magnesia, & Beryllia from $1000°$ to $1800°C$." J. Am. Ceram. Soc. 37 (2, part II) 84-8 (1954).

McRitchie, F.H., "Ceramic Materials for Application in the Design of Jet Propelled Devices." Ohio State Univ. Research Foundation Rept. 14 (July 1946), 3 pp. PB 60,669, abstracted in Bibliog. Sci. Ind. Repts. 4 (11) 965 (1947).

McVickers, R.C., Ford, S.D. & Dugdale, R.A., "Polishing & Etching Techniques for Dense Alumina." J. Am. Ceram. Soc. 45 (4) 199 (1962).

Maerker, R., "Cryolite & Its Use as a Raw Material for Enamels." Glas-Email-Keramo-Tech. 8 (4) 117-21; (5) 178-82 (1957).

Mah, A.D., "Heats of Formation of Alumina, Molybdenum Trioxide, & Molybdenum Dioxide." J. Phys. Chem. 61, 1572 (1961).

Maiman, T.H., "Stimulated Optical Radiation in Ruby." Nature 187, 493 (1960).

Maiman, T.H., "Optically Pumped Solid State Lasers." Solid State Design 4(11)17-21 (1963).

Majumder, B.L., "Young's Modulus of Elasticity of Ceramic Materials by Flexure." Trans. Indian Ceram. Soc. 11 (3) 168-81 (1952).

Makara, A.M., Slutskaya, T.M. & Mosendz, N.A., "Welding of High-Strength Steels with Fused Fluxes." Avtomat. Svarka 12, No. 10, 3-8 (1959).

Makarychev, A.R., "Service of Forsterite, Dinas, & High-Alumina Fire-Clay Brick in Open-Hearth Checkers." Ogneupory 22 (6) 268-74 (1957).

Makarychev, A.R., "Service of High-Alumina Brick in the Checkers of Open-Hearth Furnaces." Ogneupory 24 (1) 33-8 (1959).

Makhov, G., Kikuchi, C., Lambe, J., & Terhune, R.W., "Maser Action in Ruby." Phys. Rev. 109, 1399-1400 (1958).

Malitson, I.H., "Refraction & Dispersion of Synthetic Sapphire." J. Opt. Soc. Am. 52 (12) 1377-9 (1962).

Mal'tsev, A.A. & Shevel'kov, V.G., "Infrared Absorption Spectra of the Al_2O, Ga_2O, In_2O, & Al_2S Molecules." Teplofiz. Vysokikh Temperatur. Akad. Nauk SSSR 2 (4) 650-3 (1964).

Mamykin, P.S., "Highly Refractory Materials for Modern Metallurgy." Khim. Nauka i Prom 3 (1) 27-34 (1958).

Manasevit, H.M. & Simpson, W.I., "Single-Crystal Silicon on a Sapphire Substrate." J. Appl. Phys. 35 (4) 1349-51 (1964).

Mandarino, J.A., "Some Optical & Stress-Optical Properties of Synthetic Ruby." Univ. Michigan ASTIA AD 146029 (1957).

Mangsen, G.E., et al., "Hot Pressing of Aluminum Oxide." J. Am. Ceram. Soc. 43 (2) 55-9 (1960).

Mann, W., "Crushing Resistance of Synthetic Abrasives - Experiment in Relative Measurement." Ber. deut. keram. Ges. 40 (3) 179-85 (1963).

Manson, S.S., "Behavior of Materials Under Conditions of Thermal Stress." Natl. Advisory Comm. Aeronaut. Tech. Note, No. 2933, 105 pp. (1953).

Manson, S.S., "Thermal Stresses & Thermal Shock." pp. 393-418 in "Mechanical Behavior of Materials." Dorn, J.E. Editor, McGraw-Hill Book Co., New York (1961).

Manson, S.S. & Smith, R.W., "Theory of Thermal Shock Resistance of Brittle Materials Based on Weibull's Statistical Theory of Strength." J. Am. Ceram. Soc. 38 (1) 18-27 (1955).

Manuilov, K.J. & Zimbal, F.I., "Production of Highly Dispersed Alumina Suitable for Polishing." Legkie Metally 6 (10) 8-15 (1937); abstracted in Chem. Zentr. 1938, I, 4705.

Marbaker, E.E., "Coatings for Wire-Wound Resistors." J. Amer. Ceram. Soc. 28 (12) 329-42 (1945).

Margrave, J.L., "Chemistry at High Temperatures." Science 135 (3501) 345-50 (1962).

Margulis, O.M. & Kamenetskii, A.B.,, "The Application of Aluminum Phosphate for Bonding Refractory Corundum Materials & Coatings." Ogneupory 29 (7) 329-32 (1964).

Marion, S.P. & Thomas, A.W., "Effect of Diverse Anions on the pH of Maximum Precipitation of Aluminum Hydroxide." J. Colloid Sci. 1 (3) 221-34 (1946).

Markovskii, L.Ya., et al., "Micromethod for Determining the Melting Points of Highly Refractory Materials." Ogneupory 22 (1) 42-6 (1957).

Marsden, P.B. & Berstein, G., "Lubricating Oil Thickened to a Grease with a Mixture of Silica & Pyrogenic Alumina." U.S. 2,965,568, 12/20/60.

Marshall, D. & Bennett, D.G., "Effects of Precompression on the Thermal Shock Resistance of Pure Oxide Ceramics." PB Rept. 135 072, 21 pp.

Marshall, D.W., "Alumina Abrasives for Surface Grinding." U.S. 3,141,747, 7/21/64.

Masin, J.S. & Allen, T.C., "Coated Abrasive Grains & Method of Coating." U.S. 2,541,658, 2/13/51.

Massazza, F., "Preparation of Calcium & Barium Aluminates Based on Hydraulic Cements. I. Solid State System $CaO-BaO-Al_2O_2$." Ann. Chim. (Rome) 53 (7) 1002-17 (1963).

Masuch, B., "Aluminum Oxide Insulating Coatings for Vacuum-Tube Cathode Heaters." Ger. 1,017,292, 10/10/57.

Matiasovksy, K., Malinovsky, M. & Ordzovensky, S., "Electrical Conductivity of the Melts in the System $Na_3AlF_6-Al_2O_3-NaCl$." J. Electrochem. Soc. 111 (8), 973-6 (1964).

Matignon, C., "Action of High Temperatures on Some Refractory Substances." Compt. rend. 177, 1290 (1923).

Matsumoto, H., Yamauchi, S. & Nishiyama, G., "Alumina Base Cermets: I, Adherence Between Alumina & Chromium." Nagoya Kogyo Gijutsu Shikensho Hokoku 6 (2) 110-4 (1957). II. Effects of Additives on the Adhesion Between Alumina & Iron. Ibid 7 (7) 543-52 (1958).

Matsumura, T., "The Electrical Properties of Alumina at High Temperatures." Can. J. Phys. 44, 1685-98 (1966).

Matthijsen, M.J.C., "Sintered Ceramic Tool Materials: I, II." Metaalbewerking 1958, No. 22, pp. 452-7; No. 23, pp. 467-70; abstracted in J. Iron Steel Inst. (London) 192 (2) 184 (1959).

Mattox, D.M., "Metallizing Ceramics, Using a Gas Discharge." Am. Ceram. Soc. Bull. 43 (8) 586 (1964).

Mattuck, R.D. & Shandberg, M.W.P., "Optical Method for Determining the c-Axis of Ruby Boules." Rev. Sci. Instr. 30 (3) 195-6 (1959).

Matveef, M.A. and Kharatonov, F.Ya., "Mechanical Strength of Pure Oxide Ceramics." Izv. Akad. Nauk SSSR, Neorgan. Materialy 2, 395-402 (1966).

Mauer, F.A. & Bolz, L.H., "Measurement of Thermal Expansion of Cermet Components by High-Temperature X-Ray Diffraction." N.B.S. Rept. No. 5837, Suppl. 1 to WADC Tech. Rept., No. 55-473 (1957).

Mauzin, A., "Superfinish—Experimental & Scientific Study." Rev. gen. mecanique 33 (365) 5-10 (1949).

Maxwell, W.A., Gurnick, R.S. & Francisco, A.C., "Preliminary Investigation of the "Freeze Casting" Method for Forming

Refractory Powders." Natl. Advisory Comm. Aeronaut. Research Memo. E53L21, 19 pp. (1954).

May, J.E., "Growth of Alpha Aluminum Oxide Platelets from Vapor." J. Am. Ceram. Soc. 42, (8) 391-3 (1959,a.).

May, J.E., "Kinetics of High-Temperature Processes." John Wiley & Sons, New York (1959,b) pp. 30 to 37.

Mayer, D.W., "Cold Cathodes for Vacuum Tubes." U.S. 3,041,210, 6/26/62.

Mayer, W.G. & Hiedemann, E.A., "Optical Methods for Ultrasonic Determination of Elastic Constants of Sapphire." J. Acoust. Soc. Am. 30 (8) 756-60 (1958).

Mayfield, M.J., "Cerment Bodies." U.S. 2,961,325, 11/22/60.

Mazer, E.T. & Straughan, V.E., "Method for Producing Inorganic Fibers." U.S. 3,082,054, 3/63.

Maziliauskas, S., "Frits for Ceramic Bonds." Ceram. Age 74 (4) 40-4 (1958).

Maziliauskas, S., "Development in Ceramic-Metal Compositions." Ceram. Age, 75 (3) 40-1 (1960).

Meerson, G.A., "Some Problems of Powder Compacting." Soviet Powder Met. (English Transl.) 1962 (5) 315-24.

Megaw, H.D., "The Crystal Structure of Hydrargillite, Al(OH)$_3$." Z. Krist. A87, 185-204 (1934).

Megaw, H.D., Proc. Roy. Soc. (London) A142, 198-214 (1933).

Meggers, W.F., "Strongest Lines of Singly Ionized Atoms." J. Optical Soc. Amer. 31 605-11 (1941).

Mehan, R.L. and Feingold, E., "Room and Elevated-Temperature Strength of Alpha Alumina Whiskers and Their Structural Characteristics." J. Mater. 2 239-70 (1967).

Mehmel, M., "Importance of Lithium in Ceramic Bodies & Glazes." Sprechsaal 90 (4) 90-1; (5) 111-5 (1957).

Mehta, P.K., "Retrogression in the Hydraulic Strength of Calcium Aluminate Cement Structures." Mineral Process. 5 (11) 16-9 (1964).

Meiklejohn, A. & Posner, E., 'Effect of the Use of Calcined Alumina in China Biscuit Placing on the Health of the Workman." Brit. J. Ind. Med. 14 (Oct.) 229-31 (1957).

Meister, G., "Impregnated Crucible." U.S. 2,766,032, 10/9/56.

Mellor, J.W.M., "Comprehensive Treatise on Inorganic & Theoretical Chemistry." Vol. V. Longmans, Green & Co. (London) New York (1924).

Menon, P.G., "Adsorption of Carbon Monoxide on Alumina at High Pressures." J. Am. Chem. Soc. 87, 3057-60 (1965).

Meredith, N.F. & Schwartzwalder, K., "Spark Plug & Process for Making It." U.S. 2,969,582, 1/31/61.

Merry, J.D. & Vondracek, C.H., "Three Uses for Flame-Sprayed Alumina." Mater. Design Eng. 57 (2) 73 (1963).

Mesick, H.F., Jr., "Method of Making a Ceramic-to-Metal Bond." U.S. 2,776,472, 1/8/57.

Messenger, J.U., "Alumina Gel Pellets." U.S. 2,471,000, 5/24/49.

Metzger, A.J., "Lightweight Ceramics Ideal for Radomes." Ceram. Ind. 68 (6) 122-3, 135 (1957).

Metzger, A.J., "Annotated Bibliography of Articles on Lightweight Ceramics." Bull. Va. Polytech. Inst. 49, (8); Eng. Expt. Sta. Ser. No. 110, 42 pp. (1956).

Metzger, A.J., "Patents on Lightweight Ceramics." Bull. Va. Polytech. Inst. 49 (9); Eng. Expt. Sta. Ser. No. 111, 69 pp. (1956).

Meyer, E., "Investigation of Hardness Testing & Hardness." Z. Ver. Deut. Ing. 52, 17 645-54, 740-8, 835-44 (1908).

Meyer, H., "Flame Spraying of Alumina." Werkstoffe u. Korrosion 11 (10) 601-6 (1960).

Meyer, H., "Behavior of Powders in a Plasma Stream." Ber. deut. keram. Ges. 39 (2) 115-22 (1962).

Meyer, H., "The Application of Chromium Oxide Containing Aluminum in Flame Spraying." Ber. deut. keram. Ges. 40, 385 (1963).

Meyer, H., "Fusion of Powders in a Plasma Jet." Ber. deut. keram. Ges. 41 (2) 112-8 (1964).

Meyer-Hartwig, E., "Process of Manufacturing Ceramic Compounds & Metallic Ceramic Compounds." U.S. 2,982,014, 5/2/61.

Meyerson, M.R., et al., "Gauge Blocks of Superior Stability—Initial Developments in Materials & Measurement." J. Res. Natl. Bur. Stds. 64C (3) 175-207 (1960).

Michaelson, H.B., "High-Temperature Materials for Vacuum Service." Materials & Methods 38 (6) 110-5 (1953).

Michaud, M., "Emission Factors at High Temperature of Refractory Products Containing Silica & Alumina." Compt. rend. 228, 1115-6 (1949).

Michel, M., "A Calorimetric Study of the Thermal Decomposition of Alumina Trihydrate." Compt. rend. 244, 73-4, 575-6 (1957).

Michel, M. & Papee, D., "Hydrargillite." Fr. 1,367,553, 7/24/64.

Microstructure of Ceramic Materials. Proceedings of an Am. Ceram. Soc. Symp., Pittsburgh, April 1963. Natl. Bur. Std. Misc. Publ. 1964, No. 257.

Mii, H., et al., "Machinability Study of Ceramic Tools: I, Ultra-High-Speed Lathe." Nagoya Kogyo Gijutsu Shikensho Hokoku 7 (9) 638-42 (1958).

Mikhailov, K.V. & Litvinov, R.G., "Effect of the Chemical Medium of Concrete on Strength Properties of Glass Fibers." Zhelezobetonnye Konstruktsii, Teor. i Ekserim. Issled., Akad. Stroit. i Arkhitekt. SSSR, Ural'sk. Filial, Sb. Tr. 1963, 242-54.

Mikiashvili, Sh. M., Samarin, A.A. & Taylev, L.M., "Viscosity of Molten Slags of the System Manganese Oxide-Silica-Alumina." Izvest. Akad. Nauk SSSR., Otdel. Tekh. Nauk 1957, No. 1, pp. 115-22; abstracted in J. Iron Steel Inst. (London) 188 (3) 285-6 (1958).

Milewski, J.V., "Whisker Reinforced Plastics." Machine Design 37 (11) 216-25 (May 13, 1965).

Miller, A.W. & Roberts, C.W., "Drying of Gases with Activated Alumina." Ind. Chemist, 34 (397) 141-5 (1958).

Miller, E.D., Jr., "Alumina Refractories." U.S. 3,121,640, 2/18/64.

Miller, E.D., Jr., "Alumina Refractories Improved by Volatilized Silica." U.S. 3,226,241, 12/28/65.

Miller, E.D., Jr., "Cast High-Alumina Shapes." U.S. 3,220,862, 11/30/65.

Miller, F.A. & Wilkins, C.H., "Infrared Spectra & Characteristic Frequencies of Inorganic Ions—Their Use in Qualitative Analysis." Anal. Chem. 24 (8) 1253-94 (1952).

Miller, H.R., & Wheeler, W.H., "Ceramic Materials for Airframes." Ceram. Age 71 (4) 36-9 + 2 pp. (1958).

Miller, J.G. & Weir, J.V., "Alumina Desiccant & Process for Preparation Thereof." U.S. 2,841,564, 7/1/58.

Miller, M.A., Foster, L.M., & Baker, C.D., "Aluminum." U.S. 2,829,961, 4/8/58.

Miller, W.A., "Aluminum Oxide Semiconductors." U.S. 2,729,880, 1/10/56.

Miller, Wm. A. and Shott, Wm. L., "Corrosion of Refractories by Blast Furnace Slags." Am. Ceram. Soc. Bull. 47, 648-53 (1968).

Milligan, L.H., "The Mechanism of the Dehydration of Crystalline Aluminum Hydroxide & of the Adsorption of Water by the Resulting Alumina." J. Phys. Chem. 26, 247 (1922).

Milligan, L.H., "Note on Modulus of Rupture of Cylindrical Ceramic Rods When Tested on a Short Span." J. Am. Ceram. Soc. 36 (5) 159-60 (1953).

Milligan, W.O., "Recent X-Ray Diffraction Studies on the Hydrous Oxides & Hydroxides." J. Phys. & Colloid Chem. 55, 497-507 (1951).

Milligan, W.O. & McAtee, J.L., "Crystal Structure of γ-AlOOH & γ-ScOOH." J. Phys. Chem. 60, 273-7 (1956).

Milliken, T.H., Jr., Mills, G.A. & Oblad, A.G., "Chemical Characteristics & Structure of Cracking Catalysts." Dis. Faraday Soc. No. 8, 279-90 (1950).

Mills, G.A., Weller, S., Hindin, S.G., & Milliken, T., "Relation Between Dehydration & Catalytic Properties of Alumina." Z. Elektrochem. 60 (8) 823-7 (1956); abstracted in J. Appl. Chem. (London) 7 (10) ii-328 (1957).

Minamiie, Y. & Tashiro, M., "Alumina Containing No Chromium." Japan 1210, 2/10/64.

Minowa, S., Hayashi, H. & Kato, M., "Reactions Between Molten Steel & Refractories: I, Products of the Reaction Between Molten Iron & Alumina." Nogoya Kogyo Gijutsu Shikensho Hokoku 8 (5) 28-32 (1959).

Minowa, S., Kato, M. & Noguchi, C., "Reaction Between Molten Steel & Refractories: II, Corrosion of Al_2O_3-SiO_2 Refractories by Molten Steel." Nagoya Kogyo Gijutsu Shikensho Hokoku 8 (6) 18-23 (1959).

Minowa, S., Kato, M., & Mizuta, M., "Reaction Between Molten Steel & Refractories: V, Starting Temperature of Reaction Between Graphite & SiO_2 or Al_2O_3." Nagoya Kogyo Gijutsu Shikensho Hokoku 10 (3) 152-5 (1961).

Minowa, S., Kato, M., & Mizuta, M., "Reaction Between Molten Steel & Refractories: VII, Rate of Reaction Between Molten High Carbon Iron & Silica or Alumina Crucible." Nagoya Kogyo Gijutsu, Shikensho Hokoku 10 (5) 308-13 (1961).

Mischke, R.A. & Smith, J.M., "Thermal Conductivity of Alumina Catalyst Pellets." Ind. Eng. Chem. Fundamentals 1 (4) 288-92 (1962).

Misra, S.K. & Chaklader, A.C.D., "System Copper Oxide-Alumina." J. Am. Ceram. Soc. 46 (10), 509 (1963).

Misra, M.L. & Puri, I.J., "Insulating Brick from Bauxite Using Aluminum Powder." Indian Ceramics 7 (10) 233-5 (1961); reprinted in Refractories J., 37 (9) 284, 288 (1961).

Misra, M.L. & Puri, I.J., "High Alumina Insulating Brick from Bauxite." Indian Ceramics 7 (10) 235-8 (1961). Reprinted in Refractories J., 37 (9) 278-9 (1961).

Mistler, R.E. and Coble, R.L., "Comments on the Use of Log-Log Plots in Analyzing Grain Growth Data." J. Am. Ceram. Soc. 51, 472 (1968).

Mitchell, J.B., Spriggs, R.M., & Vasilos, T., "Microstructure Studies of Polycrystalline Refractory Oxides." Tech. Rept. No. RAD-TR-63-32 (Aug. 7, 1963) USDC, OTS No. AD 413,944.

Mitchell, L., "Ceramics." Ind. Eng. Chem. 46 (10) 2056-64 (1954).

Mitscherlich, A., "Investigation of Alunite, of Lowigite, & The Alumina Hydrates." J. prak. Chem. 83, 455 (1861).

Miyabe, S. & Nishigaya, I., "Amorphous Substances Present in Grains of A* Abrasives: I." J. Japan. Ceram. Assoc. 51 (609) 556-61 (1943).

Miyazawa, H., & Okada, J., "The Electric Breakdown of Alumina at High Temperature." J. Phys. Soc. Japan 6, 55-9 (1951).

Mizutani, Y., Sakaguchi, K. & Iizuka, K., "Evaluation of Adsorbents for Use in Petroleum Refining." Kogyo Kagaku Zasshi 60, 24-6 (1957).

Modi, H.J. & Fuerstenau, D.W., "Streaming Potential Studies of Corundum in Aqueous Solutions of Inorganic Electrolytes." J. Phys. Chem. 61, 640-3 (1957).

Moiseev, P.S., "Fibrous Alumina." Kolloid Zhur. 9 (1) 53-6 (1947).

Molchanov, V.S. & Prikhid'ko, N.E., "Corrosion of Silicate Glasses by Alkaline Solutions: III, Inhibitors of the Alkaline Corrosion of Glasses." Izvest. Akad. Nauk SSSR, Otdel. Khim. Nauk. 1958, No. 7, pp. 801-8.

Mong, L.E. & Donoghue, J.J., "Preparation of Zirconia & Alumina Shapes." Am. Ceram. Soc. Bull. 29 (11) 405-7 (1950).

Montgomery, E.T., et al., "Method of Coating a Metal Surface." U.S. 2,775,531, 12/25/56.

Montoro, V., "Crystalline Structure of Bayerite." Ricerca Sci. 13, 565-71 (1942).

Moorbath, S. Atomic Energy Research Establishment Report C/M 127 (August 1952).

Moore, B., "Recrystallized Alumina Products: I-III." Ind. Chemist 18 (210) 236-40; (211) 306-9; (212) 337-8 (1942).

Moore, B., "Impact Strength of Transparent & Translucent Fused Silica & Recrystallized Alumina." J. Soc. Chem. Ind. (London) 59 (6) 119-20 (1940).

Moore, C.E., "Atomic Energy Levels." Vol. I, Circular 469 U.S.D. Commerce, N.B.S. (June 15, 1949).

Moore, C.E. & Russell, H.N., "Binding Energies for Electrons of Different Types." J. Res. Natl. Bur. Stds. 48 (1) 61-7 (1952); RP 2285.

Moore, H., "Fusion-Cast Refractories: I." Ceramics 2 (20) 405-10 (1950).

Moore, N.C., "Process for Producing Metal Ceramic Compositions." U.S. 2,785,974, 3/19/57.

Moore, N.C. & Huffadine, J.B., "Friction Members Having Sintered Compositions." U.S. 2,893,112, 7/7/59.

Morgan Crucible Co., Ltd., "Insulating Firebrick & Process of Manufacture." Brit. 661,607, 9/26/51.

Morgan Crucible Co., Ltd., "Refractory Bodies & Articles." Brit. 688,992, 1/21/53.

Mori, J., "X-Ray Investigation of the Hydration Products of Quenched Compositions of the System $CaO-Al_2O_3-SiO_2$." Semento Gijutsu Nenpo 9, 54-7 (1955).

Morimoto, T., Shioma, K., & Tanaka, H., "Heat of Immersion in Water." Bull. Chem. Soc. Japan 37, (3) 392-5 (1964).

Moriya, T., et al., "Studies on Divitrification of Phosphate Glasses: I, Devitrification Process of $Al_2O_3-P_2O_5$ Glass." Yogyo Kyokai Shi 68 (774) 145-53 (1960).

Morley, J.G., "Strong Fibers & Fiber-Reinforced Metals." Proc. Roy. Soc. 282A, 43-52 (1964).

Morrison, A.R. et al., "High Modulus, High Temperature Laminates with Fibers & Flakes." Owens-Corning Fiberglas Co. AFML, WPAFB, AF 33 (616), 5802 (Feb. 1962).

Morrison, W.S., "New Methods for Purifying Water for the Ceramic Industry." Bull. Am. Ceram. Soc. 20 (7) 246-7, 250-1 (1941).

Morveau, L.B.G. de., "Alun." Eyclopedie Methodique, Dictionaire de Physique, Paris (1786).

Moscou, L. & Vlies, G.S. van der., "Electron Microscope Investigations on the Morphology of Aluminum Hydroxides." Kolloid-Z, 163 (1) 35-41 (1959).

Moses, J.H. & Michel, W.D., "Bauxite Deposits of British Guiana & Suriname in Relation to Underlying Unconsolidated Sediments Suggesting Two-Step Origin." Econ. Geol. 58 (2) 250-62 (1963).

Moteki, A., "Studies on the Injection Molding of Ceramics: III, Properties of High-Alumina Porcelain Produced by Injection Molding." Yogyo Kyokai Shi 68 (770) 23-32 (1960).

Mott, W.R. Electrochem. Ind. 2, 129 (1904).

Moulson, A.J. & Popper, P., "Assessment of Suitability of Ceramics for Valves." Spec. Ceram. Proc. Symp. Brit. Ceram. Res. Assoc. 1962, 355-77 (Pub. 1963).

Mountvala, A.J. & Murray, G.T., "Effects of Gaseous Environment on the Fracture Behavior of Alumina." J. Am. Ceram. Soc. 47 (5) 237-9 (1964).

Muan, A., "Stability of the Phase $Fe_2O_3 \cdot Al_2O_3$." Am. J. Sci., 256 (6) 413-22 (1958).

Muan, A., "Reactions Between Iron Oxides & Alumina-Silica Refractories." J. Am. Ceram. Soc. 41 (8) 275-86 (1958).

Muan, A. & Gee, C.L., "Phase Equilibrium Studies in the System Iron Oxide-Al_2O_3 in Air & at 1 atm. O_2 Pressure." J. Am. Ceram. Soc. 39 (6) 207-14 (1956).

Mueller, H.R., "Firing of Normal Corundum Disks." Ber. Deut. Keram. Ges. 42, 1-5 (1965).

Mullite Refractories Co., "Shamva Mullite." Shelton, Conn. 50 pp. free. Ceram. Abstracts, April 1940.

Murkes, J., "Absolute Hardness Scale." Tek, Tidskr. 82 (2) 37-40 (1952); Engrs. Digest 13 (4) 112-4 (1952).

Murray, L.A., Lamorte, M.F. & Vogel, F.L., "Development of Effective Lasers." RCA Engr. 8 (5) 12-15 (1963).

Murray, P., "Ceramics & Atomic Energy." Claycraft 26 (1) 2-7 (1952); (5) 210-6 (1953).

Murray, P., "Metal Ceramic Mixtures." Powder Met. 1960, No. 5, pp. 64-80.

Murray, P., Liven, D.T., & Williams, J., "Hot Pressing of Ceramics." pp. 147-71, "Ceramic Fabrication Processes," ed. W.D. Kingery. Technology Press of MIT, and John Wiley & Sons, Inc., New York (1958).

Murray, R.F. & Rhodes, D.W., "Waste Disposal & Processing." AEC Res. & Development Report TID-4500, Ed. 17, Sept. 28, 1962. IDO-14581.

Murray, R.F. & D.W. Rhodes., "Low Temperature Polymorphic Transformations of Calcined Alumina." U.S. At. Eng. Comm. IDO-14581 (1962).

Murray, P., Rogers, E.P. & Williams, A.E., "Practical & Theoretical Aspects of the Hot Pressing of Refractory Oxides." Trans. Brit. Ceram. Soc. 53 (8) 474-510 (1954).

Murthy, M.K. & Kirby, E.M., "Infrared Study of Compounds & Solid Solutions in the System Lithia-Alumina-Silica." J. Am. Ceram. Soc. 45 (7) 324-9 (1962).

Myers, C.L. & Sherby, O.D., "Creep of Sintered Aluminum Powder (SAP) Above & Below the Melting Point of Aluminum" J. Inst. Metals 90, 380-2 (1962) (Paper No. 2133).

Nabarro, F.R.N., "Deformation of Crystals by Motion of Single Ions." pp. 75-90 in Rept. Conf. Strength of Solids (Univ. Bristol) July 1947 (Published 1948).

Nachtman, J.S., "High Temperature Structural Material & Method of Producing." U.S. 2,840,891, 7/1/58.

Nachtman, J.S., "Aluminum Base Powder Products." U.S. 2,947,068, 8/2/60.

Nagai, S. & Katayama, J., "Refractory Cements & Mortars: IV." J. Japan Ceram. Assn. 46, 20-6 (1938); Chem. Zentr. 1938, I, 2936-7.

Nagai, S., Suzuki, K., & Ota, Z., "Erosion of High Alumina Brick by Cement Materials." Yogyo Kyokai Shi 65 (738) 147-53 (1957).

Nagaoka, K. & Hara, M., "Studies on Fine Grained Crystalline Glass-Ceramics—Submicrocrystallization of Glasses of the System SiO_2-Li_2O-Al_2O_3 by P_2O_5: V, Effects of Al_2O_3 Component on the Crystallization." Osaka Kogyo Gijutsu Shikensho Kiho 11 (2) 115-21 (1960). Ibid. 13 (2) 105-16 (1962).

Nagashima, H., Miyake, E. & Ito, A., "Cast ZrO_2-Al_2O_3 Refractories." Japan 14,348, 9/18/62.

Nagumo, T. & Murakoshi, M., "Process for Purifying Alumina & Recovering a Gallium Concentrate." U.S. 3,144,304, 8/11/64.

Nahin, P.G. & Huffman, H.C., "Alumina & Alumina-Supported Catalysts." Ind. Eng. Chem. 41, 2021-7 (1949).

Nakai, J. & Fukami, Y., "Investigations on Alumina." J. Soc. Chem. Ind. Japan 39 (6) 203-4B (1936).

Nakai, J. & Miyazaki, T., "The Tunneling Current Through Thin Metal Oxide Film." Technol. Rept. Osaka Univ. 13 (563-589), 321-9 (1963).

Nakamoto, K., Margoshes, M., & Rundle, R.E., "Stretching Frequencies as a Function of Distances in Hydrogen Bonds." J. Am. Chem. Soc. 77, 6480-6 (1955).

Nakayama, N., Yamada, K., Kajihara, H., Sakai, S. & Mii, H. "Machinability Study of Ceramic Tools: VII, Roughness of Machined Surface." Nagoya Kogyo Gijutsu Shikensho Hokoku 10 (8) 457-66 (1961).

Nakayama, N., et al. "Machinability Study of Ceramic Tools: VIII, Deformed Surface Layer of Workpiece." Nagoya Kogyo Gijutsu Shikensho Hokoku 10 (10) 581-8 (1961).

Nakayama, N., Yamada, K., Naoki, K., Kajihara, H., & Saki, S., "Machinability Study of Ceramic Tools: X, Hardness of Ceramic Tools at High Temperature & Temperature Distribution Near the Tool Point." Nagoya Kogyo Gijutsu Shikensho Hokoku 11 (7) 373-83 (1962).

Naray-Szabo, I., "Relation Between the Structure & the Physical Properties of Glass: I." Acta Phys. Acad. Sci. Hung. 8 (1-2) 37-64 (1957).

Nasyrov, G.Z. & Lainer, A.I., "Participation of Silica in Hydrochemical Reduction of Alumina from Reduced Alunite." Izv. Vysshikh Uchebn. Zavedenii, Tsvetn, Met. 7 (3) 79-86 (1964).

Navias, L., "Advances in Ceramics Related to Electronic Tube Developments." J. Am. Ceram. Soc. 37 (8) 329-50 (1954).

Navias, L., "Sintering Experiments on Sapphire Spheres." J. Am. Ceram. Soc. 39 (4) 141-5 (1956).

Navias, L., "Metal-Ceramic Electrical Resistors." U.S. 2,855,491, 10/7/58.

Navias, L., "Comparison Between Al_2O_3-Ta & Al_2O_3-W Reactions above 1600°C in a Vacuum." Am. Ceram. Soc. Bull. 38 (5) 256-9 (1959).

Navias, L., "The Technology of High Temperature Ceramics." Northwestern Ohio Section, Amer. Ceram. Soc., Jan. 18, 1960.

Navias, L., "Preparation & Properties of Spinel Made by Vapor Transport & Diffusion in the System MgO-Al_2O_3." J. Am. Ceram. Soc. 44 (9) 434-46 (1961).

Navias, L., "Magesia-Alumina Spinel Articles & Process of Preparing Them." U.S. 3,083,123, 3/26/63.

Neiman, R., "Precision Casting by Investment Molding Process: II." Amer. Foundryman 6 (10) 7-15 (1944).

Nekrich, M.I., "Complex Utilization of Alunites." Trudy Khar'kov. Khim. Tekhnol. Inst. im. S.M. Kirova 1945, No. 5, pp. 187-8.

Nelson, D.F. & Boyle, W.S., "Continuously Operating Ruby Optical Maser." App. Opt. 1 (2) 181-3 (1962).

Nelson, D.F. & Remeika, J.P., "Laser Action in a Flux-Grown Ruby." J. App. Phys. 35, 522-9 (1964).

Neogi, P. & Mitra, A.K., "A New Scaly Variety of Aluminum Hydroxide." J. Chem. Soc. 130, 1222-3 (1927).

Nesin, A., "Method of Making Crystalline Alumina Lapping Powder." U.S. 3,121,623, 2/18/64.

Neuhaus, A., "Ionic Colors of Crystals & Minerals (Chromium Coloring)." Z. Krist., 113, 195-233 (1960).

Neuhaus, A. & Richartz, W., "Absorption Spectra & Coordination of Natural & Synthetic Single Crystals & Crystal Powders Colored Allochromatically by Cr^{3+}." Angew. Chem. 70 (14) 430-4 (1958).

Newman, E.S. & Wells, L.S., "Effect of Some Added Materials on Dicalcium Silicate." J. Res. NBS 36 (2) 151-3 (1946). RP-1696.

Neville, A.M., "Effect of Warm Storage on Strength of Concrete Made with High-Alumina Cement." Inst. Civil Engrs. (London) 10, 185-92 (1958); abstracted in J. Appl. Chem. (London) 8 (11) ii-452 (1958).

Newnham, R.E. & de Haan, Y.M., "Refinement of the a-Al_2O_3, Ti_2O_3, V_2O_3, & Cr_2O_3 Structures." Z. Krist. 117 (2-3) 235-7 (1962).

Newsome, J.W., "Method of Converting Crystalline Alumina Hydrate to Alpha-Alumina." U.S. 2,642,337, 6/16/53.

Newsome, J.W., Heiser, H.W., Russell, A.S. & Stumpf, H.C., "Alumina Properties, Tech. Paper No. 10" (Second Revision). Aluminum Co. of America (1960).

Nicholson, J.L., "Large Area Planar and Nonplanar Aluminum Oxide Films." Electrochem. Technol. 5, 349-51 (1967).

Nielsen, T.H. & Leipold, M.H., "Thermal Expansion in Air of Ceramic Oxides to 2200°C." J. Am Ceram. Soc. 46 (8) 381-7 (1963).

Nies, B.W. & Lambe, C.M., "Movement of Water in Plaster Molds." Am. Ceram. Soc. Bull. 35 (8) 319-24 (1956).

Nieuwenburg, C.J. van & Pieters, H.A.J., "Studies on Hydrated Aluminum Silicates. I. Rehydration of Metakaolin & The Synthesis of Kaolin." Rec. trav. chim. 48 27-36 (1929).

Nikitichev, P.I., Saburenkov, E.M., & Nozdrina, V.G., "Corundum Single Crystals." U.S.S.R. 132,209, 10/5/60.

Nilsson, E.O.F., Nilsson, S.I., & Hagelin, E.G., "Method & Device for Pulverizing and/or Decomposing Solid Materials." U.S. 2,997,245, 8/22/61.

Nirmala, A. & Srivastava, S.N., "Emulsification in Nonaqueous Media." Indian J. Appl. Chem. 23, 185 (1960).

Niwa, S., "Trial of Unburned Spinel (MgO-Al_2O_3) Brick for Roofs of Open-Hearth Furnaces." Taikabutsu Kogyo 13 (65) 319-25 (1962).

Nixon, A.C. & Davis, O.L., "Aluminum Oxide Compositions." U.S. 2,450,766, 10/5/48.

Noble, R.D., Bradstreet, S.W., & Rechter, H.L., "Ceramic Coating Compositions & Articles Coated Therewith." U.S. 2,995,453, 8/8/61.

Nock, J.A., "Today's Aluminum Aircraft Alloys." SAE Trans. 61, 209 (1953).

Noda, T. & Isihara, Y., "Effect of the Addition of Salts on the Crystal Growth of Alumina." J. Soc. Chem. Ind. Japan 43 (3) 166-8 (1940).

Noguchi, C., "Synthesis of Spinel: V, Effects of Mineralizer at 1535°." J. Japan Ceram. Assoc. 56 (632) 77-82 (1948).

Nolte, H.J., "Metallizing Ceramics." U.S. 2,904,456, 9/15/59.

Nolte, J.J. & Spurek, R.F., "Metal-Ceramic Sealing with Manganese." Television Eng. 1 (11) 14-8 (1950).

Norin, R. & Magneli, A., "New Niobium Oxide Phases." Naturwissenschaften 47 (15) 354-5 (1960).

North American Coal Corp. (Clark, Ezekail L.), "Recovering Sulfuric Acid from Sulfates of Aluminum & Iron." U.S. 3,086,846, 4/23/63.

Norton Grinding Wheel Co., Ltd., "Ceramic Nuclear-Fuel Elements." Brit. 831,679, 3/30/60.

Norton Introduces New Refractory. Am. Ceram. Soc. Bull. 39 (9) 481 (1960).

Norton, C.L., "Pebble Heater." Chem. & Met. Eng. 53 (7) 116 (1946).

Norton, C.L., Jr., "Refractory." U.S. 3,135,616, 6/2/64.

Norton, C.L., Jr. & Duplin, V.J., Jr., "High-Temperature Castable Refractories." U.S. 2,527,500, 10/24/50.

Norton, C.L., Jr. & Hooper, B., "Notes on Reaction Between MgO & Various Types of Refractories." J. Amer. Ceram. Soc. 29 (12) 364-7 (1946).

Norton, F.H., "Analysis of High-Alumina Clays by the Thermal Method." J. Am. Ceram. Soc. 23 (9) 281-2 (1940). Ibid 22 (2) 54-63 (1939).

Norton, F.H., "Refractories." 3rd ed., revised. McGraw-Hill Book Co., Inc., New York, 782 pp (1949).

Norton, F.H. & Fellows, D.M., et al. "Progress Report for Jan. 1 to Mar. 31, 1950." U.S. AEC Publ. NYO-594, 22p (1950).

Norton, F.H. & Kingery, W.D., "The Measurement of Thermal Conductivity of Refractory Materials." U.S. AEC Publ. NYO 599, 32 pp. (1951).

Norton, F.H. & Kingery, W.D. AEC Tech. Infor. Service. Oak Ridge Report, January 1, 1952.

Norton, F.H., Kingery, W.D. Economos, G. & Humenik, M. Jr., "Study of Metal-Ceramic Interactions at Elevated Temperatures." U.S. Atomic Energy Comm. NYO-3144, 83 pp. (1953).

Norton, F.J., "Permeation of Gases Through Solids." J. App. Phys. 28 (1) 34-9 (1957).

Norton J.T., "Selecting Cermets for High-Temperature Applications." Machine Design 28 (8) 143-6, 148 (1956).

Novak, L.J., "Gas Plating of Alumina on Quartz." U.S. 2,989,421, 6/20/61.

Nowak, J.M., "Ceramic Materials for Leading Edges of Hypersonic Aircraft." Ceram. Age 77 (10) 109-12, 114-5 (1961).

Nowak, J.M. & Conti, J.C., "Development of Lightweight Ceramic Honeycomb Structures." Am. Ceram. Soc. Bull. 41 (5) 321-5 (1962).

Nozdrina, V.G. and Tsinobar, L.I., "Spontaneous Crystallization of Corundum under Hydrothermal Conditions." Kristallografiya 11, 475-6 (1966).

O'Connor, D.J., Johansen, P.G., & Buchanan, A.S., "Electrokinetic Properties & Surface Reastions of Corundum." Trans. Faraday Soc. 52, 229-36 (1956).

Oel, H.J., "Ceramic Materials for Breeders, Moderators, Reflectors, Control Rods, Shielding, & Construction Elements & for Thermoelectric Power Generation." Ber. deut. keram. ges. 40 (2) 73-84 (1963).

Ogle, J.C. & Weinrich, A.R., "Coating a Surface with Quartz." Abstracted in Trans. Brit. Ceram. Soc. 47 (6) 205A (1948).

Ohta, N. & Kagami, K., "Change of "Active Structure" of Alumina Hydrogel with Duration of Standing." Repts. Govt. Chem. Ind. Research Inst. Tokyo 51 381-4 (1956); abstracted in J. Appl. Chem. (London) 7 (6) i-481 (1957).

Oishi, Y., Hamano, Y., Kinoshita, M., & Takizawa, Y., "Manufacture of Oxide Cutting Tools by Hot-Pressing Technique." Osaka Kogyo Gijutsu Shinkensho Kiho 14 (4) 313 (1963).

Oishi, Y. & Hashimoto, H., "Effect of Firing Atmosphere on Sintering of Alumina." Yogyo Kyokai Shi 70, 257-63 (1962).

Oishi, Y. & Hashimoto, H., "Studies on Sintering of Alumina: II, Effect of Atmosphere on the Sintering of Alumina." Osaka Kogyo Gijutsu Shikensho Kiko 14 (1) 116 (1963).

Okhotin, M.V. & Bazhbeuk-Melikova, I.G., "Effect of Al_2O_3 on Surface Tension of Glass in a Highly Viscous Condition." Steklo i Keram. 9 (6) 3-4 (1952).

Okhotin, M.V. & Ushanova, A.V., "Effect of Fluorides on the Crystallization & Viscosity of Alkali-free High-Alumina Glasses." Steklo i Keram. 16 (2) 15-6 (1959).

Okorokov, S.D., Golynko-Vol'fson, S.L., Yarkina, T.N. & Chepik, R.A., "Interaction Between Calcium Aluminates & Gypsum at High Temperatures." Zhur. Priklad. Khim. 35 (2) 256-63 (1962).

Okuda, H., et al. "Grinding with Ball Mills: I, Mills with Rubber Lining." Nagoya Kogyo Gijutsu Shikensho Hokoku 2 (10) 27-30 (1953).

Okura, T., et al. "Effects of Some Chemicals on the Flocculation of Aluminum Hydroxide." Kogyo Josui 52, 32-5 (1963).

Olcott, J.S. & Stookey, S.D., "Glass Body Having a Semicrystalline Surface Layer & Method of Making It." U.S. 2,998,675, 9/5/61.

Olson, O.H. & Morris, J.C., "Determination of Emissivity & Reflectivity Data on Aircraft Structural Materials." WADC TR 56-222, Part II, 31 pp. (1958); Part III; (1959).

Olt, R.D., "Synthetic Maser Ruby." Appl. Optics 1 (l) 25-32 (1962).

Omley, H.A., "Metal-to-Ceramic Seal." U.S. 2,917,140, 12/15/59.

Onitsch-Modl, E.M., "Sintering Characteristics, Microstructure, & Properties of $Cr-Al_2O_3$ Cermets." Planseeber. Pulvermet. 3, 42-56 (1955); abstracted in J. Appl. Chem. (London) 6 (5) i-672 (1956).

Onitsch-Modl, E.M., "Metallurgy of Sintered Aluminum." Intern. Leichtmetalltagung, Vortraege Diskussionsbeitr. 4th Leoben, Austria 1961, 101-7 (Pub. 1962).

Ono, S., Fendo, T. & Onaka, K., "Zirconia-Alumina Cast Refractories." Japan 3896, 6/8/62.

Ono, I. & Shibata, H., "Cast Refractories Containing Zirconia & Alumina." Japan 3897, 6/8/59.

Onofrio, A.d'., "Hydration of Cement." Belg. 632,236, 9/2/63.

Oomes, L.E., Boer, J.H. de & Lippens, B.C., "Phase Transformations of Aluminum Hydroxides." Proc. Intern. Symp. Reactivity Solids, 4th, Amsterdam 1960, 317-20 (Pub. 1961).

Orbach, K.K., "Ceramic Uses of Plasma Jet." Ceram. Ind. 79 (5) 72-5 (1962).

Orcel, J., "Differential Thermal Analysis for Determination of Constituents of Clays, Laterites, & Bauxites." Congr. Internat. Mines, Met. Geol. Appl., 7e Session, Paris, 1935, Geol. 1, 359-73.

O'Reilly, D.E. & Poole, C.P. Jr., "Nuclear Magnetic Resonance of Alumina Containing Transition Metals." J. Phys. Chem. 67 (9) 1762-71 (1963).

Oreshkin, P.T., "Electrical Conductivity of Aluminum & Zinc Oxides at High Temperature." Zhur. Tekh. Fiz, 25, 2447-50 (1955).

Organic Binders & Other Additives for Glazes & Engobes. Trans. Brit. Ceram. Soc. 61, 524 (1962).

Orlova, I.G., "Deformation Mechanism of Unfired Corundum Ceramics upon Heating." Dokl. Akad. Nauk SSSR. 165, 387-90 (1965).

Orlova, I.G. & Kainarskii, I.S., "Kinetics of Corundum Deformation on Heating," Dokl. Akad. Nauk SSSR 157 (2), 331-3 (1964).

Orlova, J.G., Kainarskii, I.S., & Prokopenko, M.I., "Effect of Modifying Additives on the Strength of Corundum Ceramics" Izv. Nauk SSSR, Neorgan. Materialy 1, 804-9 (1965).

Orowan, E., "Energy Criteria of Fracture." Welding J. 34, 157-60 (1955).

Orsini, L. & Petitjean, M., "Infrared Spectrographic Study of a Boehmite & Its Dehydration Products." Compt. rend. 237 (4) 326-8 (1953).

Ortman, C.D., "Kiln Refractory." U.S. 2,895,840, 7/21/59.

Osborn, E.F., & Muan, A., "Phase Equilibrium Diagrams of Oxide Systems." Plate 9. Published by Amer. Ceram. Soc. & Edward Orton Ceramic Foundation (1960).

Osborn, E.F. & Mann, A., "Phase Equilibrium Diagrams of Oxide Systems." Plates 1-10. The Amer. Ceram. Soc., 4055 N. High St., Columbus, Ohio (1960).

Osiowski, A., "Nuclear Magnetic Resonance (N.M.R.) As a Method of Investigating the Hydration of Cement." Cements-Wapno-Gips 19 (3) 66-8 (1964).

Otto, W.H., "Relationship of Tensile Strength of Glass Fibers to Diameter." J. Am. Ceram. Soc. 38 (3) 122-4 (1955).

Otto, W.H. & Preston, F.W., "Evidence Against Oriented Structure in Glass Fibers." J. Soc. Glass Technol. 34 (157) 63-68T (1950).

Owen, A.E., "Properties of Glasses in the System $CaO-B_2O_3$: I, Dc Conductivity & Structure of Calcium Boro-aluminate Glasses." Phys. Chem. Glasses 2 (3) 87-98 (1961).

Owen, A.E., "Properties of Glasses in the System $CaO-B_2O_3-Al_2O_3$: III, Thermal Expansion in the Cabal Glasses & Its Correlation with Structure & Electrical Properties." Phys. Chem. Glasses 3 (4) 134-8 (1962).

Oxley, J.H., Browning, M.F., Veigel, N.D., & Blocker, J.M., Jr., "Microminiaturized Fuel Elements by Vapor-Deposition Techniques." Ind. Eng. Chem. Prod. Res. Develop 1, No. 2, 102-7 (1962).

Ozment, H.E., "Alumina with a Small Percentage of Sodium" Fr. 1,358,998 4/17/64.

Pace, J.H., Sampson, D.F. & Thorp, J.S., "Spin-Lattice Relaxation Times in Sapphire & Chromium-Doped Rutile at 34.6 Gc/s." Proc. Phys. Soc. (London), 77 (494) 257-60 (1961).

Pacz, A. U.S. 1,551,613, 9/1/25.

Paladino, A.E. & Coble, R.L., "Effect of Grain Boundaries on Diffusion–Controlled Processes in Aluminum Oxide." J. Am. Ceram. Soc. 46 (3) 133-6 (1963).

Paladino, A.E. & Roiter, B.D., "Czochralski Growth of Sapphire." J. Am. Ceram. Soc. 47 (9) 465 (1964).

Paladino, A.E., Swarts, E.L. & Crandall, W.B., "Unsteady-State Method of Measuring Thermal Diffusivity & Biot's Modulus for Alumina Between 1500° and 1800°C." J. Am. Ceram. Soc. 40 (10) 340-5 (1957).

Palmisano, R.R. and Drager, J., "Low-Density Aluminum-Ceramic Composite." Am. Ceram. Soc. Bull. 45, 702-5 (1966).

Palmour, H. III., "Review of High-Temperature Metal-Ceramic Seals." J. Electrochem. Soc. 102 (7) 160-4 (c) (1955).

Palmour, H. III., "Raw Materials–A Key to Structural Behavior of Oxide Ceramics." Bull. Am. Ceram. Soc. 43 (4) 333 (1964).

Palmour, H. III, Choi, D.M., Barnes, L.D., McBrayer, R.D. & Kriegel, W.W., "Deformation in Hot-Pressed Polycrystalline Spinel." Mater. Sci. Res. 1, 158-97 (1963).

Palmour, H. III., DuPlessis, J.J., & Kriegel, W.W., "Microstructural Features & Dislocations on Thermally Etched Sapphire Surfaces." J. Am. Ceram. Soc. 44 (8) 400-4 (1961).

Palmour, H. III, Waller, J., McBrayer, R.D., Choi, D.M. & Barnes, L.D., "Raw Materials for Refractory Oxide Ceramics." ML-TDR-64-110, AFML, Wright-Patterson Air Force Base, Ohio. July 1964.

Palumbo, T.R., "Electrical Resistors & Method of Making." U.S. 2,487,581, 11/8/49.

Pampuch, R., "Sintering of Pure Oxides & Oxides with Additives." Prace Inst. Hutnic 10, 333-47 (1958).

Pampuch, R., "Sintering of Pure & Activated Oxides in the Solid State." Silikattech. 10 (2) 69-77 (1959).

Panasko, G.A. & Yashunin, P.V., "Calculations of the System Na_2O-Al_2O_3-H_2O." Zh. Prikl. Khim. 37 (2) 285-9 (1964).

Panyushkin, V.T. and Mal'tsev, V.S., "Calculation of Thermodynamic Potentials of Aluminum Suboxides." Tr. Inst. Met. i Obogashch., Akad. Nauk Kaz. SSR 11, 79-82 (1964),

Papadakis, M., "Study of Industrial Ball Mills." Rev. materiaux construct. et trav. publ. 1960, No. 542, pp. 295-308.

Papee, D., Charrier, J., Tertian, R., & Houssemaine, R., "Investigation of the Constitution of Activated Alumina." Congress of Aluminum, Paris 1954.

Papee, D. & Tertian, A., "Thermal Decomposition of Hydrargillite & the Constitution of the Activated Alumina." Bull. soc. chim. France 1955, No. 7-8, pp. 983-91; abstracted in J. Appl. Chem. (London) 5 (11) ii-604 (1955).

Papee, D., Tertian, R., & Biais, R., "Research on the Constitution of Gels & Crystalline Hydrates of Alumina." Bull. Soc. Chim. Fr., 1301-10 (1958).

Papkov, V.S. & Berezhkova, G.V., "Whisker Crystals of Alumina." Kristallografiya 9 (3), 442-4 (1964).

Pappis, J. & Kingery, W.D., "Electrical Properties of Single-Crystal & Polycrystalline Alumina at High Temperatures." J. Am. Ceram. Soc. 44 (9) 459-64 (1961).

Parche, M.C., "Facts About Aluminum Oxide." Research & Development Division, The Carborundum Co., Niagara Falls, N.Y.

Park, J.L., "Method of Making Ceramic Insulators." U.S. 2,952,877, 9/20/60.

Park, J.L., Jr., "Manufacture of Ceramics." U.S. 2,966,719, 1/3/61.

Parker, E.R., "Ceramic Ductility." Ceram. Ind. 75 (6) 65-8 (1960).

Parker, E.R., Pask, J.A. & Himmel, L., "Ductile Ceramics." PB Rept. 146841, c8 pp.; U.S. Govt. Res. Repts. 34 (2) 177 (1960).

Parker, E.R., Pask, J.A. & Washburn, J., "Ductile Ceramics Research." AD-283,585, 6pp.; U.S. Govt. Res. Repts. 37 (24) 112 (1962).

Parker, R.J., Grisaffe, S.J. & Zaretsky, E.V., "Surface Failure of Alumina Balls Due to Repeated Stresses Applied in Rolling Con-

tact at Temperatures to 2000°F." May 1964, 15 p. OTS price, $0.50. (NASA Technical Note D-2274).

Parmelee, C.W. & Harman, C.G., "Effect of Alumina on the Surface Tension of Molten Glass." J. Am. Ceram. Soc. 20 (7) 224-30 (1937).

Parodi, G.P., "Siliceous Material for Coating Glass Furnaces." Ital. 479,716, 4/10/53.

Parratt, Noel J., "Efficient Production of Refractory Whiskers," Brit. 998,166. 7/14/65.

Partridge, J.H., "Refractory Glass." U.S. 2,199,856, 5/7/40.

Partridge, J.H., "Creep of Refractory Materials." Trans. Brit. Ceram. Soc. 53 (11) 731-40 (1954).

Partridge, J.H. & Busby, T.S., "Slip Cast Mullite & Zircon Tank Blocks." Atti Congr. Intern. Vetro, 3rd Congr. Venice, 1953, pp. 519-29; for abstract see Ceram. Abstr. 1955, Aug., p. 145a.

Pascal, P., "Constitution & Evolution of Precipitates of Alumina." Comp. rend. 178, 481-3 (1924).

Pashos, T., "Status of Irradiation of Compacted UO_2 Powder Fuel Rods." Symposium on Powder-Filled Uranium Dioxide Fuel Elements, Nov. 1963. Worcester, Mass., Sponsored by Norton Co.

Passmore, E.M., Moschetti, A., & Vasilos, T., "The Brittle-Ductile Transition in Polycrystalline Al_2O_3." Phil. Mag. 13, 1157-62 (1966).

Passmore, E.M., Spriggs, R.M. & Vasilos, T., "Strength-Grain Size-Porosity Relations in Alumina." J. Am. Ceram. Soc. 48 1-7 (1) (1965).

Patrick, Alton J., "Radiation Damage to Ceramics, Etc." U.S. At. Energy Comm. LA-3285, 47 pp (1965).

Patrick, W.A., Jr., "Preparing Alumina Gel., U.S. 2,503,168, 4/4/50.

Patrick, W.S. and Cutler, I.B., "Grain Growth in Sintered Alumina." J. Am. Ceram. Soc. 48 541-2 (1965).

Pattison, J.N., "The Total Emissivity of Some Refractory Materials Above 900°C." Trans. Brit. Ceram. Soc. 54 698-705 (England) (1955).

Pattison, J.N., Keely, W.M. & Maynor, H.W., "Solid State Reaction Study of Hydrated & a-Aluminas with the Nitrates of Nickel & Cobalt." J. Chem. Eng. Data 5 (4) 433-4 (1960).

Patzak, I., "Presence of Al_2TiO_5 (Tielite) in Highly Calcined Bauxite." Tonind. Ztg. Keram. Rundschau 88 (6), 126 (1964).

Pauling, L., "The Principles Determining The Structure of Complex Ionic Crystals." J. Am. Chem. Soc. 51, 1010 (1929).

Pavlushkin, N.M., "Production of High Strength Corundum." Steklo i Keram. 13 (11) 19-23 (1956).

Pavlushkin, N.M., "Strength of Sintered Corundum." Steklo i Keram 14 (7) 14-7 (1957).

Pavlushkin, N.M., "Effect of Additions of Group II Elements on the Properties of Sintered Corundum." Tr. Mosk, Khim. Tekhnol. Inst. No. 37 135-47 (1962).

Pavlushkin, N.M., "Effect of Preparation on the Properties of Sintered Corundum." Freiberger Forschungsh. B 78, 5-21 (1963).

Pavlushkin, N.M., "Effect of Dispersity & the Form of Starting Material on the Properties of Sintered Corundum." Freiberger Forschungsh. B 78, 23-42 (1963).

Pavlushkin, N.M., Zhuravlev, A.K. & Eitingon, S.I., "Enameled Alumina for Decorative Ware., Steklo i Keram. 18, No. 7, 35-7 (1961).

Payne, B.S., Jr., "Cementation Coatings for Refractory Metals." Am. Ceram. Soc. Bull. 43 (8) 567-71 (1964).

Pears, C.D. et al., "Evaluation of Tensile Date for Brittle Materials Obtained with Gas Bearing Concentricity." Final Report, Mar. 63, 92 p. incl. illus., Tables. Rept. No. ASD-TDR-63-245

Pearse, R.W.B., "Identification of Molecular Spectra." 2nd ed. John Wiley & Sons, Inc., New York, 276 pp. (1950).

Pearson, A., "Sintering of Dry Ball Milled Alumina" Master's Thesis, Washington Univ., St. Louis, Mo., June 1968.

Pearson, A., Marhanka, J.E., MacZura, G., & Hart, L.D., "Dense, Abrasion-Resistant, 99.8% Alumina Ceramic." Am. Ceram. Soc. Bull. 47, 654-8 (1968).

Pearson, J.G., "Chemical Background of the Aluminum Industry." Royal Institute of Chemistry, London, 103 pp. (1955).

Pearson, S., "Delayed Fracture of Sintered Alumina." Proc. Phys. Soc. (London) 69 (444B) 1293-6 (1956).

Pecherskaya, Yu. I., Kazanskii, V.B. & Voevodskii, V.V., Use of the Electron Spin Resonance Method in the Investigation of an Oxide Catalyst." Actes congr. intern. catalyse, 2e, Paris, 1960 2, 2121-32, discussion 2132-4 (Pub. 1961).

Pechiney, A.R. Fr. 349,709, 4/8/04.

Pechiney-Compagnie de Produits Chimiques et Electro-metallurgiques. Neth. Appl. 6,611,170. 2/13/67.

Pechiney-St. Gobain, Prod. Chim., "New Porous Products with a Basis of Alumina." Fr. 1,326,386, 1/30/62.

Pechman, A., "Ceramics for High-Temperature Applications." Ceram. Age 62 (5) 27-31, 34-35 (1953); Ceramics 6 (61) 19-20 22-27 (1954); Pacific Coast Ceram. News 3 (6) 24, 30; (7) 20; 32; (8) 23; (9) 24 (1954).

Pedersen, H., "Iron & Cement Production, etc." Brit. 232,930, 6/14/26; U.S. 1,618,105, 2/15/27.

Pellini, W.S. & Harris, W.J., Jr., "Flight in the Thermosphere: II-IV." Metal Progr. 77 (4) 113-5; (5) 89-92; (6) 83-93 (1960).

Pemberton, A., "Simple Laboratory Technique for Determining Hardness & Toughness of Abrasives." Ind. Diamond Rev. 11 (26) 116-7 (1951).

Pentecost, J.L., Davies, L.G. & Ritt, P.E., "Electrical Resistivity Measurements on Highly Refractory Electrical Insulators." Presented at 60th Annual Meeting Am. Ceram. Soc., April 29, 1958.

Pentscheff, N.P., et al., "The Composition, Structure, & Aging of Aluminum Hydroxide Prepared from Aluminate Solutions." Compt. rend. Acad. Bulgare Sci. 16 (7) 725-8 (1963).

Peppler, R.B. & Newman, E.S., "Heats of Formation of Some Barium Aluminates." J. Res. Natl. Bur. Stds. 47 (6) 439-42 (1951); RP 2269.

Peppler, R.B. & Wells, Lansing S., "System of Lime, Alumina, & Water From 50° to 250°C." J. Res. Natl. Bur. Stds 52 (2) 75-92 (1954).

Peskin, W.L., "Special Refractories for Aluminum." Modern Metals (August 1961).

Peters, D.W., "Alumina Polyphase Heterojunction." J. Am. Ceram. Soc. 48 (4) 220-1 (1965).

Peters, D.W., "Thermoelectric Power of Single-Crystal Al_2O_3." J. Phys. Chem. Solids 27, 1560-2 (1966).

Peters, D.W., Feinstein, L., and Peltzer, C., "High-Temperature Electrical Conductivity of Alumina." J. Chem. Phys. 42, 2345-6 (1965).

Peters, F.A., Johnson, P.W., & Kirby, R.C., "Methods for Producing Alumina from Clay - An Evaluation of Three Sulfuric Acid Processes." U.S. Bur. Mines Rept. Invest. 1963 No. 6229, 57 pp.

Petitjean, M. Thesis No. 79. Univ. of Lyon, March 4, 1955.

Petrescu, P., "The Nature of Exoelectron Emission Centers." Acad, rep. populare Romine, Inst. Fiz. Atomica si Inst. Fiz., Studii cercetari fiz. 11, 867-78 (1960).

Petrovic, L.J. and Thodos, G., "Evaporation from Alumina in Fixed Fluid Beds." Brit. Chem. Eng. 11 1041-4 (1966).

Petzold, A., "Ceramic Coatings for High-Temperature Protection of Metals Against Corrosion." Werkstoffe u. Korrosion 9 761-5 (1958).

Pevzner, R.L., "New Highly Refractory Material, "Thermite Mullite." Bull. acad. sci. URSS., Classes sci. Tech. 1946, 1431-8; abstracted in Chem. Zentr. 1947, I (21/22) 1037.

Pevzner, R.L., "Production of Corundum for Abrasives Without the Use of Furnaces." Doklady Akad. Nauk SSSR 67 (4) 707-9 (1949).

Pevzner, R.L., "Thermitocorundum Refractories & Characteristics of Their Applications." Legkaya Prom. 14 (8) 36-7 (1954).

Pflanz, J.E. & Muller-Hesse, H., "Solid-State Reactions in The System $BaO-Al_2O_3-SiO_2$: I, $BaO-Al_2O_3$, $BaO-SiO_2$, & $Al_2O_3-SiO_2$." Ber. deut. keram. Ges. 38 (10) 440-50 (1961).

Phelps, G.W., "Soluble Salts - Their Effect on Casting Properties in Ball Clays." Ceram. Ind. 76 (3) 68-9 (1961).

Phelps, G.W. & Maguire, S.G., Jr., "Water as a Ceramic Raw Material." Am. Ceram. Soc. Bull. 35 (11) 422-6 (1956).

Phillips, C.J. & DiVita, S., "Thermal Conditioning of Ceramic Materials." Am. Ceram. Soc. Bull. 43 (1) 6-8 (1964).

Piatasik, R.S. & Hasselman, D.P.H., "Effect of Open & Closed Pores on Young's Modulus of Polycrystalline Ceramics." J. Am. Ceram. Soc. 47 (1) 50-1 (1964).

Pichler., "Zirlite." Jb. Min. 57 (1871). (See Dana).

Pieper, A.O., "Process of Making Refractory Material." U.S. 2,770,552, 11/13/56.

Pike, R.G. & Hubbard, D., "Source of the Non-Migratable Ionic Charges Developed by High-Alumina Cement During Hydration." Highway Res. Board 37, 256 (1958).

Pincus, A.G., "Mechanism of Ceramic-to-Metal Adherence - Adherence of Molybdenum to Alumina Ceramics." Ceram. Age 63 (3) 16-20, 30-2 (1954).

Pincus, A.G., "Critical Compilation of Ceramic Forming Methods." TDR No. RTD-TDR-63-4069, AF Materials Laboratory (1963).

Pines, B. Ya., "Recent Soviet Work on the Dielectric Properties and Sintering of Alumina." J. Mater. Sci. 1, 312-13 (1966).

Pines, H. & Chen, C-T., "Alumina: Catalyst & Support. VIII. Aromatization of n-Octane-IC^{14} & Cyclooctane Over Chromia-Alumina Catalysts." Proc. Intern. Congr. on Catalysis, 2nd, Paris, 1960 1, 367-84, discussion 384-7 (Pub. 1961).

Pingard, L.C., "Activated Alumina." U.S. 2,881,051, 4/7/59.

Piriou, B., "Optical Constants of Corundum and Magnesium Oxide." Compt. Rend. 259, (Group 5), 1052-5 (1964).

Pistorius, C.W.F.T., "Phase Relations in the System $CaO-Al_2O_3-H_2O$ to High Pressures & Temperatures." Am. J. Sci. 260 (3), 221-9 (1962).

Plank, C.J., "Forming Porous Glass & Composition Thereof." U.S. 2,480,672, 8/30/49.

Plank, C.J., Adsorption of Ions from Buffer Solutions by Silica, Alumina, & Silica-Alumina Gels." Phys. Chem. 57, 284-90 (1953).

Plankenhorn, W.J., "Refractory Ceramic Base Coats for Metal.'. J. Am. Ceram. Soc. 31 (6) 145-53 (1948).

Plass, G.N., "Mie Scattering & Absorption Cross Sections for Aluminum Oxide & Magnesium Oxide." App. Opt. 3 (7) 867-72 (1964).

Pendl, J.N., & Gielisse, P.J., "Atomistic Expression of Hardness," Z. Kristallogr. 118 (5/6) 404 (1963).

Plummer, M., "The Formation of Metastable Aluminas at High Temperatures." App. Chem. (London) 8, 35-44 (1958).

Plummer, W.A., Campbell, D.E., & Comstock, A.A., "Method of Measurement of Thermal Diffusivity to 1000°C." J. Am. Ceram. Soc. 45 (7) 310-6 (1962).

Plunkett, J.D., "NASA Contributions to the Technology of Inorganic Coatings." NASA SP-5014, National Aeronautics & Space Admin., Washington, D.C. (1964).

Podszus, E., "Electrical Conductivity of High-Insulating Oxides & Nitrides at High Temperatures." Z. Elektrochem. 39 (2) 75-81 (1933).

Pointud, R. & Roger, J., "Manufacture of Refractory Ware for Use in Metallurgical Laboratories." Rev. Met. 54 (April) 283-7 (1957); abstracted in J. Iron Stell Inst. (London) 188 (1) 74 (1958).

Pole, G.R. & Beinlich, A.W., Jr., "New Refractory Compositions Resistant to Molten Rock Phosphate." J. Am. Ceram. Soc. 26 (1) 21-37 (1943).

Pole, G.R. & Moore, D.G., "Electric-Furnace Alumina Cement for High-Temperature Concrete." J. Amer. Ceram. Soc. 29 (1) 20-4 (1946).

Poluboyarinov, D.N., "High-Alumina Ceramics." Khim. Nauka i Prom. 3 (1) 8-14 (1958).

Poluboyarinov, D.N., Andrianov, N.T., Gazman, I. Ya., & Lukin, E.S., "Vaporization of Porous Oxide Ceramics. . . ." Ogneupory 31, 33-7 (1966).

Poluboyarinov, D.N. & Balkevich, V.L., "High-Alumina Refractories from Electromelted Corundum." Ogneupory 14 (12) 538-46 (1949).

Poluboyarinov, D.N. & Balkevich, V.L., "Highly Refractory Materials from Recrystallized Alumina." Ogeneupory 16 (3) 109-19 (1951).

Poluboyarinov, D.N. & Ershova, E.S., "Effect of Dispersion of Alumina on Sintering of Corundum Body." Trudy Moskov.

Khim. Tekhnol. Inst. im. D. I. Mendeleeva 1949, No. 16, Stroitel. Materialy Sbornik, No. 2, pp. 59-72.

Poluboyarinov, D.N. & Kalliga, G.P., "Deformation Under Load at High Temperatures of Alumina-Silicate Refractories with High Content of Alumina." Ogneupory 16 (6) 272-80 (1951).

Poluboyarinov, D.N. & Kirshenbaum, Ya. B., "Sintering of High-Alumina Briquettes." Qgneupory 17 (6) 243-52 (1952).

Poluboyarinov, D.N. & Popil'skii, R. Ya., "Technology of Manufacture of High-Alumina Refractories from Technical Alumina." Ogneupory 12 (6) 243-54 (1947).

Poluboyarinov, D.N. & Popil'skii, R.Ya., "Technology of High-Alumina Refractories with Technical Aluminum Oxide." Ogneupory 12 (12) 537-44 (1947).

Poluboyarinov, D.N. & Popil'skii, R. Ya., "Manufacture of High-Alumina Refractories on the Basis of the Synthesis of Mullite." Ogneupory 14 (2) 58-64 (1949).

Poluboyarinov, D.N., Popil'skii, R. Ya., & Sterlyadkina, Z.K., "Volume-Structural Changes Connected with Phase Transformations During Firing of Mullite-Corundum Refractories." Ogneupory 20 (7) 315-25 (1955).

Poluboyarinov, D.N. & Vydrik, G.A., "Sintering of Corundum Body as a Function of Prior Firing & Dispersion of Technical Alumina." Doklady Akad. Nauk SSSR 88 (2) 325-8 (1953).

Pomerantz, M.A., "Secondary Electron Emission from Oxide-Coated Cathodes." I, J. Franklin Inst. 242, 41-61 (1946).

Ponomarev, V.D., et al., "Investigation of Thermal Decomposition of Alunite (Alumstone)." Izv. Vysshikh Uchebn. Zavedenii, Tsyetn. Met. 6 (2), 94-101 (1963).

Poole, C.P., & Itzel, J.F., "Optical Reflection Spectra of Chromia-Alumina." J. Chem. Phys. 39 (12) 3445 (1963).

Poole, C.P., Jr. & Itzel, J.F., Jr., "Electrom Spin Resonance (E.S.R.) Study of the Antiferromagnetism of Chromia Alumina." J. Chem. Phys. 41 (2), 287-95 (1964).

Popil'skii, R. Ya., & Galkina, I.P., "Tests on the Casting of High Grog Bodies." Ogneupory, 25 (3) 137-42 (1960).

Popil'skii, R. Ya. & Nemets, L.M., "Use of Paraffin Bond in Pressing Alumina Shapes." Ogneupory 16 (7) 323-32 (1951).

Popil'skii, R. Ya., Pankratov, Yu. Fr., & Koifman, N.M., "Formation of A Poreless Structure in Polycrystalline Corundum." Dokl. Akad. Nauk SSSR 155 (2), 326-9 (1964).

Popov, S.K. "The Growth & Uses of Gem-Grade Corundum Crystals," from "Growth of Crystals," Shubnikov, H.V. & Sheftal', N.N. (Editors), Translation, Consultants Bureau, Inc., New York (1959).

Porai-Koshits, E.A. ed., "Structure of Glass: Vol. 2, Proceedings of the Third All-Union on the Glassy State, Leningrad, 1959." Translated from Russian, 1960. xii + 480pp. illus. Available from Consultants Bureau Enterprises, Inc. New York 11.

Porter, J.L., "Control of Phosphate in Alumina Production." U.S. 2,557,891, /619/51.

Porto, S.P.S. and Krishnan, R.S., "Raman Effect of Corundum." J. Chem. Phys. 47, 1009-11 (1967).

Pouillard, E., "The Behavior of Alumina & of Titanium Oxide with Oxides of Iron." Ann. Chim. 5 (12) 164-214 (1950).

Poulos, N.E., Elkins, S.R., et al., "High-Temperature Ceramic Structures." AD 270805, 124 pp. U.S. Govt. Res. Repts. 37 (7) 64 (1962).

Powder, E.R., "Process for Making Foamed Ceramic Products." U.S. 3,133,820, 5/19/64.

Powder Metallurgy, 1954. Iron Steel Inst. (London), Spec. Rept. 1956, No. 58, ix + 399 pp

Powell, R.H., "Control of Polishing Alumina." Chem. Age (London) 50 (1295) 378 (1944).

Powers, D.L., "Thermal & Mechanical Testing of Foam Alumina & Foam Zirconia." Ad 276983, 21 pp. illus.; U.S. Govt. Res. Repts., 37 (18) 28 (1962).

Powers, W.H. & Kappmeyer, K.K., "Investigation of Special Refractories for Steel Plant Applications." Am. Ceram. Soc. Bull. 44 (7) 561-7 (1965).

Powers, T.C., "Measuring Young's Modulus of Elasticity by Means of Sonic Vibrations." ASTM Proceedings 38 (Part II) (1938).

Preist, D.H. & Talcott, R., "Thermal Stresses in Ceramic Cylinders Used in Vacuum Tubes." Am. Ceram. Soc. Bull. 38 (3) 99-105 (1959).

Prescott, C.H. & Hincke, W.B., "High-Temperature Equilibrium Between Alumina & Carbon." J. Am. Chem. Soc. 49 2753 (1927).

Prettre, M., Imelik B., Blanchin, L., & Petitjean, M., "Mechanism of Formation of Adsorptive & Catalytically Active Aluminum Oxides." Angew. Chem. 65, 549-55 (1953).

Preusser, E., "Chemical Setting of Refractory Raw Materials with Diluted Phosphoric Acid." Silikattech. 12 (2) 81-3 (1961).

Price, D.E. & Wagner, H.J., "Preparation & Properties of Fiber-Reinforced Structural Materials." Battelle DMIC Memo 176 (Aug. 1963).

Primak, W. & Fuchs, L.H., "Radiation Damage to the Electrical Conductivities of Natural Graphite Crystals." Phys. Rev. 103 (3) 541-4 (1956).

Primak, W., et al., "Radiation Damage in Diamond & Silicon Carbide." Phys. Rev. 103 (5) 1184-92 (1956).

Primak, W. & Szymanski, H., "Radiation Damage in Vitreous Silica—Annealing of the Density Changes." Phys. Rev. 101 (4) 1268-71 (1956).

Provance, J.D. (Houze Glass Corporation). U.S. 3,044,888, 7/17/62.

Pryslak, N.E., "Sandwich-Type Metal-to-Ceramic Vacuum-Tight Seal." Ceram. Age 65 (3) 21-2 (1955).

Przibram, K., "Widely Distributed Blue Fluorescence of Organic Origin. III." Osterr. Akad. Wiss., Math.-naturw. Kl., Anz. No. 4, 165-70 (1960).

Puehler, F. & Wagner, R., "Ceramic Oxide Cutting Materials in Precision Working - Examples from Production." Ind. Anzeig. 80 (47) 673-6 (1958); Ind. Diamond Abstr. 1958, Aug., P. A118.

Pulfrich, H., "Ceramic-to-Metal Seal." U.S. 2,163,407, 6/20/39.

Pulliam, G.R., "Influence of Surface on Ceramic Mechanical Properties." PB Rept. 161309, 36 pp.; U.S. Govt. Res. Repts. 33 (5) 522 (1960).

Pupko, A.L., "Aluminum Oxide Films as Carriers". Doklady Akad. Nauk 63, 259 (1948).

Puppe, F., "Grinding Electrocorundum." Schleif. & Poliertech. 15 (10) 193-7 (1938).

Puri, J.M., "Use of Ceramic Materials for Gas Turbine Blades." Research (London) 12 (7) 246-53 (1959).

Purt, G., "Binary System BaO-Al$_2$O$_3$." Radex Rundschau 1960 (4) 201.

Pyrex. Corning Glass Works.

Quayle, J.C., et al., "Method of Making a Ceramic Body." Swed. 119,136, 10/25/45.

Quinan, J.R. & Sprague, N.W., "Coated Abrasives." U.S. 3,011,882, 12/5/61.

Radio Corp. of America, "Heating Elements of High-Melting Metals." Ger. 705,766, 4/3/41.

Raether, S., "Electronic Conductivity & Macrostructure of Aluminum Oxide Layers Formed by Anodic Oxidation." Z. angew. Phys. 11, 456-60 (1959).

Ramamurthy, L., "Elastic Constants of Single-Crystal Corundum Below Room Temperature." J. Phys. Chem. Solids, 28 363-66 (1967).

Raman, C.V. Proc. Indian Acad. Sci. 34A 61-71 (1951).

Ramaswamy, S., et al., "Bauxite as a Drying Agent for Wet Chlorine Gas from Electrolytic Cells: I." J. Sci. Ind. Res. (India) 14B (10) 527-32 (1955).

Ramke, W.G. & Latva, J.D.," Refractory Ceramics & Intermetallic Compounds." Aerospace Eng. 22 (1) 76-84 (1963).

Ramsay, W.S., "Casting of Ceramic Articles." U.S. 2,521,128, 9/5/50.

Ramsey, J.A., "Exoelectron Emission & The Optical Properties of Oxide-Coated Metal Surfaces." Nature 185, 602 (1960).

Randolph, R.L., "Injection Molded Ceramics." Materials in Design Eng. 54 (2) 10-12 (1961).

Ranganathan, T., MacKean, B.E., & Maun, A., "System Manganese Oxide-Alumina in Air." J. Am. Ceram. Soc. 45 (6) 279-81 (1962).

Rankin, G.A. & Merwin, H.E., "Ternary System CaO-Al$_2$O$_3$-MgO." J. Am. Chem. Soc. 38, 568-88 (1916).

Rao, R.V.G.S., "Elastic Constants of Alumina." Proc. Inc. Acad. Sci. 29A, 352-60 (1949).

Rao, N.S.G., "Chrome Alumina Pink Ceramic Colors." Central Glass & Ceram. Res. Inst. Bull. (India) 5 (1) 25-8 (1958).

Rao, N.S.G., "Ceramic Colors." Central Glass & Ceram. Res. Inst. Bull. (India) 5 (2) 66-75 (1958).

Rao, S.R. & Leela, M., "Magnetic Study of Corundum." Current Sci. (India) 22 72-3 (1953).

Rasch, R., "Schematic Representation of the Bonds within Crystals, with Spinels & Corundum as Examples." Sprechsaal 87, 413-5, 449-51 (1954); Chem Abstr. 49 (2) 689h (1955).

Rasmussen, J.J., Stringfellow, G.B. & Cutler, I.B., "Effect of Impurities on the Strength of Polycrystalline Magnesia & Alumina." J. Am. Ceram. Soc. 48 (3) 146-50 (1965).

Rautenberg, T.H., Jr. & Johnson, P.D., "Light Production in the Aluminum-Oxygen Reaction." J. Opt. Soc. Am. 50, 602-6 (1960).

Raychaudhuri, S.P. & Hussain, A., "Preliminary Study of the Aging of Alumina & Silica Gels & of the Precipitates Obtained from Mutual Coagulation of Alumina & Silicic Acid Sols." J. Indian Chem. Soc. 20 (6) 195-6 (1943).

Rea, R.F., "Ceramic Parts & Tooling for Mechanical Applications." Tool Engr. 1955, March, pp. 95-8.

Rea, R.F. & Ripple, J.W., "Cutting & Grinding Ceramics." Am. Ceram. Soc. Bull. 36 (5) 163-7 (1957).

Ready, D.W. and Kuczynski, G.C., "Sublimation of Aluminum Oxide in Hydrogen." J. Am. Ceram. Soc. 49, 26-9 (1966).

Rechsteiner, E.B., "Lasers: A Dynamic New Industry." Ind. Research 4 (10) 14-21 (1962).

Redwood, M., "The Absorption of Ultrasound in Solids." Me. Sci. Met. 56, 172-80 (1959).

Reed, E.L., "Stability of Refractories in Liquid Metals." J. Am. Chem. Soc. 37, 146-53 (1954).

Reed, T.B., "Growth of Refractory Crystals Using the Induction Plasma Torch." J. Appl. Phys. 32 (12) 2534-35 (1961).

Reed, L., "Heat Treatment Studies on High-Alumina Ceramics." Advan. Electron Tube Tech., Prov. Natl. Conf. 6, 15-29 (1962). (Pub. 1963).

Reed, L., "Survey of the Factors Affecting Ceramic-to-Metal Sealing." Am. Ceram. Soc. Bull. 43 (8) 584 (1964).

Reed, L. & McRae, R., "Evaporated Metalizing." Am. Ceram. Soc. Bull. 43 (8) 586 (1964).

Reese, K.M. & Cundiff, W.H., "Alumina." Ind. Eng. Chem. 47 (9) 1672-80 (1955).

"Refractories Bibliography, 1928-1947." American Iron & Steel Institute & The American Ceramic Society, Columbus, Ohio 1950, 2109 pps.

"Refractories Bibliography 1947-1956." Compiled by the Joint Refractory Committee of The American Iron & Steel Institute & the Refractories Institute. Univ. of Oklahoma Press, Norman, Okla., 1822 columns (1959).

Refractory Service in Glass Furnaces. Report of Comm. C-8 on Refractories of the ASTM: App. II Industrial Survey of Conditions Surrounding Refractory Service in Continuous Plate Glass & Window Glass Furnaces. Proc. ASTM 39, 349-56 (1939); Ind. Heating 6 (9) 831-6; (12) 1171-2 (1939).

Regester, R., "The Strength & Perfection of Aluminum Oxide Whiskers." Masters Thesis, Univ. Penn. (1963).

Regester, R., Gorsuch, P.D., & Girifalco, L., "Structure and Properties of Aluminum Oxide Whiskers." Mater, Res. Stand. 7, 203-6 (1967).

Reheis Co., Inc., "Aluminum Chloride Hydroxides & Aluminum Hydroxide." Brit. 954,088, 4/2/64.

Reich, H.F. & Panda, J.D., "Investigating The Slagging Resistance of Refractories—Critical Review of the Use of the Heating Microscope." Tonind.-Ztg. u. Keram. Rundschau 85 (8) 186-90; (10) 223-9 (1961).

Reichelt, W. & Schmeken, H., "Coating." U.S. 2,903,544, 9/8/59.

Reichertz, P.P. & Yost, W.J., "Crystal Structure of Synthetic Boehmite." J. Chem. Phys. 14, 495 (1946).

Reichmann, R., "Spark Plug." U.S. 1,799,225, 4/7/31.

Reichmann, R., "Insulators for Spark Plugs, etc." U.S. 1,803,355, 5/5/31.

Reichmann, R. & Kohl, H., "Testing Metal Oxides for Use in Making Cast Ceramic Articles." Ger. 609,345, 2/13/35.

Reichmann, R., "Aluminum Oxide." Ger. 560,575, 2/8/35.

Reichmann, R., "Shaped Bodies of Non Plastic Metallic Oxides Such as Park Plug Insulators." U.S. 2,031,129, 2/18/36.

Reichmann, R., "Method of Examining the State of Crystallization of Calcined Aluminum Oxide." U.S. 2,058,178, 10/20/36.

Reichmann, R. & Kohl, H., "Method of Molding Nonplastic Metallic Oxides." U.S. 1,934,091, 11/7/33.

Reinhart, F., "Advances in Ceramics During 1955." Glas-Email-Keramo-Tech. 6 (10) 348-52; (12) 432-5 (1955). Ger. 929,113, Ger. 929,178.

Reinhart, F., "New Types of Refractory Materials." Glas-Email-Keramo-Tech. 9 (9) 327-9 (1958).

Reinhart, F., "Joining of Ceramic to Ceramic & of Ceramic to Glass." Glas-Email-Keramo-Tech. 12 (1) 10-2 (1961).

Remeika, J.P., "Growth of Single Crystals of Corundum & Gallium Oxide." U.S. 3,075,831, 1/29/63.

Remeika, J.P., "Process for Growing Single Crystals." U.S. 3,079,240, 2/26/63.

Remmey, G.B., "New Alumina-Silica Refractories." Chem. Eng. Progress 44 (12) 943-6 (1948).

Rempes, P.E., Weber, B.C. & Schwartz, M.A., "Slip Casting of Metals, Ceramics, & Cermets." Am. Ceram. Soc. Bull. 37 (7) 334-9 (1958).

Rengade, E., "Effect of Blending Portland & Aluminous Cements." Rev. Mat. Constr. Trav. Pub. No. 300, pp. 255-6 (1934); translated in Concrete (Cement Mill Ed.), 44 (3) 40-1 (1936).

Renkey, A.L. & Reardon, D.E., "Refractory Spheres, Heavy-Duty Ceramics for Pebble Heaters." IEEE (Inst. Elec. Electron Engrs.) Trans. Aerospace AS-1 (2), 1411-6 (1963).

Reshetnikov, M.A., "Heat Capacity of Corundum from 0 to 2000°K." Zh. Neorgan. Khim. 11, 1489-96 (1966).

Rettner, E.S., et al., "Mechanism of Operation of the Barium Aluminate Impregnated Cathode." J. Appl. Phys. 28 (12) 1468-73 (1957).

Reynolds, F.H. and Elliot, A.B.M., "Etching Corundum with Silicon." Phil. Mag. 13, 1073-4 (1966).

Rhines, F.N., "Powder Metallurgy." ed. Leszynski, 31-51 Interscience Publishers, New York (1961).

Rhoads, J.L. & Berg, M., "Method of Sealing a Metal Member to a Ceramic Member." U.S. 3,006,069, 10/31/61.

Rhodes, W.H., "Contamination from Al$_2$O$_3$ Furnace Tube." J. Am. Ceram. Soc. 44 (6) 300 (1961).

Rhodes, W.H., Sellers, D.J., Vasilos, T., Heuer, A.H., and Duff, R., "Microstructure Studies of Polycrystalline Refractory Oxides." Contract NOw-66-0506.

Ribbe, P.H. & Alper, A.M., "Inverse Segregation of MgO & MgAl$_2$O$_4$ in Fusion-Cast Bodies." J. Am. Ceram. Soc. 47 (4) 162-7 (1964).

Richards, R.G. & White, J., "Phase Relations of Iron Oxide-Containing Spinels." Trans. Brit. Ceram. Soc. 53, 233-70 (1954).

Richardson, H.M., et al., "Action of Ferrous Oxide on Alumino-silicate Refractories." Gas Res. Board Commun. GRB 32, pp. 38-45 (1947).

Richardson, H.M., et al., "Action of Mixtures of Calcium & Ferrous Oxides on Aluminosilicate Refractories." Gas Res. Board Commun. GRB 41, pp. 27-39 (1948).

Richardson, K.D., "Ceramics for Aircraft Propulsion Systems." Bull. Am. Ceram. Soc. 33 (5) 135 (1954).

Richt, A.E., "Metallographic Examination of Irradiated Cermet Fuel Elements." U.S. At. Energy Comm. CF-58-8-35, 9 pp. (1958).

Richter, W., "Plasticizing Corundum Raw Materials." Fr. 1,294,519, 5/25/62.

Richter, W. & Kammerich, G., "Investigations on Ceramic & Mixed-Ceramic Materials for Cutting Tools." Silikattech. 6 (12) 528-34 (1955).

Ricker, R.W., "Castable Refractory." U.S. 2,912,341, 11/10/59.

Riddle, F.H., "Spark-Plug Insulator." U.S. 2,337,930, 12/28/43.

Riddle, F.H., "Alumina & Silicon Carbide Refractory." U.S. 2,388,080, 10/30/45.

Riddle, F.H., "Ceramic Spark Plug Insulators." J. Am. Ceram. Soc. **32** (11) 333-45 (1949).

Riddle, F.H., "Alumina Insulating Material & Method of Making." U.S. 2,524,601, 10/3/50.

Ridgway, R.R., "Resistor & Method of Making." U.S. 2,252,981, 8!19/41.

Ridgway, R.R., "Method of Purifying Crystalline Alumina & An Abrasive Material made Thereby." U.S. 2,301,706, 11/10/42.

Ridgway, R.R., "Crystalline Alumina & Method & Apparatus for Making." Can. 463,699, 3/14/50.

Ridgway, R.R. & Bailey, B.L., "Article of Self-Bonded Granular Material & Method of Making." U.S. 2,091,569, 8/31/37.

Ridgway, R.R., Ballard, A.H., & Bailey, B.L., "Hardness Values for Electrochemical Products." Trans. Am. Electrochem. Soc. **63**, 23 (1933).

Ridgway, R.R., & Glaze, J.B., "Aluminous Abrasive." U.S. 1,719,131, 7/2/29.

Ridgway, R.R., Klein, A.A. & O'Leary, W.J., "Preparation & Properties of So-Called "Beta Alumina." Electrochem Soc. Preprint 70-9, 16pp. (1936).

Riedel, J.D. Ger. 386,614, 12/13/23; Ger. 424,701, 1/29/26.

Riegert, R.P., "For a Better Definition of Ceramics." Ceram. News **12** (2) 11, 18; (8) 20 (1963).

Rieke, J.K. & Daniels, F., "Thermoluminescence." J. Phys. Chem. **61**, 629-33 (1957).

Ries, K.G., et al., "Resin-Bonded Abrasive Article & Method of Manufacture." U.S. 2,559,664, 7/10/51.

Riesmeyer, A.H., "Removing Sodium from Alumina." U.S. 2,411,807, 11/26/46.

Riesmeyer, A.H. & Gitzen, W.H., "Purification of Alumina." U.S. 2,411,806, 11/26/46.

Riesz, C.H. "Mechanism of Wear of Nonmetallic Materials." PB Rept. 161955, 25 pp.; U.S. Govt. Res. Repts. **34** (6) 780 (1960).

Riesz, C.H., & Weber, H.S., "Mechanism of Wear of Nonmetallic Materials." AD 275361, 23 pp., illus.; U.S. Govt. Res. Repts. **37** (16) 33 (1962).

Riesz, C.H. and Weber, H.S., "Friction and Wear of Sapphire." Mech. Solid Friction, Papers Conf. Kansas City, Mo. **1963**, 67-81 (1964).

Rigby, G.R. (with a foreword by A.T. Green), "Thin-Section Mineralogy of Ceramic Materials." Published by British Refractories Research Association, 179 pp. 10 fig. (1948).

Rigterink, M.D., "Ceramic Electrical Insulating Materials." J. Am. Ceram. Soc. **41** (11), part II, 501-6 (1958).

Rindone, G.E., Day, D.E. & Caporali, R., "Relative Acidities of Glasses Containing Al_2O_3 & TiO_2 as Determined by the Oxygen Electrode." J. Am. Ceram. Soc. **43** (11) 571-7 (1960).

Ringel, C.M., "Effects of Firing Atmospheres and Magnesia Additions in the Sintering of Aluminum Oxide." Univ. Microfilms Order No. 66-2132.

Rinne, F., "Morphological & Phisicochemical Studies of Synthetic Spinels as an Example of Nonstoichiometric Compounds." Neues Jahrb. Mineral. Geol, Beil-Bd., **58A**, 43-108 (1928).

Rishel, P.A., Infield, J.M. and Kirchner, H.P., "Leaching and Machining of Polycrystalline Alumina." Ceram. Bull. **47**, 702-6 (1968).

Riskin, V. Ya. & Goncharov, V.V., "Research on the Pressing of Refractories at High Temperatures." Ogneupory **22** (4) 186-8 (1957).

Ristic, M.M. and Afgan, N., "Interrelation of the Kinetics and Mechanism of Sintering." Bull. Boris Kidric Inst. Nuc. Sci. **16**, 267-9 (1965).

Roach, T.L. & Himmelbau, D.M., "Adsorption of Calcium, Strontium, & Thallium Ions from Molten Salts by Silica & Alumina." J. Inorg. & Nuclear Chem. **17** (3-4) 373-81 (1961).

Robert, L., "Relation Between Heat of Wetting & Adsorbability of Liquid Organic Compounds." Compt. rend. **233**, 1103-4 (1951).

Roberts, E.J. & Jukkola, W.W., "Catalytic Treatment of Alumina in Fluidized Beds." U.S. 2,833,622, 5/6/58.

Roberts, I.L. U.S. 683,000,9/17/01.

Roberts, J.P., "Ceramics & Glass: 3, Microscopical Investigation of Mechanical Breakdown in Sintered Aluminum Oxide, Including Observations on Reflected Light Examination." Selected Govt. Res. Repts. (Gt. Brit.) **10**, 15-29 (1952).

Roberts, J.P. & Watt, W., "Axial Loading of Sintered Corundum Tensile Test Pieces." Trans. Brit. Ceram. Soc. **50** (3) 122-43 (1951).

Roberts, J.P. & Watt, W., "Ceramics & Glass: 5, Mechanical Properties of Sintered Alumina." Selected Govt. Research Repts. (Gt. Brit.) **10**, 51-6 (1952).

Robijn, P. & Angenot, P., "Emissivity of Refractory Materials." Verres Refractories **17** (1) 3-10 (1963).

Robinson, Mcd., Pask, J.A. & Fuerstenau, D.W., "Surface Charge of Alumina & Magnesia in Aqueous Media." J. Am. Ceram. Soc. **47** (10), 516-20 (1964).

Robinson, S.P., "Heat Exchange Pebbles." U.S. 2,644,799, 7/7/53; U.S. 2,672,671, 3/23/54; U.S. 2,680,278, 6/8/54; U.S. 2,685,528, 8/3/54; U.S. 2,630,617, 3/10/53; U.S. 2,630,616, 3/10/53; U.S. 2,635,950, 4/21/53; U.S. 2,633,622, 4/7/53; U.S. 2,618,566, 11/8/52; U.S. 2,680,692, 6/8/54.

Robson, H.E. & Broussard, L., "Use of X-Ray Spectrometer Data in Radial Distribution Analysis of the Diffraction Patterns for Amorphous Materials." Advan. X-Ray Anal. **4** 108-16 (1961).

Robson, J.T. & Tucker, H.C., "Method of Coating Concrete Articles with Vitreous Coatings & Resulting Article." U.S. 2,708,172, 5/10/55.

Robson, T.D., "Castable Refractories in Electric Furnace Manufacture." Ceramics **4** (44) 134-8 (1952).

Robson, T.D., "High-Alumina Cements & Concretes." John Wiley & Sons, Inc., New York (1962).

Rochow, E.G., "Electrical Conduction in Quartz, Periclase, & Corundum at Low Field Strength." J. App. Phys. **9** (10) 664-9 (1938).

Rodigina, E.N. & Gomel'skii, K.Z., "Heat Content of Corundum Modification at High Temperature." Zhur. Fiz. Khim. **29**, 1105-12 (1955).

Rodman Laboratory, "Minutes of Symposium on Ceramic Cutting Tools." PB 111757. U.S. Dept. of Commerce, Office of Technical Services, Washington 25, D.C., 88 pp.

Roe, F.C. & Schroeder, H.S., "Abrasion-Resistant Refractory Materials as Applied to Blast Furnace Operations." Blast Furnace Steel Plant **40** (4) 429-35, 442, 458, 460 (1952).

Rogener, H., "Direct-Current Resistance of Ceramic Materials." Z. Elektrochem. **46** (1) 25-7 (1940).

Rogers, E.J. & Mooney, E.L., "Molding of Ceramics." U.S. 2,694,245, 11/16/54.

Rolin, M. and Pham, H.T., "Phase Diagrams of Mixtures Not Reacting with Molybdenum." Rev. Hautes Temp. et Refract. **2**, 175-85 (1965).

Rontgen, P., Winterhager, H., & Kammel, R., "Constitution & Properties of Slags from Smelting Processes: I, Viscosity of Slags of the System $Feo-Al_2O_3 \cdot SiO_2$." Z. Erzbergbau u. Metallhuttenw **9** (May) 207-14 (1956); abstracted in J. Iron Steel Inst. (London) **187** (1) 61-2 (1957).

Rogers, E.J. & Mooney, E.L., "Molding of Ceramics." U.S. 2,694,245, 11/16/54.

Rollason, E.C., et al., "Characteristics of Metallographic Polishing Powders." J. Iron Steel Inst. **162**, 265-70 (1949); Brit. Ceram. Abstracts **49** (1) 10a (1950).

Roller, P.S., "Plasticity." Chem. Industries **43** (4) 398 (1938).

Rooksby, H.P., "X-Ray Identification & Chrystal Structure of Clay Minerals." ed. G.W. Brindley. The Mineralogical Society Brit. Museum, London (1951).

Rooksby, H.P., "The Preparation of Crystalline Delta Alumina." J. App. Chem. **8** (1) 44-9 (1958).

Rooksby, H.P. & Rooymans, C.J.M., "Formation & Structure of Delta Alumina." Clay Minerals Bulls. **4** (25) 234-8 (1961).

Rooymans, C.J.M. (Communication to H.P. Rooksby). J. Inorg. & Nuclear Chem. **11**, 78-9 (1959).

Rose, G., "Hydrargillite." Pogg Ann. **48**, 564 (1839).

Rose, H.E. & Sullivan, R.M.E., "Ball, Tube, & Rod Mills." Chemical Publishing Co., Inc., New York, 258 pp., 114 illus. (1958).

Rose, H.E. & Trbojeric, M.D., "Grinding at Supercritical Speeds in Ball Mills." Nature **183** (4664) 813-7 (1959).

Rose, K., "How Glass Reflectors are Made Through Vaporization of Aluminum." Materials & Methods 28 (6) 85-7 (1948).

Ross, J.F., "Alumina Ceramics." Fr. addn. 82,775, 4/16/64. Addn. to 1,314,617.

Rossett, A.J. de., Finstron, C.G. & Adams, C.J., "Adsorption of H_2S on Alumina at Low Coverages." J. Catalysis 1 (3), 235-43 (1962).

Rossi, R.C., "Kinetics and Mechanism of Final-Stage Densification in the Vacuum Hot-pressing of Fine-Grained Aluminum Oxide." Univ. Microfilms Order No. 65-3076.

Rossi, R.C., & Fulrath, R.M., "Epitaxial Growth of Spinel by Reaction in the Solid State." J. Am. Ceram. Soc. 46 (3) 145-9 (1963).

Rossi, R.C. and Fulrath, R.M., "Improving the Density of Hot-Pressed Refractory Oxide Powder Bodies." U.S. 3,343,915.

Rossi, R.C. and Fulrath, R.M., "Final-Stage Densification in Vacuum Hot-Pressing of Alumina." J. Am. Ceram. Soc. 48, 558-64 (1965).

Rossini, F.D., Wagman, et al., "Selected Values of Chemical Thermodynamic Properties." N.B.S. (US) Circular No. 500, Feb. 1, 1952.

Roth, R.S. & Hasko, S., "Beta-Alumina-Type Structure in the System Lanthana-Alumina." J. Am. Ceram. Soc. 41 (4) 146 (1958).

Roth, W.A., "The Structure of Electrolytically Prepared Aluminum Oxide." Z. anorg. allgem. Chem. 244 48-56 (1940).

Roth, W.A., Wirths, G. & Berendt, H. Z. Elektrochem 48, 264-7 (1942).

Roth, W.A. & Wolf, U., "Heats of Formation of Calcium Aluminates." Z. Elektrochem 46 (3) 232-3 (1940).

Roth, W.A., Wolf, U. & Fritz, O., "Heats of Formation of Al_2O_3 & La_2O_3." Z. Elektrochem 46 (1) 42-5 (1940).

Rotherham, L., et al., "Ceramics for Gas Turbines." Iron Steel Inst. (London) Spec. Rept. 1952, No. 43, pp. 273-80.

Roudabush, N.W., "Bricks." U.S. 2,861,793, 11/25/58.

Roy, D.M., Roy, R., & Osborn, E.F., "Subsolidus Reactions in Oxide Systems in the Presence of Water at High Pressures." J. Am. Ceram. Soc. 36 (5) 147-51 (1953).

Roy, D.M., Roy, R., & Osborn, E.F., "System $MgO-Al_2O_3-H_2O$ & Influence of Carbonate & Nitrate." Am. J. Sci. 251, 337 (1953).

Roy, R., Hill, V.G. & Osborn, E.F., "Polymorphs of Alumina & Gallia." Ind. Eng. Chem. 45 (4) 819-20 (1953).

Roy, R., and Osborn, E.F., "System $Al_2O_3-SiO_2-H_2O$." Am. Mineralogist 39, (11/12) 853-5 (1954).

Roy, S.K. and Coble, R., "Solubility of Hydrogen in Porous Polycrystalline Aluminum Oxide." J. Am. Ceram. Soc. 50 435-6 (1967).

Roy, S.K. and Coble, R.L., "Solubilities of MgO, TiO_2, and $MgTiO_3$ in Aluminum Oxide." J. Am. Ceram. Soc. 51, 1-6 (1968).

Roy, S.S., "Origin of Laterite: A Review." J. Mines, Metals, & Fuels 15, 218-19, 288 (1967).

Royal Worcester Industrial Ceramics, Ltd., "Grinding Ceramics to Close Tolerances." Am. Ceram. Soc. Bull. 41 (10) 606-7 (1962).

Rozen, A.M., Karpacheva, S.M. & Shevelev, Ya.V., "Mobility of Oxygen in Oxides & Kinetics of Oxygen Exchange." Problemy Kinetiki i Kataliza, Akad. Nauk SSSR., Inst. Fiz. Khim., Soveshchanie, Moscow, 1956 9, 251-63 (Pub. 1957).

Rubin, G.A., "Determination of the Thermal Shock Resistance of Ceramic Materials." Ber. deut. keram. Ges. 40 (1) 13-5 (1963).

Rudkin, R.L., Parker, W.J., & Jenkins, R.J., "Thermal Diffusivity Measurements on Metals & Ceramics at High Temperatures." AD-297,836, 20 pp.; U.S. Govt. Res. Repts. 38 (11) 51 (1963).

Ruff, O. & Konschak, M., "Vapor Pressure Measurements on Cu, Au, Al_2O_3 et al." Z. Elektrochem 32, 515-25 (1926).

Ruh, E. & McDowell, J.S., "Thermal Conductivity of Refractory Brick." J. Am. Ceram. Soc. 45 (4) 189-95 (1962).

Ruh, E. & Renkey, A.L., "Thermal Conductivity of Refractory Castables." J. Am. Ceram. Soc. 46 (2) 89-92 (1963).

Ruh, E. & Wallace, R.W., "Thermal Expansion of Refractory Brick." Am. Ceram. Soc. Bull. 42 (2) 52-6 (1963).

Rundle, R.E. & Parasol, M., "O-H Stretching Frequencies in Very Short & Possibly Symmetrical Hydrogen Bonds." J. Chm. Phys. 20, 1487-8 (1952).

Rushmer, R.H. & Elsey, C.L., "Intaglio Stones - Pioneers in Fine Finish." Grinding Finishing 10 (7) 30-1 (1964).

Russell, A.S. & Cochran, C.N., "Surface Areas of Heated Alumina Hydrates." Ind. Eng. Chem. 42 (7) 1336-40 (1950).

Russell, A.S., Edwards, J.D., & Taylor, C.S., "Solubility & Density of Hydrated Aluminas in Sodium Hydroxide Solutions." J. Metals (8) 1123-8 (1955).

Russell, A.S. & Lewis, J.E., "Abrasive Characteristics of Alumina Particles." Ind. Eng. Chem. 46, 1305-19 (1954).

Russell, R.J. & Hardinge, H., "Means & Methods of Supplying Heat to Grinding Mills." U.S. 3,078,048, 2/19/63.

Rustambekyan, S.F., "The Production of Fused Cast Mullite." Ogneupory 29 (2) 67-71 (1964).

Rutman, D.S., "More on Refractories for the Refining Zone of Rotary Kilns." Ogneupory 24 (5) 198-200 (1959).

Rutman, D.S., et al., "Thermally Stable High Alumina Ladle Brick & Stoppers of Mullite-Corundum Composition." Ogneupory 22 (12) 546-9 (1957).

Ryan, J.R. & Ruh, E., "Emission Spectroscopy for Analysis of Ceramic Materials." Ceram. Age 80 (5) 14-6 (1964).

Rybnikov, V.A., "New Refractories." Steklo i Keram. 7 (2) 13-6 (1950).

Rybnikov, V.A., "Use of Hydrate of Alumina in the Production of High Alumina Refractories." Ogneupory 21 (5) 233-4 (1956).

Rbynikov, V.A., Volynskii, E.A. & Vodop'yanov, G.V., "Service of High-Alumina Brick in the Checkers of the Regenerators of Open-Hearth Furnaces." Ogneupory 23 (3) 109-11 (1958).

Rybnikov, V.A. & Volynskii, E.A., "Service of Chrome-Alumina Brick in the Checkers of the Regenerators of Open-Hearth Furnaces." Ogneupory 24 (4) 171-2 (1959).

Rybnikov, V.A., "Properties of Magnesite & High-Alumina Refractories." Ogneupory 21 (1) 37-9 (1956).

Ryshkewitch, E., "Tensile Strength of Several Ceramic Bodies of the One-Component System." Ber. Deut. Keram. Ges. 22, 363-71 (1941).

Ryshkewitch, E., "The Compressive Strength of Pure Refractories." Ber. deut. Keram. Ges. 22 54-65 (1941).

Ryshkewitch, E., "Modulus of Elasticity of Sintered Oxide Refractories." Ber. deut. keram. Ges. 23, 243-69 (1942).

Ryshkewitch, E. Ber. deut. keram. Ges. 25, 95-112 (1944).

Ryschkewitsch, E., "One Component Oxide Ceramics on the Basis of Physical Chemistry." (Oxydkeramik der Einstoffsysteme vom Standpunkt der Physikalischen Chemie). Springer-Verlag, Berlin 1948, vi P 280 pp.

Ryschkewitsch, E., "Rigidity Modulus of Some Pure Oxide Bodies - 8th Communication to Ceramography." J. Am. Ceram. Soc. 34 (10) 322-6 (1951).

Ryschkewitsch, E., "Ceramic Cutting Tool." U.S. 2.270,607, 1/20/42.

Ryshkewitch, E., "Compression Strength of Porous Sintered Alumina & Zirconia - 9th Communication to Ceramography." J. Am. Ceram. Soc. 36 (2) 65-8 (1953).

Ryshkewitch, E., "Ceramic & Logic." Brit. Clayworker 65 (775) 223-4 (1956).

Ryshkewitch, E., "Are Ceramics Really Brittle." Ceram. Ind. 69 (6) 116-7 (1957).

Ryshkewitch, E., "Cutting Tools of Sintered Alumina." Ber. deut. keram. Ges. 34 (1) 3-5 (1957).

Ryshkewitch, E., "Bend Strength of Sintered Alumina." Trans. Brit. Ceram. Soc. 55 (9) 565-70 (1956).

Ryshkewitch, E., Strott, A.J., & Uts, D.L., "Pure Refractory Oxide Bodies Having High Density." U.S. 3,226,456. 12/28/65.

Ryschkewitsch, E. & Taylor H., "Ceramic Tool Material." U.S. 2,979, 414, 4/11/61.

Saalfeld, H., "Transformation of $a-Al_2O_3$ into $\beta-Al_2O_3$." Z. anorg. u. allgem. Chem. 286, 174-9 (1956); abstracted in Brit. Aluminium Co., Ltd., Light Metals Bull. 18 (24) 963-4 (1956).

Saalfeld, H., "Transformation of $a-Al_2O_3$ by Cryolite-Like Substances." Z. anorg. u. allgem. Chem. 291 (1-4) 117-21 (1957).

Saalfeld, H., "The Dehydration of Gibbsite & The Structure of a Tetragonal $\gamma-Al_2O_3$." Clay Minerals Bull. 3 249-56 (1958).

Saalfeld, H., "Single Crystal Investigations on the Dehydration of Hydrargillite." Z. Krist. 112 (1) 88-96 (1959).

Saalfeld, H., "The Structures of Gibbsite & of the Intermediate Products of Its Dehydration." N. Jb. Miner. Abh. 95 1-87 (1960).

Saalfeld, H., "Structure Phases of Dehydrated Gibbsite." Proc. Intern. Symp. Reactivity Solids, 4th, Amsterdam 1960, 310-6 (Pub. 1961).

Saalfeld, H., "Structural Investigations in the System Al_2O_3-Cr_2O_3." Z. Krist. 120 (4-5), 342-8 (1964).

Saalfeld, H. & Jagodzinski, H., "Unmixing in Al_2O_3-Supersaturated Mg-Al Spinels." Z. Krist 109 (2) 87-109 (1957).

Saalfeld, H. and Mehrotra, B.B., "The Structure of Nordstrandite." Naturwissenschaften 53, 128-9 (1966).

Sable, A., "Method & Apparatus for the Manufacture of Alumina." U.S. 2,659,660, 11/16/53.

Safford, H.W. & Silverman, A., "Alumina-Silica Relationship in Glass." Presented at meeting of Glass Division, American Ceramic Society, Conneaut Lake Park, Pa., Sept. 1941; abstracted in Ceram. Ind. 37 (4) 48 (1941).

Sainte-Clair, D. Ann. Phys. Chim. 46 (3) 196) (1856).

Sainte-Clair, D. Ann. Chim. Phys. (3) 61, 309 (1861).

St. Pierre, P.D.S., "Constitution of Bone China: I, High-Temperature Phase Studies." Can. Dept. Mines & Tech. Surveys, T.P. 2, p. 55 (1953).

St. Pierre, P.D.S., "Note on the System CaO-Al_2O_3-P_2O_5." J. Am. Ceram. Soc. 39 (10) 361-2 (1956).

St. Pierre, P.D.S. & Gatti, A., "Process for Producing Transparent Polycrystalline Alumina." U.S. 3,026,177. 3/20/62.

Saito, T., "The Thermal Decomposition of Some Inorganic Hydrates by the Measurement of Proton Magnetic Resonance Absorption. II. Thermal Dehydration of $Al(OH)_3$ & Rehydration of the Dehydration Products." Tokyo Kogyo Shikensho Hokoku 57 (5), 187-93 (1942).

Sait, T., "Proton Magnetic Resonance Studies on the Thermally Dehydrated Products of Alumina Trihydrate." Bull. Chem. Soc. Japan 33, 1626-7 (1960).

Sakamoto, K., "Mechanism of the Decomposition of Sodium Aluminate Solutions." Ext. Met. Aluminum 1, 175-89 (1963).

Salkind, M.J., "Whisker & Whisker Strengthened Composite Materials." Watervliet Arsenal Tech. Rept. RR-6315 (Oct. 1963).

Salmang, H., "Importance of Surface Tension in the Slip-Casting Process." Ber. deut. keram. Ges. 33 (3) 65-72 (1956).

Salmang, H., "Glass Manufacture - Physical & Chemical Principles. Springer-Verlag, Berlin, 1957., 354 pp., 115 illus.

Salmang, H., "Physical & Chemical Fundamentals of Ceramics." (Die Keramik - Physikalishe und Chemische Grundlagen) 4th ed. Springer-Verlag, Berlin, 351 pp., 124 illus. (1958).

Salmassy, O.K., Duckworth, W.H. & Schwope, A.D., "Mechanical Property Tests on Ceramic Bodies." WADC Tech. Rep. No. 53-50.

Salmassy O.K., Bodine, E.G., Duckworth, W.H., & Manning, G.K. WADC Tech. Rep. No. 53-50, Part 2.

Salmassy, O.K., Schwope, A.D. & Duckworth, W.H., "Statistical Significance of the Strength of Brittle Materials." Am. Ceram. Soc. Bull. 33 (8) 240-3 (1954).

Samaddar, B.N., Kingery, W.D. & Cooper, A.R., "Dissolution in Ceramic Systems II Dissolution of Alumina, Mullite, Anorthite, & Silica in a Calcium-aluminum-Silicate Slag." J. Am. Ceram. Soc. 47 (5) 249-54 (1964).

Samaddar, B. & Lahiri, D., "Investigations on the Activation of Slags of CaO-Al_2O_3-SiO_2 System by DTA." Trans. Indian Ceram. Soc. 21 (3) 75-85 (1962).

Samsonov, G.V., "Handbooks of High Temperature Materials: No. 2, Properties Index." Plenum Press, New York, xii + 418 pp. (1964).

Sandford, F. & Ericsson, E., "Effect of Composition of Kiln Atmosphere in the Firing of Refractory Oxides." J. Am. Ceram. Soc. 41 (12) 527-31 (1958).

Sandmeyer, K.H., "Fused Cast Zirconia-Alumina Articles." U.S. 2,903,373, 9/8/59.

Sandmeyer, K.H., "Fused Cast Refractory Articles & Method of Making Them." U.S. 2,911,313, 11/3/59.

Sandmeyer, K.H. & Miller, W.A., "A Fused Cast Alumina Refractory." Am. Ceram. Soc. Bull. 44 (7) 541-4 (1965).

Sang, H.J., "High-Strength Dielectric Materials for Very Fast Aircraft." PB Rept. 128738, 41 pp.; U.S. Govt. Res. Repts. 29 (2) 89 (1958).

Sanlaville, J., "Adsorption of Water Vapor by Activated Alumina & the Dynamic Drying of Gases." Genie chim. 78, 102-22 (1957).

Sano, M. & Taniguichi, T., "Extraction of Alumina Containing Low Silica." Japan 9456, 7/27/62.

Sargent, E.C., "Nuclear Ceramics: I." Ceram. Age 69 (5) 28-29, 32-33 (1957).

Sargent, L.B., & Kipp, E.M., "Activated Alumina for Maintenance of Gas-Engine Crank Case Oil." Iron Steel Eng. 31, 73-6 (1954).

Sarver, J.F., "Comment on 'Vegard's Law and the System Alumina-Chromia'." J. Am. Ceram. Soc. 46 (1) 58 (1963).

Sasvari, K. & Zalai, A., "The Crystal Structure & Thermal Decomposition of Alumina and Alumina Hydrates as Regarded from the Point of View of Lattice Geometry." Acta Geol. Acad. Sci. Hung. 4, 415-66 (1957).

Sasvari, K. & Hegedus, A.J., "X-Ray & Thermoanalytical Study on Thermal Dissociation of Aluminum Oxide Hydrates." Naturwiss. 42, 254 (1955).

Sata, T., "Measurement of the Melting Point of Pure Oxides by Using Small Amounts of Sample." Rev. Int. Hautes Temp. Refract. 3, 337-41 (1966).

Sata, T. & Kiyoura, R., "Fabrication of Cermets of the System Alumina-Chromium by Hot Pressing." Yogyo Kyoaki Shi 67 (760) 116-23 (1959).

Sato, T., "Dehydration of Alumina Trihydrate." J. Appl. Chem. (London) 9 (6) 331-40 (1959).

Sato, T., "Hydrothermal Reaction of Alumina Trihydrate." J. Appl. Chem. (London), 10 (10) 414-7 (1960).

Sato, T., "Thermal Transformation of Alumina Trihydrate, Bayerite." J. Appl. Chem. (London) 12 (12) 553-6 (1962).

Satterfield, W.R., "Refractory & Method." U.S. 2,846,324, 8/5/58.

Saucier, H., "Rapid Orientation Method for Crystals of Synthetic Corundum." Bull. soc. franc. mineral. 76, 480-2 (1953).

Saunders, A.C., "Refractory Articles." U.S. 2,943,008, 6/28/60.

Saunders, L.E., "Preparing Commercially Pure Alumina from Aluminous Materials." U.S. 960,712, 6/7/20; U.S. 1,269,224, 6/11/18.

Savinelli, E.A., "Removal of Fluoride Ion from Water with Activated Alumina." J. App. Chem. 6, 9 (1956).

Savitskii, K.V., Ilyushchenkov, M.A., et al., "Thermovacuum Etching of Al_2O_3 and SiC Crystals." Kristallografiya 11, 341-4 (1966).

Sawamura, H., "Electron-Microscopic Observation of Aluminum Hydroxide." J. Sci. Research Inst. (Japan) 46 15-6 (1952).

Scandrett, H.F. & Porter, J.L., "Process for Treating Aluminous Ores." U.S. 2,852,343, 9/16/58.

Schaefer, C.F. & Schwartzwalder, K., "Method of Firing Alumina Ceramics." U.S. 2,867,888, 1/13/59.

Schaefer, C.F. & Stoia, J.Z. U.S. 3,034,191, 5/15/62.

Schaffer, P.S., "Vapor-Phase Growth of Alpha Alumina Single Crystals." J. Am. Ceram. Soc. 48, 508-11 (1965).

Schairer, J.F., "System CaO-FeO-Al_2O_3-SiO_2: I, Results of Quenching Experiments on Five Joins." J. Am. Ceram. Soc. 25 (10) 241-74 (June 1942).

Schairer, J.F. & Bowen, N.L. "System K_2O-Al_2O_3-SiO_2." Am. J. Sci. 253, 681-746 (1955).

Schairer, J.F. & Bowen, N.L. "System Na_2O-Al_2O_3-SiO_2." Am. J. Sci. 254 (3) 129-95 (1956).

Schairer, J.F. & Yagi, K. "System Feo-$Al_2O_3SiO_2$." Am. J. Sci. Bowen Vol. Part II, pp. 471-512 (1952).

Schatt, W. & Schulze, D., "Electron Microscope Studies of Fracture on Sintered Bodies of Alumina & Chromium." Silikattech 8 (12) 524-32 (1957).

Schatt, W. & Schulze, D., "Electron Microscope Observations of the Free Surface of Sintered Corundum." Ber. deut. keram. Ges. 36 (11) 364-7 (1959).

Schatz, E.A., "Spectral Reflectance of Sintered Oxides." J. Am. Ceram. Soc. 51, 287-9 (1968).

Schawlow, A.L., "Optical Masers." Sci. American 204 (6) 52-61 (1961).

Schawlow, A.L. & Townes, C.H., "Infrared & Optical Masers." Phys. Rev. **112**, 1940 (1958).

Scheetz, H., "Investigation of the Theoretical & Practical Aspects of the Thermal Expansion of Ceramic Materials." **AD 270178**, 63 pp., illus.; U.S. Govt. Res. Repts. 37 (7) 59 (1962).

Schell, D.C. & Neff, J.M., "Refractory Oxide Bodies & Coatings for Aircraft Power Plants." U.S. Air Force, Air Materiel Command, AF Tech. Rept. No. 6081, 12 pp. (June 1950).

Scheuplein, R. & Gibbs, P., "Surface Structure in Corundum: I, Etching of Dislocations." J. Am. Ceram. Soc. **43** (9) 458-72 (1960).

Scheuplein, R. & Gibbs, P., "Surface Structure in Corundum: II, Dislocation Structure & Fracture of Deformed Single Crystals." J. Am. Ceram. Soc. **45** (9) 439-52 (1962).

Schiavo, J. and Janus, T.P., "Ultrathin Al_2O_3 Films fo Electron Microscopy." Rev. Sci. Instrum. **38**, 691-2 (1967).

Schifferli, L.M., Jr., "Slip Casting." Metal Progr. **76** (4) 99-102 (1959).

Schillbach, H., "Vacuum-Tight Sealing of Ceramic to Ceramic or Ceramic to Metallic Bodies." Ger. 700,840, 11/28/40.

Schilling, A.H. & Smith, R.B., "Information on Ceramic & Water-Cooled Turbine Blading for Gas Turbines Obtained from E. Schmidt (LFA)." CIOS File XXX-66, Item 5, 31. **PB L 58,374**, 3 pp. (Sept. 1945); abstracted in Bibliog. Sci. Ind. Repts. 5(3) 205 (1947).

Schlanger, S.O. "Nordstrandite ($Al_2O_3 \cdot 3H_2O$) from Guam." Am. Mineralogist **50**, 1029-37 (1965).

Schlechtweg, H., "Fracture of Rotating Brittle Disks." Ingenieur Arch **6** (5) 365-72 (1935); abstracted in Physik. Ber. **17** (3) 277 (1936).

Schloemer, H. and Mueller, F., "Effect of Etching on Dislocations in Orientated Basal Flakes of Corundum Crystals." Ber. Deut. Keram. Ges. **45**, 11-18 (1968).

Schlotzhauer, L.R., Hutchins, J.R., "Corrosion Resistance of Al_2O_3-ZrO_2 SiO_2 Refractories." Glass Ind. **47**, 26-9, 46, 48, (1966).

Schlotzhauer, L.R. & Wood, K.T., "Method of Making a Refractory Body & Article Made Thereby." U.S. 2,842,447, 7/8/58.

Schmacher, G. & Dintera, H., "Sintered Mixtures of Titanium Oxide & Alumina for Tools." Ger. 1,149,288, 5/22/63.

Schmah, H., "Simple Preparation of Well-Crystallized Bayerite." Z. Naturforsch 1, 323-4 (1946).

Schmellenmeier, H., "Simple Method for the Manufacture of Solid Ceramic Shapes for Temperatures up to 1800°C." Z. Tech. Physik **24** (9) 217-8 (1943).

Schmidt, J., "Various Aspects of Corundum Trichites." Compt. rend. **258** (18) 4480-1 (1964).

Schmidt, K.H., Brodt, R. & Lampatzer, K., "High-Temperature, Heat-Resisting Protective Layers on Metal Parts, Especially Steel Sheet & Strip." Ger. 1,115,555, 10/19/61.

Schmitter, E., "Curves for Differential Thermal Analyses Obtained from a Study of Bauxites, Bauxitic Clays & Other Minerals." Univ. Nacl. Autonoma Mex., Inst. Geolo., Bol. No. 63, 1-57 (1962).

Schneider, A., "Artificial Abrasives—Silicon Carbide & Electrocorundum." Radex Rundschau 1959, No. 1 pp. 421-33.

Schneider, A. & Gattow, G., "Heat of Formation of Aluminum Oxide." Z. anorg. u. allgem. Chem. **277** (1-2) 41-8 (1954).

Schneider, R., "Aluminum Hydroxide for Manufacture of Opaque Glass." Glashutte **56** (36) 703-4 (1926). Glass Ind. 8 (3) 62-3 (1927).

Schneider, S.J., "Effect of Heat-Treatment on the Constitution & Mechanical Properties of Some Hydrated Aluminous Cements." J. Am. Ceram. Soc. **42** (4) 184-93 (1959).

Schneider, S.J., "Compilation of Melting Points of the Metal Oxides." U.S. Nat. Bur. Std. Monograph 68 (1963).

Schneider, S.J. and McDaniel, C.L., "Melting Point of Alumina in Vacuum." Rev. Int. Hautes Temp. Refract. **3**, 351-61 (1966).

Schneider, Sam J. and McDaniel, C.L., "Effect of Environment on the Melting Point of Al_2O_3." J. Res. Nat. Bur. Stds. A71, 317-33 (1967).

Schneider, S.J. & Mong, L.E., "Thermal Length Changes of Some Refractory Castables." J. Res. NBS (US) **59** (1) 1-8 (1957) RP 2768.

Schneider, S.J. & Mong, L.E., "Elasticity, Strength, & Other Related Properties of Some Refractory Castables." J. Am. Ceram. Soc. **41** (1) 27-32 (1958).

Schofield, H.Z., Lynch, J.F., & Duckworth, W.H., "Fundamental Studies of Ceramic Materials." Final Summary Report ATI-74493 (March 31, 1949).

Scholder, R. & Mansmann, M., "Compounds of the β-Alumina Type." Z. anorg. allgem. Chem. **321** (5-6) 246-61 (1963).

Scholz, S., "New Views on the Hot-Pressing of Refractory Compounds." Soviet Powder Met. Metal Ceram. (English Transl.) **1963**, No. 2, pp. 170-5. .

Scholze, H., "Aluminum Borates." Z. anorg. u. allgem. Chem. **284**, 272-7 (1956): abstracted in Brit. Aluminium Co. Ltd. Light Metals Bull. 18 (21) 824 (1956).

Schreiber, E. and Anderson, O.L., "Pressure Derivatives of the Sound Velocities of Polycrystalline Alumina." J. Am. Ceram. Soc. **49** 184-90 (1966).

Schrewelius, N.G., "Constitution & Microhardness of Fused Corundum Abrasives." J. Am. Ceram. Soc. **31** (6) 170-5 (1948).

Schrewelius, N.G., "Constitution & Microhardness of Fused Corundum Abrasives—Correction." J. Am. Ceram. Soc. **31** (10) 216 (1948); Ibid. (6) 170-75.

Schrewelius, N.G., "Manufacturing Corundum." Swed. 125,868, 2/11/48.

Schrewelius, N.G., "Abrasive Corundum from Chamotte." Swed. 132,872, 5/19/47.

Schroder, A., "Isomorphous Lamination in Corundum & Its Importance for the Separation of Natural from Synthetic Corundum." Zentr. Mineral. Geol. A, No. 5, pp. 129-35 (1936).

Schroeder, H.S. & Logan, Ian M., "Bonded Alumina Refractories." U.S. 2,569,430, 9/25/51.

Schulz, H., "Grinding & Polishing: II." Glas-Email-Keramo-Technik 3 (7) 249-50 (1952).

Schurecht, H.G., "Reactions of Slag with Refractories: II, Refractory Coatings Produced with Metallic Aluminum." J. Am. Ceram. Soc. **22** (11) 384-8 (1939); for Part I see Ibid. (4) 116-23.

Schurecht, H.G., "Spark Plug Insulators Containing ZnO." U.S. 2,804,392, 8/27/57.

Schurecht, H.G., "Spark Plug Insulators Containing Stannic Oxide." U.S. 2,917,394, 12/15/59.

Schurecht, H.G., "Electrically Semiconducting Ceramic Body." U.S. 3,037,140, 5/29/62.

Schuster, D.M. & Scala, E., "The Mechanical Interaction of Sapphire Whiskers with a Birefringent Matrix." Trans. AIME **230**, 1635-40 (1964).

Schwab, G.M., "Semiconductor Properties of Aluminum Oxide." Z. Angew. Phys. **14**, 763-5 (1962).

Schwab, G.M., & Konrad, A., "Influence of Radiation on Alumina as a Para Hydrogen Catalyst." J. Catalysis 3 (3) 274-9 (1964).

Schwarz, H. & Hempel, S., "Tamped Tunnel Kiln Cars Made of Refractory Concrete Based on Alumina Cement." Silikattech. **11** (5) 233 (1960).

Schwartz, B., "Thermal Stress Failure of Pure Refractory Oxides." J. Am. Ceram. Soc. **35**, (12) 325-33 (1952).

Schwartz, M.A., White, G.D., & Curtis, C.E., "Crucible Handbook. A Compilation of Data on Crucibles." U.S. Atomic Energy Comm. ORNL-1354, 28 pp. (1953).

Schwartzwalder, K., "Spark-Plug Insulator." U.S. 2,332,014, 10/19/43.

Schwartzwadker, K., "Injection Molding of Ceramic Materials." Bull. Am. Ceram. Soc. 28 (11) 459-61 (1949).

Schwartzwalder, K., "Spark Plug Electrode Seal." U.S. 3,139,553, 6/30/64.

Schwartzwalder K. & Barlett, H.B., "Ceramic Composition & Process for Making." U.S. 2,760,875, 8/28/56.

Schwartzwalder, K. & Fessler, A.H., "Ceramic Body & Method of Making." U.S. 2,308,115, 1/12/43.

Schwartzwalder, K. & Somers, A.V., "Sealing Cement for Spark Plug." U.S. 2,829,063, 4/1/58.

Schweig, B., "Secondary Glass Constituents: VII, Aluminum Oxide." Glass 22 (10) 263-5 (1945).

Schwerin, B., "Manufacture of Aluminum Hydroxide." U.S. 1,216,371, 2/20/17.

Schwerin, G.B. Ger. 274,039, 276,244, 1910.

Schwiersch, H., "Thermal Dissociation of the Natural Hydroxides of Aluminum & Trivalent Iron." Chemie der Erde Vol. VIII, 252-315 (1933).

Schwiete, H.E., Granitzki, K.E. & Karsch, K.H., "Thermal Conductivity of Refractory Materials of the System Al_2O_3-SiO_2 at 200° to 1600°C." Ber. deut. keram. Ges. 38 (12) 529-34 (1961).

Schwiete, H.E., Mueller-Hesse, H., & Pflanz, J.E., "Investigation of the BaO-Al_2O_3-SiO_2 System with the Aid of Infrared Spectroscopy." Forschungsber. Landes Nordrhein-Westfalen 1961, No. 998, 170 pp. DM-49.

Scott, H.F. & Emblem, H.G., "Some Applications of Ethyl Silicate in Refractories." Refractories J. 27 (7) 286-9 (1951).

Scott, J., "Effect of Seed & Temperature on the Particle Size of Bayer Hydrate." Ext. Met. Aluminum 1, 203-18 (1963).

Scott, T.R., "New Acid Alumina Process." Research (London) 14 (2) 50-4 (1961).

Searcy, A.W., "High-Temperature Inorganic Chemistry. Progress in Inorganic Chemistry." Vol. III, pp. 49-128. Edited by F.A. Cotton. Interscience Publishers, New York (1962).

Sears, G.W., "High Strength Crystals." U.S. 2,813,811, 11/19/57.

Sears, G.W. & DeVries, R.C., "Nucleation of Alumina at High Supersaturations & Subsequent Recrystallization." J. Chem. Phys. 32 (1) 93-5 (1960).

Sears, G.W. & DeVries, R.C., "Morphological Development of Aluminum Oxide Crystals Grown by Vapor Deposition." J. Chem. Phys. 39 (11), 2837-45 (1963).

Sears, G.W., DeVries, R.C. & Huffine, C., "Twist in Alumina Whiskers." J. Chem. Phys. 34, 2142-3 (1961).

Sears, G.W. & Navias, L., "Evaporation of Aluminum Oxide." J. Chem. Phys. 30 (4) 1111-2 (1959).

Sedalia, B.M., "Bauxite Versus Diaspore Clay in High Alumina (70%) Refractory Brick." M.Sc. Thesis (unpublished) School of Mines & Metallurgy, Univ. Missouri, Rolla, Mo.

Sedlacek, R., "Tensile Strength of Brittle Materials." Tech. Doc. Rept. No. ML-TDR-64-49 (March 1964). AF Materials Lab. Res. & Tech. Div. AF.SC W-P AFB, Ohio.

Seeger, K., "Physics & Technique of the Afteremission of Electrons (Exoelectrons)." Angew. Chem. 68, 285-9 (1956).

Seigle, L.L., "Role of Grain Boundaries in Sintering." Kinet. High-Temp. Processes, Conf., Dedham, Mass. 1958, 172-8 (Pub. 1959).

Seki, I.K., "Electrocast Refractories." Repts. Imp. Ind. Research Inst. Osaka, Japan 18 (8) 131 pp. (1937).

Sekiguchi, J., "Magnesia-Alumina Refractories." Japan 8585, 7/18/62.

Sellers, D.J., Heuer, A.H., Rhodes, W.H., & Vasilos, T., "Alumina Crystal Growth by Solid-State Techniques." J. Am. Ceram. Soc. 50, 217-18 (1967).

Selsing, J., "Ceramic Products." U.S. 2,898,217, 8/4/59.

Selwood, P.W., "Magnetochemistry, 2nd Ed." Interscience Pub. Inc., New York, xii + 435 pp. (1956).

Senarmont, H. de. C.R. Acad. Sci., Paris 32, 763 (1851).

Shaffer, P.T.B., "Handbook of High-Temperature Materials." No. 1. Materials Index. Plenum Press, New York, xx + 740 pp. (1964).

Shackelford, J.F. and Scott, W.D., "Relative Energies of [1100] Tilt Boundaries in Aluminum Oxide." J. Am. Ceram. Soc. 51 688-92 (1968).

Shakhtakhtinskii, G.B., et al., "Concentration of Vanadium in Caustic Solutions During Production of Alumina from Alunite." Azerb. Khim. Zh. 1963, (4), 109-13.

Shalabutov, Yu. K., "The Origin of Nonstationary Processes in Aluminum Oxide Coatings of Vacuum-Tube Heaters." Fiz. Tverdogo Tela 1, 296-306 (1959); Soviet Phys. Solid State 1, 266-74.

Shalnikova, N.A. & Yakovlev, Y.A., "X-Ray Determinations of Lattice Constants & Coefficients of Thermal Expansion for Leucosapphire & Ruby." Kristallografiya 1, 531-3 (1956).

Shapland, J.T. & Livovich, A.F., "Evaluation of Five Commercial Calcium Aluminate Cements." Am. Cera. Soc. Bull. 43 (7) 510-3 (1964).

Sharma, R.A. and Das, T.P., "Crystalline Fields in Corundum-Type Lattices." J. Chem. Phys. 41, 3581-91 (1964).

Sharma, R.A. & Richardson, F.D., "Activities in Lime-Alumina Melts." J. Iron Steel Inst. (London) 198 (4) 386-90 (1961).

Shaw, G.S., "Manufacture of Glass Containers." Brit. 571,310, 8/29/45.

Shaw, M.C. & Smith, P.A., "Workpiece Compatibility of Ceramic Cutting Tools." ASLE (Am. Soc. Lubrication Engrs.) Trans. 1, 336-44 (1958).

Sheheen, S.J. & Mooney, E.L., "Semiconducting Ceramic Spark-Producing Devices." U.S. 2,861,014, 11/18/58.

Sheindlin, A.E., et al., "Enthalpy & Heat Capacity of Corundum Melt up to a Temperature of about 2800°K." Inzh.-Fiz. Zh., Akad. Nauk Belorussk. SSR 7 (5) 63-5 (1964).

Sherby, O.D., "Factors Affecting the High-Temperature Strength of Polycrystalline Solids." Acta Met. 10, 135-47 (1962).

Shevlin, T.S., "Development, Properties, & Investigation of a Cermet Containing 28% Alumina & 72% Chromium." U.S. Air Force, Air Res. & Development Command, WADC Tech. Rept. No. 53-17, 52 pp. (Dec. 1952).

Shevlin, T.S., "Cermets, Their Nature & Uses." Ceram. Age 70 (5) 22 (1957).

Shevlin, T.S. & Hauck, C.A., "Fundamental Study & Equipment for Sintering & Testing of Cermet Bodies: VII, Fabrication, Testing, & Properties of 34 Al_2O_3-66 Cr-Mo Cermets." J. Am. Ceram. Soc. 38 (12) 450-4 (1955).

Shimizu, Y., et al., "Aging of Aluminogel." Kogyo Kagaku Zasshi 67 (5) 788-97 (1964).

Shiraki, Y., "Slip Casting: I." J. Japan Ceram. Assoc. 57 (636) 37-41 (1949).

Shiraki, Y., "Slip Casting: III, Physical Properties of Sintered Alumina Porcelain." J. Ceram. Assoc. Japan 58 (643) 9-14 (1950).

Shiraki, Yoichi, "Slip Casting: IV, Casting of Amphoteric Materials, Especially Aluminum Oxide." J. Ceram. Assoc. Japan 58 (644) 41-5 (1950).

Shirikov, Yu. G. & Kirillov, I.P., "Magnetic Properties of Impregnated & Mixed Nickel-Alumina Catalysts." Izv. Vysshikh Uchebn. Zavedenii, Khim. I Khim. Tekhnol. 6 (6) 945-51 (1963).

Shishakov, N.A., "Electronographic Study of Hydroxide Films on Metals." Zhur. Fiz. Khim. 26, 106-11 (1952).

Shomate, C.H. & Naylor, B.F., "High-Temperature Heat Contents of Aluminum Oxide. Aluminum Sulfate, Potassium Sulfate, Ammonium Sulfate, & Ammonium Bisulfate." J. Amer. Chem. Soc. 67 (1) 72-5 (1945).

Shook, W.B., Author, "Critical Survey of Mechanical Property-Test Method for Brittle Materials." ADS Tech. Doc. Rept. 63-491, July 1963. AF Materials Lab., Aeronautical Systems Div., AF Systems Command W-PAFB, Ohio.

Shreve, O.D., "Infrared, Ultraviolet, & Raman Spectroscopy—Application to Analysis of Complex Materials." Anal. Chem. 24 (10) 1692-9 (1952).

Shternberg, A.A. & Kuznetsov, J.A., "Corundum, Crystallization of, on a Seed from the Gas Phase." Kristallografiya 9 (1), 121-3 (1964).

Shubnikov, A.V., Klassen-Neklyudova, M.V., & Grum-Grzhimailo (Editors), "Physical Properties of Synthetic Corundum—A Symposium." Trudy Inst. Krist. Adad. Nauk SSSR 8, 356 pp. (1953).

Shul'man, A.R., "Technique of Studying Electrical Conductivity of Semiconductors at High Temperatures." Zhur. Tekh. Fiz. (USSR) 9, 389-98 (1939).

Shul'man, A.R., "Electrical Conductivity of Alumina at High Temperatures." Zhur. Tekh. Fiz. 10, 1173-82 (1940).

Shul'man, A.R., Makedonskii, V.L., & Yaroshetskii, I.D., "Secondary Electronic Emission for Aluminum Oxide at Different Temperatures." Zhur. Tekh. Fiz. 23, 1152-60 (1953).

Shul'man, A.R. & Rozentsveig, I.Y., "Secondary Electron Emission of Alumina." Doklady Akad. Nauk SSSR 74, 497-500 (1950).

Shulz, I.H., "Grinding & Polishing, I." Glas-Email-Keramo Technik 3, 151-4 (1952).

Sibley, L.W., Mace, A.E., Grieser, D.R., & Allen, C.M., "Characteristics Governing the Friction & Wear Behavior of Refractory Materials for High-Temperature Seals & Bearings." PB Rept. 171010, 54 pp.; U.S. Govt. Res. Repts. 34 (6) 737 (1960).

Sicka, R.W., et al., "Reinforcement of Nickel Chromium Alloys with Sapphire Whiskers." Final Rept. on Navy Bureau of Weapons Contract Now 64-01256 (Oct. 27, 1964).

Siede, A. & Metcalfe, A.G., "Cermet Preparation by Reactions in the Iron-Aluminum-Oxygen System." PB Rept. 131820, 43 pp.; U.S. Govt. Res. Repts. 30 (2) 80-1 (1958).

Siegrist, P.F., "Apparatus for Making Thin Ceramic Plates." U.S. 2,715,256, 8/16/55.

Siekmann, H.J. & Sowinski, L.A., "What Angles are Best for Ceramic Tools?" Metal-Working Production 101 (Sept. 6) 1573-5 (1957); abstracted in J. Iron Steel Inst. (London) 189 (3) 287 (1958).

Sieverts, A. & Moritz, H., "Manganese & Hydrogen." Z. physik. Chem. A-180, 249-63 (1937).

Siewert, G. & Jungnickel, H., "Alkaline Reaction of Commercial Al_2O_3. Adsorption Analysis with Al_2O_3." Ber. 76B, 210-3 (1943).

Sifre, G., "La Oxide Containing Sintered Al_2O_3." Fr. 1,330,309, 6/21/63.

Silverman, W.B., "Effect of Alumina on Devitrification of Soda-Lime-Silica Glass." J. Am. Ceram. Soc. 22 (11) 378-84 (1939).

Silverman, W.B., "Effect of Alumina on Devitrification of Sodium Oxide-Dolomite-Lime-Silica Glasses." J. Am. Ceram. Soc. 23 (9) 274-81 (1940).

Simpson, H.E., "Ten Years of Progress in the Glass Industry, 1948-1958." Glass Ind. 40 (11) 606-27, 661-9; (12) 698-701, 728-29 (1959).

Sinclair, E.L., "How to Select Abrasives for Grinding Ceramics." Ceram. Ind. 67 (5) 92-3, 95 (1956).

Singer, F., "Colored Bodies." Pottery Gaz. 71 (1) 27-34 (1946); abstracted in J. Soc. Glass Technol. 30 (138) 85 (1946).

Singer, F., "Protection of Refractory Material & Metals in Contact with Molten Aluminum." Brit. 580,916, 10/9/46.

Singer, F., "Sinter Alumina as Engineering Material for Cutting Tools & Turbine Blades." Ceramics 1 (5) 215-25; (6) 279-88 (1949).

Singer, F. & Singer, S.S., "Production of Sinter Alumina Cutting Tools." Ind. Diamond Rev. 17 (200) 126-31 (1957).

Singer, F. & Singer, S.S., "Sintered Alumina as A Cutting Tool Material." Keram. Z. 9 (10) 532-6 (1957).

Singer, F. & Singer, S.S., "Sinter Alumina for Cutting Tools." Refractories J. 34, 456-62 (1958).

Singer, F. & Singer, S.S., "Industrial Ceramics." Chapman & Hall Ltd, London, 1455 pp. (1963).

Singer, F. & Thurnauer, H., "Sinter Alumina." Metallurgia 36, 237-42, 313-5 (1947).

Sinha, K.P. & Sinha, A.P.B., "Vacancy Distribution & Bonding in Some Oxides of Spinel Structure." J. Phys. Chem. 61 758-61 (1957); Chem. Abs. 51 (19) 14362i (1957).

Skaupy, F., "Metallic Conductivity & Lattice Distortion." Die Technik 2, 77-9 (1947).

Skaupy, F. & Hoppe, H., "Crystal Radiation & Grain-Boundary Radiation of Non-metallic Substances (Oxides)." Z. tech. phys. 13, 226-8 (1932).

Skaupy, F. & Liebmann, G., "The Temperature Radiation of Non-Metallic Bodies, Especially Oxides." Z. Elektrochem. 36, 784-6 (1930).

Skobeev, I.K., "Processing of Aluminosilicates into Alumina Using Ammonium Sulfate." Nauchn. Tr., Irkutskii Politekhn. Inst. 1963 (19) 229-34.

Skunda, M., "Method for Shaping Ceramic Articles." U.S. 2,726,433, 12/13/55.

Slonin, A.E. (ed.), "Seventh Symposium on Electromagnetic Windows." Ohio State Univ. RTD (June 1964).

Slyh, J.A., et al., "Refractory Oxides as Liners for Combustion Chambers of Rocket Motors." Battelle Memorial Inst. Rept. 8 (Oct. 1946). 7 pp. PB 60, 655; abstracted in Bibliog. Sci. Ind. Repts. 4 (11) 966 (1947).

Slyh, J.A. & Bixby, H.D., "Ultrasonics in Ceramics." Am. Ceram. Soc. Bull. 29 (10) 346-8 (1950).

Smalley, A.K., Riley, W.C. & Duckworth, W.H., "Al_2O_3-Clad UO_2 Ceramics for Nuclear-Fuel Applications." Am. Ceram. Soc. Bull. 39 (7) 359-61 (1960).

Smirnov, M.N., Eliseeva, A.A. & Lobanova, E.V., "Granular Activated Aluminum Oxide." U.S.S.R. 136,336, 3/14/61.

Smirnov, G.M. & Mchedlov-Petrosyan, O.P., "Products of Interaction of $BaSO_4$ with Al_2O_3 at 1200° to 1400°C." Doklady Akad. Nauk. SSSR 64 (2) 223-4 (1949).

Smith, A.E. & Beeck, O.A., "Beta-Alumina Catalysts." U.S. 2,454,227, 11/16/48.

Smith, B.R., "Method of Making Metallized Ceramic Bodies." U.S. 3,074,143, 1/22/63.

Smith, E.F., "Alumina Compositions." U.S. 2,847,387, 8/12/58.

Smith, G.H. & Yenni, D.M., "Shaping Unicrystalline Bodies of Material such as Corundum & Spinel." U.S. 2,537,165, 1/9/51.

Smith, R.K. & Metzner, A.B., "Rates of Surface Migration of Physically Adsorbed Gases." J. Phys. Chem. 68 (10) 2741-7 (1964).

Smith, R.L. & Sandland, G.E., "Some Notes on the Use of a Diamond Pyramid in Hardness Testing." J. Iron Steel Inst. 111, 285-94 (1925).

Smoke, E.J., "Thermal Endurance of Some Vitrified Industrial Compositions." Ceram. Age 54 (3) 148-9 (1949).

Smoke, E.J., "Spinel As Dielectric Insulation." Ceram. Age 63 (5) 13-6 (1954).

Smoke, E.J., "Fabrication of Dense Ceramic Radomes by Casting." Ceram. Age 69 (4) (1957).

Smoke, E.J., "Thermal Conditioning of Dense Ceramics." Ceram. Age 69 (4) 32-3 (1957).

Smoke, E.J., Callahan, J.P., et al., "Improved Ceramics." PB Rept. 138939, 76 pp.; U.S. Govt. Res. Repts. 32 (6) 730 (1959).

Smoke, E.J., Illyn, A.V., & Koenig, J.H., "Effect of Thermal Conditioning on High-Frequency Ceramic Dielectric Insulation." Nat. Acad. Sci.-Nat. Res. Council Publ. 396 34-6 (1955).

Smoke, E.J., & Koenig, J.H., "Thermal Properties of Ceramics." Rutgers Univ. Eng. Res. Bull. 1958, No. 40, 53 pp.

Smoke, E.J. & Koenig, J.H., "Development of Refractory Ceramics That Can Be Processed at Temperatures Considerably Lower Than Their Maximum Use Temperature." AD-287, 914, 42 pp. illust; U.S. Govt. Res. Repts. 38 (4) 64-5 (1963).

Smoke, E.J. & Koenig, J.H., "Development of Refractory Ceramics that can be Processed at Temperatures Considerably Lower than Their Maximum Use Temperature." AD 275787, 43 pp. illus.; U.S. Govt. Res. Repts., 37 (16) 23-4 (1962).

Smoke, E.J., et al., "High-Temperature Materials." U.S. Dept. Com. Office Tech. Serv. AD 287,500 142 pp. (1962).

Smoke, E.J., et al., "Development of Refractory Ceramics That Can be Processed at Temperatures Considerably Lower than Their Maximum Use Temperature." AD 264776, 41 pp.; illus.; U.S. Govt. Res. Repts. 37 (1) 58 (1962).

Smoot, T.W. and Ryan, J.R., "Dense Refractory Slip-Cast Shapes of High Purity." U.S. 3,235,923. 2/22/66.

Smothers, W.J., et al., "Bibliography of Differential Thermal Analysis." Univ. Arkansas, Inst. of Science & Technol., Res. Ser. No. 31, 44 pp. (1951).

Smothers, W.J. & Reynolds, H.J., "Sintering & Grain Growth of Alumina." J. Am. Ceram. Soc. 37 (12) 588-95 (1954).

Smutny, Z. & Tomshu, F., "Question of Selection of Refractories for Lining Refining Furnaces." Ogneupory 23 (4) 182-8 (1958).

Smyth, H.T., "Physical Principles in Ceramic Radome Design." Ceram. Age 69 (4) 33, 35 (1957).

Snow, R.B., "Equilibrium Relationships on the Liquidus Surface in Part of the $MnO-Al_2O_3-SiO_2$ System." J. Am. Ceram. Soc. 26 (1) 11-20 (1943).

Snyder, P.E. & Seltz, H., "Heat of Formation of Aluminum Oxide." J. Am. Chem. Soc. **67** (4) 683-5 (1945).

Society of Glass Technology, Refractories Committee, "Survey of Refractories Used in Glass Tank Furnaces." J. Soc. Glass Technol. **42** (209) 63-99P (1958).

Soden, R.R. and Monforte, F.R., "Etching Reagent for Oxides." J. Am. Ceram. Soc. **48**, 548 (1965).

Soga, N., Schreiber, E., & Anderson, O.L., "Estimation of Bulk Modulus and Sound Velocities of Oxides at Very High Temperatures." J. Geophys. Res. **71**, 5315-20 (1966).

Sokolov, V.A., Banashek, E.I., & Rubinchik, S.M., "Enthalpy of Corundum at 678-1330°K." Zh. Neorgan. Khim. **8**, 2017-20 (1963).

Solomin, N.V., "Microstructure of Baddeleyite-Corundum Electrofused Refractory Before & After Service in a Glass-melting Furnace." Ogneupory **21** (6) 168-74 (1956).

Solomin, N.V., "Ceramic Compositions for Spark-Plug Insulators." U.S.S.R. 120,760, 6/19/59.

Solomin, N.V. & Gladina, N.M., "Corrosion of Refractories by Glass with High Alkaline Earth Content." Steklo i Keram. **13** (5) 1-3 (1956).

Solomin, N.V. & Galdina, N.M., "Production & Testing of Baddeleyite-Corundum Electrofused Refractories." Steklo i Keram. **13** (1) 1-5 (1956).

Soltis, P.J., "Anisotropic Mechanical Behavior in Sapphire (Al$_2$O$_3$) Whiskers." Naval Air Eng. Center, Philadelphia. AML Rept. 1831, April 1964.

Somers, A.V., "Alumina Ceramic Bodies Containing Niobia." U.S. 3,088,832, 5/7/63.

Somers, A.V. & Schwartzwalder, K., "Cement for Spark Plugs, Etc." U.S. 2,906,909, 9/29/59.

Sommer, A. & Brady, L.J., "Refractory Article & Method of Making." U.S. 2,874,067, 2/17/59.

Somov, A.I., et al., "The Growing of Single Crystals of Corundum by the Czochralski Technique." Izv. Akad. Nauk SSSR Neorgan. Materialy **1**, 1049-50 (1965).

Sono, K., "Some Experiments on Vanadium Colors." Nagoya Kogyo Gijutsu Shikensho Hokoku **2** (2) 29-31 (1953).

Sorg, H.E., U.S. 2,740,067, 3/27/56.

Sorg, H.E., "Ceramic Vacuum Tube." U.S. 2,754,445, 7/10/56.

Souza Santos, P., Vallejo-Freire, A., Parsons, J. & Watson, J.H.L., "The Structure of Schmidt's Aluminum Hydroxide Gel." Experimenta **14**, 318-20 (1958).

Souza Santos, P., Vallejo-Freire, A., & Souza Santos, H.L., "Electron-Microscope Studies on the Ageing of Amorphous Colloidal Aluminum." Koll. Z. **133**, 101-7 (1953).

Souza Santos, P., Watson, J.H.L., Vallejo-Freire, A., & Parsons, J. "X-Ray & Electron Microscope Studies of Aluminum Oxide Trihydrates." Koll. Z. **140**, 102-12 (1955).

Speil, S., et al., "Differential Thermal Analysis, Its Application to Clays & Other Aluminous Minerals." U.S. Bur. Mines Tech. Paper 664, 81 pp. (1945). Supt of Documents, Washington, D.C.

Spencer, F.E.V. & Turner, D., "Properties & Applications of High Alumina Ceramics in Industry." J. Inst. Production Engrs. **37** (9) 548-59, 567 (1958); abstracted in Brit. Aluminium Co. Ltd., Light Metals Bull. **20** (24) 837 (1958).

Spencer, L.F., "Abrasive Sheets & Belts." Metal Ind. (London) **94** (12) 223-4 (1959).

Spinner, S. & Valore, R.C., Jr., "Comparison of Theoretical & Empirical Relations Between the Shear Modulus & Torsional Resonance Frequencies for Bars of Reactangular Cross Section." J. Res. Natl. Bur. Stds. **60** (5) 459-64 (1958).

Spoerry, M., "Production of Corundum." Swiss 250,894 7/1/48.

Spriggs, R.M., "Expression for Effect of Porosity on Elastic Modulus of Polycrystalline Refractory Materials, Particularly Aluminum Oxide." J. Am. Ceram. Soc. **44** (12) 628-9 (1961).

Spriggs, R.M., "Effect of Open & Closed Pores on Elastic Moduli of Polycrystalline Alumina." J. Am. Ceram. Soc. **45** (9) 454 (1962).

Spriggs, R.M. & Bender, S.L., "Vegard's Law & the System Alumina-Chromia." J. Am. Ceram. Soc. **45** (10) 506 (1962).

Spriggs, R.M. & Brissette, L.A., "Expression for Shear Modulus & Poisson's Ratio of Porous Refractory Oxides." J. Am. Ceram. Soc. **45** (4) 198-9 (1962).

Spriggs, R.M., Mitchell, J.B., & Vasilos, T., "Mechanical Properties of Pure, Dense Aluminum Oxide as a Function of Temperature & Grain Size." J. Am. Ceram. Soc. **47** (7) 323-7 (1964).

Spriggs, R.M. & Vasilos, T., "Effect of Grain Size on Transverse Bend Strength of Alumina & Magnesia." J. Am. Ceram. Soc. **46** (5) 224-8 (1963).

Spriggs, R.M. & Vasilos, T., "Functional Relation Between Creep Rate & Porosity for Polycrystalline Ceramics." J. Am. Ceram. Soc. **47** (1) 47-8 (1964).

Stackhouse, R.D., Guidotti, A.E. & Yenni, D.M., "Plasma-Arc Plating. A New Production Process for High-Temperature Design." Prod. Eng. **29** No. 50, 104-6 (1958).

Stanworth, J.E. & Turner, W.E.S., "Effect of Small Additions of Alumina on Reactions in the Mixture 6SiO$_2$+Na$_2$CO$_3$ + CaCo$_3$." Jour. Soc. Glass Tech. **21** (85) 299-309 (1937).

Starokadomskaya, E.L., Tsitovskii, I.L., & Klepkova, E.N., "Investigation of Materials for High-Temperature Thermocathode Preheaters." Trudy Moskov. Khim.-Tekhnol. Inst. im. D.I. Mendeleeva **1960**, No. 31, 84-91.

Staudinger, H., "Ceramic Cutting Materials." Werkstattstechnik **50** (5) 254-62 (1960). Ind. Diamond Abstr. **17**, A175 (1960).

Stavorko, A.P., "Producing a Trial Batch of High-Alumina Brick from Latnenski Clay & Technical Alumina." Ogneupory **25** (1) 7-8 (1960).

Stavrolakis, J.A. & Norton, F.H., "Measurement of the Torsion Properties of Alumina & Zirconia at Elevated Temperatures." J. Am. Ceram. Soc. **33** (9) 263-8 (1950).

Stead, H.J., "Finishing Lenses." U.S. 2,554,070, 5/22/51.

Steele, B.R., Rigby, F. & Hesketh, M.C., "The Modulus of Rupture of Sintered Alumina Bodies." Proc. Brit. Ceram. Soc. **6**, 83-94 (1966).

Steggerda, J.J., "The Forming of Active Aluminum Oxide." Thesis, Delft Dec. 7, 1955.

Steijn, R.P., "Wear of Sapphire." J. Appl. Phys. **32** (10) 1951-58 (1961).

Steimke, F.C., Jr., "Fused Cast Refractory." U.S. 2,919,994, 1/5/60.

Steindorff, E., "German Developments in the Production of Synthetic Sapphire & Spinel." Gemmologist **16** (186) 28-9 (1947).

Steinheil, A., "The Structure & Growth of Thin Surface Films on Metals on Oxidation in Air." Ann. Physik **19**, 465-83 (1934).

Steinhoff, E., "Investigations on the Causes of the Destruction of Checker Brick in Glass-Melting Furnaces." Glastech. Ber. **28** (7) 265-72 (1955).

Steinhoff, E., "Cause of Breakdown of Chamotte & High Alumina Recuperator Tubes." Glastech. Ber. **36** (4) 109-20 (1963).

Stephens, D.L. & Alford, W.J., "Dislocation Structures in Single-Crystal Al$_2$O$_3$." J. Am. Ceram. Soc. **47** (2) 81-6 (1964).

Stern, W., "Directional Hardness & Abrasion Resistance of Synthetic Corundum." Ind. Diamond Rev. **12** (140) 137-40 (1952).

Sterry, J.P., "Testing Ceramic-Metal Brazes." Metal Progr. **79** (6) 109-11 (1961).

Sterry, J.P., "Ceramic Metal Brazing for Missile Structures." Ceram. Ind. **81** (5) 69 (1963).

Stett, M.A. and Fulrath, R.M., "Chemical Reaction in a Hot-Pressed Aluminum Oxide-Glass Composite." J. Am. Ceram. Soc. **50**, 673-6 (1967).

Stevanovic, M. and Elston, J., "Effect of Fast Neutron Irradiation in Sintered Alumina and Magnesia." Proc. Brit. Ceram. Soc. **7**, 423-37 (1967).

Stevels, J.M., "Networks in Glass & Other Polymers." Translated in Glass Ind. **35** (12) 657-62 (1954).

Stewart, G.H. ed., "Science of Ceramics:" Vol 1 – Proceedings of a Conference Held at Oxford, June 1961. Academic Press, New York, 334 pp.

Stewart, M.M., "Method for Preparing Beta-Alumina Trihydrate." U.S. 3,096,154, 7/2/63.

Stirland, D.J., et al., "Observations on Thermal Transformations in Alumina." Trans. Brit. Ceram. Soc. **57** (2) 69-84 (1958).

Stock, D.F. & Dolph, J.L., "Refractories for Aluminum Melting Furnaces." Am. Ceram. Soc. Bull. **38** (7) 356-9 (1959).

Stofel, E. & Conrad, H., "Fracture & Twinning in Sapphire (a-Al_2O_3 Crystals)." Trans. AIME **227** (5) 1053-60 (1963).

Stokes, C.A. & Secord, R.N., "Production of Finely Divided Aluminum Oxide from Bauxite." U.S. 3,007,774, 11/7/61.

Stone, E.L. & Albertin, L., "Surface Coatings Protect Parts During Heat Treatment." Metal Progr. **79** (6) 103-6 + 4 pp. (1961).

Stone, P.E., Egan, E.P. Jr., & Lehr, J.R., "Phase Relationships in the System $CaO-Al_2O_3-P_2O_3$." J. Am. Ceram. Soc. **39** (3) 89-98 (1956).

Stookey, S.D., "Method of Making Ceramics & Product Thereof." U.S. 2,920,971, 1/12/60.

Stookey, S.D., "Ceramic Body & Method of Making It." U.S. 2,971,853, 2/14/61.

Stookey, S.D., "Glass Ceramics." Chem. Eng. News **39** (25) 116-25 (1961).

Stotko, H., "New Method for Testing Loose Abrasive Grains: I, II." Radex Rundschau **1960**, No. 4, pp. 203-20; No. 5, pp. 262-91.

Stott, V.H., "Gas-Tight Sintered Alumina Ware." Trans. Ceram. Soc. **37** (8) 346-54 (1938).

Stowe, V.M., "Purification of Aluminum Hydrate." U.S. 2,405,275, 8/6/46.

Stowe, V.M., "Activated Alumina: Heat of Wetting by Water." J. Phys. Chem. **56**, 484-6, 487-9 (1952).

Strain, H.H., "Chromatographic Adsorption Analysis." Interscience Publishers, Inc., New York (1941).

Straka, R.C., Jr., "Product Forms of Alumina-Silica Ceramic Fibers." Am. Ceram. Soc. Bull. **40** (8) 493-5 (1961).

Streng, A.G. & Grosse, A.V., "Storing Gaseous Ozone Under Pressure." Ind. Eng. Chem. **53**, No. 5, 61A-64A (Intern. Ed. 50A-53A) (1961).

Stringfellow, A.E., "Ceramic Arcing Plate Material." U.S. 2,864,919, 12/16/58.

Strokov, F.N., et al., "Carbonation of Aluminate Solutions with an Increase in the Alkali Content & The Nature of the Alkali in the Hydrated Alumina." Sbornik Rabot Gosudarst. Inst. Priklad. Khim. **1940**, No. 32, pp. 166-79; Khim. Referat. Zhur. **4** (2) 84 (1941).

Stroup, P.T., "Carbothermic Smelting of Aluminum." The 1964 Extractive Metallurgy Lecture. Trans. AIME **230** (3), 356-72 (1964).

Stumpf, H.C., Russell, A.S., Newsome, J.W. & Tucker, C.M., "Thermal Transformations of Aluminas & Alumina Hydrates." Ind. Eng. Chem. **42**, (7) 1398-1403 (1950).

Stutzman, R.H., Salvaggi, J.R., & Kirschner, H.P., "Investigation of the Theoretical & Practical Aspects of the Thermal Expansion of Ceramic Materials: Vol. I, Literature Survey." PB Rept. **161826**, 162 pp.: U.S. Govt. Res. Repts. **34** (3) 309 (1960).

Suits, C.G., "New Substance Hard as Diamond." Am. Ceram. Soc. Bull. **36** (4) 152 (1957).

Sulc, V., "Manufacturing Balls from Corundum or its Solid Solutions." Czech. 115,328. 7/15/65.

Sully, A.H., Brandes, E.A. & Waterhouse, R.B., "Some Measurements of the Total Emissivity of Metals & Pure Refractory Oxides & The Variation of Emissivity with Temperature." Brit. J. Appl. Physics **3** (3) 97-101 (1952).

Sun, K-H, & Callear, T.E., "Fluoride Glasses." U.S. 2,466,506, 4/5/49.

Sun, K-H., "Beryllo-Aluminate Glass." U.S. 2,466,508, 4/5/49.

Sutton, W.H., "Apparatus for Measuring Thermal Conductivity of Ceramic & Metallic Materials to 1200°C." J. Am. Ceram. Soc. **43** (2) 81-6 (1960).

Sutton, W.H., "Development of Composite Structural Materials for Space Vehicle Applications." Amer. Rocket Society, Jour. **32**, 593-600 (1962).

Sutton, W.H., "Investigation of Whiskers-Reinforced Metallic Composites." AD 601-266, p. 712-27 (Nov. 1963).

Sutton, W.H. "Investigation of Bonding in Oxide Fiber Reinforced Metals." Final Rept. on Contract DA36-034-ORD-37682 (June 30, 1964). **AD-455770.**

Sutton W.H. & Chorne, J., "Development of High-Strength Heat-Resistant Alloys by Whisker Reinforcement." Metals Eng. Quarterly **3**, 44-51 (1963).

Sutton, W.H & Feingold, E., "Bonding in Ni/Ti, Ni/Cr, Ni/Zr-Sapphire Systems." Am. Ceram. Soc. Bull. **43** (8) 585 (1964).

Sutton, W.H. & Feingold, E., "Investigation of Bonding in Oxide Fiber Reinforced Metals." 1st Quarterly Rept. for Period July 1 to Sept. 30, 1964, **AD-458220.**

Sutton, W.H., Rosen, B.W., & Flom, D.G., "Whisker Reinforced Plastics for Space Applications." S.P.E. Jour. **20** (11) 1203-9 (1964).

Suzuki, S., "Electron Microscope Studies on the Aging of $Al(OH)_2$." Kolloid-Z. **156** (1) 67-70 (1958).

Suzuki, S., "Magnesia Spinel." J. Japan Ceram. Assn. **47** (558) 302-9 (1939).

Suzuki, S. & Fujita, Y., "Refractory Materials for Light-Metal Metallurgy: I." J. Japan Ceram. Assoc. **49** (584) 466-73 (1941).

Svec, J.J., "Alumina Scores in Computer Assemblies." Ceram. Ind. **87**, (3) 54-5, 98-9 (1966).

Swan, R.J., "Polishing of Metallurgical & Geological Specimens for Microscopic Examination." Metallurgia **68** (405) 47-50 (1963).

Swanson, H.E. & Fuyat, R.K., "Standard X-Ray Powder Diffraction Patterns." Nat. Bur. Stds. Circ. 539, 2 (1953), 3 (1954).

Swanson, H.E., Cook, M.L., Isaacs, T. & Evans, E.H., "Standard X-Ray Powder Diffraction Patterns." Natl. Bur. Stds. Circ. 539, 9 (1960).

Swift, H.R., "Effect of Magnesia & Alumina on Rate of Crystal Growth in Some Soda-Lime-Silica Glasses." Presented at Meeting of Am. Ceram. Soc., Buffalo, N.Y., April-May 1946; abstracted in Amer. Ceram. Soc. Bull. **25** (4) 132 (1946).

Symposium on Ceramics & Ceramic-Metal Seals. Rutgers Univ., April 21, 1953 (See Ceramic Age Feb-Sept. 1954).

Syritskaya, Z.M., & Yakubi, V.V., "Glass Region in the System $P_2O_5-Al_2O_3-ZnO$." Steklo i Keram. **17** (2) 18-21 (1960).

Sysoev, L.A. & Obukhovskii, Ya. A., "Preparation of Single Crystals of High-Melting Oxides." U.S.S.R. 164-016, 7/30/64.

Szabo, Z.G., Csanyi, L.J., & Kavai, M., "Determination of the Solubility Products of Metal Hydroxide Precipitates." Z. anal. chem. **146**, 401-14 (1955).

Tabor, D., "Hardness of Solids." Endeavour **13** (49) 27-32 (1954).

Tacvorian, S., "Study of the Thermal Shock Resistance of Certain Sintered Refractory Materials." Bull Soc. Franc. ceram. **1955**, No. 29, pp. 20-40.

Tacvorian, S. & Levecque, M., "Material Consisting of Sintered Alumina & Metal Mixtures." Ger. 963,766, 5/9/57.

Tacvorian, S., et al., "Sintered Refractory Material." U.S. 2,829,427, 4/8/58.

Takahashi, K., et al., "Phosphate Glasses of High Chemical Durability." J. Ceram. Assoc. Japan **61** (690) 621-4 (1953).

Talaber, J., "Durability of Aluminum Cements." Coloq. Intern. Durabilite Betons, Rappt. Final. Prague 1961, 109-14 (Pub. 1962).

Tallan, N.M., "Dielectric Losses Resulting from Dislocations in Sapphire." State Univ. of New York College of Ceramics at Alfred Univ. Ceramic Res. Dept., microfilm available from Dissertation Abstracts, 313 N. First St., Ann Arbor, Mich.

Talsma, H. (to E.I. du Pont de Nemours), "Porous Alumina or Silicate Refractories Bonded by Al_2O_3..." U.S. 3,255,027. 6/7/66.

Tamele, M.W., "Chemistry of the Surface & the Activity of Al_2O_3-SiO_2 Cracking Catalyst." Faraday Discussions 8, 270-9 (1950).

Tamura, Y., et al., "Symposium on Glass Melting 1959." Union Scientifique Continental du Verre. Charleroi, Belgium. 963 pp., illus. 800 Bfr. "Volatilization at High Temperatures of Alkali or Alkaline Earth Borate Glasses."

Tanabe, Y. & Sugano, S., "Absorption Spectra of Ruby." J. Phys. Soc. Japan **12** (5) 556 (1957).

Tanaka, T., "Hydraulic Properties of Granulated Blast-Furnace Slags: IV, Strength Development of Quenched Glassy Slags in the System CaO-Al₂O₃-SiO₂." Semento Gijutsu Nenpo 6, 49-56 (1952).

Tanaka, T., Takemoto, K., & Kikuchi, H., "Differential Thermal Analysis of Quenched Glassy Slags in the System CaO-Al₂O₃-SiO₂." Semento Gijutsu Nenpo 6 (56-65) (1952).

Tanaka, T. & Watanabe, Y., "Expansive Cements." Semento Gijutsu Nenpo 8, 192-202 (1954).

Tangerman, E.T., "What We Should Know About Ceramic Tools." Am. Machinst 100 (6) 153-74 (1956).

Tanigawa, H. & Tanaka, H., "Studies on the Manufacture of Low Expansion Ceramics by Addition of Lithia-Containing Glasses: I, Sintering of Natural Petalite." Osaka Kogyo Gijutsu Shikensho Kiho 14 (3) 266 (1963).

Tar, I., et al., "The Colloidal Properties of Al(OH)₂ Gels." Magy, Kem. Folyoirat 68, 413-7 (1962).

Tarte, P., "New Applications of Infrared Spectrometry for Crystallochemical Problems." Silicates Ind. 28 (7/8), 345-54 (1963).

Tatousek, R.J. & Peters, A.T., "Evaluation of Cast Refractory Hot-Top Linings." Open Hearth Proc. 40, 158-66 (1957).

Taylor, C.S. & Edwards, J.D., "Some Reflection & Radiation Characteristics of Aluminum." Heating, Piping, Air Conditioning (Journal Section) 11, 59-63 (1939).

Taylor, C.S., Tucker, C.M. & Edwards, J.D., "Anodic Coatings with Crystalline Structure on Aluminum." Trans. Electrochem. Soc. 88, 325-33 (1945).

Taylor, H.D., "Microscopy Technique for Alumina Ceramics." Am. Ceram. Soc. Bull. 42 (3) 104-5 (1963).

Taylor, K.M., "Abrasive Cutoff Wheel." U.S. 2,564,217, 8/14/51.

Taylor, R.E., "Thermal Conductivity & Expansion of Beryllia at High Temperatures." J. Am. Ceram. Soc. 45 (2) 74-8 (1962).

Taylor, R.J., "The Activation of Alumina." J. Soc. Chem. Ind. (London) 68, 23-6 (1949).

Tchoubar, C. & Oberlin, A., "Transformation of Albite by Action of Water." J. Micros. 2, 415-32 (1963).

Tefft, W.E., "Elastic Constants of Synthetic Single-Crystal Corundum." J. Res. Natl. Bur. Stds., A70, 277-80 (1966).

Teichner, S., Juillet, F. & Arghiropoulos, B., "A New Colored & Nonstoichiometric Form of Alumina." Bull. soc. chim. France 1959, 1491-5.

Tentarelli, L.A. & White, J.M., "Low Temperature Ceramic-Metal Seals." Am. Ceram. Soc. Bull. 43 (8) 586 (1964).

Terminasov, I. & Kharson, L., "Controlling Quality of Abrasives by X-Rays." Vestnik Metalloprom. 17 (9) 2-5 (1937).

Terry, J.H., "New Uses for Ceramics in Motors & Transformers." Am. Ceram. Soc. Bull. 36 (12) 454-6 (1957).

Tertian, R., "The Constitution & Crystal Structure of Activated Alumina (γ-Al₂O₃). Compt. rend. 230, 1677 (1950).

Tertian, R. & Papee, D., "Thermal & Hydrothermal Transformations of Alumina." J. chim. phys. 55, 341-53 (1958).

Tertian, R., Papee, D., & Charrier, J., "X-Ray Studies of Transition of Anhydrides of Alumina." Compt. rend. 238, 98-9 (1954).

Teter, J.W., Gring, J.L. & Keith, C.D., "Alumina Catalyst Base." U.S. 2,838,375, 2,838,444, 2,838,445, 6/10/58.

Tewari, S.N., Dey, A.K. & Ghosh, S., "Aging of Hydroxides." Z. anorg. Chem. 271, 150-2 (1953).

Tewari, S.N. & Ghosh, S., "Amphoteric Nature of Hydrated Aluminum Oxide & the Determination of its Isoelectric Point." Proc. Natl. Acad. Sci. India 21A, Symposium on Chem. Hydrous Oxides 41-51 (1952).

Thermal Syndicate, Ltd., "Mullite Combustion Tubes." Chem. Age (London) 57 (1461) 60 (1947).

Thery, J. and Briancon, D., "Structure and Properties of the Aluminates of Sodium." Rev. Hautes Temp. et Refract. 1, 221-27 (1964).

Thibault, N.W. & Nyquist, H.L., "The Measured Knoop Hardness of Hard Substances & Factors Affecting Its Determination." Trans. Am. Soc. Metals 38 271-330 (1947)

Thibon, H., Charier, J., & Tertian, R., "Thermal Decomposition of Alumina Hydrates." Bull. soc. chim. France 1951, 384-92.

Thiele, K.H., Schwartz, W., & Dettmann, L., "Preparation of Very Pure, Fine Aluminum Oxide from Triethylaluminum." Z. Anorg. Allg. Chem. 349, 324-7 (1967).

Thielke, N.R., "Application of Crystal Chemistry to the Search for New Refractories." J. Am. Ceram. Soc. 33 (10) 304-9 (1950).

Thielke, N.R., "Refractory Materials for Use in High-Temperature Areas of Aircraft." U.S. Air Force, Air Res. & Dev. Command, WADC Tech. Rept. No. 53-9, 48 pp. (Jan. 1953).

Thiessen, P.A. & Thater, K.L., "Pure Aluminum Ortho-Hydroxide in Gelatinous & Finely Powdered Form." Z. anorg. allg. Chem. 181, 417 (1929).

Thilo, E. & Jander, J., "The System Al₂O₃-Cr₂O₃ & The Red Color of Ruby." Forschungen u. Forschr. 26, 35 (1950).

Thilo, E., Jander, J. & Seemann, H., "Color of Ruby & The (Al,Cr)₂O₃ Solid Solutions." Z. anorg. u. allgem. Chem. 279 (1-2) 2-17 (1955).

Thilo, E., Jander, J., Seemann, H. & Sauer, R., "The Color of Ruby." Naturwissen. 37, 399 (1950).

Thomas, A.G. & Jones, H.J., "Hot Pressing of Ceramic Powders." Powder Met. 1960, No. 6, pp. 160-9.

Thomas, F., "One Grinder Conquers Nose Cone." Grinding Finishing 9 (9) 32-5 (1963).

Thomas, W.F., "An Investigation of the Factors Likely to Affect the Strengths & Properties of Glass Fibers." Phys. & Chem. Glasses 1 (1) 4-18 (1960).

Thompson, J.J., "Forming Thin Ceramics." J. Am. Ceram. Soc. Bull. 42 (9) 480-1 (1963).

Thompson, J.F., "Producing Alumina." U.S. 2,469,088, 5/3/49.

Thompson, J.G. & Mallett, M.W., "Preparation of Crucibles from Special Refractories by Slip Casting." Jour. Res. Nat. Bur. Stds. 23 (2) 319-27 (1939). R.P. 1236.

Thomson, R.F., "Sintered Powdered Copper Base Bearing." U.S. 2,831,243, 4/22/58.

Thomson, R.F., "Sintered Powdered Metal Piston Ring." U.S. 2,855,659, 10/14/58.

Thomson-Houston Company, Ltd., "Improvement in & Relating to Wear-Resisting Bodies." Brit. 4,887, 2/27/13.

Thosar, B.V., "Fluorescent Ion of Chromium in Ruby." Phil. Mag. 26 (175) 380-9 (178) 878-87 (1938).

Thurnauer, H., "Review of Ceramic Materials for High-Frequency Insulation." J. Am. Ceram. Soc. 20 (11) 368-72 (1937).

Thurnauer, H., "Electrically Conductive Ceramic Thread Guide." U.S. 2,369,266, 2/13/45.

Tiede, E.L., "Glass Composition." U.S. 2,571,074, 10/9/51.

Tighe, N.J., "Jet Thinning Device for Preparation of Al₂O₃ Electron Microscope Specimens." Rev. Sci. Inst. 35 (4) 520-1 (1964).

Timofeeva, V.A. "Growth Conditions of Al₂O₃, ZnO, and Ga₂O₃ Crystals." Rost. Kristollov, Akad. Nauk SSSR, Inst. Kristallogr. 6, 86-92 (1965).

Timofeeva, V.A. & Voskanyan, R.A., "Growing Crystals of Corundum from Solutions of Molten Lead Fluoride." Kristallografiya 8, 293-6 (1963).

Timofeeva, V.A. & Yamzin, I.I., "Corundum (& Spinel) from Gas Phases." Trudy Inst. Krist. Akad. Nauk SSSR 1956, No. 12, 67-72.

Tinklepaugh, J.R., Truesdale, R.S. Swica, J.J. & Hoskyns, W.R., "Grain Size Effects on the Thermal Conductivity of Ceramic Oxides." O.T.S. 171346, U.S. Dept. of Comm., Washington, D.C. (Feb. 1961).

Tinklepaugh, J.R. & Crandall, W.B. (eds.). "Cermets." Reinhold Publishing Co., New York, vi + 239 pp. (1960).

Tinklepaugh, J.R., Truesdale, R.S., Swica, J. J. & Hoskyns, W.R., "Grain Size Effects on the Thermal Conductivity of Ceramic Oxides." Feb. 1961. 77 p. incl. illus. (Project 7022; Task 70634). Contract AF 33(616)-7661.

Topchieva, K.V., et al., "The Effect of Aluminum Oxide Dehydration on Its Catalytic Activity." Kataliz v Vysshei Shkole, Min. Vysshego i Srednego Spets. Obrazov. SSSR, Tr. 1-go (Pervogo) Mezhvuz. Soveshch, po Katalizu 1958, No. 1, Pt. 1, 241-7 (Pub. 1962).

Torkar, K., "Property Relationships of Porous Materials." Ber. deut. keram. Ges. 31 (5) 148-56 (1954).

Torkar, K. & Krischner, H., "Aluminum Hydroxides & Oxides. IV. Description of Two New Forms of Aluminum Oxide Found in the Hydrothermal Reaction with Metallic Aluminum." Monatsh. Chem. 91, 658-68 (1960).

Torkar, K. and Krischner, H., "Comment on 'A New Alumina Hydrate, Tohdite'." Bull. Chem. Soc. Japan 39, 1356 (1966).

Torkar, K. et al., "Production of Amorphous Aluminas." U.S. 3,038,784, 6/12/62.

Torkar, K. & Worel, H., "Phase Diagram of the System $Al_2O_3-H_2O$." Monatsh, Chem. 88 743 (1957).

Toropov, N.A. & Galakhov, F. Ya., "System Alumina-Silica." Doklady Akad. Nauk SSSR 78, 299-302 (1951).

Toropov, N.A. & Galakhov, F.Y., "The Mullite Problem." Voprosy Petrograf. i Mineral, Akad. Nauk SSSR, 2, 245-55 (1953).

Toropov, N.A. & Galakhov, F. Ya. "Solid Solutions in the System $Al_2O_3-SiO_2$." Izvest. Akad. Nauk SSSR., Otdel, Khim. Nauk 1958 No. 1, pp. 8-11.

Toropov, N.A. & Sirazhiddinov, N.A., "The System $MgAl_2O_4-LaAlO_3$." Zh. Neorgan. Khim. 9 (5) 1300-8 (1964).

Toropov, N.A. & Stukalova, M.M., "Replacement of Sodium in Crystals of β-Alumina with Calcium, Strontium, & Barium." Compt. rend. Acad. Sci. USSR 27 (9) 974-7 (1940); Khim. Referat. Zhur. 4 (2) 80 (1941).

Toropov, N.A. & Volkonskii, B.V., "Hydraulic Activity of Granulated Slags." Doklady Akad. Nauk SSSR 66 (1) 95-7 (1949).

Torrey, J., New York Med. Phys. J. 1 68 (1822).

Tosterud, M., "Aluminum Hydrate of Low Water Content & Process of Producing Same." U.S. 1,953,201, 4/3/34.

Towner, R.J., "APM Alloys." Metals Eng. Quarterly 1 (1) 24 (1961).

Townsend, M.G., "Cobaltous Ion in Alumina." J. Phys. Chem. 68 (6) 1569-72 (1964).

Trambouze, Y. & Perrin, M., "Variation of the Protonic Acidity of Silica-Alumina Gels." Compt. rend. 236, 1261 (1953).

Tran-Huu, T. & Prettre, M., "Temperature Dependence of the Dehydration Reactions of Hydrargillite." Compt. rend. 234 (13) 1366-8 (1952).

Treadwell, W.D. & Terebesi, L., "Energy of Formation of Alumina from the Elements." Helv. Chim. Acta 16, 922-39 (1933).

Treffner, W.S. & Williams, R.M., "Heat Evolution Tests with Calcium Aluminate Binders & Castables." J. Am. Cer. Soc. 46 (8) 399-406 (1963).

Trent, E.M. & Comins, D.B., "Process for Preparing Hard Aluminium Oxide Sinter Bodies." Ger. 1,098,427, 1/26/61.

Tresvyatskii, S.G., "Role of Closed Pores in Sintering Pure Highly Refractory Oxides." Ogneupory 25 (3) 130-2 (1960).

Tretyachenko, G.H., "Criteria of the Thermal Stability of Cermets." Soviet Powder Met. Metal Ceram., 1963, No. 1, pp. 44-51.

Tripp, H.P. & King, B.W., "Thermodynamic Data on Oxides at Elevated Temperatures." J. Am. Ceram. Soc. 38 (12) 432-7 (1955).

Trippe, P., "Russians Put Emphasis on Ceramics." Metalworking Production 101 (March 17) 443-7 (1957); abstracted in J. Iron Steel Inst. (London) 188 (1) 87 (1958).

Trofimov, A.K. & Tolkachov, S.S., "Investigation of the Gamma to Alpha Transformation by the Luminous Spectra." Doklady Akad. Nauk SSSR 104, 154-5 (1955).

Trombe, F., "Superrefractory Oxides." Bull. Soc. Franc. Ceram. 1949, No. 3, pp. 18-26.

Trombe, F., "Treatments & Reactions at High Temperatures." Verres et refract. 3, 83-96 (1949).

Trombe, F. & Foex, M., "Use of Infrared Radiation at High-Energy Illumination for the Treatment & Study of Refractive Substances at Elevated Temperature." Congr. intern. chim. pure et appl. 16^e Paris 1957, Mem. sect. chim. minerale 585-92 (Pub. 1958).

Tromel, G., et al., "System Silica-Alumina." Ber. deut. keram. Ges. 34 (12) 397-401 (1957).

Trostel, L.J., Jr., "Stability of Alumina & Zirconia in Hydrogen." Am. Ceram. Soc. Bull. 43 (4) 309 (1964).

Trostel, L.J., Jr., "Stability of Al_2O_3 and ZrO_2 in Hydrogen." Am. Ceram. Soc. Bull. 44, 950-2 (1965).

Trout, O.F. Jr., "Exploratory Investigation of Several Coated & Uncoated Metal, Refractory, & Graphite Models in a 3800°F Stagnation Temperature Air Jet." NASA Tech. Note, 190, 73 pp. (1960).

Truesdale, R.S., Swica, J.J. & Tinklepaugh, J.R., "Metal Fiber-Reinforced Ceramics." WADC TR 58-452, ASTIA Doc. No. 207079, PB 151610. (1958, Dec.).

Tseung, A.C.C. & Carruthers, T.G., "Refractory Concretes Based on Pure Calcium Aluminate Cement." Trans. Brit. Ceram. Soc. 62 (4) 305-20 (1963).

Tsigler, V.D., et al., "High Alumina Lightweight & Its Use." Ogneupory 25 (7) 299-307 (1960).

Tsitsishvili, G.V. & Sidamonidze, Sh.I., "The Effect of Irradiation on the Adsorption & Catalytic Properties of Alumina." Soobshch. Akad. Nauk Gruz. SSR 31 (3) 569-76 (1963).

Tucker, S., "Method for Obtaining Iron-Free Aluminum Compounds from Clays." U.S. 2,847,279, 8/12/58.

Tucker, R.N. & Gibbs, P., "Impurity Penetration Along Dislocation Lines in Alpha Alumina." J. App. Phys. 29, 1375-6 (1958).

Tuleff, J., "Development of Pores in Ceramic Products as a Function of Their Firing Temperature." Bull. Soc. Franc. Ceram. 1961, No. 50, pp. 17-24.

Tulovskaya, Z.D. & Segalova, E.E., "Thermographic Study of the Process of Hydration of Monocalcium Aluminate at Different Temperatures." Zh. Prikl. Khim. 37 (2) 267-75 (1964).

Tumanov, V.I., Funke, V.F. & Belen'kaya, L.I., "Wettability of Aluminum Oxide & of Carbides by Metals of the Iron Group." Zh. Fiz. Khim. 36, 1574-7 (1962).

Tunison, D. & Burdick, V., "The Fracture Analysis of Alumina Using the Electron Microscope." U.S. Dept. Comm., Office Tech. Serv., AD 269,145 35 pp. (1961).

Turkevich, J., & Hillier, J., "Electron Microscopy of Colloidal Systems." J. Anal. Chem. 21, 475-85 (1949).

Turnbaugh, J.E. and Norton, F.H., "Low-Frequency Grain Boundary Relaxation in Alumina." J. Am. Ceram. Soc. 51, 344 (1968).

Turnock, A.C. & Lindsley, D.H., "Fe-Al & Fe-Ti Spinels & Related Oxides." Carnegie Inst. Washington Year Book 60, 157 (1961).

Twells, R., "Producing Ceramic Gauges," U.S. 2,525,324, 10/10/50.

Uchikawa, H., Tsumagari, A. & Koike, H., "Mechanism of the Formation of Calcium Aluminates." Semento Gijutsu Nenpo 17, 50-9 (1963).

Ueda, I. & Okada, J., "Electric Breakdown of Alumina." Proc. Phys. Soc. Japan 4, 8_9 (1949).

Ueltz, H.F.G., "Refractory Composition for Journals, Bearings, etc." U.S. 2,745,763, 5/15/56.

Ueltz, H.F.G., "Castable Aluminum Oxide Mixture & Articles Made Therefrom." U.S. 2,965,506, 12/20/60.

Ueltz, H.F.G., "Abrasive Grain." U.S. 3,079,243, 2/26/63.

Ulrich, G., "Notice Regarding the Crystal Structure of the Corundum-Hematite Group." Norsk. Geol. Tidsskrift 8, 115-22 (1925).

Umblia, E.J., "Method of Producing a Solderable Metallic Coating on a Ceramic Body & of Soldering to the Coating." U.S. 2,835,967, 5/27/58.

Universal Oil Products Co., "Inorganic Oxide Particles of Spherical Shape & Their Manufacture." Brit. 596,591, 1/21/48.

Unmack, A., Second International Congr. Crystalog. Inorganic Structures 8, Section G, June 27 to July 5, 1951.

Urbain, G. and Rouannet, M., "Secondary Reference Points in the Practical Temperature Scale. Experimental Determination of the Melting Point of Al_2O_3." Rev. Int. Hautes Temp. Refract. 3, 363-9 (1966).

Urban, R.F., "Checker-Brick Trials (Superduty & High-Alumina)." Open Hearth Proc. 38, 103-8 (1955).

Use of Abrasive Grain in the Manufacture of Coated Abrasives: I. Anon. Grinding Finishing 9 (7) 33-6 (1963).

U.S. Atomic Energy Commission., "Granulation of Ceramic Material." Brit. 832,309,4/6/60.

Van Arkel, A.E. & Fritzius, C.P., "Infrared Absorption of Hydrates." Rec. Trav. chim. Pays Bas. 50 1035 (1931).

Van Der Beck, R.R., et al., "How Oxide Cutting Tools are made & What They do." Ceram. Ind. 69 (6) 113-5, 128-31 (1957).

Van Houten, G.R., "Survey of Ceramic-to-Metal Bonding." Am. Ceram. Soc. Bull. 38 (6) 301-7 (1959).

Van Nordstrand, R.A., Hettinger, W. & Keith, C., "A New Alumina Trihydrate." Sinclair Research Labs., Harvey, Ill. Nature 177, 713-4 (1956).

Van Olphen, H., "Internal Mutual Flocculation in Clay Suspensions." J. Colloid Sci. 19 (4) 313-22 (1964).

Van Ooosterhout, G.W., "Morphology of Synthetic Submicroscopic Crystals of Alpha and Gamma FeOOH." Acta Cryst. 13, 932 (1960).

Van Ooosterhout, G.W. & Rooymans, C.M.J., "A New Superstructure in γ Ferric Oxide." Nature 181, 44 (1958).

Varian Associates Instrument Division. "NMR and EPR Spectroscopy." Pergamon Press, Inc., New York, 320 pp. (1961).

Vasilos, T. and Gannon, R.E., "Ablation Properties of Ceramics." Am. Ceram. Soc. Bull. 44, 971-4 (1965).

Vasilos, T. & Harris, G., "Impervious Flame-Sprayed Ceramic Coatings." Am. Ceram. Soc. Bull. 41 (1) 14-7 (1962).

Vasilos, T. & Spriggs, R.M., "Pressure Sintering—Mechanisms & Microstructures for Alumina & Magnesia." J. Am. Ceram. Soc. 46 (10) 493-6 (1963).

Vasilos, T., Spriggs, R.M. & Mitchell, J., "Microstructure Studies of Polycrystalline Refractory Oxides." AD-288,146, 19 pp., illus.; U.S. Govt. Res. Repts. 38 (4) 68 (1963). AD-296-269, 18 pp. illus., Ibid. (10) 36.

Vassiliou, B.E., "Simple Laboratory Method for Making Crucibles & Other Shapes." Trans. Brit. Ceram. Soc. 56 (10) 516-8 (1957).

Vatter, H. Ger. 645,871, 4/35.

Vauquelin, L.N., Ann. chim. Phys. 42 (1) 113 (1802).

Vecchi, G., "Nickel in Ceramics." Faenza 38 (1) 16-7 (1952).

Velikin, B.A., "Gun Lining of Metallurgical Furnaces." Ogneupory 24 (8) 354-60 (1959).

Venable, C.R. Jr., "Erosion Resistance of Ceramic Materials for Petroleum Refinery Applications." Am. Ceram. Soc. Bull. 38 (7) 363-8 (1959).

Ventriglia, U., "Plasticity." Ricerca sci. 20 (3) 310-9 (1950).

Vergnon, P., Juillet, F., Elston, J. & Teichner, S. J., "Flash Sintering of Aluminas." Rev. Hautes Temp. Refractaires 1 (1), 27-39 (1964).

Verhoogen, J., "Physical Properties & Bond Type in Mg-Al Oxides and Silicates." Am. Mineralogist 43 (5-6) 552-79 (1958).

Verneuil, M.A., "Memoir on the Artificial Reproduction of Rubies by Fusion." Ann. chim. Phys. 3, 20-48 (1904).

Verwey, E.J.W., "The Crystal Structure of Gamma Iron Oxide & Gamma Alumina." Z. Krist. 91, 65-9 (1935).

Verwey, E.J.W., "The Structure of the Electrolytic Oxide Layer on Aluminum." Z. Krist 91, 317-20 (1935).

Verwey, E.J.W., "Incomplete Atomic Arrangement in Crystals." J. Chem. Phys. 3, 592-3 (1935).

Verwey, E.J.W. & van Bruggen, M.G., "Electrical Resistance & Method of Making." U.S. 2,524,611, 10/3/50.

Vest, R.W., "Electrical Behavior of Refractory Oxides." AD 255079, 8 pp. illus.; U.S. Govt. Res. Repts. 36 (1) 64 (1961).

Vest, R.W., "Electrical Behavior of Refractory Oxides." AD-283,667, 10 pp., illus.; U.S. Govt. Res. Repts. 37 (24) 67-8 (1962).

Vielhaber, "Alumina in Enamels." Emailwaren-Ind. 17 (27-8) 73-4 (1940).

Vincent, G.L., "How Avco Ceramic Re-entry Vehicles are Tested & Flown on Martin's Titan ICBM." Ceram. Ind. 75 (5) 88-93 (1960).

Vines, R.F., Semmelman, J.O., Lee, P.W., & Fonvielle, F.P. "Mechanisms Involved in Securing Dense Vitrified Ceramics from Preshaped Partly Crystalline Bodies." J. Am. Ceram. Soc. 41 (8) 304-9 (1958).

Vinogradova, L.V., et al., "Trial Production of Thermally Stable Mullite-Corundum Brick at the Podol'sk Refractories Plant." Ogneupory 21 (4) 178-9 (1956).

Viola, N. & McQuarrie, M., "Note on Relation Between Unfired & Fired Densities of Aluminum Oxide Compacts." J. Am. Ceram. Soc. 42 (5) 261-2 (1959).

Viro, S.E., "Grinding Ceramic Materials in Ball Mills in a Vacuum." Keram. Sbornik No. 15, pp. 34-40 (1941).

Vishnevskii, V.N., Gnyp, R.G., et al., "X-Ray and Thermal Luminescence of Pure and Doped Corundum Single Crystals." Ukr. Fiz. Zh. 11, 991-7 (1966).

Vogel, W., "Directed Crystallization of Glasses." Sixth Conference of Silicate Industry (Budapest, 1961) Abstract. Phys. & Chem. of Glasses 4 (3) June 1963.

Voitsekhovskii, R.V. & Vovnenko, A.M., "Effect of the Dispersed Phase on the Measurements of the Activity of the Hydrogen Ion in Salt Solutions." Kolloid. Zhur. 20, 697-704 (1958).

Volchek, I.Z., "Hydraulic Properties of High-Alumina Slags as a Function of Their Phase Characteristics." Doklady Akad. Nauk SSSR 90 (3) 437-40 (1953).

Volf, M.B., "Kavalier Glass." Czechoslovak Glass Rev. 2 (1/2) 12-5 (1947).

Volk, H.F. and Meszaros, F.W., "Oxygen Permeation Through Alumina" Ceramic Microstructures, ed. Fulrath and Pask, Wiley, N.Y.

Vollmer, J., "Light Guide Radiation Pyrometry." J. Opt. Soc. Am. 49 (1) 75-7 (1959).

Volosevich, G.N. & Poluboyarinov, D.N., "Methods of Controlling the Microstructure of Corundum Ceramics." Doklady Akad. Nauk SSSR 113 (1) 152-5 (1957).

Volosevich, G.N. & Poluboyarinov, D.N., "Methods for Regulating the Crystallization & Properties of Corundum Ceramics." Sovrem. Metody Issled. Silikatov i Stroit. Materialov, Vses. Khim. Obshchestvo, Sb. Statei 1960, 214-21.

Voltz, S.E. & Weller, S.W., "Alumina Stabilized by Thoria to Resist a-Alumina Formation." U.S. 2,810,698 10/22/57.

Voltz, S.E. & Weller, S.W., "Alumina Stabilized by Hafnia to Resist a-Alumina Formation." U.S. 2,810,699, 10/22/57.

Vondracek, C.H., "Evaluation of Inorganic Potting Compounds." Mater. Symp., Natl. SAMPE (Soc. Aerospace Mater. Process Engrs.) Symp., 7th, Los Angeles 1964 (26), 20 pp.

Vorus, T.A., Jewett, R.P. & Accountius, O.E., "Direct Observation of Dislocations in Sapphire." J. Am. Ceram. Soc. 46 (9) 459-60 (1963).

Vydrik, G.A., "Effect of Comminution of Technical Alumina on the Sintering of Corundum Bodies." Steklo i Keram. 16 (10) 35-9 (1959).

Wachtman, J.B. Jr. & Lam, D.G. Jr., "Young's Modulus of Various Refractory Materials as a Function of Temperature." J. Am. Ceram. Soc. 42 (5) 254-60 (1959).

Wachtman, J.B., Jr. & Maxwell, L.H., "Plastic Deformation of Ceramic Oxide Single Crystals." J. Am. Ceram. Soc. 37 (7) 291-9 (1954).

Wachtman, J.B., Jr. & Maxwell, L.H., "Plastic Deformation of Ceramic Oxide Single Crystals: II." J. Am. Ceram. Soc. 40 (11) 377-85 (1957).

Wachtman, J.B. & Maxwell, L.H., "Factors Controlling Resistance to Deformation & Mechanical Failure in Polycrystalline (glass-free) Ceramics." PB Rept. 131623, 76 pp. U.S. Govt. Res. Repts. 29 (5) 274 (1958).

Wachtman, J.B. & Maxwell, L.H., "Strength of Synthetic Single Crystal Sapphire & Ruby as a Function of Temperature & Orientation." J. Am. Ceram. Soc. 42 (9) 432-3 (1959).

Wachtman, J.B., Jr Scuderi, T.G., & Cleek, G.W., "Linear Thermal Expansion of Aluminum Oxide & Thorium Oxide from 100° to 1100°K." J. Am. Ceram. Soc. 45 (7) 319-23 (1962).

Wachtman, J.B., Tefft, W.E., & Lam, D.G. Jr., "Young's Modulus of Single Crystal Corundum." Mechanical Properties of Engineering Ceramics, Kriegel, W.W. & Palmour, H. III. Interscience Publishers (1961).

Wachtman, J.B., Tefft, W.E., Lam, D.G. & Stinchfield, R.P., "Elastic Constants of Synthetic Single Crystal Corundum at Room Temperature." J. Res. Natl. Bur. Stds. 64A (3) 213-28 (1960).

Wade, W.R., "Measurements of Total Hemispherical Emissivity of Several Stably Oxidized Metals & Some Refractory Oxide Coatings." NASA Memo., 1-20-59L, 30 pp. (1959).

Wade, W.R. & Hackerman, N., "Heats of Immersion." J. Phys. Chem. 64, 1196 (1960).

Waeser, B., "Sintering of Metals & Metal-Ceramic Combinations (patent Review)." Kolloid-Z, **140** (2/3) 159-64 (1955).

Wagner, H.J., "Filaments & Whiskers in Reinforced Structural Materials." Battelle Tech. Rev. **12** (12) 8-14 (1963).

Wagner, H.J., "Fiber Reinforced Metals." Battelle DMIC Review of Recent Developments (Dec. 11, 1964).

Wagner, H.W., "New Concepts of the Properties of Abrasives." Mech. Eng. **72** (3) 225-6 (1950); Ind. Diamond Rev. **10** (114) 133-5 (1950); Grits & Grinds **41** (3) 9-13 (1950).

Wagner, P. & Coriell, S.R., "High Temperature Compatibility of Cesium Gas with Some Dielectrics." Rev. Sci. Instr. **30**, 937-8 (1959).

Wagner, W., "Abrasive Material with Porous Grains." Ger. 939,377, 6/21/56.

Wahl, N.E., "Rain Erosion of Ceramics." Ceram. Age. **69** (1) 28 (1957).

Wainer, E. & Beasley, "Fiber Forming Process." U.S. 3,082,051, 3/63.

Wainer, E. & Cunningham, A., "Sapphire Fibers." U.S. 3,077,380, 2/12/63.

Wainer, E. & Mayer, E.F., "Method of Making Inorganic Fibers." U.S. 3,096,144, 7/2/63.

Wakao, Y. & Hibino, T., Effect of the Addition of Metallic Oxides on the α-transformation of Alumina." Nagoya Kogyo Gijutsu Shikensho Hokoku 11, 588-95 (1962).

Wakao, Y. & Hibino, T., "Sintering of Alumina. VII. Alumina Powder for the Production of Sintered Alumina." Nagoya Kogyo Gijutsu Shikensho Hokoku 10, 121-6 (1961).

Walkenhorst, W., "Simple Method for the Preparation of Structureless Carrier Layers of Aluminum Oxide." Naturwissenschaften 34 (12) 373 (1947).

Walker, R.F., "Mechanism of Material Transport During Sintering." J. Am. Ceram. Soc. **38** (6) 187-97 (1955); Ibid (12) 466.

Walker, R.F., Efimenko, J. & Lofgren, N.L., "Rate of Vaporization of Refractory Substances." Planetary Space Sci. 3, 24-30 (1961).

Wallace, R., Florio, J. & Polanyi, T., "Investigation of the Chemical Reaction Between Tungsten & Aluminum Oxide." PB Rept. 171373, 73 pp.; U.S. Govt. Res. Repts. **35** (4) 408 (1961).

Wallmark, S. & Westgren, A., "X-Ray Analysis of Barium Aluminates." Arkiv Keni, Mineral Geol. **12B** (35) 4 pp. (1937; abstracted in Mineralog. Abs. 8 (11) 366 (1943).

Walton, J.D., Jr & Bowen, M.D., "The Evaluation of Ceramic Materials Under Thermal Shock Conditions." Mech. Properties Eng. Ceramics Proc. Conf., Raleigh, N.C. 1960, 149-73 (Pub. 1961).

Walton, J.D. & Harris, J.N., "High-Temperature Insulation for Wire: I." PB Rept. 131812 36 pp.; U.S. Govt. Res. Repts. **30** (2) 59 (1958).

Walton, J.D., Jr & Poulos, N.E., "Cermets from Thermite Reactions." J. Am. Ceram. Soc. **42** (1) 40-9 (1959).

Walton, J.D., Jr., Poulos, N.E. & Mason, C.R., "Use of Thermite Reactions to Produce Refractory Cermets." Ceram. Age **75** (6) 39-45 (1960).

Walton, S.F., "Refractory." U.S. 2,246,226, 6/17/41.

Walton, S.F. & Bassett, L.B., "Surface Coated Abrasive Grain." U.S. 2,527,044, 10/24/50.

Walworth, C.B., "Ceramic Fiber—Seven Forms & How to Use Them." Materials in Design Eng. **46** (5) 124-9 (1957).

Warburton, R.S. & Wilburn, F.W., "Application of Diff. Therm. Anal. & Thermo-grav. Anal. to the Study of the Reactions Between Glassmaking Materials: IV. $CaCO_3$-SiO_2-Al_2O_3 Systems." Phys. Chem. Glasses 4 (3) 91-8 (1963).

Ward, D. & Passmore, M.E., "Nose Cones—Study in Grinding Technique." Machine & Tool Blue Book **55** (6) 131-7 (1960).

Warde, J.M., "Materials for Nuclear Power Reactors." Materials & Methods **44** (2) 121-44 (1956).

Warde, J.M., "Refractories for Nuclear Energy." Refractories Inst. Tech. Bull., No. 94, 18 pp. (Feb. 1956).

Warman, M.O. and Budworth, D.W., "Effects of Residual Gas on the Sintering of Alumina to Theoretical Density in Vacuum." Trans. Brit. Ceram. Soc. **66**, 265-71 (1967).

Warshaw, S.I. & Norton, F.H., "Deformation Behavior of Polycrystalline Aluminum Oxide." J. Am. Ceram. Soc. **45** (10) 479-86 (1962).

Wartenberg, H.V., "Alumina." Z. anorg. allgem. Chem. **269**, 76-85 (1952); **270**, 328 (1952).

Wartenberg, H.V. & Eckhardt, K., "Melting-Point Curves of Inert Oxides. VIII. Systems with CeO_2." Z. anorg. allgem. Chem. **232**, 184 (1937).

Wartenberg, H.V. & Moehl, H., "The Reduction of Alumina & Other Oxides by Tungsten at High Temperatures." Z. physik Chem. **128**, 439-44 (1927).

Wartenberg, H.V. & Prophet, E., "Conductivity of Corundum." Z. Elektrochem. **38** (11) 849-50 (1932).

Wartenberg, H.V. & Reusch, H.J., "Melting Points Curves of Highly Refractory Oxides." Z. Anorg. allgem. Chem. **207**, 1-20 (1932).

Wartenberg, H.V., Wehner, G., & Saran, E., "Surface Tension of Molten Aluminum & Lanthanum Oxides." Nachr. Ges. Wiss. Gottingen Math-physik, Kl (Fachgr. II) **2**, 65-71 (1936).

Watkins, D.B., et al. "Effect of Alumina on Tensile Strength of Glass." Glass Ind. **31** (1) 19-21, 50 (1950).

Watson, D.R., Hrishikesan, K., & Deutscher, J.S., "Preparation of Finely Milled Alumina for Dense Sintered Bodies." U.S. 3,274,311. 9/20/66.

Watson, D.R., Lippman, A. Jr., & Royce, D.V., Jr., "Method for Reducing the Soda Content of Alumina." U.S. 3,106,452, 10/8/63.

Watson, G.R., "Molding Hollow Articles from Refractory Materials Under Heat & Pressure." U.S. 2,535,180, 12/26/50.

Watson, J.H.L., Vallejo-Freire, A., Souza-Santos, P., & Parson, J., "Fine Structure & Properties of Fibrous Alumina." Kolloid-Z. **154** (1) 4-15 (1957).

Watt, W., et al., "Specialized Ceramic Materials with Particular Reference to Ceramic Gas Turbine Blades." BIOS Final Rept. 470, Item 21, 22. PB L 63,820, 33 pp. (Aug. 1945); abstracted in Bibliog. Sci. Ind. Repts. 5 (6) 509 (1947).

Watts, A.S., "Control of the Properties of Glazes by the Aid of Eutectics: II, The Alkali-Alumina-Silica & Lead-Alumina-Silica Systems Separately & in Combination." J. Am. Ceram. Soc. **41** (7) 249-53 (1958).

Way, K., "Nuclear Data." Nat. Bur. Std. Circ. 499 (Sept. 1, 1950).

Webb, W.W., Dragsdorf, R.D. & Forgeng, W.D., "Dislocations in Whiskers." Phys. Rev. **108**, 498-9 (1957).

Webb, W.W., & Forgeng, W.D., "Growth & Defect Structure of Sapphire Microcrystals." J. Appl. Phys. **28** (12) 1449-54 (1957).

Webb, W.W., Wissler, W.A., & Forgeng, W.D., "Process for Preparing Virtually Perfect Alumina Crystals." U.S. 3,011,870, 12/5/61.

Weeks, J.L. & Seifert, R.L., "Note on the Thermal Conductivity of Synthetic Sapphire." J. Am. Ceram. Soc. **35** (1) 15 (1952).

Weertman, J.R., "Theory of Steady-State Creep Based on Dislocation Climb." J. App. Phys. **26** 1213-7 (1955).

Wefers, K., "The Structure of Aluminum Trihydroxides." Naturwissenschaften **49**, No. 9, 204-5 (1962).

Weibull, W., "A Statistical Theory of the Strength of Materials." Ing. Vetenskaps. Akard. Handl. 151 Stockholm (1939).

Weil, N.A., "Studies of Brittle Behavior of Ceramic Materials." ASD Tech. Doc. Rept. No. ASD-TR-61-628 (April 1962), 495 pp. Part II, April 1963, 683 pp.

Weil, N.A., Bortz, S.A. & Firestone, R.F., "Factors Affecting The Statistical Strength of Alumina." Materials Science Research Vol. 1. Plenum Press, New York, 335 pp. (1963).

Weill, D.F., "Relative Stability of Crystalline Phases in the Al_2O_3-SiO_2 System." Univ. Microfilms (Ann Arbor, Mich.) Order No. 63-5466, 105 pp. Dissertation Abst. 24 1140 (1963).

Weinig, S., "What's New in Ceramic Ductility." Ceram. Ind. **72** (6) 96 (1959).

Weiser, H.B., "X-Ray Studies of the Hydrous Oxides. I. Alumina." J. Phys. Chem. **36**, 3010 (1932).

Weiser, H.B., "Inorganic Colloid Chemistry, Vol. II. The Hydrous Oxides & Hydroxides. John Wiley & Sons, New York, p. 90-120 (1935).

Weiser, H.B. & Milligan, W.O., "X-Ray Studies on the Hydrous Oxides. VI. Alumina Hydrate." J. Phys. Chem. **38**, 1175-82 (1934).

Weiser, H.B. & Milligan, W.O., "The Constitution of Inorganic Gels." Advances in Colloid Science I. Interscience Publishers, New York (1952).

Weiser, H.B., Milligan, W.O. & Purcell, W.R., "Alumina Floc." Ind. Eng. Chem. 33 (5) 669-72 (1941).

Weissenberg, G. & Meinert, N., "Optical Glass & Its Production." U.S. 2,763,559, 9/18/56.

Weissenberg, G. & Ungemach, O., "Borate Optical Glass." U.S. 2,764,492, 9/25/56.

Welch, J.H., "Observations on Compositions & Melting Behavior of Mullite." Trans. Internat. Ceram. Congr. 7th London, Eng. 1960, 197-206 (Pub. 1961).

Welling, C.E., "Method of Preparing Alumina & Hydroforming Catalyst Thereon." U.S. 2,852,473, 9/16/58.

Wells, L.S., Clarke, W.F. & McMurdie, H.F., "Study of the System $CaO\text{-}Al_2O_3\text{-}H_2O$ at Temperatures of 21° & $90^\circ C$." Jour. Res. Nat. Bur. Stds. 30 (5) 367-409 (1943); RP 1539.

Wendell, C.B., Jr & Engelson, F.E., "Process of Making Aluminum Oxide." U.S. 2,693,406, 11/2/54.

Wenger, L.S. & Knapp, W.J., "Prestressing & Thermal Shock of Ceramics: I, II." Pacific Coast Ceram. News 4 (1) 23-4; (2) 20-1 (1955).

Wenzel, K., "Effect of Titanium Compounds on the Thermal Expansion of Standard Electrocorundum." Silikattech., 11 (4) 164-6 (1960).

West, R.R. & Gray, T.J., "Reactions in Silica-Alumina Mixtures." J. Am. Ceram. Soc. 41 (4) 132-6 (1958).

Westbrook, J.H., "Metal-Ceramic Composites: I." Am. Ceram. Soc. Bull. 31 (6) 205-8 (1952).

Westerholm, R.S., "Development & Application of Rokide Pure Oxide Coatings." Proc. Porcelain Enamel Inst. Forum 21, 8-12 (1959).

Westerholm, R.J., McGeary, T.C., Huff, H.A. & Wolff, R.L. "Flame-Sprayed Coatings." Machine Design 33 (18) 82-92 (1961).

Westmoreland-White, B., "Hard Compositions for Tools, Dies. Etc." Brit. 575,753, 3/13/46.

Weyl, W.A., "Acid-Base Relationship in Glass Systems: II, A New Acid-Base Concept Applicable to Aqueous Systems, Fused Salts, Glasses, & Solids." Glass Ind. 37 (6) 325-31, 336, 344, 346, 350 (1956).

Weyl, D., "Influence of Internal Strains on the Texture & Mechanical Strength of Porcelains." Ber. deut. keram. Ges. 36 (10) 319-24 (1959).

Whalen, T.J., McHugh, C.O., & Humenik, M. Jr., "Dispersion-Strengthened Ceramics." Proc. Sagamore Army Mater. Res. Conf. Conf. 12th, Raquette Lake, N.Y. 1965 219-27 (Pub. 1966).

Whalley, E. & Winter, E.R.S., "Studies of the Exchange Reactions of Solid Oxides: II, Exchange of Oxygen Between Water Vapor & Certain Metallic Oxides." J. Chem. Soc. 1950 1175-7.

Wheeler, W.H. & Olivitor, J.P., "Low Density Refractory Oxide." U.S. 2,992,930, 7/18/61.

Wheeler, W.H., "Heat Shields, Ceramic Radiative." Ceram. Age 79 (10) 161 (1963).

Wheildon, W.M. Jr., "Coating Metals & Other Materials with Oxide & Articles Made Thereby." U.S. 2,707,691, 5/3/55.

Wheildon, W.M., "Oxide Coatings–Resistant to Temperatures in Excess of 3000°F. and to Erosion of Supersonic Velocities: I." Pacific Coast Ceram. News 4 (11) 21, 25 (1955).

Wheildon, W.M., "Oxide Coatings–Resistant to Temperatures in Excess of 3000°F and to Erosion of Supersonic Velocities: II." Pacific Coast Ceram. News 4 (12) 20-1 (1955).

Wheildon, W.M., "Wear Resistant Ceramic Tooling in the Fabrication of Ceramic Shapes." Am. Ceram. Soc. Bull. 42 (5) 308-11 (1963).

White, A.E.S., "Cutting Tools for Hard, Abrasive Materials." Brit. 837,013, 6/9/60.

White, A.E.S., et al. "Metal-Ceramic Bodies." Iron Steel Inst. (London), Spec. Rept. 1956 No. 58, 311-4.

White, E.A.D., "New Technique for Production of Synthetic Corundum." Nature 191 (4791) 901-2 (1961).

White, E.A.D., "Corundum Crystals." Brit. 935-390, 8/28/63.

White, H.E., "Practical Advantages in the Use of Special Refractories." Ind. Heating 5 (10) 945-52; (11) 1061-3 (1938).

White, J., "Some General Considerations on Thermal Shock." Trans. Brit. Ceram. Soc. 57 (10) 591-623 (1958).

White, J., "Changing-Pattern in Refractories Technology." Refractories J. 39 (4) 126-39 (1963).

White, J.F., "Review of Ceramic Material Requirements for Nuclear Reactor Use." Am. Ceram. Soc. Bull. 32 (9) 301-5 (1953).

White, J.F. & Clavel, A.L., "Extrusion Properties of Non-Clay Oxides." Am. Ceram. Soc. Bull. 42 (11) 697-702 (1063).

Whitehead, S., "Dielectric Breakdown of Solids. Oxford & The Clarendon Press, Oxford (1951).

Whiteway, S.G., "Measurement of Low Permeability in Ceramic Test Pieces." Am. Ceram. Soc. Bull. 39 (11) 677-9 (1960).

Whitmore, D.H., "Excitation Processes in Ceramics & Anomalous Increase in Thermal Conductivity at Elevated Temperatures." J. Appl. Phys. 31 (6) 1109-12 (1960).

Whitney, L.F., et al. "Evaluation of Hot Hardness Measuring Techniques & The Development of a Hot Microhardness Tester." PB Rept. 126492, 122 pp.; U.S. Govt. Res. Repts. 29 (3) 134 (1958).

Whittemore, O.J., Jr., "Properties & Uses of Pure Oxide Heavy Refractories." J. Am. Ceram. Soc. 32 (2) 48-53 (1949).

Whittemore, O.J., Jr., "Extreme Temperature Refractories." J. Can. Ceram. Soc. 28, 43-8 (1959).

Whittemore, O.J., "Process of Making a Nuclear Fuel Element." U.S. 3,081,249, 3/12/63.

Whittemore, O.J., Jr. & Ault, N.N., "Thermal Expansion of Various Ceramic Materials to $1500^\circ C$." J. Am. Ceram. Soc. 39 (12) 443-4 (1956).

Whittemore, O.J., Jr., et al. "Molded Alumina Cutting Tools." U.S. 3,093,498, 6/11/63.

Wicken, O.M. & Birch, R.E., "Refractories Selection for Modern Rotary Kilns." Pit Quarry 56 (1) 162-4, 175, 181 (1963).

Wickersheim, K.A. & Lefever, R.A., "Optical Properties of Synthetic Spinel." J. Opt. Soc. Am. 50 (8) 831-2 (1960).

Widmann, H., "Manufacture of (Alumina) Cutting Tools." Glas-Email-Keramo-Tech. 15 (6) 185-8 (1964).

Wilcox, P.D. and Cutler, I.B., "Strength of Partly Sintered Alumina Compacts." J. Am. Ceram. Soc. 49, 249-52 (1966).

Wilder, D.R., "Progress in Sintering." Electrochem. Technol. 1 (5-6) 172-9 (1963).

Wilder, D.R. & Fitzsimmons, E.S., "Further Study of Sintering Phenomena." J. Am. Ceram. Soc. 38 (2) 66-71 (1955).

Wilkins, R.G., Lodwig, R.M. & Greene, S.A., "The Chemical Composition of Metallized Flames." Symp. Combust., 8th, Pasadena, Calif. 1960, 375-88 (Pub. 1962).

Williams, A.E., "Refractory Concrete." Ceramics 3 (31) 378-83 (1951).

Williams, A.E., "Castable Refractories–Properties & Applications of Aluminous Cement in Metallurgical Work." Metal Ind. (London) 92 (1) 3-7 (1958).

Williams, J.C. & Nielsen, J.W., "Wetting of Original & Metallized High-Alumina Surfaces by Molten Brazing Solders." J. Am. Ceram. Soc. 42 229-35 (1959).

Williams, L.S., "Stress Endurance of Sintered Alumina." Trans. Brit. Ceram. Soc. 55 (5) 287-312 (1956).

Williams, L.R. & Stamper, J.W., "Bauxite." U.S. Dept. Interior, Bur. Mines Preprint (1963).

Williams, R., "Ceramics for Vacuum Tube Envelopes." Ceram. Age, July 1954, p. 41-54.

Williams, R.M. & Simpson, H.E., "Effect of Alumina & Lime on the Solubility of Lithium Silicates." Ceram. Ind. 60 (4) 137-8, 165 (1953).

Wills, H.J., "Securing Fine Surfaces by Grinding." Machinery (N.Y.) 49 (8) 202-4 (1943).

Wills, H.J., "Securing Fine Surfaces by Grinding: III." Machinery (N.Y.) 49 (10) 171-3 (1943).

Wills, H.J., "Securing Fine-Surfaces by Grinding: V." Machinery (N.Y.) 50 (1) 172-4 (1943).

Willstatter, R. & Kraut, H., " Aluminum Hydroxide I. Hydrates & Hydrogel." Ber. 56, 149-62 (1923).

Willstatter, R. & Kraut, H., "An Alumina Gel of the Formula Al(OH)₃ I. Hydrate & Hydrogel." Ber. 56B, 1117-21 (1923).

Willstatter, R., Kraut, H. & Erbacher, O., "Hydrates & Hydrogels VII. Isomeric Hydrogels of Alumina." Ber. 58, 2448-58 (1925).

Willstatter, R., Kraut, H. & Lobinger, L., "The Simplest Silicic Acids." Ber. 61B, 2280-93 (1928).

Wilsdorf, H.G.F., "Structure of Amorphous Aluminum Oxide Films." Nature 168, 600-1 (1951).

Winchell, A.N., "Spinel Group." Amer. Mineralogist 26, 422-8 (1941).

Winkler, E.R., Sarver, J.F., & Cutler, I.B., "Solid Solutions of Titanium Dioxide in Aluminum Oxide." J. Amer. Ceram. Soc. 49, 634-7 (1966).

Winter, E.R.S., "Studies of the Exchange Reactions of Solid Oxides: I, Exchange of Oxygen Isotopes Between Gaseous Oxygen & Certain Metallic Oxides." J. Chem. Soc. 1950, pp. 1170-5.

Winzer, R., "Corrosion in Crucibles made from Pure Oxide." Angew. Chem. 45, 429-31 (1932).

Wisely, H.R., "Resistivity Characteristics of Some Ceramic Compositions Above 1000° F." Am. Ceram. Soc. Bull. 36 (4) 133-6 (1957).

Wisely, H.R., "Investigation for the Development of Ceramic Bodies for Electron Tubes." AD-277,842, 79 pp., illus.; U.S. Govt. Res. Repts. 37 (20) 19 (1962).

Wislicenus, H. Koll. Z. 2, 11 (1908).

Wisnyi, L.G., "High-Alumina Phases in the System Lime-Alumina." Dissertation Abstr. 15, 2498-9 (1955); abstracted in J. Appl. Chem. (London) 6 (7) ii-36 (1956).

Wisnyi, L.G. & Taylor, K.M., "Materials in Nuclear Applications." Am. Soc. Testing Materials Spec. Tech. Pub. 1960, No. 276, iv + 344 pp. illus.

Woerner, J.J., "Ceramic Metalizing Process." U.S. 3,051,592, 8/28/62.

Wolf, R.F.A., "Production of Aluminum Powder." U.S. 2,930,686, 3/29/60.

Worel, H. & Torkar, K., "Process for the Production of α-Alumina from Aluminum Fluoride." U.S. 3,027,231, 3/27/62.

Workman, G.M., "Assessment of Ladle Firebrick Quality & Performance." Trans. Brit. Ceram. Soc. 57 (9) 551-72 (1958).

Worrall, W.E., "Flow Properties of Acid-Deflocculated Alumina Slips." Trans. Brit. Ceram. Soc. 62 (8) 659-72 (1963).

Wozniczek, H., "Investigation of Activated Al₂O₃ Hydrates by Dispersion Analysis According to the Figurowskii Method." Poznan. Towarz. Przyjaciol. Nauk Wydzial Mat. Przyrod Prace Komisji Mat. Przyrod. 10, 139-47 (1962).

Wygant, J.F., "Elastic & Flow Properties of Dense, Pure Oxide Refractories." J. Am. Ceram. Soc. 34 (12) 374-80 (1951).

Wygant, J.F. & Bulkley, W.L., "Refractory Concrete for Refinery Vessel Linings." Am. Ceram. Soc. Bull. 33 (8) 233-9 (1954).

Wygant, J.F. & Crowley, M.S., "Curing Refractory Castables—It Isn't the Heat, It's the Humidity." Am. Ceram. Soc. Bull. 43 (1) 1-5 (1964).

Wygant, J.F. & Kingery, W.D., "Stability of Ceramic Raw Materials." Bull. Am. Ceram. Soc. 31, 251-5 (1952).

Yamada, K., et al ., "Machinability Study of Ceramic Tools: III, Tool Wear in Ultra-High-Speed Cutting." Nagoya Kogyo Gijutsu Shikensho Hokuku 7 (2) 74-80 (1958).

Yamada, K., et al. "Machinability Study of Ceramic Tools: IV. Performance of Ceramic & Carbide Tools in Cutting Alloyed Cast Iron at Ultrahigh Speed." Nagoya Kogyo Gijutsu Shikensho Hokoku 9 (3) 1-7 (1960).

Yamada, K., et al., "Machinability Study of Ceramic Tools: V, Performance of Carbide & Ceramic Tools in Cutting Carbon Steel." Nagoya Kogyo Gijutsu Shikensho Hokoku 9 (8) 1-11 (1960).

Yamada, K., et al., "Machinability Study of Ceramic Tools: VI, Effects of Tool-Tip Chamfering on Ultra-High-Speed Cutting of Steel & Cast Iron." Nagoya Kogyo Gijutsu Shikensho Hokoku 10 (2) 63-75 (1961).

Yamaguchi, Goro & Nippon Gaishi Kaisho, Ltd., "Large Single Crystals of Corundum." Jap. 737('65).

Yamaguchi, G., "Utilization of Spinel with Lower Valent Al Ion as a Reducing Agent for Glass." J. Ceram. Assoc. Japan 61 (689) 549-53 (1953).

Yamaguchi, G., "Spinel Produced by the Sintering Reactions Between Basic Oxides & Alumina." J. Ceram. Assoc. Japan 61 (690) 594-9 (1953).

Yamaguchi, G., et al., "New Alumina Hydrate, "Tohdite." Bull. Chem. Soc. Japan 37 (5) 752-4 (1964).

Yamaguchi, G. & Sakamoto, K., "Crystal Structure of Bayerite." Bull. Chem. Soc. Japan 31, 140 (1958).

Yamaguchi, G. & Tanabe, H., "Fatty Abrasive of Alumina." J. Ceram. Assoc. Japan 62 (691) 34-8 (1954).

Yamaguchi, G., Tanabe, H. & Tomiura, K., "Researches on Spinel Pigments." J. Ceram. Assoc. Japan 62 (693) 191-6 (1954).

Yamaguchi, G. & Tomiura, K., "Pink Pigments of MnO-P₂O₃-Al₂O₃ & Cr₂O₃-P₂O₃ Systems." J. Ceram. Assoc. Japan 62 (692) 111-4 (1954).

Yamaguchi, G. & Tanabe, H., "Research On Artificial Emery." J. Ceram. Assoc. Japan 62 (693) 208-12 (1954).

Yamaguchi, G. & Yanagida, H., "Relation Among γ-,η-, & δ-Al₂O₃ under Hydrothermal Conditions." Bull. Chem. Soc. Japan 35, 1896-7 (1962).

Yamaguchi, G., Yanagida, H., & Ono, S., "Reply to Torkar & Krischner's Comment on 'A New Alumina Hydrate, Tohdite'." Bull. Chem. Soc. Japan 39, 1356 (1966).

Yamaguchi, G., Yanagida, H. & Soejima, S., "Solubility and the Velocity of Dissolution of Corundum Under Hydrothermal Conditions." Bull. Chem. Soc. Japan 35, 1789-94 (1962).

Yamamoto, H., et al. "Studies on Alumina Used for Radio Receiving Tubes." J. Ceram. Assoc. Japan 61 (687) 419-25 (1953).

Yamanouchi, S. & Kato, S., "Glass-melting Crucible Consisting of β-Alumina." J. Japan, Ceram. Assoc. 54 (622) 29-32 (1946).

Yamauchi, T. & Kato, S., "Transition of β-Alumina by Heating." J. Japan Ceram. Assoc. 51 (602) 71-7 (1943).

Yamauchi, T. & Kato, S., "Studies on β-Alumina: IX, Existence of β-Alumina in the Bayer Alunina Calcined at Low Temperature." J. Japan. Ceram. Assoc. 52 (617) 194-6 (1944).

Yamauchi, T. & Kondo, R., "Slip Casting of Alumina." J. Japan Ceram. Assoc. 57 (642) 160-3 (1949).

Yamauchi, T. & Kondo, R., "Electrical Resistance of Alumina Porcelain." J. Ceram. Assoc. Japan 58 (645) 82-7 (1950).

Yanada, H., Suwa, S., & Yamaguchi, G., "Minute Corundum Crystal." Japan 7,750, 7/10/62.

Yanagida, H. and Kroger, F.A., "The System Al-O." J. Am. Ceram. Soc. 51, (12) 700-6 (1968).

Yanagida, H. & Yamaguchi, G., "Thermal Effects On the Lattices of η- and γ-Aluminum Oxide." Bull. Chem. Soc. Japan 37 (8) 1229-31 (1964).

Yanagida, H., Yamaguchi, G., & Kubota, J., "Two Types of Water Contained in Transient Aluminas." Bull. Chem. Soc. Japan 38, 2194-6 (1965).

Yenni, D.M. & Runck, W.A., "Process of Making Spark Plug Electrode Structures." U.S. 2,998,632, 9/5/61.

Yokokawa, T., & Kleppa, O.J., "A Calorimetric Study of the Transformation of Some Metastable Modification of Alumina to Alpha Alumina." J. Phys. Chem. 68 (11) 3246-9 (1964).

Yopps, J.A. & Fuerstenau, D.W., "Zero Point Charge of Alpha Alumina." J. Colloid Sci. 19 (1) 61-71 (1964).

Yoshiteru, H. and Makoto, K., "Exaggerated Grain Growth of Alumina in Hot-Pressing." Proc. Jap. Congr. Test. Mater. 10 154-6 (1967).

Young, C.G., "Investigations of Radiation Effects in α-Al₂O₃ by Electron-Spin Resonance." Univ. Microfilms (Ann Arbor, Mich.), Order No. 61-5438, 78 pp.; Dissertation Abstr. 22, 2849 (1962).

Young, G.H., Wilkins, C.H.T., et al. "High Temperature Printed Circuitry." PB Rept. 149143, 31 pp.; U.S. Govt. Res. Repts. 34 (5) 572 (1960).

Young, G.H., Owen, C.J., et al. "High-Temperature Printed Circuitry." PB Rept. 149144, 149145, & 149146, 120 pp. U.S. Govt. Res. Repts. 34 (5) 572-3 (1960).

Young, L., "Transients in the Formation of Anodic Films." Can. J. Chem. 37, 1620-1 (1959).

Young, Y.R., "Dissipation of Energy by 2.5 to 10 k.e.v. Electrons in Al_2O_3." J. Appl. Phys. 28 (5) 524-5 (1957).

Yudin, B.F. and Karklit, A.K., "Evaporation of Refractory Oxides. I Thermodynamics of Evaporation of Refractory Oxides at Elevated Temperatures." Zl. Prinkl. Khim. 39, 537-44 (1966).

Yuille, E.C., "Preparation of Dispersions Containing Hydrous Aluminum Oxide." U.S. 3,112,265, 11/26/63.

Yurchak, R.M., "Method of Determining Bending Strength of Fine Ceramic Shapes." Steklo i Keram. 7 (7) 15-7 (1950).

Yuzhaninov, I.A., "Loss of Alumina as Dust in Fluidized-Bed Furnaces." Izv. Vysshikh Uchebn. Zavedenii, Tsvetn. Met. 7 (3) 100-4 (1964).

Zachariesen, W.H., "The Atomic Arrangement in Glass." J. Am. Chem. Soc. 54 (10) 3841 (1932).

Zaionts, R.M., "Production of Spark Plug Insulators for Automobile Engines in Czechoslovakia." Steklo i Keram. 15 (8) 43-6 (1958).

Zalar, S., "Production of Alumina by the Continuous Extraction of Bauxite in Towers." Nova Proizvodnja 1952, No. 4-5, pp. 337-51; abstracted in Brit. Aluminium Co., Ltd. Light Metals Bull. 20 (5) 163 (1958).

Zaporoshtseva, O.K. & Gorbai, Z.V., "Use of Aluminum Hydroxide in a Plate Glass Batch." Steklo i Keram. 18 No. 4. 32-3 (1961).

Zechmeister, L. & Cholnoky, L. "Principles & Practices of Chromatography." John Wiley & Sons, Inc., New York (1941).

Zeemann, F.A.O.G., "Refractory Plates for Porcelain Firing." Ceramica (Sao Paulo) 3 (11) 100-3 (1957).

Zemlicka, J., "Synthesis of Large Corundum Crystals." Ber. Geol. Ges. Deut. Demokrat. Rept. Gesamtgebiet Geol. Wiss. 7, 492-4 (1962).

Zen, E., "Validity of "Vegard's Law." Am. Mineralogist 41 (5/6) 523-4 (1956).

Zet, F., "Production of Extruded Profiles from Nonplastic Corundum Bodies." Sklar. Keram. 11 (2) 50-3 (1961).

Zhikharevich, S.A., Getman, I.A., & Kozyreva, L.A., "Technology of Dense Volume-Stable High-Alumina Refractories for Blast Furnace Linings." Ogneupory 23 (9) 385-95 (1958).

Zhilin, A.I. & Ignat'eva, L.P., "Cast Lining Blocks from Alumina Slag for Cement Kilns." Tsement 6 (6) 41-5 (1939).

Zhorov, G.A., Kovalev, A.I., & Sivakova, E.V., "Thermal Conductivity and Emissivity of Alumina Coatings at High Temperatures." Teplofiz. Vys. Temp. 4, 643-8 (1966).

Ziegler, K. & Ruthard, K., "Alumina Insulating Coat on Copper." Ger. 1,025,955, 3/13/58.

Ziese, W., et al., "Production of Colloidally Soluble Aluminum Hydroxide." U.S. 2,907,634, 10/6/59.

Zimens, K.E., "Magnetic Measurements on Active Aluminum Oxide & Hydroxide." Svensk. Kem. Tid. 52, 205-22 (1940).

Zimmerann, K. & Burton-Banning, L., "Use of Various Clays in Abrasive Wheel Bonds." Sprechsaal 94 (10) 245-58 (1961).

Zimmerman, W.F., "Development of a Foamed Alumina Cement." Am. Ceram. Soc. Bull. 38 (3) 97-8 (1959).

Zimmerman, W.F., & Allen, A.W., "X-Ray Thermal Expansion Measurements of Refractory Crystals." Am. Ceram. Soc. Bull. 35 (7) 271-4 (1956).

Zimmerman, W.F. & Haeckl, R.S., "Cellular Lightweight Alumina Ceramic." U.S. 2,966,421, 12/27/60.

Zintl, E., Morawietz, W., & Gastinger, E., "Boron Monoxide," Z. anorg Allgem. Chem. 245, 8-11 (1940).

Zintl, E. & Morawietz, W., "Mixed Crystals of Cryolite with Alumina." Z. anorg. & Allgem Chem. 240 (2) 145-9 (1938).

Zschacke, F.H., "Effectiveness of Opacifiers in Glasses, Glazes, & Enamels." Keram. Rundschau 49 (20) 197-201 (1941).

Zwilsky, K.M. & Grant, N.J., "Metal-Metal Oxide Composites for High Temperature Use." Metal Progr. 80 (2) 108-11 + 2 pp. (1961).